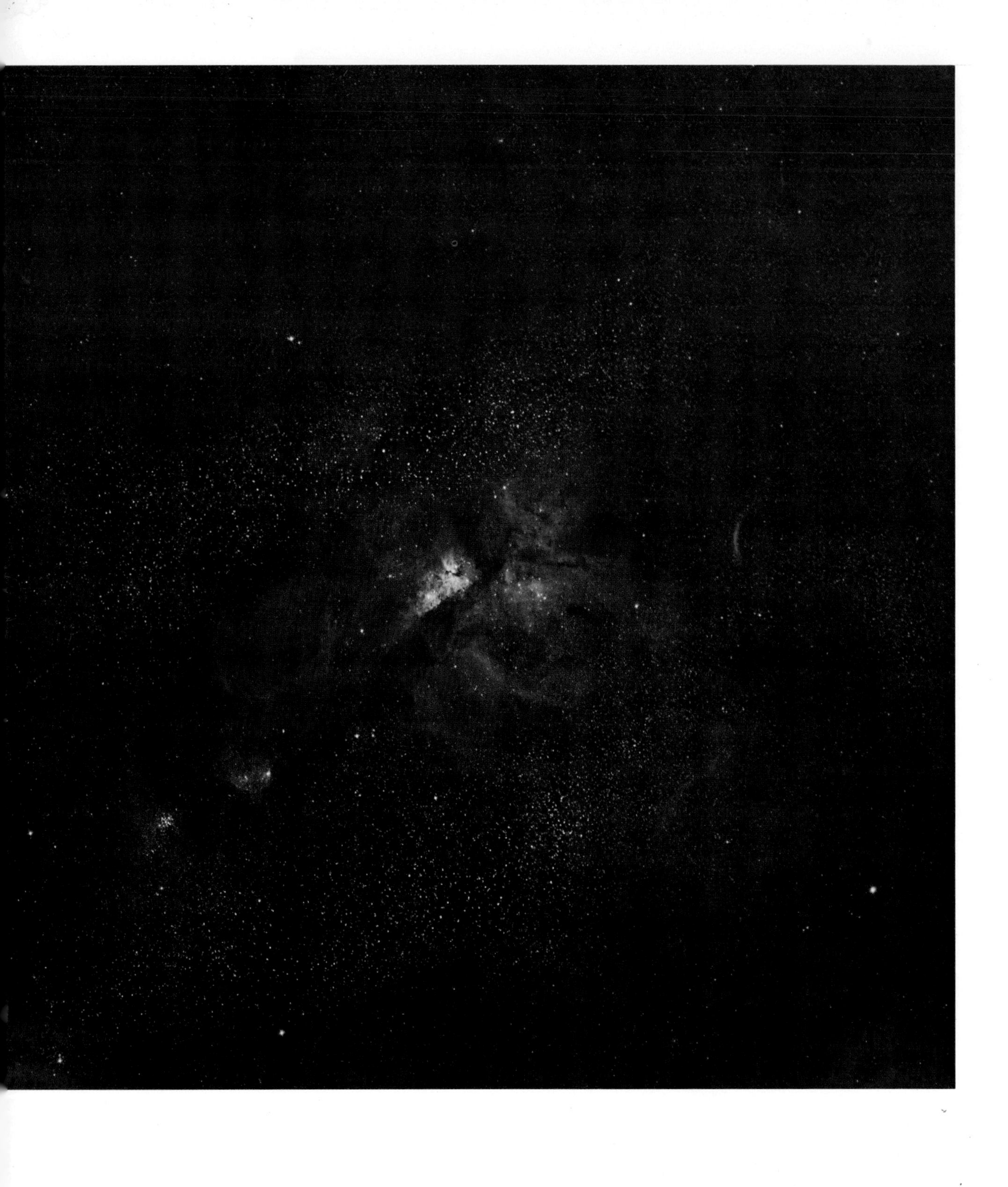

DRAMA OF THE UNIVERSE

George O. Abell
University of California,
Los Angeles

HOLT, RINEHART AND WINSTON
New York · Chicago · San Francisco · Atlanta · Dallas
Montreal · Toronto · London · Sydney

Acknowledgments

Photograph Credits

COVER PHOTOGRAPHS:

FRONT: Dumbbell nebula (*Hale Observatories*); Trifid nebula (*Kitt Peak National Observatory*); BACK: Elliptical galaxy NGC 5128 in Centaurus (© *Association of Universities for Research in Astronomy, Inc. The Cerro Tololo Inter-American Observatory*)

HALF-TITLE PHOTOGRAPH: Mars landscape photographed by Viking 2. (*NASA*)

FRONTISPIECE PHOTOGRAPH: The gaseous nebula, Eta Carinae, photographed with the 61-cm Schmidt telescope. Cerro Tololo, Chile. (© *1975 by the Association of Universities for Research in Astronomy, Inc. The Cerro Tololo Inter-American Observatory*)

OPENER PHOTOGRAPHS

Prologue: *Alinari Art Reference Bureau*

Act I and Epilogue: *NASA*

Act II: *Cornell University*

Act III: *Hale Observatories*

Acts IV and VI: *Kitt Peak National Observatory*

Act V: *Jet Propulsion Laboratory*

Act I—Scene 1: *Yerkes Observatory;* Scene 2: *Istituto e Museo di Storia della Scienza di Firenze;* Scene 3: *Crawford Collection, Royal Observatory Edinburgh;* Scene 4: *NASA/JPL;* Scene 5: *Hale Observatories;* Scene 6: *Lick Observatory*

Act II—Scene 1: *Yerkes Observatory;* Scene 2: *California Institute of Technology;* Scene 3: *Hale Observatories;* Scene 4: *Courtesy of the Archives, California Institute of Technology*

Act III—Scene 1: *NASA/JPL;* Scenes 2 and 5: *Hale Observatories;* Scene 3: *UCLA Observatory;* Scene 4: *Institution for Astronomy, Lund University, photograph courtesy Griffith Observatory*

Act IV—Scene 1: *High Altitude Observatory, a division of the National Center for Atmospheric Research, sponsored by the National Science Foundation;* Scene 2: *National Accelerator Laboratory;* Scenes 3, 4, and 7: *Hale Observatories*

Act V—Scenes 1, 2, and 3: *NASA/JPL;* Scene 4: *NASA/Ames Research Center;* Scene 5: *Daniel Brody/Editorial Photocolor Archives*

Act VI—Scenes 1, 2, and 3: *Kitt Peak National Observatory*

Additional credits are included in captions on pertinent pages of the text.

The photographs credited to the Hale Observatories in the color sections of this book are copyrighted by California Institute of Technology and Carnegie Institute of Washington.

Reproduced by permission from the Hale Observatories.

Library of Congress Cataloging in Publication Data

Abell, George Ogden 1927-
Drama of the universe.
Bibliography: p. ii
Includes index.
1. Astronomy. I. Title.
QB45.A15 520 77-22338

ISBN 0-03-022401-2

Printed in the United States of America

8901 039 987654321

PREFACE

To Jaques the World's a stage, and we are the players. I think, though, that the cosmos itself is a far grander arena, and that the planets, stars, and galaxies, which act out the almost eternal *Drama of the Universe*, provide some of the finest theater we can hope to experience.

At first it may seem that this universal play lacks the unity of time, place, and action required of classical drama. But wait! Time and space, as we shall see, are inextricably interrelated, and even at this moment we observe in remote parts of the universe events that occurred in the dim past—many thousands of millions of years ago! And as for action, not only is matter everywhere made of the same kinds of atoms, but so far as we know, everywhere in space and everywhere in time that matter obeys and has obeyed the same natural laws that govern events on earth today.

It is to emphasize the inherent drama and the true excitement of astronomy that I have cast this book in the format of a play, with six acts, a prologue and an epilogue, and some scenes. The material is not presented topic by topic, as if the sun, meteorites, and superclusters of galaxies are unrelated topics to be studied separately. While trying not to omit important aspects of astronomy, I have organized the text around ideas, emphasizing the unity of concepts, whether they are applied to the satellites of Mars or to the rotation of a galaxy.

I suppose this approach will not appeal to all teachers, but that is why I am revising my other texts that have more traditional approaches: *Exploration of the Universe*, Third Edition, and the brief nonmathematical version, *Realm of the Universe*. There are many advantages of the traditional approach; for example, salient facts about Jupiter are found in one place, as are those pertaining to the measurement of stellar parallaxes. But today lots of teachers want to try something new—an approach less concerned with teaching students astronomy than with teaching them *about* astronomy; an approach that relates science in general and astronomy in particular to our culture, which shows that astronomy is really one of the humanities as well as a science, which describes the universe as we understand it but from the viewpoint of how it concerns us, not just for its own sake.

That is why I have written *Drama*. It is not a book intended for science students (although I hope science students will enjoy reading it). It is intended for those who wonder what astronomy is all about, but who are dissuaded from further study by too abstract or pedantic an approach; it is a book intended for poets. I hope my mother would have enjoyed reading it. Anyway, I enjoyed writing it.

Act I deals with Newtonian gravitational theory—how it developed, what it is, and what it has done for us. Ironically, astrology played a role in the history of gravitation, and I have taken the liberty of discussing the subject at some length. I believe there is a great beauty to the rational development of the Newtonian structure, and wish to contrast it with the murky occult, which has received such widespread and, I think, unconscious involvement of all too many of our students.

The second act is about the new science of the twentieth century, and how it has expanded the realm of science beyond that reached by Newton. The act opens with the nature of radiation and its relation to electricity and magnetism, which leads, I think naturally, to the quantum theory and relativity.

The third act shows how we measure or estimate distances in space, from the moon to the most remote galaxies, and how we have thus built up our present picture of the large-scale structure of the universe.

Act IV brings the universe alive, in a sense, by showing the drama of the birth, evolution, and death of stars—and of the solar system. Life in the universe is the theme of Act V; here we discuss the worlds of the solar system, the possibilities of life on any of them, the chances of life elsewhere in the universe, and how we might know about it. Here, too, I try to put interstellar travel and UFOs into perspective. The act closes with some comments about the future of life on earth. Act VI, the final act, concerns the grand questions about the origin and evolution of the universe itself.

I realize that some other books on the market get right to the heart of modern astrophysics, and start off with pulsars, black holes, galaxies, and quasars. Those are among the hot topics in present-day astronomy, and the authors of the books I refer to seem to feel that these topics should be emphasized at the expense of classical topics, such as Kepler's laws. I agree that the forefront of modern astronomy must be emphasized, but I don't believe the uninitiated reader can fully appreciate a black hole without some understanding of the science on which the idea is based. He can be given a "gee whiz" survey of hot topics and a lot of unrelated facts, but he can get that kind of science from the newspaper. I cannot bring myself to pull the idea of gravitation right out of the air, as though it were a bit of self-evident and ancient trivia that do not warrant development. I want to show the logical structure—how it all hangs together. It may be traditional to start with Newtonian theory, but that tradition makes good pedagogical sense to me. So heeding the advice of Hans Sachs, I have honored the traditional as well as the new.

I have done everything without mathematics. I think we can give students a feeling for the concepts without training them to work specific arithmetic problems. Although the text is somewhat sophisticated in the level of many of the ideas presented, I think it is accessible, and I hope it is interesting, to any intelligent reader. Incidentally, the exercises, inserted inside the text where they suggest themselves, are intended to be solved without algebra or numerical computation that cannot be done in the head.

Finally, because the book is a try at something new, I have taken the liberty of being a little personal in places. I have related some experiences of my own and of many of my colleagues, and I have felt free to express opinions; I hope it is always clear which is opinion and which is not.

As in any new book, it is inevitable that this one will have typos and slips. I am grateful to the many people who have called my attention to such errors in my other texts, and similarly will most appreciate hearing about the ones in this book, so that they can be removed in future printings.

A number of people have read part or all of the manuscript, and have offered very many helpful suggestions. None, however, has seen the final version, so I alone must assume responsibility for any goofs that have slipped through. These good readers include Mr. Harold Corwin, and Professors Morris S. Davis, William D. Davis, Owen Gingerich, Miroslav Plavec, Duane H. D. Roller, Paul E. Trejo, Thomas A. Weber, and Donat G. Wentzel. I particularly want to single out Professor Dimitri Mihalas, whose very careful reading and extensive comments were enormously valuable. He has offered to let me buy him a drink of his choice in return, and I am waiting for the chance to pay up!

The art was done by Eric G. Hieber. He worked on some of my other books; I like his product, and asked for him again. The same is true of my copy editor, Sara Boyajian; but Sara is far more than just an editor. She has been a combination mother-nurse to the book, who has guided it in all details from the earliest manuscript to the final cover design. I don't believe one could find a harder working, more faithful, or more dependable editor!

Finally, no one suffers over a new manuscript more than the author's spouse. My own dear wife, Phyllis, is no exception! Without her tremendous help, encouragement, and understanding, there would have been no need for this Preface.

Encino, California
December 1977

G. O. Abell

PROGRAM

Prologue 1

The beginnings: the universe; the solar system; life

Act I • **Order, Not Chaos**

The unification of celestial and terrestrial science

SCENE 1 *The Fixed and Wandering Stars* 12

Motions in the sky; the seasons; phases of the moon and eclipses

SCENE 2 *The Age of Astrology* 28

Motions of the planets; the horoscope and the ancient belief in astrology; precession; the Ptolemaic system

SCENE 3 *The Copernican Revolution* 40

The heliocentric hypothesis; the science of Copernicus, Brahe, Kepler, Galileo, and Newton

SCENE 4 *Universal Gravitation* 56

The law of gravitation and its proof with the moon and the apple; artificial satellites and space travel

SCENE 5 *More Proof of the New Order* 67

The minor planets; precession and tides explained; comets as gravitational probes; discovery of Neptune

SCENE 6 *How Universal Is Universal Gravitation?* 77

Gravitation beyond the solar system; contrast between modern science and the ancient occult; some limitations of Newtonian theory

Act II • Light, Space, and Time

The new physics of the twentieth century

SCENE 1 *Light and the Telescope* 88

Properties of light; geometrical optics

SCENE 2 *Visible and Invisible Light* 97

Color; interference; the Doppler effect; the electromagnetic spectrum; radio telescopes; radiation laws; quantum theory of light; particles and waves

SCENE 3 *Atoms and Radiation* 115

Atoms; charge; electricity and magnetism; Maxwell's unification of electromagnetic phenomena; quantum theory; the uncertainty principle and the statistical nature of the behavior of matter

SCENE 4 *Special Relativity—Meaning of Space and Time* 130

Speed of light; simultaneity; length and time in relativity; equivalence of mass and energy

SCENE 5 *General Relativity; A New Meaning of Gravitation* 142

Principle of equivalence; spacetime; warping of spacetime by matter; tests of general relativity

Act III • The Depths of Space

The distance ladder and surveying space

SCENE 1 *Mapping Our Own Neighborhood* 156

Time and place on earth; distances in the solar system

SCENE 2 *The Search for Stellar Parallaxes* 168

Stellar parallax; aberration of starlight; binary stars; stellar distances and motions; the Milky Way and Herschel's "grindstone"; stellar magnitudes

* SCENE 3 *Analyzing Starlight* 186

Stellar distances from the inverse-square law of light; stellar spectra; the Hertzsprung-Russell diagram; stellar properties

* SCENE 4 *The Milky Way Galaxy—The Old Universe* 197

Shapley's studies; the structure of the Galaxy; radio mapping of the spiral arms; our modern view of the Galaxy

SCENE 5 *Beyond the Milky Way—The New Universe* 209

The "nebulae"; the controversy over the nebulae; Hubble's resolution of the controversy; the extragalactic distance ladder

Act IV • **The Birth, Life, and Death of a Star**

The evolution and fates of stars, and the role of the atomic nucleus

SCENE 1 *The Anatomy of a Star* 224

The sun and its properties; solar phenomena; the solar interior; early ideas of the sun's energy source

* SCENE 2 *The Atomic Nucleus* 242

Cosmic rays; subatomic particles; the quark theory; nuclear forces; nuclear fusion; the problem of the solar neutrinos

* SCENE 3 *Stellar Evolution and Nucleogenesis* 254

Models of stars; explanation of the main sequence; evolution of stars to red giants; formation of the heavy elements in stellar interiors; pulsating stars; mass ejection from stars; stellar explosions

* SCENE 4 *Between the Stars* 267

Gas and dust in interstellar space; bright and dark nebulae; dynamics of the interstellar medium; radio waves from interstellar space; interstellar molecules

SCENE 5 *Origin of the Solar System* 282

Age of the solar system; the solar nebula; formation of the planets and their satellites; the minor planets, comets, and meteoroids

SCENE 6 *Checking Out the Theory—Evolution of M2001* 294

Comparison of the predictions of stellar evolution theory with real star clusters of different ages

SCENE 7 *The End* 303

The end states of stellar evolution: white dwarfs, neutron stars, and black holes; pulsars; predicted properties of black holes

Act V • **The Search for Life**

Are we alone?

SCENE 1 *The Living Earth* 324

Structure of the earth and its atmosphere; plate tectonics and continental drift; origin of life on the earth

SCENE 2 *The Moon: A Resident Alien* 337

Kinetic theory and the moon's lack of an atmosphere; the moon's motions; the lunar surface; origin of the moon and its features

SCENE 3 *The Other Worlds* 350

The other planets in the solar system and the chances of life on them; satellites of the planets; studies of the planets with space probes; the Martian dream

SCENE 4 *Is There Intelligent Life Elsewhere in the Galaxy?* 372

Estimate of the number of habitable planets in the Galaxy; estimate of the number of civilizations; interstellar travel; UFOs; interstellar communication

SCENE 5 *Is There Intelligent Life on Earth?* 384

A sober appraisal of the rate at which we are fouling our own nest with our gluttony, our wastes, and our own number

Act VI • **The Grand Questions**

The origin and evolution of the universe

SCENE 1 *The Galaxian Zoo* 396

The other galaxies and their properties; the Hubble law;

PROLOGUE

"In the beginning . . . " So opens the Bible, with its account of the origin of the universe. To many, the Biblical story, in one or another of its various interpretations, is the absolute truth.

Science neither confirms nor refutes the Genesis story of the creation. Science is concerned with an interpretation of nature—not an explanation. Science begins with observations, or the results of experiments, and attempts to find a model (or hypothesis, or theory) that predicts those—*and other*—observations and experimental results as consequences of that model's basic postulates. A scientific hypothesis, however, can never claim to be absolutely true, because ultimately every absolute truth must be based on premises of faith—a subject science does not—and cannot—deal with.

This is not to say that most of us, including virtually all scientists, do not really *believe as truths* many of the most basic scientific hypotheses. For example, I have no inner doubts whatsoever that the earth revolves about the sun. These beliefs, however, are not really verifiable within the means of science; indeed, I cannot *prove* that anything in the universe—except me—exists! Yet our understanding of the motion of the earth, and of the sun's rising tomorrow, is so well founded that most of us would be shaken to our psychological roots if the whole concept should be shown to be completely wrong. Technically, a belief such as that the sun will rise tomorrow is a religious, rather than a scientific, belief; nevertheless most of us really believe in what we regard to be well-established scientific laws. It is different, though, with our ideas about the beginning of the universe.

We do not yet have a complete and well-tested theory of the origin of the universe itself. Still, I am often asked, how do *I* think it all began—what theory of cosmology do *I* believe in? Frankly, I do not know, nor do I even have any particular opinion, other than that our most-favored views on the subject today are probably wrong. Mind, I have no evidence that the current views of most cosmologists are in error; it's just that our experience has shown that most often new ideas at the frontier of science turn out to be incorrect. It is only after an incredible amount of hard work, wrong turns up blind alleys, heated arguments by experts, and failures of beautiful theories

FIGURE P.1
Detail from Ghiberti's gold plates depicting the story of *Genesis* on the doors of the Bapistery in Florence.

that a picture emerges that is generally agreed to fit all the observations and to withstand the most crucial tests. We have not yet reached that step in cosmology, so I suspect that the currently favored view, which I am about to describe, will not even be remembered by those teaching astronomy a few decades from now.

But it *is* fun to speculate. The most conservative form of speculation is to suppose that all our current understanding of the laws of physics is basically correct—that the universe is real, that it is the same everywhere, and that it behaves according to the most straightforward and simple interpretation of those laws, which we assume to be known. In other words, what do we find if we extrapolate our best current knowledge of nature backward in time to "the beginning"—whatever that may mean?

We cannot really know the real "beginning" (what came before?), but if we do extrapolate backward in time, with the best knowledge we have (quite possibly wrong, as I have said), we come to a time some 15 to 25 thousand million years ago when all the matter in the universe was compressed to an enormously high density and at an incredibly high temperature—at least 10 thousand million degrees. Even then the universe was infinite—or at least may as well have been for all we can ever tell—and occupied no particular place; that is, there was no special "point" in space from which the universe began. The stuff of the universe was not that which exists today. Indeed, the atoms that compose our bodies had not yet even been formed. All that existed were certain basic particles, such as protons, antiprotons, electrons, positrons, neutrons, antineutrons, and neutrinos (all of which we shall discuss later).

In any event, this mass, according to present thinking, being very hot and very dense, expanded rapidly, and as it expanded, it cooled. Do not be concerned about the fact that it could have been infinite and still have expanded into greater infinities; after all, if it were finite, we could never picture what was out there beyond it anyway.[1] (Later we will see that, at least philosophically, we can avoid infinities with curved spacetime, but it is sort of begging the question.) We must, nevertheless, imagine that the universe expanded, thinned out, and cooled.

As the universe expanded and cooled, for a period of several minutes its material passed through temperatures in the range of from a few tens of millions to a few hundreds of millions of degrees. Then the various subnuclear particles fused together to form the nuclei of various kinds of atoms. However, the many collisions between the particles in the hot mass broke up the nuclei of the more complex atoms almost as quickly as they formed and, by the the end of the period when nuclear fusion could occur, only the simplest kinds of atomic nuclei survived. (There were no atoms yet—it was still too hot; the electrons that would belong to the atoms at room temperature were all bouncing about by themselves as free particles.) Most of the nuclei—over 90 percent, were those of hydrogen, that is, they were just protons. Something less than 10 percent were the nuclei of the next simplest kind of atom, helium. A tiny smidgen, perhaps one in a hundred thousand or less, consisted of nuclei of heavy hydrogen (consisting of one proton and one neutron), and there were only the barest insignificant traces—virtually none—of the nuclei of the atoms of other kinds of elements. Essentially, then, the universe was a great (infinite) ball of hydrogen and helium nuclei and electrons. There were no stars, no planets, no dinosaurs, not the atoms of carbon or oxygen—not even the *nuclei* of those atoms! It was a simple, pure kind of universe (the kind theoreticians like). Now, whereas we may speak lightly of it, this may very well have been the situation—at least that is what our presently known laws of physics would suggest.

Meanwhile, that infinite ball of tremendously hot, ionized gas went on expanding, and as it did so, it continued to cool. After nearly a million years it had cooled down to only about 3000°K.[2] By that time the density of the gas was very low—an extremely high vacuum on our laboratory standards. Still, it was *opaque,* in that the material absorbed light and other radiant energy very efficiently. Hot gas radiates, so this *primeval fireball* was filled with light; if you could have been inside, it would have appeared extremely bright. Because it was so opaque, however, you wouldn't have been able to see your hand in front of your eyes. It would have been, until then, like the interior of a star.

At this point the atomic nuclei began to combine with electrons to make ordinary atoms (of hydrogen and helium), and very quickly the universe changed from a hot ionized gas to a neutral gas. The neutral atoms no longer radiated and no longer absorbed radiation, and were no longer dominated by it. Then the gas cooled even more rapidly, and in those regions where its density was ever so slightly greater than average, gravitation pulled the atoms toward each other. Thus the matter of the universe gradually coalesced into vast tenuous clouds.

[1] Galileo (read his *Two New Sciences*) was worried about infinities and greater infinities. For example, the largest number we can imagine is infinity; yet every number has a square, and a cube, and a 47th power—still greater infinities!

[2] In astronomy we always measure temperature from absolute zero (−273°C) and in Celsius (or centigrade) degrees (See Appendix 4). An absolute temperature of 3000°K is about 4900° Fahrenheit.

Within the great clouds of matter, there formed subcondensations of enormous size. Today these subcondensations contain almost all (perhaps *all*) the matter of the universe; we call them *galaxies*. We suspect that the galaxies began as clouds of gas (within the larger original clouds), but most of the material of the galaxies fragmented again into smaller clouds, which gradually condensed into relatively small balls of gas. These "small" gas balls were typically a million kilometers in diameter, and contained hundreds of thousands of times as much matter as is contained in the entire earth. These spheres of gas held themselves together under the force of their own self-gravitation, which compressed the matter inside them to extremely high densities and pressures, and thereby raised the temperatures of those internal gases to millions of degrees. In other words, the matter of the universe, which had cooled off and condensed into galaxies, and within them into small balls of gas, was now in the deep interiors of those balls reheated—in fact, reheated to temperatures high enough that the atoms, again stripped of their electrons, underwent nuclear fusion once more. This time, the nuclei of the hydrogen atoms fused to form more atomic nuclei of helium; in the process, enormous amounts of energy were released (we shall see how later), which maintained the high temperatures and made those spheres of gas self-luminous. We call them stars.

Remember that these original stars, the *first-generation stars*, were composed purely of hydrogen and helium (assuming, of course, that this picture of the early history of the universe is correct). There were not yet any of the kinds of atoms in existence that would be required to make a solid planet like the earth. There were probably large numbers of *double stars*—pairs of stars revolving about each other—and doubtless many *clusters* of stars, but no "rocky" stuff. Typical galaxies contained tens of thousands of millions to hundreds of thousands of millions of these stars of hydrogen and helium, all widely separated from one another. Also, at least in many galaxies, a small

FIGURE P.2
The galaxy, M81, in *Ursa Major*. (*Lick Observatory*)

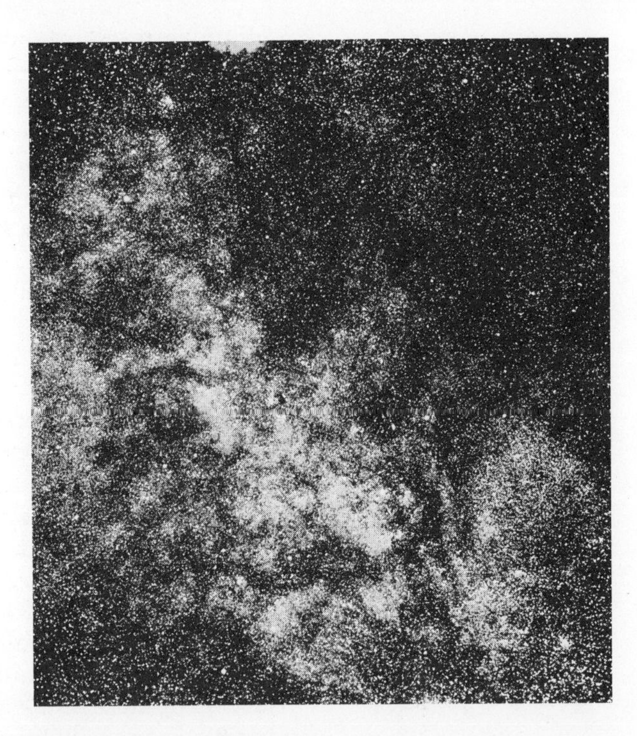

FIGURE P.3
The Milky Way in *Sagittarius.* (*Yerkes Observatory*)

fraction of the original gas did not condense into stars, but remained tenuously strewn about in interstellar space. Still, there was only matter composed of hydrogen and helium; no atoms to make people.

Then where did *we* come from? From atoms "cooked up" in the stars themselves! We estimate that it took about a thousand million years from the time of the big bang that started the expansion of the universe to make those first stars. Once formed, the stars changed things relatively rapidly. Many, perhaps most, of those first-generation stars were so hot inside that the nuclear conversion of hydrogen to helium progressed at great speed, and in less than a million years some of the stars had used up the available hydrogen in their central, hottest regions. Later we shall see how changing the composition of the gas in the center of a star forces the star to change its whole structure. As a result of the nuclear reactions that give the star the energy by which it shines, it evolves through stages of various sizes, densities, and temperatures.

Soon after its central hydrogen is exhausted, the outer parts of a star distend until the star is a giant many times its former size. Meanwhile, its internal gases compress and heat up still more, until its central temperature reaches hundreds of millions of degrees. With this much higher temperature, new kinds of nuclear fusion occur; helium nuclei fuse into heavier elements, building up carbon, and later oxygen, nitrogen, and other familiar elements —even iron. At advanced stages in their evolution many of the stars eject large fractions of their gaseous constituents into space, as great shells of gas. Others explode violently, spewing their matter into space. In the heat of those explosions, we can now understand how at least small amounts of even the heaviest elements can be formed, including all the kinds of atoms known in nature.

Thus, within a few millions of years after a galaxy has formed into stars, many of those stars are ejecting much of their material into the interstellar medium, carrying into space not only the original hydrogen and helium but newly formed atoms of helium and heavier elements as well. The gas of the interstellar space, once pure hydrogen and helium, is now "contaminated" with other kinds of atoms. Meanwhile, that interstellar gas condenses, here and there, into new stars, *second-generation stars,* formed of a richer assortment of chemical elements. These new stars repeat the process, first undergoing fusion of hydrogen into helium, then forming more heavier elements, and finally ejecting some of those back into space, to enrich further the interstellar material from which third and future generations of stars will form.

In this way, over the thousands of millions of years, the material that makes up the stars and interstellar material of a galaxy evolves in composition. It is still *mostly* hydrogen and helium, but a small percentage (from 1 to 4 percent, perhaps) is a mixture of the various other chemical elements. So it was in a particular galaxy a little under 5 thousand million years ago. At that time, quite recently in the lifetime of that galaxy, a cloud of interstellar gas began to fall together under the influence of its own gravitation. That cloud of gas was slowly rotating, and as it contracted to ever smaller sizes, it spun faster and faster to conserve its *angular momentum*—just as a figure skater spins faster when she pulls in her outstretched arms close to her body. We call that particular spinning cloud the *solar nebula.*

Because of its rotation the solar nebula flattened, much like a pizza, into a disk. Half or more of the material managed to condense into a central sphere of gas that eventually became our sun. The remainder was left behind in that rotating disk. The sun, at the center of the nebula, became hot, and began shining by nuclear fusion of hydrogen to helium. Today, 4.6 thousand million years later, the sun still derives its energy by turning hydrogen to helium, and it will be another 5 thousand million years before the hydrogen is used up at its center.

FIGURE P.4
The Crab nebula in *Taurus,* remnant of the supernova outburst of 1054. *(Hale Observatories)*

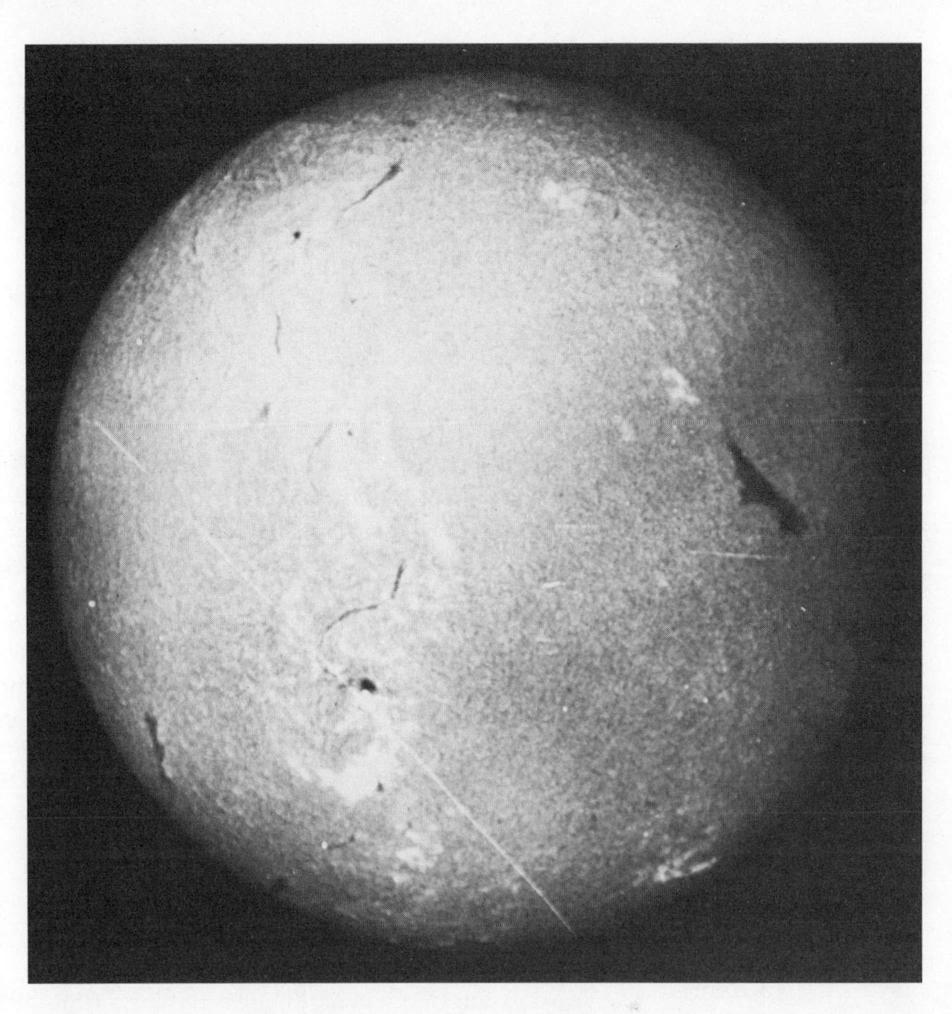

FIGURE P.5
The sun, a garden variety star. *(Hale Observatories)*

Meantime, while the sun was forming, some of the atoms of gas in the flat disk—especially those of the heavier elements—began to condense into tiny solid grains. Far out in the disk, it was quite cold, and most of the grains consisted of ice and substances such as frozen carbon dioxide, but there were some rocky and metallic grains as well. In the inner part of the disk, the radiation from the young sun heated up the gas, so that it was too hot for ices, and only metallic and rocky stuff condensed. These grains of dust collided with each other in the rotating solar nebula, and began to collect, or *accrete,* into larger bodies. Before long, most of the solid grains had either collected into a small number of fairly sizable bodies revolving about the sun or had been lost to the solar nebula. These bodies became planets.

Close to the sun, the planets formed mainly from rocky and metallic materials. Farther out, where it was cooler, the planets ended up with a large fraction of icy material. The largest of them, in addition, attracted a considerable amount of the gases of the nebula to their rocky and icy cores. In fact, hydrogen and helium captured from the nebular disk, now compressed to a solid or liquid state, make up most of the mass of the planets Jupiter and Saturn. Even when they were completely formed, the combined mass of the planets was scarcely one-tenth of 1 percent of the mass of the sun. The rest of the material in the solar nebula has long since been lost to the solar system, probably by the action of radiation pressure of sunlight.

FIGURE P.6
Jupiter. *(Hale Observatories)*

We now turn our attention to one of the tiny planets—our earth—the third from the sun. Originally that little object had no atmosphere, because it had too little gravitation to capture the warm gases in the solar nebula. However, some of its rocks contained hydrates, carbonates, and nitrogen compounds. They also contained some radioactive elements that release energy as they slowly decay. Beneath the surface, that energy was trapped by the surrounding rocks, and heated them until they released the water, carbon dioxide, and nitrogen they held in chemical combination. Those substances, in the form of hot gases, escaped to the surface, mostly through vulcanism. The water, on reaching the cooler surface, recondensed to form the oceans. The carbon dioxide was able to recombine with surface rocks. This left the nitrogen, the major constituent of the earth's atmosphere today.

Most experts, however, believe that for a time other gases existed as well. Some of the molecules of water and carbon dioxide could have been dissociated by ultraviolet solar radiation. This process released some hydrogen gas, and then ammonia and methane (compounds of hydrogen, nitrogen, and carbon) could easily form. Thus the early environment at the surface of our earth was very different from that of today. There was what biologists call a "soup" containing water, hydrogen, ammonia, and methane. Laboratory experiments show that when such a soup is subjected to electrical discharge (such

FIGURE P.7
Vulcanism was a major source of the earth's atmosphere. *(U.S. Biological Survey)*

FIGURE P.8
The Andromeda galaxy.
(Hale Observatories)

as lightning) or irradiated with ultraviolet radiation (lots of it comes from the sun), organic compounds including amino acids, porphyrins, and sugars are produced in large quantities. These are *not life,* but are the basic building blocks of all living cells. We do not yet know how to produce living organisms, but we do know how the chemical units of which they formed would naturally arise in the expected environment of the primordial earth. Somehow, at least here on earth, some of those units found their way into replicable biological organisms. Then began the long road of biological evolution.

We do not know when the first life appeared on earth. The earliest fossils found so far are of blue-green algae about 3.5 thousand million years old. Evidently life got started during the first thousand-million-year history of our planet. It was many hundreds of millions of years after that before complex creatures evolved and less than one thousand million years ago when land animals emerged. It is the green plant life, mostly in the oceans, that is responsible for the oxygen in our present atmosphere. It is estimated that 600 million years ago the oxygen content of the atmosphere was less than 1 percent. Photosynthesis by green plants has built it up to its present 21 percent. The giant dinosaurs lived very recently, only a few hundred million years ago. The age of great mammals is more recent yet.

We have now seen that the matter in the universe is concentrated in galaxies of stars. Most galaxies, if not all, belong to clusters. Our own Galaxy is part of a small cluster called the *Local Group.* The Local Group is tremendously insignificant on the cosmic scale. We would never even notice a comparable group at a distance of even 1 percent that of the most remote objects we can observe. It is really our own, very local, cosmic neighborhood. One of the neighboring galaxies in our Local Group, the Andromeda galaxy, is barely visible, from the United States and Europe, to the unaided eye on a dark, moonless autumn night. Light, traveling at the highest possible speed of 300,000 km/s, takes about 2 million years to reach us from that very near galaxian neighbor. We see it, in other words, as it was when light left it, starting its journey to us, 2 million years ago.

Try to look at the Andromeda galaxy some evening. Failing good weather or a dark enough sky, settle for one of its photographs in this book. Then contemplate what was happening about the time the light you see left its source: man was first emerging on the earth.

We shall now turn back through only a tiny part of that brief period during which man has been around – in fact only through the final 0.1 percent of the 2 million years – to begin our drama, the drama of man's exploration of his realm – the universe.

ACT I

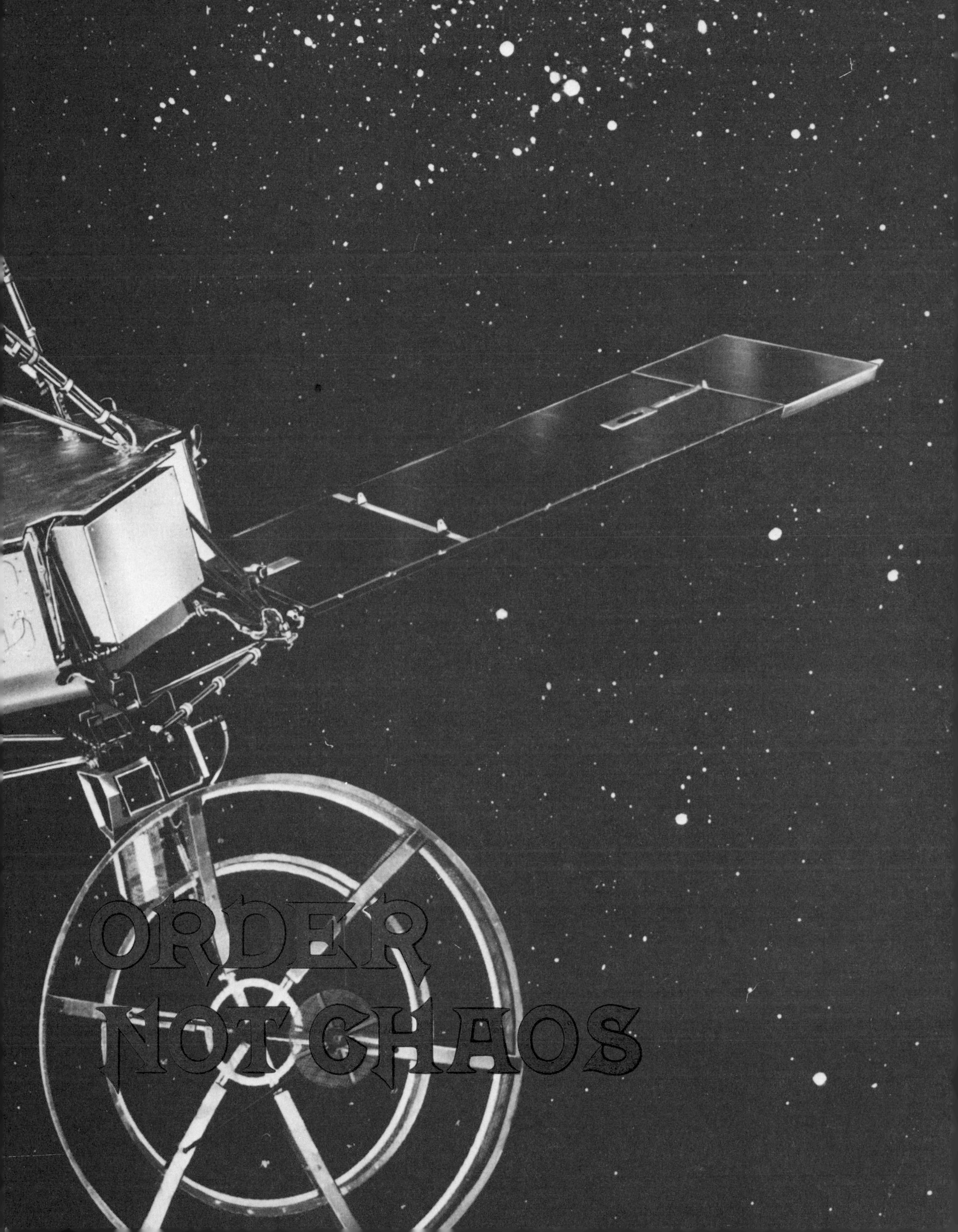

ORDER NOT CHAOS

SCENE 1

The Fixed and Wandering Stars

Among prehistoric human traces we find evidence that the advantages of some practical knowledge of celestial events, especially of time and the seasons, had not been entirely lost to those prehistoric people. For example, petroglyphs depicting the sun, moon, and stars exist in caves. Particularly impressive are the ruins of those monuments of stone constructed by Stone Age people living in what is now Great Britain and some parts of northern Europe. The most famous is Stonehenge, on the Salisbury Plain of southern England, only one of nearly a thousand such monuments known. Many have been carefully surveyed, and the alignments of the stones suggest that some may have been intended to mark the directions of the rising and setting of the sun at different seasons. In any event, man's interest in the heavenly bodies and their motions predates modern civilization by many millennia.

Every civilization was involved to some extent with the study of the heavens. Most had calendars, and kept track of the seasons. The sophisticated calendar of the Maya is particularly interesting, but the Maya also observed and kept records of the motions of some of the planets. The Polynesians evidently navigated by the stars. The Chinese maintained records of unusual celestial events. The Egyptians observed the stars to keep track of such occurrences as the annual flooding of the Nile. The Hindus, the Arabs, the Babylonians, and the Greeks all devoted energies to the understanding of the heavens.

It would be nice to think that all these people were interested in new knowledge for its own sake; some probably were. For the most part, though, a knowledge of astronomy was very practical. Crops, for example, had to be planted and harvested seasonally. And then, for many early peoples, especially those in Babylonia, and later in Greece, the bodies in the sky became inextricably involved with their religions. It is not hard to understand why this should be so. In a life where earthly events seem so random and unpredictable, it is difficult not to be impressed by the magnificent regularity and dependability of the motions in the sky. In a world of chaos the heavenly bodies must have seemed to provide a sense of familiarity and security.

The "Fixed" Stars

The stars themselves are in rapid motion, moving with respect to us at speeds of many kilometers per second. But they are so far away that their motions are not discernible to the unaided eye, even throughout a lifetime; in

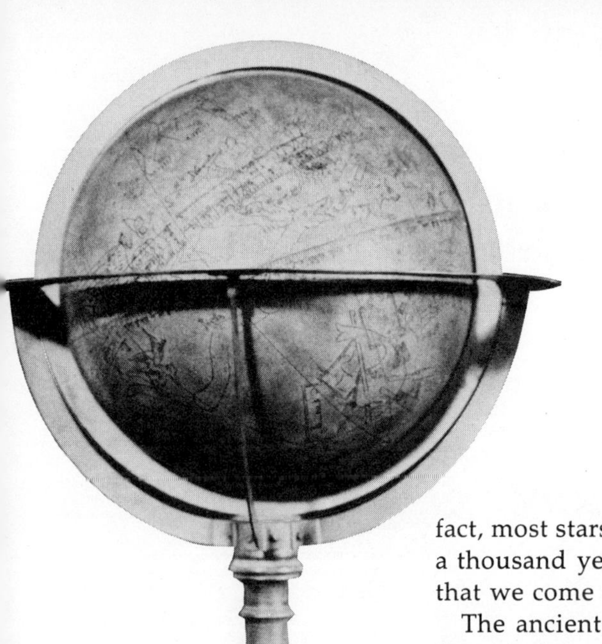

FIGURE I.1
An Arabic celestial globe. *(Istituto e Museo di Storia della Scienza di Firenze)*

fact, most stars would appear in very nearly the same relative positions after a thousand years. Thus their apparent patterns become familiar landmarks that we come to recognize and know throughout our lives.

The ancients saw these same star patterns, and in their legends named them in honor of people, animals, or things, much as we name states and cities after our heroes. These star groups are called *constellations.* Modern astronomers still refer to constellations, but now to designate particular areas of the sky, and the boundaries between modern constellations are somewhat analogous to political boundaries on the earth. The ancients, though, actually assigned individual stars in a constellation to particular parts or places in the person or animal for which it was named; thus *Betelgeuse* is the right shoulder of Orion, the Hunter, and *Regulus* is the heart of Leo, the Lion.

The stars are, of course, all at different distances from us, and the constellations are not real groups or clusters of stars (although a few small groups of stars that can be seen with the unaided eye are true associations; for example, the *Pleiades*). Nevertheless, the stars certainly give the impression of being fixed to the inner surface of a great hollow bowl centered on the earth. Most ancients believed in the literal existence of such a surface, and called it the *celestial sphere;* we still call it that today. Even though we know that the sphere is only imaginary, it is a useful concept for denoting directions of objects in the sky—we say they have such and such "positions" on the celestial sphere. The celestial sphere is not just a bowl or a hemisphere; it is a complete globe surrounding us. From the surface of the earth we see only half of it above our horizon at any one time, but it takes very little imagination to picture the other half extending down beneath us.

The ancient observers saw that the celestial sphere appears to turn about us each day, as though pivoted on a great axis passing right through the earth. As it turns, stars appear to rise in the east, move across the sky, and set in the west. Roughly half a day later, those same stars which had set in the west are seen to rise again in the east. But as they do so, the stars maintain the same fixed patterns with respect to one another; for example, the constellation Orion always looks the same—just as the map of Australia on a globe of the earth always looks like Australia, even though you may spin the globe around on its stand. That is why the ancients called the stars the *fixed stars.*

Today, nearly everyone knows that the rotation of the celestial sphere is illusory; it is just a reflection of our own motion caused by a rotation of the earth about its polar axis. This means that the pivot points of the celestial sphere, the *north* and *south celestial poles,* must lie exactly along an imaginary line through the poles of the earth. Halfway between the celestial poles, and

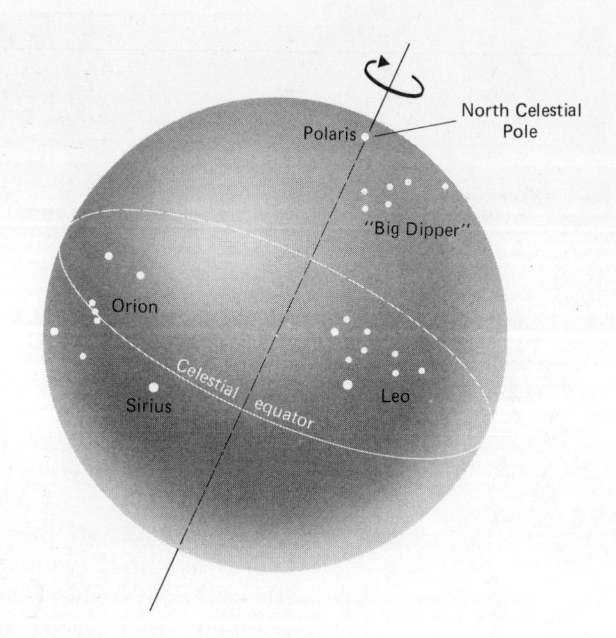

FIGURE I.2
The celestial sphere.

dividing the sky into Northern and Southern hemispheres, is the *celestial equator*, exactly analogous to the equator of the earth.

The Babylonians invented the convention of dividing a circle into 360°, and each degree into 60 smaller parts called minutes of arc ($'$), and each minute can be divided into 60 seconds of arc ($''$). Thus $60'' = 1'$ and $60' = 1°$. The distance along any circle (such as a meridian or the equator) makes a certain angle at the center of that circle (see Figure I.3); we say that the arc distance *subtends* that angle. We also use degrees, minutes, and seconds to measure the sizes of angles. Thus 90° is a right angle. The smallest angle that can be perceived by the unaided eye is about one minute (1′). The sun and moon each subtend an angle of about ½° (30′). Naturally most readers already know about latitude and longitude as coordinates on the surface of the earth. We denote the positions of places on earth according to how many degrees they are north or south of the equator, and east or west of Greenwich, England. We use very similar coordinates to specify the positions of stars or other objects on the celestial sphere. These celestial coordinates were invented by the ancients centuries before Christ. In the second century B.C. Hipparchus measured the angular coordinates of about 850 stars and prepared a catalog of those positions.

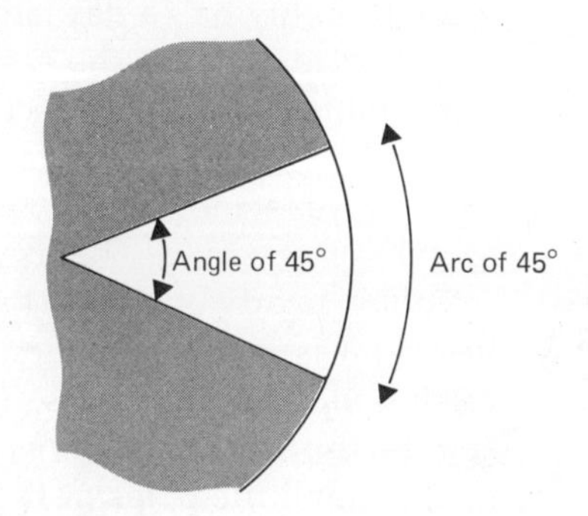

FIGURE I.3
Angular and arc measure.

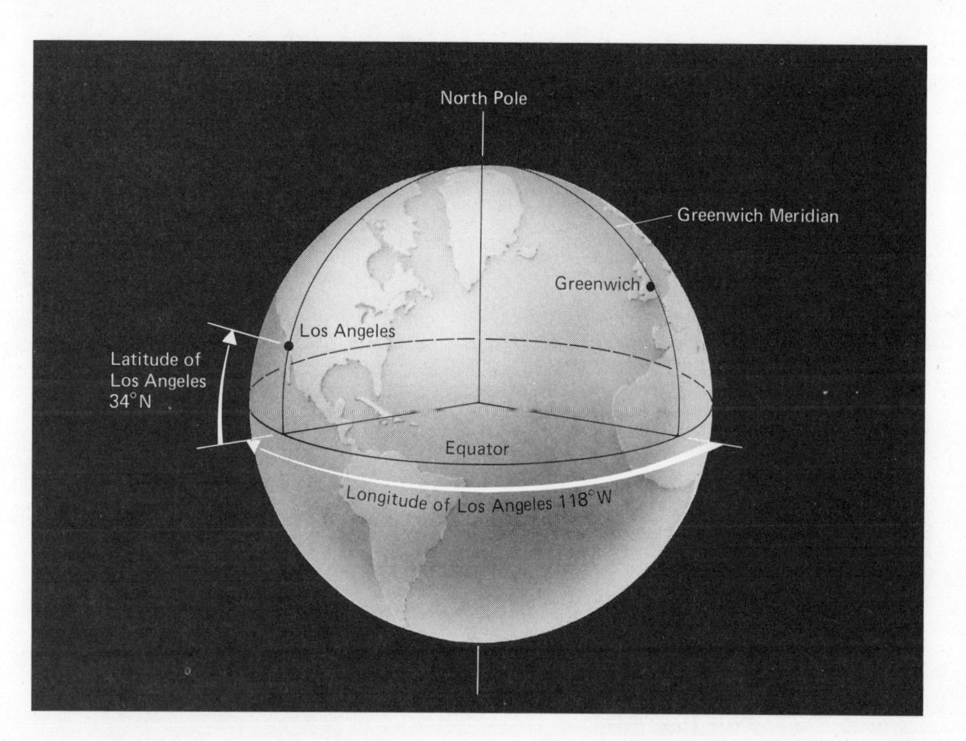

FIGURE I.4
Latitude and longitude.

---

1. How many degrees are there in a right angle?
2. How many seconds of arc are there in a semicircle?
3. The circumference of a circle is 2π times its radius. How many degrees are subtended at the center of a circle by a distance laid out along the circumference of length π times the radius?
4. If you are not already familiar with latitude and longitude on the earth, consult an appropriate reference. Then give the latitude and longitude of
   a) the north pole;
   b) the south pole;
   c) the point where the Greenwich meridian intersects the equator.
5. Now consult a good atlas, and find the latitude and longitude of
   a) London;
   b) Sydney, Australia;
   c) New Delhi;
   d) Tahiti;
   e) Los Angeles.
6. As seen by an observer at the earth's equator, where does the north celestial pole appear to be in the sky?

How the sky appears to turn overhead depends on where you are on earth. For example, to a hypothetical observer standing at the earth's north or south pole, the sky would appear to rotate around a point overhead, and nothing could rise or set (Figure I.5). On the other hand, to an observer at the equator of the earth, everything appears to rise and set during the day (Figure I.6).

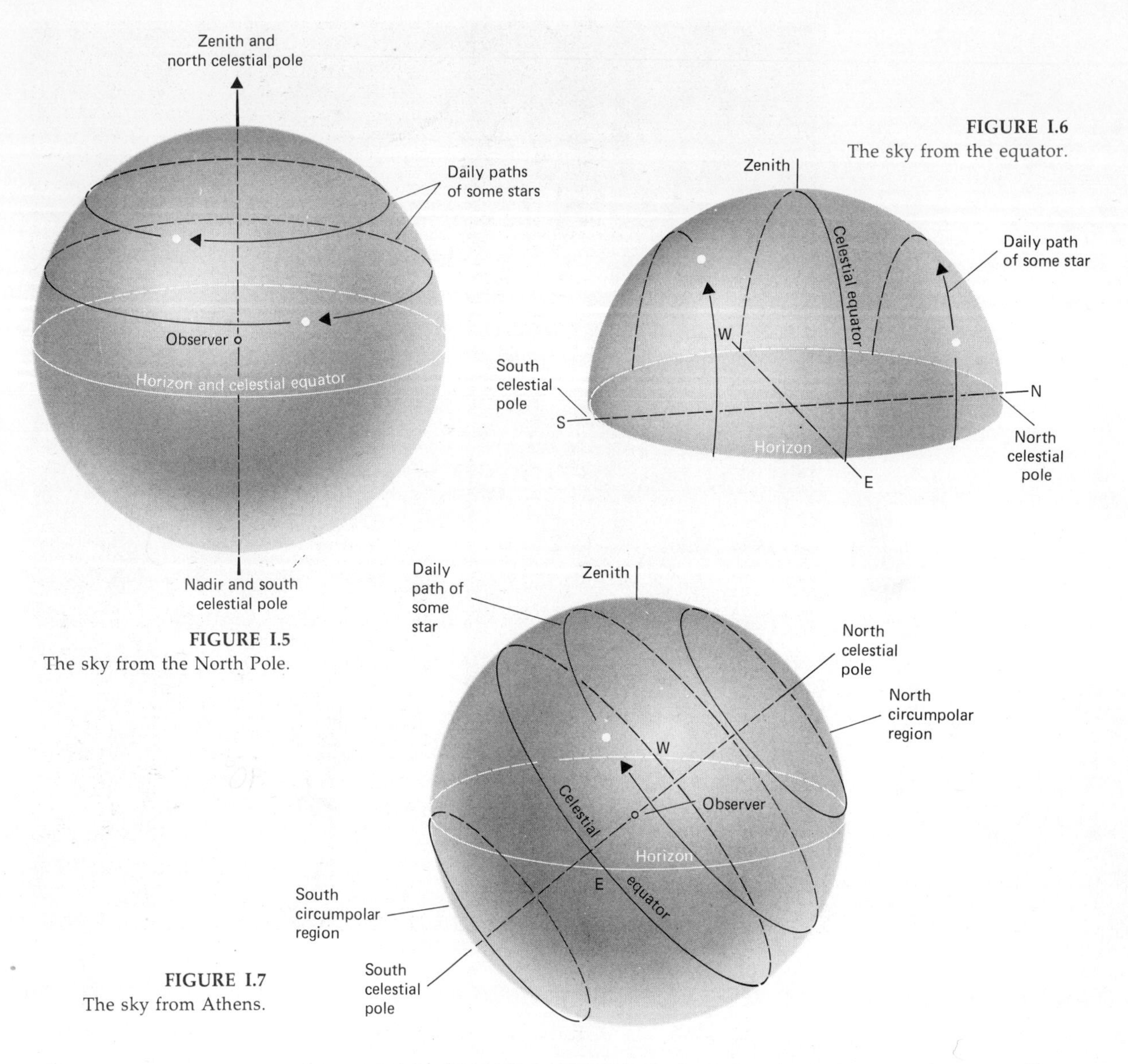

FIGURE I.5
The sky from the North Pole.

FIGURE I.6
The sky from the equator.

FIGURE I.7
The sky from Athens.

The overhead point, or *zenith,* for a person at Athens is shown in Figure I.7 as a point at the top of the celestial sphere. Of course different observers at different places on the earth would consider "up" to be at other places on the celestial sphere; everybody has his own private zenith. Halfway between the zenith and the *nadir* (straight down) is the circle we call the *horizon,* which divides the half of the sky we see from the half hidden by the earth. Anyway, our observer at Athens is at a latitude of 38°N, that is, 38° along the curved surface of the earth from the equator to the north pole. Thus the north celestial pole will lie neither on the horizon, as at the equator, nor at the zenith, as at the north pole, but 38° of the way up from the horizon to the zenith—at an *altitude* of 38°.

---

7. To an observer at the north pole, where does the celestial equator appear to be in the sky?

FIGURE I.8
Polar star trails.
(Yerkes Observatory)

If you could actually see the north celestial pole in the sky, you could tell your latitude by its altitude. One way to locate it is to point a camera in that general direction, and take a long time exposure; the turning sky will cause the star images to trail around the celestial pole in concentric circles. In the mid-twentieth century there happens to be a fairly bright star very near the north celestial pole (about one degree away); we call it the *North Star,* or *Polaris.* Because of a slow change in orientation of the earth's axis in space (which we shall come to in the next scene), Polaris was not always so near the pole. Nevertheless, the ancients knew the location of the celestial pole in the sky, and navigators from early times knew that its altitude indicated their latitude. In fact, Aristotle pointed out that the apparently changing orientation of the sky to travelers moving north or south on the earth was one proof that the earth itself is spherical.

While the sky appears to pivot about the north celestial pole, stars near it do not rise or set, but simply circle around it perpetually above the horizon; they are said to be *circumpolar.* An equal area of the sky surrounding the south celestial pole lies perpetually below the horizon. From the United States, Canada, and most of Europe, the Big Dipper (actually part of the constellation of Ursa Major) is circumpolar above the horizon, while the Southern Cross never rises. The situation is reversed for those in Australia, New Zealand, and South Africa.

Stars outside the circumpolar zone all rise and set, but spend unequal times above and below the horizon. In general, for a northern observer, a star, or other object, north of the equator will be up more than half the time, and one south of the equator will be up less than half the time; it is just the other way around for a southern observer. A point *on* the celestial equator is always up exactly half the time. These details are important, as we shall see, because they account for the seasons.

Now that we have described the celestial sphere, we should be in a position to appreciate its simple, but strikingly regular, motion in the sky. It was a concept understood by all civilizations, and is basic to the development of interest in astronomy, worship of the heavenly bodies, and ultimately our modern technology. But let us go on!

The "Wandering" Stars

We all know that the sun's bright light, scattered about by the molecules of our atmosphere, makes a bright blue glare that hides the stars in daytime. But the ancients knew the stars were there all the time, day and night, just as we have described. How could they know? Because the sun does not occupy a fixed place on the celestial sphere, but moves about. It happens that the sun moves pretty regularly, describing a circular path around the sky (on the celestial sphere) each year.

Now even the Greek philosopher Aristotle, of the fourth century B.C., pointed out that the apparent annual revolution of the sun about the sky could be caused by a motion of the earth about the sun, so that we see the sun in different directions, projected in front of different stars at different times.

However, Aristotle cited a pretty good reason for thinking the earth did not move in an orbit about the sun. If it did, he said, and if the sun occupied the center of the celestial sphere, we would be passing nearest different parts of the sphere at different times, and the apparent sizes and shapes of the constellations should appear to change. Because they do not, we must be in the central position.

Now we must understand that the celestial sphere turns about us pretty rapidly, once a day, so that all things—stars and sun included—rise and move across the sky and set each day. The independent motion of the sun is gradual. The earth takes a year to revolve about the sun, so the sun appears to take a year to pass eastward among the stars, all the way around the celestial sphere. Thus the sun rises and sets with the stars each day, but each day it rises a little later—about four minutes later—than do the stars it rose with the previous day.

Six other objects also were known to the ancients to have regular, or at least semiregular, motions, independent of the fixed stars. The most conspicuous of these is the moon. The others are objects that *look* like stars to the unaided eye, although for the most part they are somewhat brighter, except that they change positions from night to night—some of them only *very* gradually—among the stars. The whole lot, all seven of them (including the sun and moon), were called *planets* by the ancients; *planet* is the Greek word for "wanderer." The ancients could have had no idea what the sun, moon, and planets really are. But they all deviate from the simple regular motion of the fixed stars. With this information, could one deny the significance of all these "wandering stars" to our own lives and destiny? So began the religion of antiquity that, in its various perversions, has survived to the present day.

---

8. As the moon revolves about the earth each month, the earth revolves about the sun, and both in the same sense—from west to east. As a consequence, after the moon has passed the sun in the sky, the time required before it passes the sun again is different from the time it takes the moon to revolve once about the sky with respect to the background of distant stars. Which period is greater, and why?

ELLIPTICAL ORBIT

It should not surprise us that to the ancients these planets became gods, or abodes of the gods, or at least symbolic of the gods. They were given the names of the Babylonian gods, and each planet assumed the attributes of the god whose name it bore. The Babylonians believed that the positions of the planets among the constellations in the celestial sphere must be important to their society, just as the position of the sun, and the corresponding seasons, was of obvious importance. They developed the religion of *mundane astrology*, which was based on the idea that the lots of kingdoms were dependent on the motions of the planets in the sky. This was around 1000 B.C.. Later the Greeks adopted astrology from the Babylonians, and developed it to a highly complex system.

The Sun and Seasons

The earth revolves about the sun in a path that is nearly circular and which lies in a flat surface, or *plane*. Consequently, during the year the sun appears to move about the sky in a path that appears as a circle. However, the plane of the earth's revolution about the sun is not in the plane of its own equator.

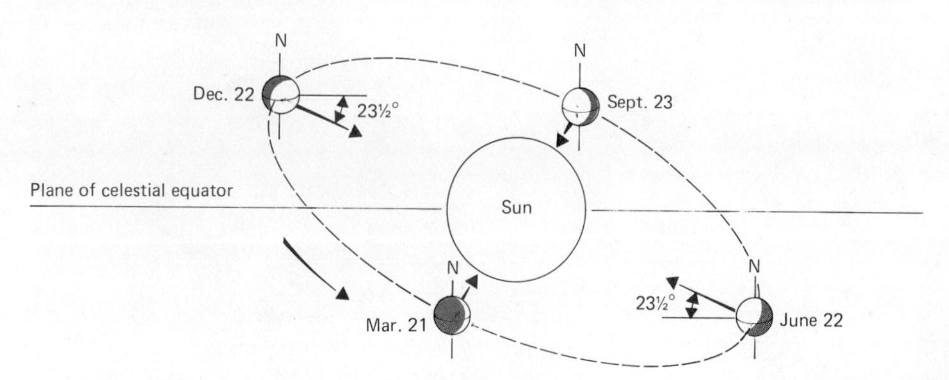

FIGURE I.9
The cause of the seasons.

That is to say, its axis of rotation is tipped to the perpendicular to its orbit (Figure I.9). That angle of tip, 23½°, is called the *obliquity of the ecliptic.* The result is that the sun's apparent path of motion about the celestial sphere, called the *ecliptic*, does not lie along the celestial equator, but is tipped to it by 23½°.

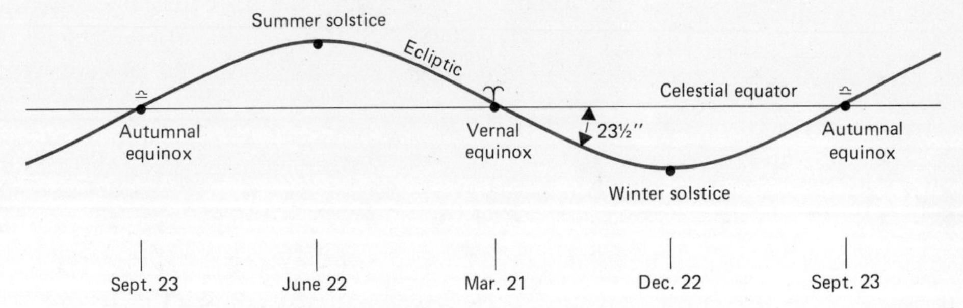

FIGURE I.10
Relation of the ecliptic to the celestial equator.

The sky is mapped on a flat plane, as in a Mercator projection, in Figure I.10. On March 21 the sun appears to cross the celestial equator in the sky. On this date it is at the position marked by the ♈ in Figure I.10—the symbol for the constellation Aries, the Ram, because 2 thousand years ago, when these things were given their names, the sun was in that constellation when it crossed from the south half of the celestial sphere to the north across the celestial equator. It is then that spring begins, and today we call that point marked by ♈ the *vernal equinox.* At that time the sun is on the celestial equator and is thus above the horizon exactly half the time; the length of the day then equals the length of the night. Literally, *equinox* means "equal night."

About three months later, the sun is as far north of the equator as it can get, at the *summer solstice,* and summer begins. The sun is then 23½° north of the celestial equator, and for people in northern latitudes it is above the horizon more than half the time (see Figure I.11). The sun is also higher in the sky when it crosses the *celestial meridian,* that circle which divides the east half of the sky from the west half. Summer is warmer than winter because the sun is up longer in the day and is higher in the sky. The hottest weather may lag the date of the solstice (about June 22) a month or two, because during the spring much of the sun's heat is used in melting snows; excess accumulation of

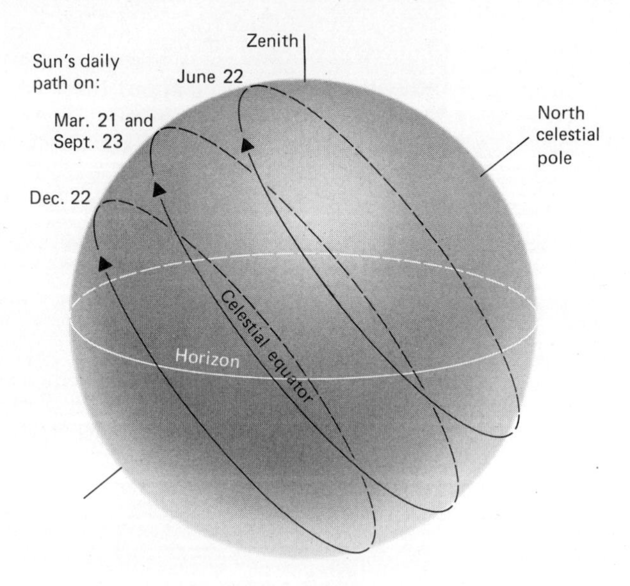

FIGURE I.11
The sun's path in the sky in different seasons.

Armillary Sphere of Antonio Santucci delle Pomarance, made for the Grand Duke Ferdinando I Medici in 1593. *(Istituto e Museo di Storia della Scienza di Firenze)*

The full moon, photographed by Apollo 17 astronauts. (*NASA*)

Telescopes donated by Galileo to the Grand Duke Ferdinando II and to his brother the Prince Leopoldo. The longest has a wooden tube covered with paper, a focal length of 1.33 meters and an aperture of 26 mm. (*Istituto e Museo di Storia della Scienza di Firenze*)

The eclipsed moon. (*Celestron International*)

The geometry of a lunar eclipse (not to scale).

The solar eclipse of 23 October 1976, seen from Ballaret, Victoria, Australia. (*Photographed by Charles David Long*)

The solar eclipse of 7 March 1970. (*NASA*)

A sequence of photographs of the solar eclipse of 20 June 1974. Photographed by Stephen Schiller. The final frame is the totally eclipsed sun photographed with a special infrared film; the different colors represent different infrared wavelengths.

sunlight generally continues to build up in northern latitudes until August or September.

Then, on or about September 23 the sun again crosses the equator, this time passing from north to south. On this day of the *autumnal equinox* there are again 12 hours of sunshine and 12 during which the sun is set. Henceforth the days grow shorter, because the sun, now south of the equator, is up less than half the time. The days shorten until about December 22, when the sun reaches the *winter solstice*, the days are at their shortest, and the sun is lowest in the sky. After December 22, the sun starts northward again, and the days grow longer. In ancient times this was a period of celebration, and early Christians chose this time to celebrate Christmas, which is why our modern

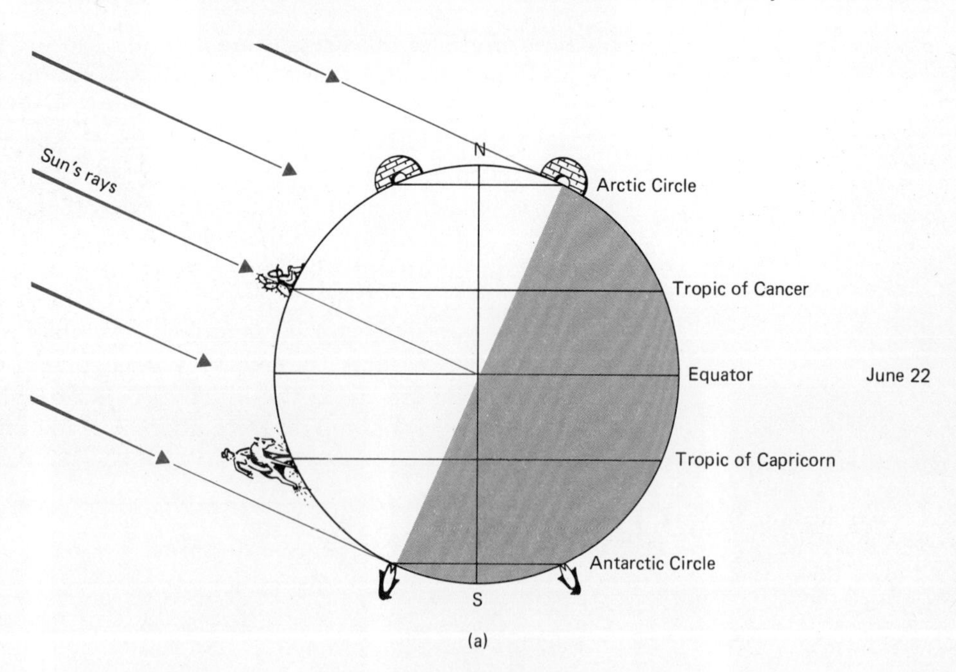

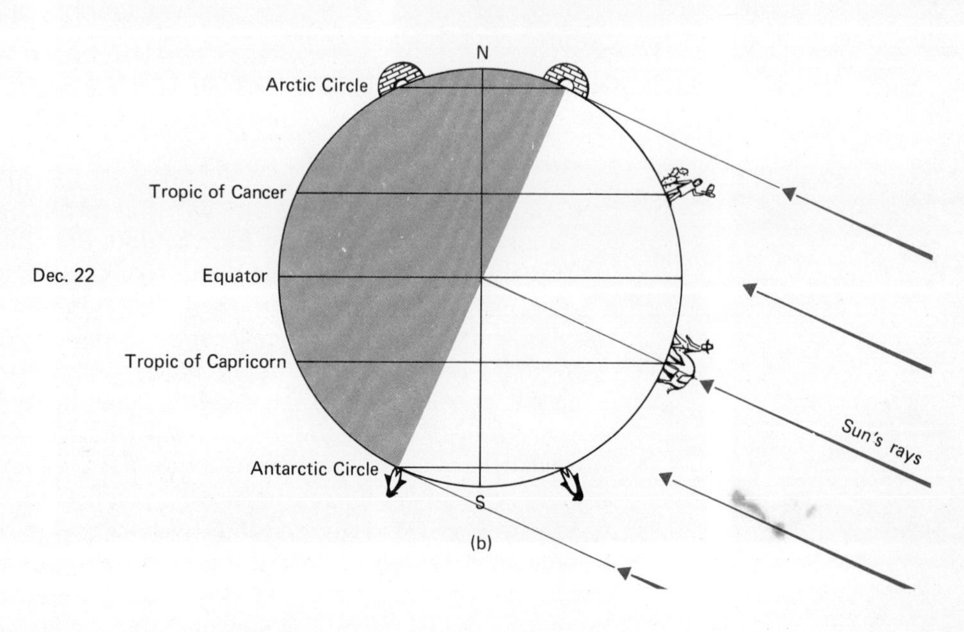

FIGURE I.12 (*a*) The seasons on June 22; (*b*) The seasons on December 22.

Christmas is so near the time of the winter solstice. Even the new year was intended to begin at about the time of the winter solstice. The slight disparity of dates between these events in our present calendar has come about through technical circumstances in the history of the calendar.

Naturally all the events of the previous paragraphs are reversed for Southern Hemisphere people. Their spring begins on September 23, and their summer on December 22. The reason that they are obliged to celebrate Christmas in the middle of their summer is because the history of this holiday began in the Northern Hemisphere. Nevertheless, my astronomical colleagues in the Southern Hemisphere still call the point in the sky where the sun is on December 22 (if they call it anything, which they are not disposed to do), the *winter* solstice.

These effects of the seasons are summarized in Figure I.12. Part (a) of the figure shows the direction of the sun's rays on the earth on June 22. Notice how direct they are on the ground at noon at northern latitudes from 15° to 30°. In fact, at 23½°N latitude the sun is overhead at noon. That parallel of latitude is called the *tropic of Cancer.* Notice, too, how the sun's rays shine over the north pole and illuminate the entire part of the earth within 23½° of the pole. That circle, 23½° from the pole, which delineates the region of 24-hour sunlight on the first day of summer, is called the *arctic circle.* On the arctic circle, the sun has still not quite set, even at midnight (the *midnight sun*). Part (b) of the figure shows the situation on December 22. Here we clearly see the reversal of the seasons. The latitudes analogous to the tropic of Cancer and the arctic circle are called the *tropic of Capricorn* and *antarctic circle,* which has the 24-hour sunshine on December 22, while the arctic circle has no sun at all.

---

9. From what place on earth does the sun, on December 22, appear to circle around the sky parallel to the horizon and 23½° above it?

10. From what places on earth does the sun appear to pass through the zenith at noon on June 22?

11. From where on earth is the sun seen to set but once a year—on September 23?

The Zodiac

So the sun, and the ecliptic, along which it travels, were of great importance to the ancients. Moreover the earth and all the other planets revolve about the sun in very nearly the same plane, like marbles rolling around on a table top (because, as we have seen, they all condensed from the same flat solar nebula). Thus the ancients, a few centuries before Christ, gave special significance to the belt around the sky, centered on the ecliptic, through which the planets all appear to move. That belt is called the *zodiac,* which means, literally, the "zone of the animals." The name comes from the fact that most of the constellations through which the zodiac passes are named for animals: Leo, the lion; Aries, the ram; Taurus, the bull; and so on. So important did the zodiac seem, that it was divided into twelve equal parts, called *signs,* and they provide another way of describing the location of each planet at any time. The twelve signs were numbered eastward from the vernal equinox, where the sun is situated on the first day of spring, and are named for the constellations that most nearly corresponded with them at that time (a few centuries B.C.). In order, to the east, they are Aries, Taurus, Gemini, Cancer, Leo, Virgo, Libra, Scorpio, Sagittarius, Capricorn, Aquarius, and Pisces. We shall return to these signs in the next scene.

To the ancients, the seven planets were the sun, the moon, and the five objects visible to the unaided eye that we call planets today: Mercury, Venus, Mars, Jupiter, and Saturn. Uranus, Neptune, and Pluto had not yet been discovered, and the earth was not realized to be an astronomical body, similar in physical structure to the other planets. In fact, the earth, in antiquity, was considered base, earthly stuff, of earth, air, water, and fire. The planets were heavenly, godly stuff, unlike that which we find here. Because of their motions the sun and moon were more akin to the other planets than to stars. The sun, moon, and planets were the wandering stars; their independent motions gave them something in common with one another. The fact that the sun is brighter than the rest (and the moon second only to the sun) indicated only a difference in degree, but not in substance. The sun obviously influences us humans, so why not the rest of the planets.

The Moon's Motions

From antiquity it was recognized that the moon shines because it is illuminated by the sun, and that the daylit side of the moon is always that hemisphere directed toward the sun. Aristotle's writings give a clear discussion of this and a correct explanation for the phases of the moon.

Look at Figure I.13. The moon was known to be much nearer to the earth than the sun, and to revolve about us in a path small compared to the sun's distance. Now if the moon is in the same general direction as the sun (*A* in the figure), its illuminated, daylight, side is turned away from us. We don't see the moon at all then, and call it *new moon.* Since it takes only about a month for the moon to go around the earth (this revolution is what is meant by the word "month," or "moonth"), it takes only a couple of days for the moon to be far enough in direction from the sun to see it in the evening sky,

FIGURE I.13
Phases of the moon.

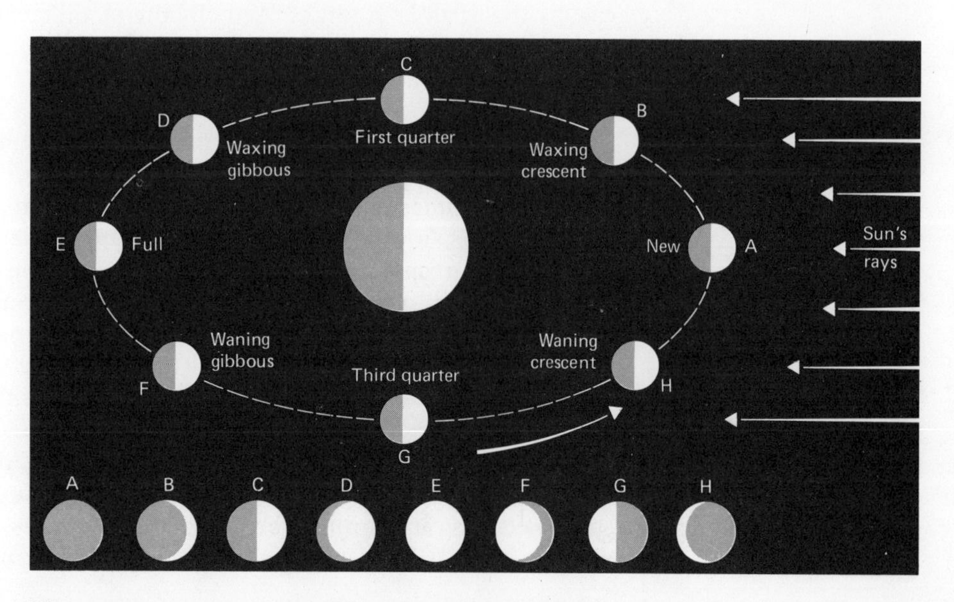

setting a couple of hours after the sun. Most of its daylight side is still turned away from us (*B* in the figure), but we do see a sliver of it, and the moon appears as a thin crescent—we call this crescent *after* new moon the *waxing* crescent.

Then, about a week after new moon, the moon is a quarter of the way around the sky from the sun, and is said to be at *first quarter*. Half of its daylight side is still turned from our view, but the half we do see appears as a half moon (C in the figure). At that time the moon appears on our celestial meridian at sunset, and it finally sets about midnight. Over the next few days, the moon continues to move further around the sky in its direction from the sun, and we see more and more of its daylight hemisphere. It then appears as a *gibbous* moon—a *waxing* gibbous during those days—and it sets later and later each succeeding night.

About two weeks after new moon, the moon is halfway around the sky from the sun (E in the figure). Then we look in one direction to see the sun and in the opposite direction to see the moon; the moon rises as the sun sets, is above the horizon all night, and does not set until the sun rises. Its full daylight side is turned to our view, and we see a *full* moon. During the following two weeks, the moon goes through the same phases again, in reverse order as it moves in its orbit around toward the direction of the sun again. For a week it passes through *waning gibbous* phases. Three weeks after new moon, it is at *third quarter,* and again appears as a half moon, this time rising at midnight and setting at noon, when the sun is on the meridian. (By the way, the moon can usually be seen easily in broad daylight; it is most instructive to watch it in the sky, changing its phases through the month, in the daytime as well as at night.) In the final week before new moon, the moon is a *waning crescent,* rising later and later, and is last seen at night just a day or two before new moon, when it rises just an hour or two before sunup. Finally, after about 29½ days from the last new moon, it has completed its cycle of phases, and again is in the sun's direction, rising with the sun in the morning, and not visible to our view.

If the moon's true path were exactly in the plane of the earth's orbit about the sun, at every new moon the moon would have to pass directly in front of the sun, and at every full moon it would have to pass directly into the shadow cast by the earth in sunlight. Actually the moon's path is inclined at about 5° to the earth's orbit, so it usually escapes this fate. All this behavior of the moon was well understood by the ancients, except that they thought the sun, as well as the moon, moved around the earth.

Twice a year, however, the places where the moon crosses the plane of the earth's orbit are in a line with the earth and sun. A new moon occurring at one of these times *does* pass in front of the sun, and the full moon at those times *does* pass into at least a part of the earth's shadow. These are *solar* and *lunar eclipses,* respectively. More often than not, the instant of the full-moon phase is at such a time that it misses passing through the dark part of the earth's shadow, although every now and then (typically once every year or two) it does happen (Figure I.14). Then everyone on the night side of the earth can see the earth's shadow on the moon. As the moon passes into our shadow, more and more of it is covered, as if the earth's shadow cuts a successively bigger and bigger bite out of the moon. At these times, the earth's shadow is always observed to be round; Aristotle pointed out that only a sphere *always* casts a round shadow, and cited this as further proof that the earth is round. Even when the moon passes completely into the earth's

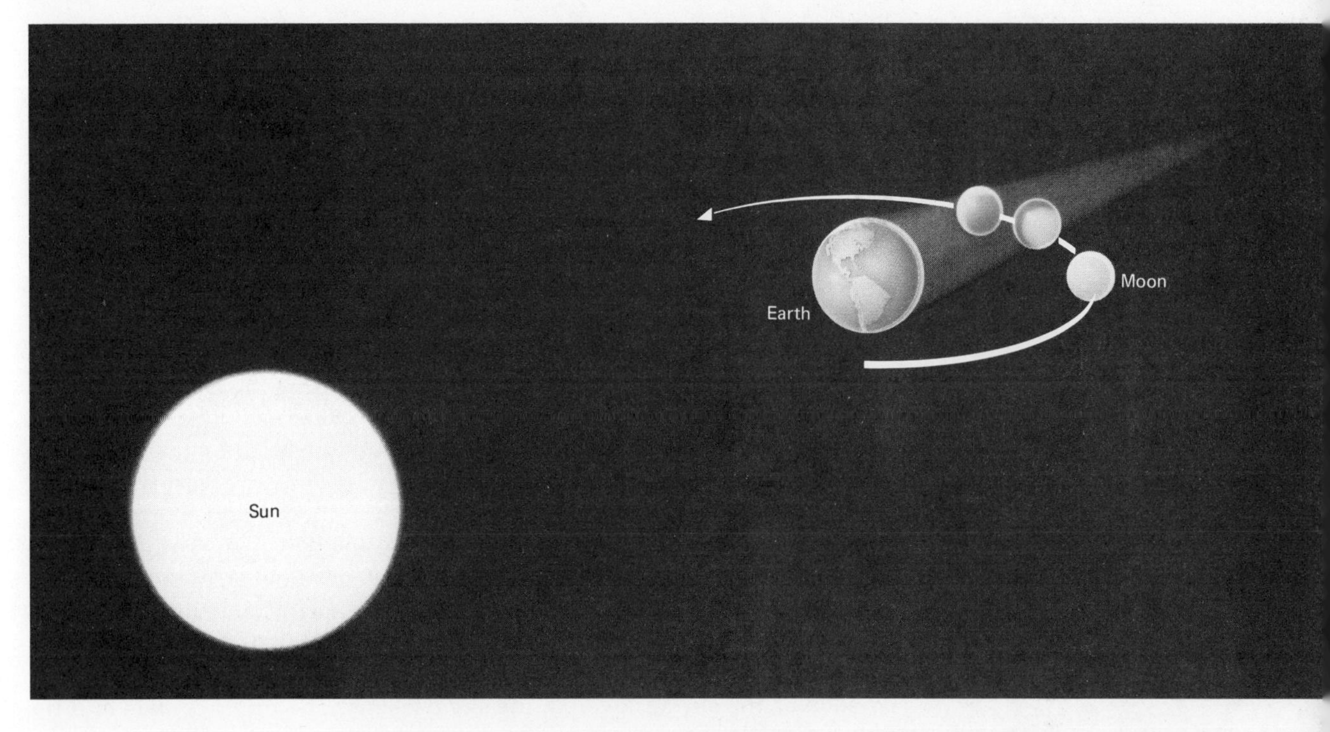

FIGURE I.14
The geometry of a lunar eclipse.

FIGURE I.15
The geometry of a total solar eclipse.

shadow, it is usually still faintly visible, as a dull coppery red disk, because the shadow is never completely dark; some sunlight passing through the atmosphere of the earth is bent by the air into the shadow and strikes the moon. Still, an eclipse of the moon is an interesting phenomenon to watch.

But nothing like a total eclipse of the sun! This happens only at those new moons when the moon happens to pass directly in front of the sun. The trouble is that the only people who can witness the phenomenon (see Figure I.15) are those at the location on the earth where the tip of the moon's

shadow strikes the ground. Remember that the sun and moon appear to be almost exactly the same size in the sky; this means that the moon's shadow can just barely reach the earth. More than half the time, in fact, it doesn't quite make it (because the moon's orbit is not quite circular and its distance from us varies slightly). Then the best anyone can see is the moon silhouetted in front of the sun, with a thin ring of sunlight still showing. This "ring"-type eclipse is called an *annular* eclipse—interesting but not spectacular (Figure I.16).

Sometimes, however, the moon is near enough for its shadow to reach the ground, and total eclipses occur that are visible from someplace on earth. Appendix 9 lists the principal solar eclipses visible during the next half century. The moon's shadow sweeps across the earth in a very thin line, at most only about 270 km across, and to see the sun completely covered by the moon, you must be in that path. Even then, the sun is covered only a very short time, never more than about 7 minutes. About an hour before the eclipse is total, you see the silhouette of the moon just barely encroaching the sun, and as the hour progresses you see more and more of the sun covered—the moon appearing to take a larger and larger bite. During these *partial* phases, you must take care to look only through certain carefully prepared filters to avoid being blinded by the bright light from the uncovered part of the sun.

As soon as the sun is covered, the sky darkens. The brighter stars and the planets are visible. Animals behave as though it were nightfall. The temperature drops, and breezes come up. And the beautiful, faint outer atmosphere of the sun, its *corona,* is clearly visible to the unaided eye. It is an extremely awe-inspiring spectacle, and no one who has an opportunity to witness it should pass it up. Total solar eclipses occur at any one place on earth at an average frequency of only once every 360 years. Large parts of the earth, for thousands of kilometers on either side of the path of totality, are presented the phenomenon of a *partial* eclipse, in which part of the sun appears cov-

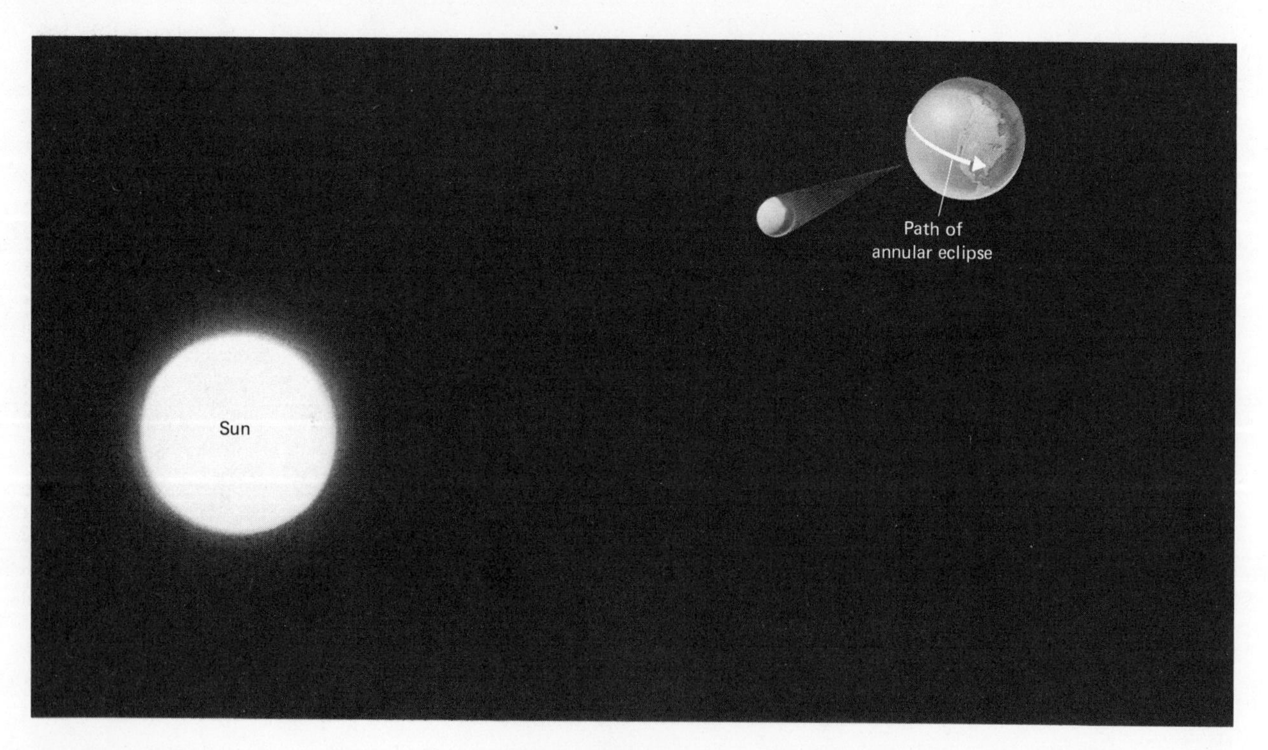

FIGURE I.16
The geometry of an annular eclipse.

FIGURE I.17
The eclipsed sun, 8 June, 1918. *(Hale Observatories)*

ered, but a partial eclipse is only moderately interesting. The ancients knew about eclipses and why they occurred. Associating the sun and moon with gods, as they did, it must have been quite a religious experience!

---

12. If the moon is on the meridian at sunrise, what is its phase?
13. When the moon is full, where is it in the sky at noon?
14. What time (about) does the moon rise on a date on which it is totally eclipsed?
15. At what time will the moon be on the meridian during a solar eclipse?

Thus many aspects of the sky were well understood by our ancestors more than 60 generations ago. If they had some rather strange notions, it was understandable, considering the state of knowledge at the time. The remarkable thing was how many right ideas they had!

For example, they knew the earth is round, and even about how large it is. The common "knowledge" that until Columbus' time people thought the earth was flat cannot be correct—at least not among those who were well educated or who understood navigation. (Columbus wasn't trying to prove the earth is round—he knew that; he was trying to establish trade routes with the Far East by going the other way around.) Of course there must have been many uneducated people who believed the earth was flat in those days. There are people today, in the most highly developed nations, who still believe that the earth is flat; I have a file of correspondence with some of them! But many millions of our fellow countrymen hold other, equally foolish ideas—ideas abandoned by educated people centuries ago. We shall be exploring some of them.

The Age of Astrology

The sun and moon move through the zodiac in a relatively simple manner. The other planets move in nearly circular orbits around the sun, as does the earth, but we see them from our own moving platform. Thus they have rather erratic-appearing motions, sometimes moving eastward through the zodiac and sometimes (as the earth passes one up in moving between it and the sun) backing up and moving westward for a while. From several centuries B.C. until the time of Kepler, in the seventeenth century, a major effort of astronomy was to find a scheme that would describe accurately the apparently perverse motions of the planets.

1. Do the planets Mercury and Venus show retrograde (east-to-west) motion in the sky? If so, why and when? If not, why not?

The reason is that the planets played the roles of gods to the ancients. The Greeks, whose fertile imaginations gave us the beginnings of democracy—the epics of Homer, the stirring speeches of Pericles, the geometry of Euclid, and the philosophy of Plato—blended their mythology, and that of the Babylonians and the Egyptians, with their concept of the majestic regularity of the heavenly motions, albeit complicated with the almost human perversity of the motions of the planets.

Planets and Gods

The Greek gods were immortal, but otherwise they had the same attributes of anger, happiness, jealousy, rage, and pleasure as those of us humans, and these same attributes were assigned to the planets that either were the gods, were their abodes, or at least represented them. Each god, and thus each planet, was a center of force, but how that force prevailed depended on how it was tempered by the effects of other gods.

If the gods themselves were capricious, at least the planets were potentially predictable in their movements. How natural to attempt to understand the whims of the gods by understanding the motions of the planets. Because our own lot in life is unpredictable, it must be purely at the mercy of the gods. But if the gods are the planets, or at least are somehow associated with them, we have only to learn the rules of the motions of the planets to understand the whims of the gods and how they shape our own lives.

As Greek scholars learned more and more about the regularities of the motions of the sun, moon, and planets, therefore, they felt they were learning more and more about the ruling forces in their own lives. How beautifully complete and typically logical of the Greek mind—armed with the geometry of Euclid—to suppose that our lives should be so ordered. The Greeks had the prophetic wisdom to suppose that the motions of the planets are indeed governed by precise laws of nature—transcending even the will of their humanesque gods—and thus, by inference, they presumed that our lives are similarly preprogrammed by the preset motions of the planets. What, then, can determine our individual lots? Only the moment that we happen to enter the world and fall into step with the eternal and predestined movements of the heavens.

Thus the belief developed that each of our lives is preset by the precise configurations of all the planets in the sky at the moment of birth; all the motions of the planets thereafter follow the laws of nature, and hence the influence of the planet-gods must similarly be constrained by their predictable relations with other planet-gods in the years to come. Thus the key to the future of an individual was the map of the heavens, showing where each planet was in the sky as seen from the precise time and place of that person's birth. This is the religion of *natal astrology*, actually believed in by ten of millions of Europeans and Americans today.

The Horoscope

Of particular importance to a person's future was what was rising when he was born. Perhaps this notion goes back to the time when the earth was believed flat, and that objects in the sky existed only temporarily, dissolving in the west as they set. Consider the effect of a planet just forming as it is about to rise in the east at the moment of a child's birth. Can we imagine the importance of that influence, permeating the atmosphere inhaled in that infant's first breath!? In any case, the ancient astrologers regarded as particularly important what was rising at one's birth. The zodiacal sign on the eastern horizon is still called the *rising sign*, and any sign, planet, or star that is rising or about to rise is said to be in the *ascendancy*. Originally, the word *horoscope* meant the *hour* to *observe* what was ascending or rising. Later, it meant a map or chart of the heavens, showing the position of each sign and each planet in the sky at some specific time. This is the present meaning of the word. In particular, a *natal* horoscope for an individual is such a chart prepared for the moment of his birth. The first natal horoscope was charted, or *cast*, about the first century B.C.

One thing a horoscope shows is where each planet is in the zodiac, through which it eternally moves. We have seen how the zodiac was divided into 12 equal parts or signs, beginning with the vernal equinox, and proceeding eastward about the sky. Thus a horoscope specifies, in part, the location of each planet according to what part of which sign it occupies. In may natal

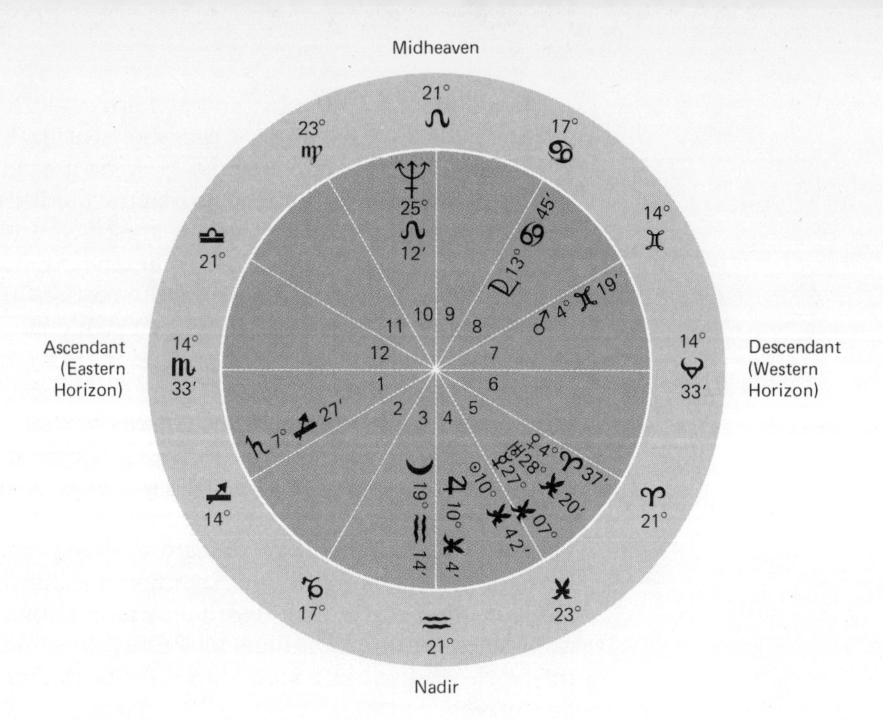

FIGURE I.18
The horoscope of a skeptical astrologer.

horoscope (Figure I.18) Saturn (symbolized by ♄) was 7°27′ into (east of the beginning of) the sign of Sagittarius (♐). The last sign of the zodiac is Pisces (♓), which the sun (⊙) traverses each year from February 19 to March 20, on the average. Since I was born on March 1, the sun was in Pisces in my natal horoscope; Pisces is my *sun sign*, and in the jargon of the occultists I am known as a *Pisces*.

Also important in astrology is where things are in the sky with respect to the horizon. Thus a horoscope depends not only on the date, which specifies where each planet is among the signs of the zodiac, and on the location on the earth from which the sky is viewed, which specifies where the celestial pole is in the sky, but also on the *time*, which finally fixes the angle through which the celestial sphere has rotated, and thus where each part of it is with respect to the horizon.

Modern astronomers orient the celestial sphere by specifying what is called *sidereal time*, which tells where the vernal equinox is in the sky as seen from the particular place in question on the earth. (Ancient astronomers did likewise.) At 0 hour sidereal time, the vernal equinox is on the meridian; at 6 hours, it is setting. It is as low as it can get in the sky at 12 hours sidereal time, and rises at 6 hours. Since the signs of the zodiac start at the vernal equinox, knowing where it is tells where all the signs and planets are with respect to the horizon.

---

2. What is the sidereal time when the autumnal equinox is on the meridian?

3. On what date does 0 hour solar time (the beginning of a day) coincide with 0 hour sidereal time (that is, when do solar and sidereal times agree)?

4. Six months after solar and sidereal times coincide, where is the sun on the ecliptic? Then, by how many hours do solar and sidereal times differ?

The ancient astronomers (and astrologers) further specified the matter by dividing the sky into twelve *houses,* each occupying a fixed place with respect to the horizon—that is, with respect to the ground where the individual concerned is born. The *first house* is that part of the sky containing objects that will rise within approximately the next two hours; hence the first house is that part of the sky immediately below the eastern horizon. The second house lies beneath the first, the third beneath the second, and so on, around the lower part of the celestial sphere. The sixth house contains objects that have set within the past couple of hours; the seventh those which will set within the next couple of hours; and so on. Objects that lie in the tenth house will be carried, by the rotation of the celestial sphere, across the meridian during the next two hours. Those in the twelfth house have just risen. Objects in the first house are thus said to be in the ascendancy (or soon will be), and those in the tenth, in *culmination* (they will soon cross the meridian).

Actually the foregoing discussion of houses has been slightly vague, because there has never been agreement among astrologers (even to this day) on the proper way to define the precise boundaries between houses. The houses are the numbered pie-shaped divisions in my horoscope (Figure I.18). The zodiac itself is represented by the circular ring at the periphery, and the symbols and numbers in it represent the signs and the number of degrees into each sign at which the various house boundaries occur; for the purpose of preparing this horoscope, I used the definition of houses attributed to the astrologer Placidius, but at least a dozen alternate definitions exist. Although the differences between the various house definitions are technical in nature, they do result in slightly different horoscopes, and can often put one or more of the planets in different houses.

So a natal horoscope is a chart of the sky showing the locations of celestial bodies at the time of a person's birth. The astrologers of antiquity attempted to interpret a person's physical and emotional character, personality, talents, strengths and weaknesses, propensity for various diseases, and the events of his life in terms of his natal horoscope.

---

5. Can you think of any reason or reasons why the sign or object in the ascendancy at the moment of your birth could influence your personality? What are they?

6. Whether or not your natal horoscope has anything to do with your personality, what other factors do you think can play a role, and why?

Astrological Doctrine

It would be out of place here to attempt to detail all the rules by which such analyses were (and still are) carried out; anyway even astrologers will not often agree with one another very well on their readings of a particular horoscope. I can cite many conflicting analyses of my own to prove that point! The interested reader will find countless volumes of instructions in his neighborhood bookstore. Just a few quick examples will suffice to give the idea.

One of the supposedly important criteria to understanding a person is his sun sign. Nearly all astrologers will agree, however, that the knowledge of a person's sun sign alone is far too little information, and any advice based on it, they say, is worthless. (Yet they continue to publish daily horoscopic ad-

FIGURE I.19
The neighborhood bookstore often contains books with "solutions" to a wide range of human problems. *(Photograph by Mary Anne Kramer; courtesy of* Mercury *magazine)*

vice in more than 2000 newspapers in the United States alone, based on just the sun sign—advice read and possibly heeded by tens of millions).

According to astrologers, a person's entire horoscope must be examined. The sign that is rising at his birth is of great importance, as are the planets that are also soon to rise. But the influence of each planet is amplified or weakened by whether it is in its *own* sign (for each planet rules a sign) or in one sympathetic with the sign it rules. *Aspects* are important, too; for example, what planets are *trine* (about 120° away from each other), or in *opposition* (opposite in the sky), or *squared* (90° away), and so on.

Each house has a certain role in a person's makeup; so the planets (and signs) in the various houses play key roles. For example, the first house (just below the eastern horizon) controls temperament and personality. Mars, the aggressive god of war, in that house might dispose one to an aggressive career (perhaps the military or athletics), especially if Aries, the sign that Mars rules, is also rising, and hence is in that first house. The second house is supposed to relate to one's wealth and fortune, the third to his siblings, the fourth to his parents, and so on. The planet and sign in the eighth house, which deals with death, might well tell the astrologer how the subject will die.

Of course, as the person goes on living, the earth goes on turning and the planets go on moving through the zodiac. The astrologer, however, keeping track of these motions and always relating them to the client's natal horoscope, believes he can foretell times of significant events in the subject's life, what times are happy ones for the subject, what ones good for important journeys, and (if the astrologer is confident in himself) even when the subject may suffer calamities or death.

Moreover, each sign of the zodiac is presumed to relate to a given part of the body; thus Aries rules the head, Leo the heart, Cancer the stomach, Scorpio the genitals, and Pisces the feet. Mars in the sign of Aries in a natal horoscope might dispose the poor subject to a tendency toward headaches all his life, and Uranus in Cancer might plague him with stomach cramps. In the Middle Ages most physicians believed in and practiced according to this *medical astrology,* and even today some doctors are reported to seek the advice of astrologers in making diagnoses.

The zodiacal signs were associated with hot and cold, wet and dry, and with the assumed elements: earth, water, air, and fire. The planets were associated with various metals: the sun with gold, the moon with silver, Mercury with quicksilver, and so on. Even nations were thought to be ruled by signs and planets, and not only were individual characteristics of people, such as stature, color of hair and eyes, attributed to details of their horoscopes, but also these characteristics of entire races, according to the signs and planets assigned to them.

There is, of course, far more—far too much even to touch on here. Suffice it to say that in antiquity it was widely believed that the stars ruled not only one's soul, but also all aspects of his life—his physical and emotional characteristics, his occupation, his success, his family, his friends and enemies, his good and bad fortune, up to his death. Astrologers could not claim then, nor can they now, to understand the planets well enough to predict all these factors with complete precision, but they tried. They had amassed an extremely elaborate set of rules to aid them. These rules were not based on centuries of data analysis, nor were they tested in controlled experiments. They were based on what we would call a magical correspondence between the gods, the planets whose names they bore, and the mythological people and creatures for which the zodiacal signs were named. That is to say, the supposed influence of the planets and signs were just those to be expected from the corresponding associations of the gods and animals of the same names.

To verify this last assertion, one need only read the *Tetrabiblos* of Ptolemy, one of the greatest astronomers, and the most important astrologer, of antiquity. The *Tetrabiblos,* written in the second century A.D., is the closest thing that exists to a bible of astrology. There Ptolemy codified the many rules by which horoscopes are interpreted, as well as explained the construction of the horoscope itself. There it can be seen how the effect of Mars in the heart of the Lion (Leo) differs from that when Mars approaches the Lion's tail.

---

7. How might one determine whether or not there is anything to the belief that the planets influence one's life in the way supposed by astrology? If you have an idea about how the astrological hypothesis might be tested, explain it in detail, and explain how psychological factors can be ruled out.

8. (For the ambitious student) Find a directory of famous people in some particular profession. Look up their birth dates and figure out from them the sun signs of those people (consult a newspaper for exact dates). Do their births favor any particular sign? Consult a statistics book, and perform an objective test to see whether any irregularities are significant. If the birth signs of the people in your sample are not randomly distributed, does this prove anything? What does it prove? If it proves nothing, why not?

Ptolemy, the scientist, shows through somewhat as well; he must have wondered how it could be that the planets could have these impressive effects, and, at least to me, seems to have been rationalizing the thing in terms of cause and effect—a sort of hardheaded twentieth century attitude. He explains, for example, that the moon, the nearest "planet" to earth, soaks up moisture from the earth, and so has a dampening influence, whereas Mars, just beyond the sun (as assumed at the time), is hot, and has an arid, drying influence. (Actually, the moon is bone dry, and Mars is now known to have considerable water—although presently frozen because of the planet's low temperature.) Also Ptolemy does not attribute *everything* to astrology; he acknowledges that there are three forces affecting man: heredity, environment, and astrology.

In any event, it should now be clear that at least one important reason that the Greeks had for studying astronomy was to understand better the motions of the planets in order to prepare better horoscopes.

The result was that the Greeks advanced astronomy to a point unsurpassed for a millennium and a half. They measured the size of the earth, the size and distance of the moon, the length of the year, and the inequalities in the lengths of the seasons. They kept track of the motions of the planets, and worked out excellent models to predict their motions, and they cataloged the stars, and noted their apparent brightnesses.

The Size of the Earth, Measured by Eratosthenes

It is easy to understand how Eratosthenes, in the third century B.C., measured the size of the earth. Eratosthenes knew that the city of Alexandria (in northern Egypt) is 5000 stadia north of Syene, the site of the modern Aswan. (A *stadium* was a Greek unit of length, based on the Olympic playing field; it was about one-sixth of a kilometer.) He also knew that on the first day of summer the sun shone from overhead at noon as seen from Syene, but at Alexandria, on the same day and at the same time, the sun was not

FIGURE I.20
Eratosthenes' method of measuring the size of the earth.

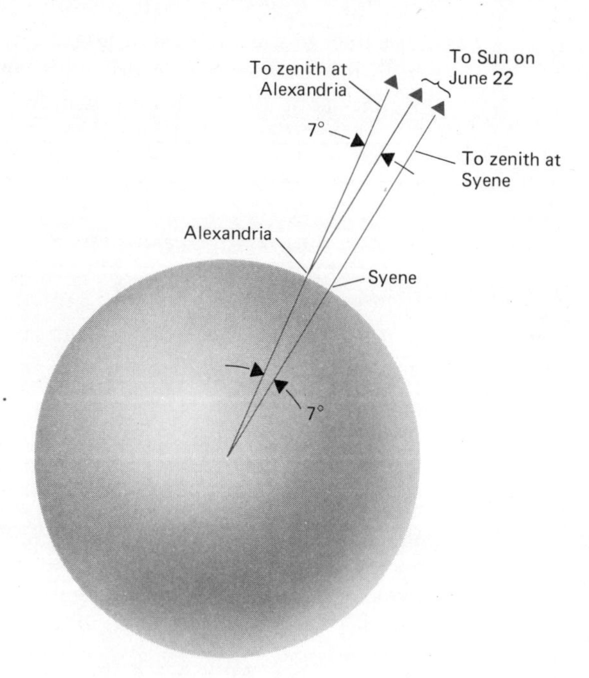

overhead, but 1/50 of a circle (about 7°) south of the zenith. Since the sun was known to be very distant, compared to the size of the earth, it was clear that observers at Syene and Alexandria would have to look in essentially the same parallel directions to see the sun (see Figure I.20). Thus if the sun was overhead at Syene, but 7° away from overhead at Alexandria, the angle at the center of the earth between those two places would also have to be 7°, or 1/50 of a circle. So it follows that the entire circumference of the earth is 50 times the distance from Alexandria to Syene, or 250,000 stadia (Eratosthenes' calculation was approximately correct).

---

9. You are in communication by telephone with a friend 10,000 km (by air) away from you. He reports that Mars is at his zenith, while at the same instant you observe Mars to be setting. What is the circumference of the earth (or perhaps the fictitious planet you are living on)?

Precession

A century later, Hipparchus, in preparing his star catalog, found that the stars are very gradually shifting positions with respect to the axis of rotation of the celestial sphere. He interpreted the phenomenon in terms of the entire sphere slowly shifting with respect to the axis about which it rotates; as a consequence, the north celestial pole describes a small circle in the northern sky, of about 23½° radius, and takes about 26,000 years to complete one cycle of this motion. (Of course, the south celestial pole is doing the same thing down south.) The discovery was remarkable for its time because the stars change position exceedingly gradually—typically 50″ per year.

The effect is called *precession*. The real reason for precession is that the axis about which the earth rotates is describing a slow conical motion in space, taking that 26,000 years for one cycle. It is similar to the way the axis of a top moves when it is spinning, but not standing quite upright. The top and the

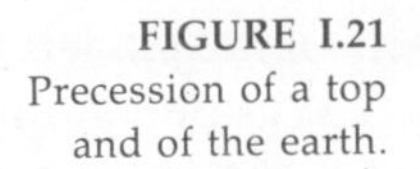

FIGURE I.21
Precession of a top and of the earth.

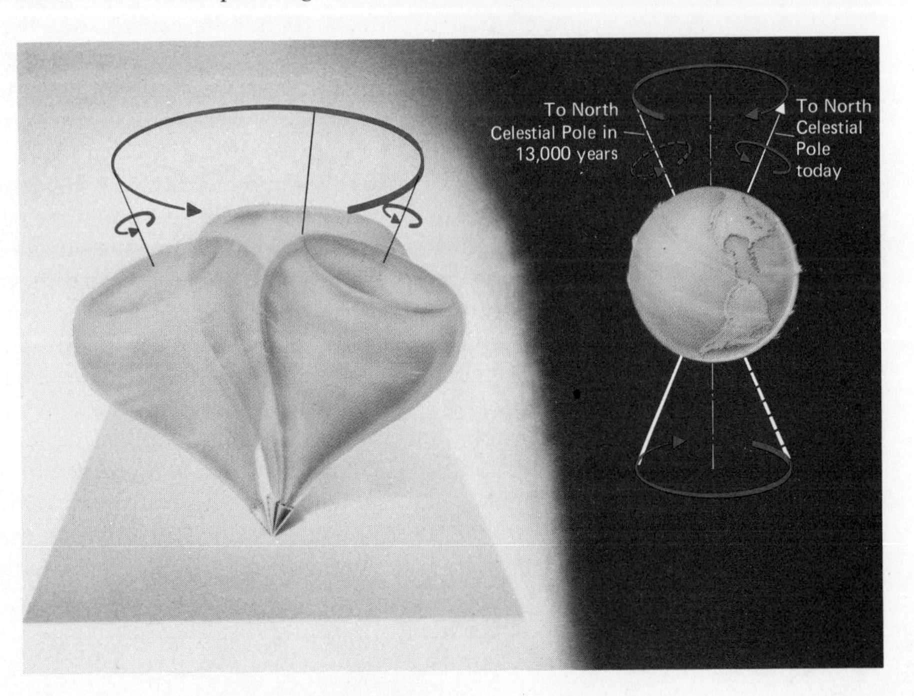

earth both precess for the same reason—gravitation. The earth's gravity tries to topple the top, and the gravitational pull of the moon and sun try to pull the earth into a position such that its axis of rotation is perpendicular to the ecliptic, rather than being tipped that $23^1/_2°$; neither body (top nor earth) gives in the expected way under the influence of the gravitational force, but describes that conical motion. The whole thing was completely explained by Newton when he introduced his gravitational theory, as we shall see later.

Of course Hipparchus and the other Greek astronomers of antiquity did not understand *why* precession occurred nor even what was doing the moving. But they still correctly observed how the sky was changing as a result of precession. If the axis of rotation of the earth (or sky) is changing, then so also must the plane of the equator of the earth (or the celestial equator), which must always be perpendicular to that axis of rotation, change with respect to the fixed ecliptic. The upshot is that the equinoxes, where the equator and ecliptic intersect, slide westward around the ecliptic, making a complete circuit of the sky in that 26,000-year period.

Fans of Rudyard Kipling's *Just So Stories* will recall how the Elephant's Child set off on his way to the Great Grey Green Greasy Limpopo River, all set about with Fever trees, while the precession of the equinoxes was precessing according to precedent. *Kipling* knew about precession, too!

But precession has something profound to do with astrology! Remember that the signs of the zodiac start, with the sign of Aries, at the vernal equinox. But if the vernal equinox is sliding around the sky every 26,000 years, then all the signs must be moving, too. Now, 2000 years ago the signs were named for the constellations they lined up with. Today, two millennia later, the signs have slipped nearly one-twelfth of the way around the sky, and are now displaced nearly a complete constellation with respect to the star groupings they are named for.

It isn't fair to discount astrology because of precession, because the astrologers who developed it to the complex form that has survived to this time were the same astronomers who discovered precession, and they thoroughly understood its effects (if not its cause). But then how do we get around Ptolemy's comments in the *Tetrabiblos* about the effects of the planets in various *parts* of the signs—for example, the Lion's heart? Was he talking about the *constellation,* for which it might make some kind of make-believe sense to refer to a limb or an organ of one of the animals? Or was he talking about the *signs,* which can hardly have hearts and tails, and anyway are moving with respect to the constellations whose names they bear? Well, he was talking about the moving signs, even though the effects of the planetary influences depended on where the planets were according to the constellations that, at the time, lined up with the signs. Confusing? Some modern astrologers have explained that each sign simply remembers the influence of the constellation of the same name when the two did coincide. (I have never heard their explanation of why the signs do not also remember the influence of other constellations they passed in previous millennia.)

Today, most astrologers still base their interpretations on the slowly moving signs. That branch of astrology is called *tropical astrology.* One justification is that, after all, the seasons, an obvious manifestation of at least one heavenly body, depend on where the sun is with respect to the equinoxes, and hence on the moving signs, not on the fixed stars. There is, however, an alternate school of astrology, *sidereal astrology,* which is based on the fixed constellations.

So today, if you are an *Aries,* meaning you were born when the sun was passing through the sign of Aries, you realize that the sun was really in the constellation of Pisces, next one to the west of Aries. Rather soon now, the vernal equinox will have passed all the way westward through Pisces and into the constellation of Aquarius. That is what is meant by the *Age of Aquarius.* Exactly when that age begins depends on exactly where you consider the boundary to lie between the constellations. Ancient star maps are noncommittal on the matter, and probably most astrologers would not readily accept the boundaries arbitrarily drawn up by the International Astronomical Union in 1928.

The Ptolemaic System

The important thing to us is that the Greeks carried on a tradition of observing the positions of the planets on the celestial sphere, and of trying to find geometrical schemes that described their motions as accurately as possible. The most famous and successful such scheme is that of Ptolemy, last of the great Greek astronomers. Various parts of Ptolemy's model had been invented by previous scholars, notably Hipparchus, but Ptolemy had the genius to synthesize existing observations and theory into a system for the planetary motions that was not seriously challenged until the time of Copernicus.

Now, we shall see later that earth, moon, and planets do not really move on perfectly circular paths, but on slightly flattened orbits, that are, technically, ellipses. Moreover, the orbital speeds of the planets vary slightly with their distances from the sun. Although the Greek astronomers never guessed the true shape of the orbits, the resulting slightly variable motions of the planets in the sky did not escape their notice. Hipparchus found that he could represent the motions of the sun and moon to sufficient accuracy by supposing that they move uniformly on circular orbits which are not quite centered on the earth, so that as seen by us they seem to move fastest in the sky when they are nearest us. Furthermore, he had to have the center of the moon's orbit revolve slowly about the earth to account for the moon's somewhat more complex behavior.

Ptolemy incorporated these and other such ideas to account for the motions of the planets. Consider how he treated a particular planet—Mars (Figure I.22). He supposed that Mars revolved each year on a small circular orbit, called an *epicycle,* whose center, in turn, revolved about the earth about once every two years. The circular path of the center of Mars' epicycle is called a *deferent.* In Ptolemy's scheme the deferent was not centered on the earth, but was somewhat to one side—it was thus an *eccentric* circle. Even this scheme, however, cannot account for the complex motion of Mars (as seen from the moving earth) to the accuracy of observations available in Ptolemy's time. Thus, as a further complication Ptolemy had the center of the epicycle revolve through equal angles in equal times, not as seen from the earth or at the center of the deferent, but as seen at another point, on the opposite side of the center of the deferent from the earth, called the *equant.* Clearly, this is beginning to sound complicated, and so it is. We shall soon see that the whole thing is really quite simple, but the new simplicity had not been discovered by the ancients. They represented all heavenly orbits with circles, but the planets do *not* move in circles, and attempts to represent their behavior with circular motions necessarily led to complications.

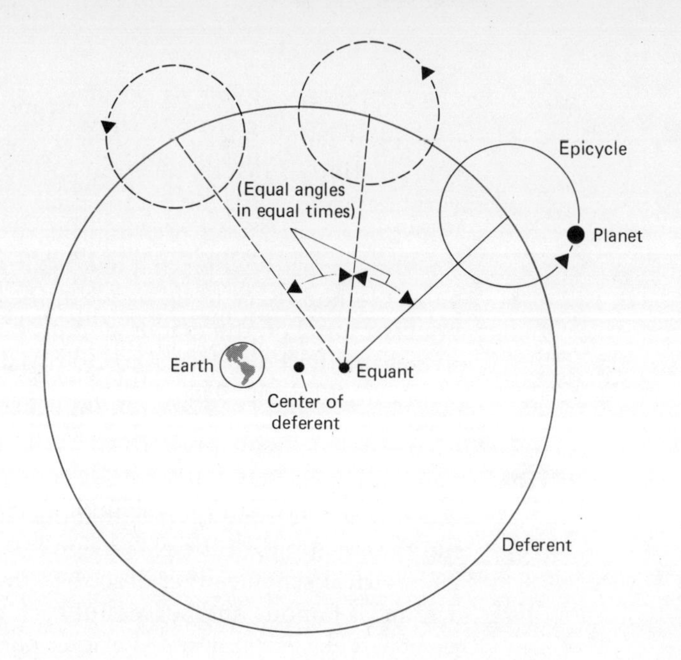

FIGURE I.22
Ptolemy's scheme of the epicycle, eccentric, and equant.

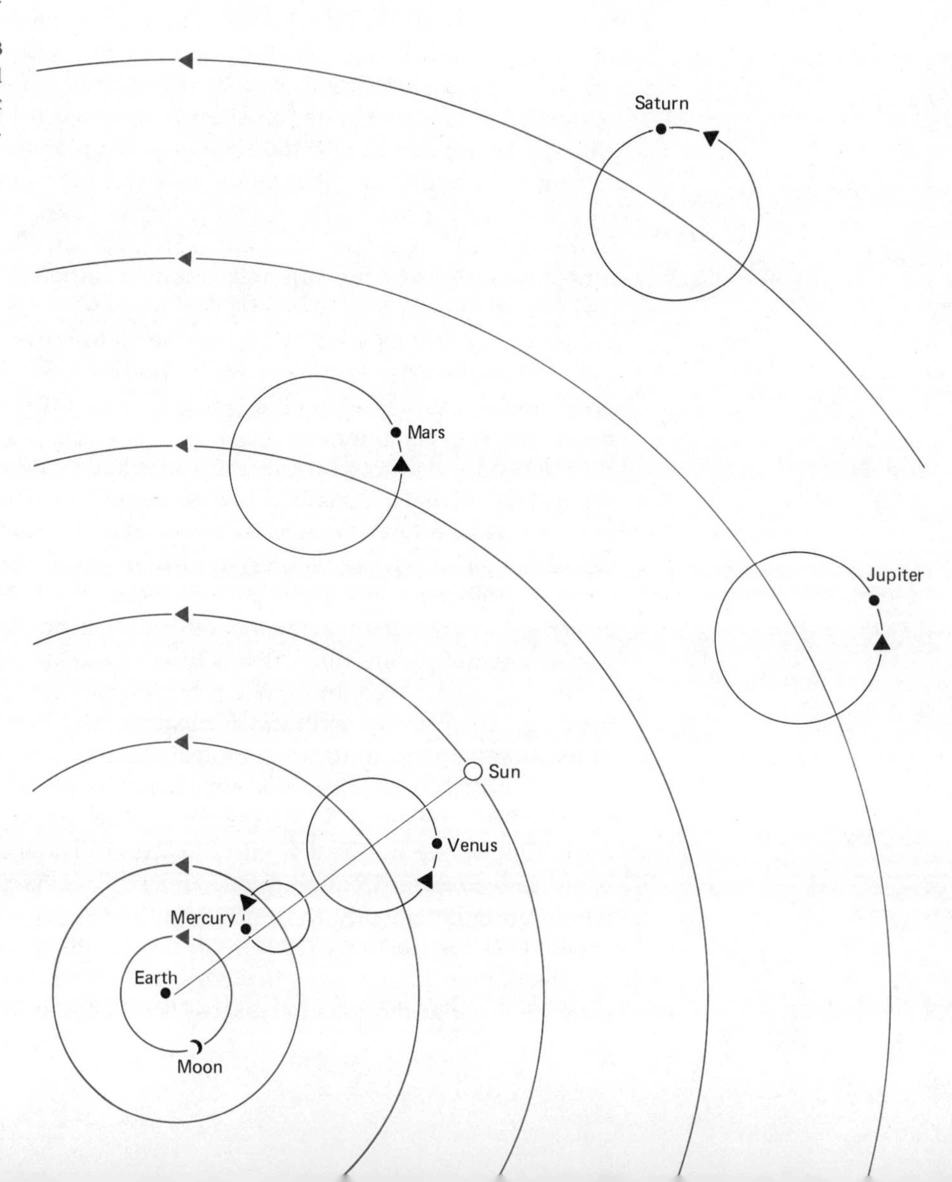

FIGURE I.23
A somewhat oversimplified diagram of the Ptolemaic system.

Nevertheless, Ptolemy's scheme was a great tribute to his genius as a mathematician. Figure I.22 shows the system of epicycle, deferent, eccentric, and equant that he adopted for each planet, and Figure I.23 shows an oversimplified view of the Ptolemaic idea of the whole solar system. We should not hold the scheme in low esteem because it is wrong. On the contrary, it is not known whether Ptolemy himself believed that it described reality. It was, at the very least, a geometrical formula, or mathematical model, that allowed astronomers to calculate where each planet would be in the sky.

Ptolemy's model worked quite well for a few hundred years, but not for fifteen hundred. Eventually there were observations good enough, and accumulated over a long enough time, to show that things could not be working as Ptolemy had prescribed. The hunt was open for a new and better order. When it was found, it far exceeded the expectations of the hunters, with profound philosophical implications—at least for a while. But we are getting ahead of our story.

SCENE 3

The Copernican Revolution

We have a tendency to reduce to the work of a few heroes the history of the great achievements of mankind. In reality it is seldom so simple. It is often difficult to establish all those to whom credit is due, and anyway it makes a long and clumsy account. Thus we think of the big stars: Copernicus, Galileo, and Newton, for example.

Not to say that these men were not very great indeed. Newton *did* write the *Principia;* Galileo, *Dialogo dei Due Massimi Sistemi;* and Copernicus, *De Revolutionibus.*

Nicholas Copernicus (1473–1543) was born in Torun on the Vistula in Poland. He studied medicine and law, but had always been interested in astronomy, and had, in fact, written several early astronomical papers. Now by the time of Copernicus some of the planets were noticeably far away from where the Ptolemaic tables predicted them to be. Over the hundreds of years small errors in Ptolemy's system had accumulated. It was a little like a pretty good clock running slightly slow or fast; the clock keeps good enough time for a while, but eventually the daily errors add up until it is an hour or more off. Like the slow clock, Ptolemy's scheme could have been reset, and would then have worked pretty well again for a few hundred years. But Copernicus introduced a very different idea.

His model put the earth, along with the other planets, in motion around the sun. This simple transformation eliminated the need for the large epicycles in the Ptolemaic scheme. Now as we have said, the planets do not have exactly circular orbits, so their motions are not precisely uniform. Ptolemy took care of these irregularities by introducing eccentrics and equants. Copernicus rejected the equants but retained the eccentrics, and he still represented all planetary motions with circles or combinations of circles. As we see, the Copernican scheme was not at all like our modern view of the solar system.

Copernicus was not the first to introduce a planetary scheme with the earth in motion. Among others, Aristarchus in the third century B.C. is said to have proposed such a view. Nor did Copernicus *prove* that the earth moved. Indeed, in its time the Copernican system was accepted not as a system of *cos-*

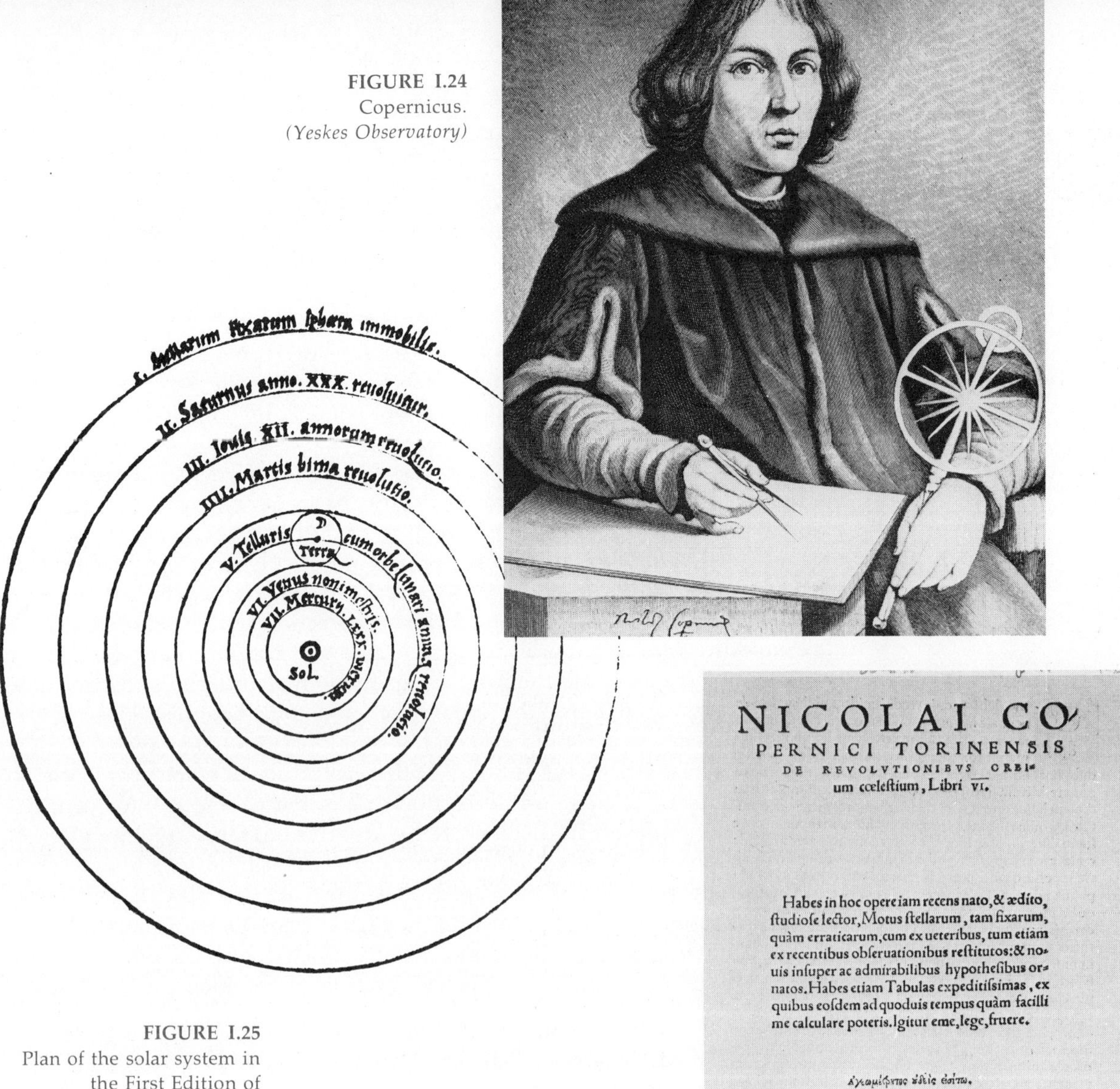

FIGURE I.24 Copernicus. *(Yeskes Observatory)*

FIGURE I.25 Plan of the solar system in the First Edition of Copernicus' *De Revolutionibus.* *(Crawford Collection, Royal Observatory Edinburgh)*

FIGURE I.26 Title page from the First Edition of *De Revolutionibus.* *(History of Science Collections, University of Oklahoma Libraries)*

mology (a physical model of the universe), but as a geometrical model—a scheme for calculating positions of the planets in the sky. Some historians of science suggest that Copernicus himself held this view. At any rate it was well over a century before the heliocentric view of the solar system was seriously considered by scientists in general as a description of reality.

Nevertheless, Copernicus eventually produced an impact on the development of science. He had described the details of his system in his great book *De Revolutionibus Orbium Celestium (On the Revolutions of the Celestial Spheres)*, which was published in 1543, the year of his death. *De Revolutionibus* was widely circulated, and had very considerable influence. It was read, for example, by Galileo, Brahe, and Kepler. And it also made the Roman Catholic *Index of Forbidden Books.*

Tycho Brahe's Observations

Tycho Brahe was a Danish nobleman, born in 1546, three years after the death of Copernicus. He made fine astronomical instruments, and devoted most of his life to careful observations of the heavens. The astronomical telescope had not yet been invented; his instruments did not magnify the planets, but were used to carefully measure their directions at different times, and hence their celestial coordinates (analogous to latitude and longitude on earth). Brahe observed a "new star" in 1572, and noted that it rivaled Venus in brilliance; he then watched it fade for 16 months until it disappeared from naked-eye visibility. (That star is now believed to have been a supernova.) He also observed a number of comets, including a rather bright one in 1577.

We must now interrupt our account of Tycho (as he is usually known) to explain the idea of *parallax.* Parallax is any apparent change in the direction of an object due to a change in the position of the observer. For example, hold a pencil in front of your face. Now look at it alternately with one eye and then with the other. Notice how the pencil seems to jump back and forth in front of the far wall as you see it from a different direction with each of your eyes. Next hold the pencil farther away, and notice that the shift back and forth is less. If a friend across the room holds up the pencil and you similarly wink your eyes at it, you will find the apparent change in direction still less. In fact, try the same experiment outdoors on the twig of a tree 100 m away, and you will see no effect at all. This is how depth perception (or stereoscopic vision) works; the apparent change in direction of an object as seen from your different eyes depends on its distance. If the object is too far away, though, your two eyes see it in so nearly the same direction that your brain is unable to detect the angle between those directions—they are sensibly parallel: thus depth perception exists only for objects within a few tens of meters.

If the distance between our eyes were very much greater, depth perception would work to much greater distances. A surveyor can measure the distance

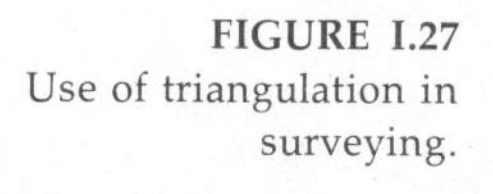

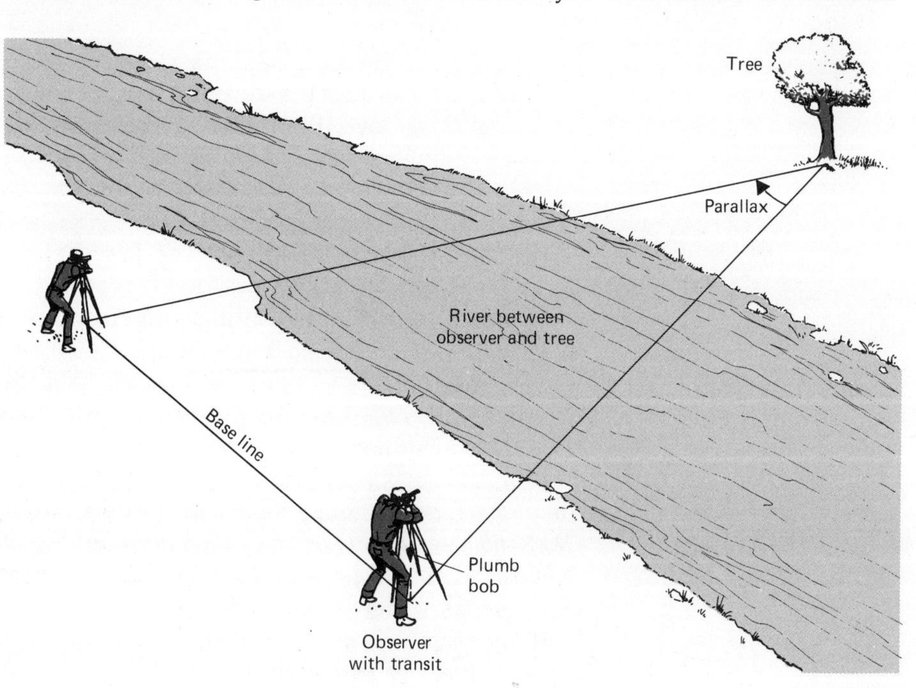

FIGURE I.27
Use of triangulation in surveying.

to a rather remote object by measuring its direction from both sides of a much larger distance, or *baseline*. In depth perception, the distance between our eyes is a baseline of only a few centimeters; surveyors may use baselines of several kilometers. The angle between the directions of the remote object as seen from the two ends of the baseline is the parallax (see Figure I.27); the farther away the object, the smaller the parallax, and the greater the baseline needed to detect it.

Brahe was interested in measuring parallaxes of objects in the sky. He used the whole earth for a baseline. If you assume the earth is turning on its axis, you realize that it carries you from one side to the other in about 12 hours; at the equator, that is over a baseline of more than 12,000 km. The moon is near enough that if it didn't move in its orbit, it would appear to shift in direction by up to 2 degrees between rising and setting, with respect to the far more remote stars. Brahe, after disentangling the effects of the moon's motion, also measured its parallax and thus found the distance of the moon. But when he tried it on the new star he had observed and on a comet, he could observe no parallax and concluded, correctly, that they must be much farther away than the moon. This was philosophically important, because until then comets were believed to be vapors (and evil omens) in the earth's atmosphere, and changes in the sky, like new stars, were believed to be impossible.

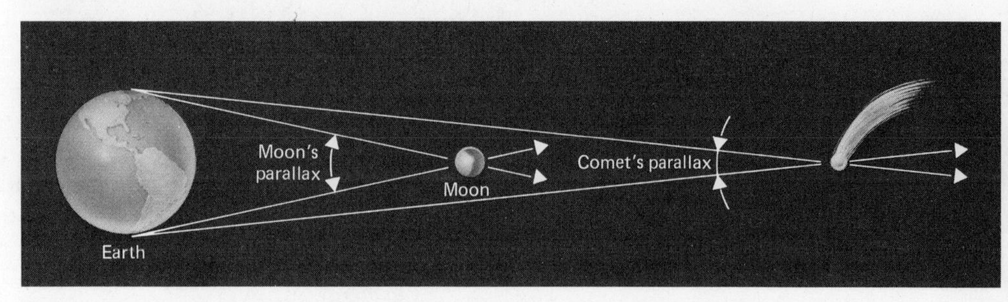

FIGURE I.28
The parallax of the comet is less than that of the moon.

---

1. In the Copernican system the celestial sphere would have to be very large in order that the annual parallax of the stars not be apparent to the unaided eye. Perhaps so large a sphere is hard to imagine, but what about the alternative in the Ptolemaic system? Discuss the conceptual problems with the celestial sphere in both systems, as regards its size, motion, and anything else you think of.
2. From the information given here and earlier in the text, plot the sizes of the earth and moon and their separation, all to scale. You will need a sharp pencil and a large sheet of paper.

Brahe tried to measure the parallaxes of the stars as well, this time by using the entire orbit of the earth for a baseline. He argued that if the earth revolved about the sun, the stars would have to show parallaxes, unless they were extremely far away. However, Brahe thought he could observe the angular sizes of some stars—that is, that they appeared as tiny disks. (What he actually observed was the effect of the earth's atmosphere smearing around the light from the stars, a phenomenon modern astronomers call *seeing*.) Brahe was not able to observe the parallaxes of any stars; yet if they were far enough away not to exhibit parallax, he calculated that to display the apparent (angular) sizes he thought he observed for them, they would actually have to have sizes far larger than the earth's orbit, an idea that seemed absurd. Thus Tycho Brahe, for a pretty good reason, discarded the Copernican idea of a moving earth. Instead, he set out for himself to discover what the real system of planetary motion is.

---

3. Stellar parallaxes were finally observed in the nineteenth century. How would the parallax of a star differ, compared with how it appears from earth, as seen from
a) Venus,
b) Jupiter.
Explain.

To this end, Brahe spent about 20 years making the most careful observations of the positions of the planets so that he could chart their motions in the sky. He obtained especially complete observations of Mars. Venus and Mercury, being nearer the sun than the earth, are never seen at large angles from the sun in the sky, and thus only in the west after sunset or in the east before sunrise. The more remote planets, Jupiter and Saturn, can, like Mars, be up all night, but because of their great distances from the sun, they move in their orbits more slowly than Mars and change positions in the sky very gradually. Mars, being close to the earth, appears to move rapidly, and had, during Tycho's observations of it in fact, made about ten trips around the sky.

Fortunately, Brahe was an arrogant cuss, and made enemies. His nose was even cut off in a duel, and he wore an artificial one made of some metal, variously claimed to be gold and silver. (Some years ago when his body was exhumed, green stains were found on the facial bones, so maybe the nose was neither gold nor silver!) Moreover Brahe was extravagant, and thus was unpopular with the Danish government, which had been supporting him in his research, and he ended up leaving Denmark in exile. I say "fortunately" because Brahe went to Prague, to become the court astronomer of Bohemia. It was there that he met Kepler.

Johannes Kepler (1571–1630), a Protestant, was from southern Germany, then as now heavily Catholic. At that time a Protestant was less than welcome, and Kepler was forced to leave his mathematics-teaching job at Graz. Being also deeply interested in astronomy, he sought out Brahe to work with him in Prague, then a scientific and cultural mecca. Because Brahe needed Kepler's expertise as a mathematician to help him discover the secrets of the universe, in 1600 he took Kepler on as an assistant. Jealous of Kepler's expertise as a mathematician, Tycho made sure not to let Kepler have in his possession too many of the planetary observations at any one time. Thus Kepler really didn't make any substantial progress in interpreting these observations until after Brahe's death in 1601.

Tycho Brahe's death is an interesting story in itself. According to Arthur Koestler in *The Sleepwalkers,* the Bohemian emperor deeply frowned on anyone leaving his dining room while he was in attendance. Evidently Brahe, while at a state dinner, was obliged to hold his water rather than relieve himself at the Royal men's room, as would have been his wont. Perhaps Tycho had partaken of too much of that good Pilsner beer. Anyway, when he finally left, he had an overfull bladder, and it burst during a carriage accident on his

FIGURE I.29
Kepler. *(The Bettman Archive)*

way home. Tycho became infected, and died a few days later after a very painful illness.

Kepler's Laws

The silver lining in that black cloud is that Kepler finally had possession of the records of Brahe's 20 years of observations of Mars and the other planets. Kepler had been an early convert to the Copernican hypothesis, and tried hard to fit the Brahe observations to various models of planetary revolution. After many trials and false starts, he finally broke part of the code and discovered some of Nature's treasured secrets. In 1609 he published the first two of his famous three laws of planetary motion in his book *The New Astronomy*, or *Commentaries on the Motion of Mars.*

For two thousand years astronomers had bumbled around in trying to represent the motions of planets with combinations of circular motions. As we have seen, they partly succeeded, but never in a wholly satisfactory way. The reason is that they had been wedded to a notion—namely, that circles are somehow the only fundamental curves of nature. (What notions frustrate our modern attempts to understand the current frontier problems?). Kepler's genius, intellectual honesty, perseverance, and patience all helped him to the discovery that the planets do not move in circles at all, but in *ellipses.*

Ellipses were well known to the ancient Greeks. In fact, Archimedes studied their properties. Ellipses are examples of *conic sections.* Imagine a cone (Figure I.30). Suppose the cone is sliced through by a plane. The curve of intersection between that plane and the surface of the cone is a conic section. As can be seen from the figure, there are two general possibilities: either the

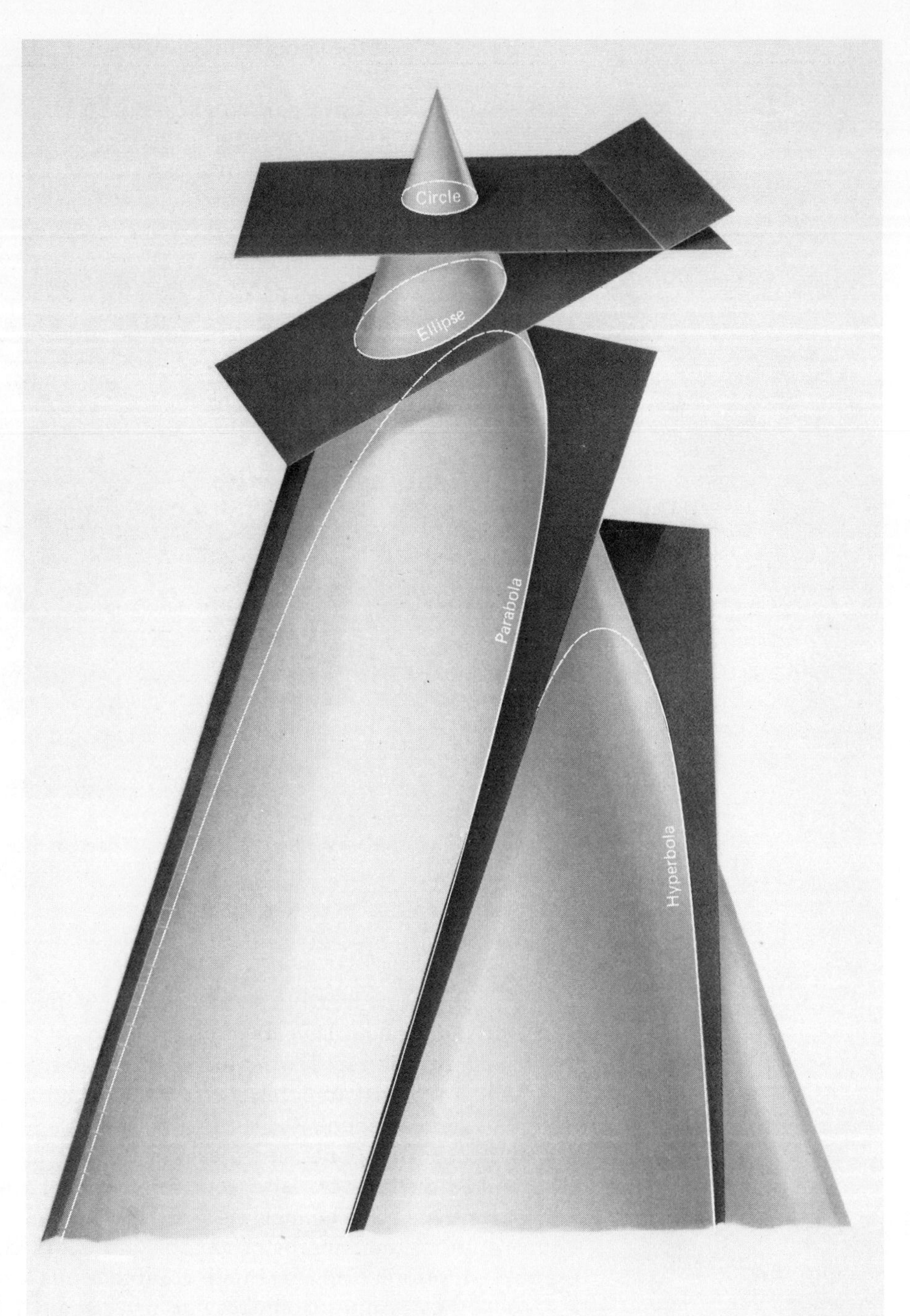

FIGURE I.30
Conic sections.

plane cuts all the way through the cone, so that the curve of intersection is closed, or the plane is so oriented that it never cuts completely through the cone, no matter how far the cone and plane may be extended, and the curve of intersection is open. In the former case, the curve of intersection is an ellipse; in the latter, the open curve is a *hyperbola*. A circle is just an ellipse that is obtained when the plane is precisely perpendicular to the axis of the cone. Finally, the curve that is the boundary between the closed family of ellipses (corresponding to planes at various angles but cutting all the way through) and the open family of hyperbolas is called a *parabola*. For the plane to cut a parabola in the cone, it must lie exactly parallel to a line in the face of the cone.

An interesting property of an ellipse is described by one of the ways to draw one. Place a sheet of paper on a drawing board and push two tacks through the paper (Figure I.31). Connect a string loosely between the tacks. Now consider the curve that results from pulling a pencil against the string and moving it around the tacks. From every point on that curve, the distance from the pencil tip to one tack plus the distance from it to the other tack is a constant—the length of the string. The points inside the curve marked by the tacks are called the *foci* of the ellipse. In a circle the two foci coincide.

FIGURE I.31
One way to draw an ellipse.

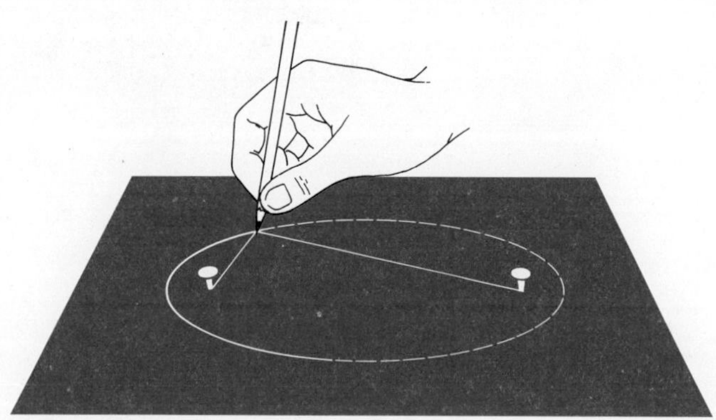

---

4. Draw an ellipse by the procedure described, using a string and two tacks. Arrange the tacks so that they are separated by one-tenth the length of the string. Comment on the appearance of your ellipse. This (if you have been careful in your construction) is approximately the shape of the orbit of Mars.

What Kepler discovered, and what is today known as Kepler's *first law,* is that the orbit of each planet around the sun is an ellipse and that the sun is at one *focus* of the ellipse. That nature should behave in such a way is indeed a beautiful and remarkable thing. But soon we shall see that it follows mathematically from an even simpler and more beautiful rule of nature.

We have already mentioned that planets vary their orbital speeds. Kepler's *second law* describes the also remarkable way in which a planet changes speed. Imagine a line connecting a planet and the sun. As the planet moves in its orbit, that line sweeps out a pie-shaped area in the plane of the planetary orbit. Kepler found that in equal intervals of time the areas so swept out by a particular planet are always the same (see Figure I.32). In other words, when a planet is close to the sun, the line from it to the sun is short, and the planet must move quickly so that the line sweeps out a fat piece of pie in order to be equal in area to the long skinny piece the sun-planet line sweeps out when the planet is far from the sun and moves slowly.

---

5. If the orbital speed of a hypothetical planet varies during its revolution between 15 km/s and 30 km/s, how many times as distant from the sun is the planet at its farthest as it is at its nearest?

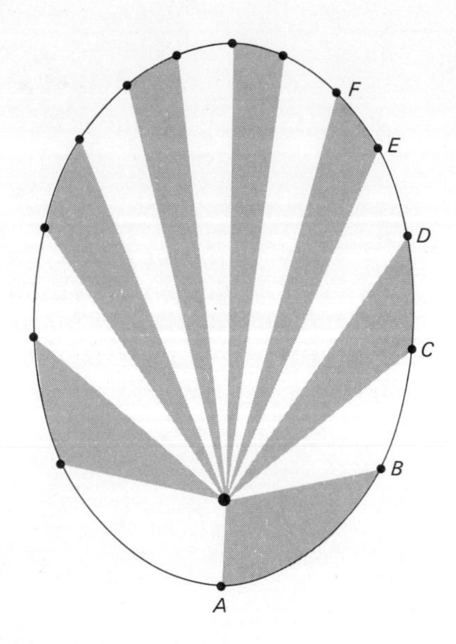

FIGURE I.32
Graphical demonstration of Kepler's second law: In equal times, the line between a planet and the sun sweeps out equal areas in the orbital plane.

Kepler thought that some kind of force must be acting on each planet to make it move as it does around the sun. He correctly deduced that the force must be aimed directly toward the sun, and quite naturally felt that the sun had something to do with it. He guessed that the force was some kind of magnetism. We shall see that this guess, although wrong, was still an astute one; Kepler had no way of knowing what magnetism is anyway.

Kepler was intrigued with the notion of a harmony in nature, and was especially delighted to discover what is now called Kepler's *third law,* which describes a simple algebraic relation between the distance of a planet from the sun and the time it takes to complete the revolution. Specifically, he found that the cube of the longest dimension of the elliptical orbit of a planet (that is, that length multiplied by itself twice) is proportional to its period of revolution squared (multiplied by itself once). The third law is sometimes called Kepler's *harmonic law,* and it does express a sort of harmony in the motions of the planets. It is stated in his book *Harmony of the Worlds* published in 1619. Kepler may have carried the harmony idea too far; he imagined that the planets, in a figurative sense, sang notes of music as they moved in their

HARMONICIS LIB. V. 207

CAP. VI

mnia (infinita in potentiâ) permeantes actu: id quod aliter à me non potuit exprimi, quam per continuam seriem Notarum intermedia-

rum. Venus ferè manet in unisono non æquans tensionis amplitudine vel minimum ex concinnis intervallis.

Atqui signatura duarum in communi Systemate Clavium, & formatio sceleti Octavæ, per comprehensionem certi intervalli concinni,

FIGURE I.33
Detail from Kepler's *Harmony of the Worlds.* (Crawford Collection, Royal Observatory Edinburgh)

paths and the book is filled with staves of music. Kepler had sent copies of various of his books to his Italian contemporary, Galileo. One need only thumb through *Harmony of the Worlds* to have a pretty good idea why Galileo never read them.

Galileo and the Sidereal Messenger

Galileo Galilei (1564–1642) was born in Pisa, the son of Vincenzo Galilei, a composer of some note. Like Copernicus, Galileo began studying for a medical career, but his true interest was in mathematics, physics, and astronomy. He held posts as professor of mathematics and astronomy first in Pisa, then at Padua, until 1610, when he went to Florence, as mathematician of the Grand Duke of Tuscany.

He had heard of the invention of the telescope in Holland, and in 1609 obtained some lenses and built one for himself. Subsequently, he built several other telescopes, all of low quality by modern standards, and the most powerful had a magnification of only 30 diameters. Galileo may not have been the first to look at objects in the sky with the telescope, but he did shake the world with his observations by describing them methodically in the little book *Sidereal Messenger (Sidereus Nuncius)*, published in 1610.

FIGURE I.34
Galileo, from a famous painting.

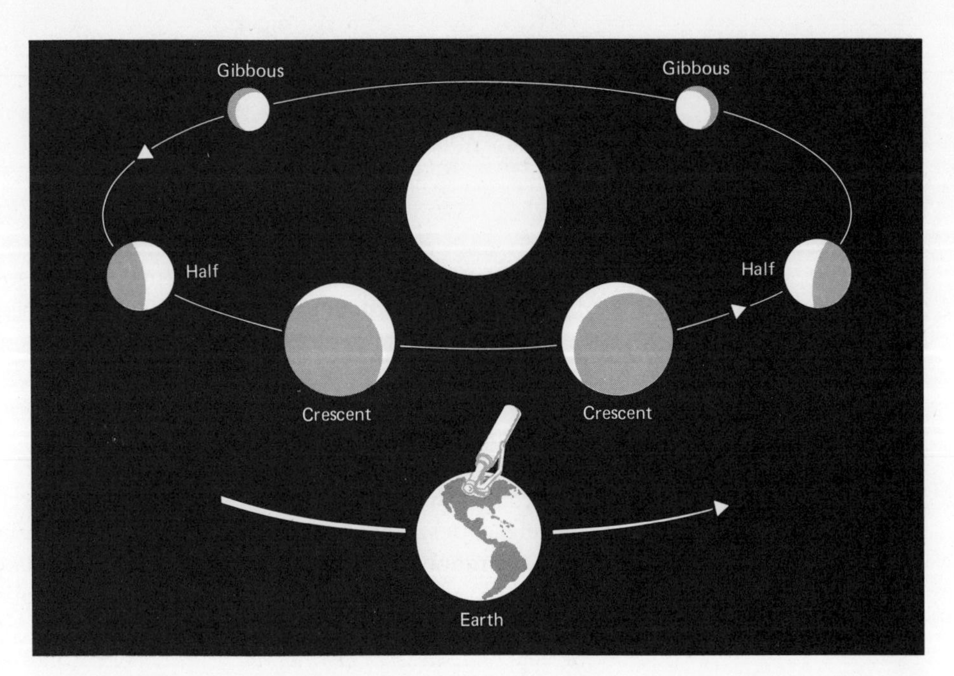

FIGURE I.35
How Venus passes through phases.

Galileo saw that Venus goes through all phases, as does the moon, which shows that Venus must revolve directly about the sun (Figure I.35), and must not always be between the earth and sun, as in Ptolemy's system. The motion of Venus about the sun did not, however, prove that the earth also so revolves. On the other hand, Galileo did make a discovery with his telescope that removed one of the objections to a moving earth; namely, that Jupiter has four moons or satellites revolving about it (at least nine additional Jovian satellites have been discovered subsequently). It had been argued that since the moon was known to revolve about the earth, the earth could not be moving or it would have long since left the moon behind. Yet here was Jupiter, certainly in motion, taking its moons along with it. In other words, a center of motion *can* itself be in motion.

---

6. Describe how the phases of Venus would appear in the Ptolemaic system, in which Venus revolves about a point always on a line connecting the earth and sun, and in which Venus is never as far away as the sun is.

Galileo's observations also shattered the notion that the heavens are perfect, immutable, and unchanging (save for their regular motions). He found the sun to be "blemished" with spots, and he showed that these sunspots (which previously had occasionally been seen with the unaided eye) must be on the sun's surface, or at least very close to it, and he even timed the sun's rotation by noting how long the spots took to move across the disk of the sun. He saw mountains and valleys on the moon, suggesting that it was a world, not a smooth crystalline orb. He found many stars not visible to the unaided eye, and clusters of stars, and that the Milky Way itself is actually the light from a myriad of individual stars. All in the heavens was, then, not as it was prescribed by ancient wisdom.

In short, Galileo's telescopic observations, and those of others who quickly followed him, although not proving the Copernican hypothesis, at least made it more credible. Other planets (at least Venus) were now shown to revolve about the sun; other planets (at least Jupiter) were now known to have moons, like the earth; other celestial worlds (at least the moon) had features suggestive of these on earth. It all fits into the grand picture that the earth was one of the planets.

Galileo's Dialogue

Of course it should now be obvious that Galileo himself firmly believed in the heliocentric theory. And he said so often enough to arouse public attention, and displeasure from the Church. Thus in 1616 the Church issued a decree forbidding anyone to hold or defend the odious Copernican hypothesis. But decree or not, the scientific community was gradually taking interest in this new idea, and finally Galileo convinced the authorities that he should be allowed to publish a book that at least impartially explored the alternative theories.

The book, entitled *Dialogue on the Two World Systems (Dialogo dei Due Massimi Sistemi),* appeared in 1632 and was written in Italian, rather than in Latin, to reach a large audience. The *Dialogue* is a hypothetical conversation between three philosophers: Salviati, through whom Galileo expresses his own views; Sagredo, who begins by believing in the Ptolemaic system, but is won over by Salviati's brilliant arguments; and Simplicio, so steeped in the ancient traditions (and perhaps a little dim) that he never sees the light.

FIGURE I.36
The three philosophers, as illustrated in Galileo's *Dialogue on the Two Systems of the World.*

FIGURE I.37
Galileo's house, and final prison, in Florence. *(Courtesy, Professor G. Godoli, Arcetri Observatory, Florence)*

In all, the *Dialogue* is a magnificent and nearly unanswerable argument for Copernican astronomy, notwithstanding the thinly cloaked irony of the preface, which reminds the reader that the arguments to follow are merely a mathematical fantasy and that divine knowledge assures us of the immobility of the earth. Galileo had made enemies among influential men in the Church hierarchy, who were neither convinced by the arguments nor assured by the preface. Galileo was charged before the Roman Inquisition of believing and holding doctrines that are false and contrary to the Divine Scriptures. He was found guilty and was sentenced to life imprisonment in his own home. He was then nearly seventy. The *Dialogue,* meanwhile, joined Copernicus' *De Revolutionibus* on the *Index of Prohibited Books,* and remained there until 1835.

Galileo is most famous for the *Dialogue* and for his trial before the Inquisition, and he doubtless played an important role in influencing scientific thought. His studies of mechanics, however, may have been equally important. He argued that motion is as natural a state as rest, and that a force is required to stop or change the direction of a moving body, just as much as to start a stationary one in motion. He also studied the properties of bodies rolling down inclined planes and falling freely, and proposed that a freely falling body accelerates (picks up speed) at a uniform rate and at the *same* rate irrespective of its weight. The fact that light and heavy objects fall together was dramatically demonstrated on the moon by an Apollo 15 astronaut, David Scott. In the lunar vacuum, where no air could interfere with the falls, he dropped a hammer and a feather and they fell together. Of course, if Galileo had been wrong on this point, we would never have gotten the Apollo vehicle to the moon in the first place.

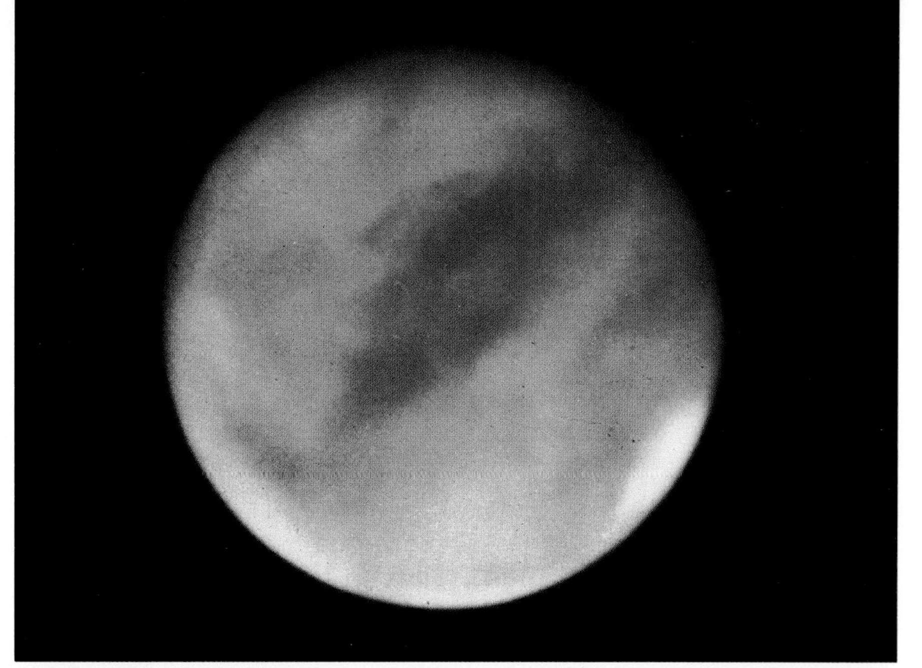

Mars, photographed with the 5-m telescope on Palomar Mountain. (*Hale Observatories*)

Saturn, photographed with the 3-m telescope of the Lick Observatory. (*Lick Observatory*)

Jupiter, photographed with the Palomar 5-m telescope. (*Hale Observatories*)

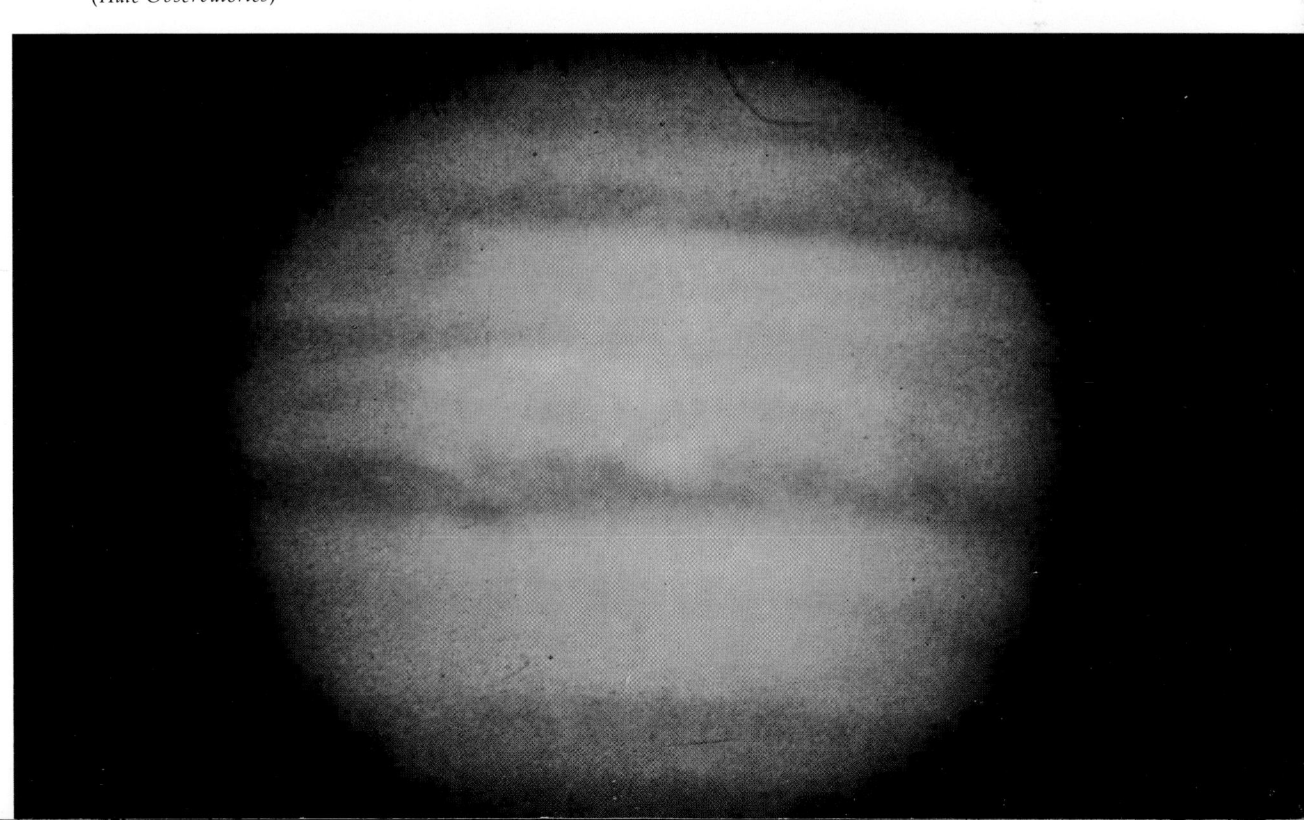

First editions of some books of great historical interest. (*Crawford Library; Courtesy, Astronomer Royal for Scotland, Royal Observatory Edinburgh*)

The earth, photographed by the Applications Technological Satellite (ATS III) from an altitude of 36,000 km. (*NASA*)

Newton's birthplace at Woolsthorpe. The apple tree in the foreground grew from the stump of the one standing in Newton's time. (*Photograph by the author*)

The laws of mechanics, such as those described by Galileo, were thus found to be the rules by which bodies on earth behave when they are put in motion and forces are exerted on them. They are terrestrial analogues of the laws of planetary motion discovered by Kepler. At the time of Galileo's death in 1642 it was not obvious that the laws of laboratory mechanics and of planetary motions are simply different manifestations of even more simple, beautiful, and fundamental laws of nature. That great synthesis is attributed to Isaac Newton.

Newton's Mechanics

Newton was born at Woolsthorpe, in Lincolnshire, on 4 January 1643. He entered Trinity College at Cambridge in 1661, and eight years later was appointed Lucasion Professor of Mathematics, a post he held during most of his productive career. Late in his life he went into government service. He died in 1727.

By the middle of Newton's life there was no longer serious question that the earth revolves about the sun, as one of the planets. The question was whether there are unifying laws of nature that apply on earth as well as throughout the solar system. Newton was able to formulate such a description of nature; he published it in his great book *Philosophiae Naturalis Principia Mathematica*, usually known simply as the *Principia*.

The key to the Newtonian system is the definition of three concepts. He formulated them in the form of his three laws of motion. The first concept is *momentum*, a measure of the state of motion of a body. If the body is at rest, it has zero momentum; if it is moving, its momentum is proportional to its speed. Newton's first law, in a sense, defines momentum as being conserved if the body is not subject to any external force. If the body is at rest, its momentum remains zero. If it is in motion, it stays in motion at the same speed, and *in the same direction*.

The second concept is *force*. Newton defined force as that which changes momentum. In the absence of a force, momentum is conserved, but if an external force is applied to a body, it changes the body's state of motion. Newton defined the magnitude, or "strength," of the force as being equal to the *rate* at which the body changes momentum. Unless the body loses some of its matter, or gains some somehow, the only way it can change momentum is to change speed or direction. Thus a force is required to start a body moving, speed it up, slow it down, stop it, or change its direction. (Note what this implies about the motions of planets in curved paths.) A specification of the speed and direction of motion of an object is called its *velocity*. A change of velocity (either speed or direction) is called *acceleration*. Thus a force produces an acceleration. The planets are changing direction; therefore they are being accelerated, and a force acts on them.

---

7. How many accelerators are there on a standard passenger car? What are they?

The third concept is *mass*. We all have an intuitive feeling for mass; it is generally what we mean when we say "weight." Technically weight is the force with which the earth's gravitation pulls on us, but when we talk about

FIGURE I.38 The momentum must be conserved in an automobile crash. *(United Press International)*

our weight, we usually are thinking about the bulk of our matter, not the pull of gravity on us, which would virtually disappear if we were removed to a point far away in space. Mass is that constant of proportionality between momentum and velocity; that is, the momentum of an object is its mass times its velocity. Thus if an automobile and a bicycle each move 20 km/hr, the automobile has the greater momentum.

Newton's third law gives us a way of measuring mass. Newton expressed it by saying that for every action there is an equal and opposite reaction. What this means is that in an isolated system, if one body exerts a force on another, the second exerts an equal and opposite force on the first; the forces are mutual. Both bodies accelerate, suffering equal changes in momentum, but in opposite directions. The recoil of a fired rifle is an example. Another is a jet airplane or rocket: fuel is ejected in one direction, and hence is given momentum. The vehicle must then experience an equal change in momentum in the opposite direction; that's what makes it go! Since the change in momentum is the same (but opposite) for the rocket and the fuel, the one that gets the greater acceleration is the one of smaller mass. Thus we have a means of comparing different masses. We need only choose one as a standard, and compare others to it by noting the relative accelerations produced by the same force, namely, a mutual force between them such as a compressed spring allowed to expand, or to a small explosion.

We have already alluded in the Prologue to *angular momentum* as that which is responsible for the rapid rotation of the solar nebula and for the revolution of the planets. From Newton's laws it can easily be shown (but we will not do so here) that if a body is rotating, or moving about an exterior point, so long as there is no force on it other than one directed exactly toward or away from the center of rotation or revolution, its angular momentum is conserved about that center of motion. (By *rotation* we mean a turning about an internal axis; by *revolution,* we mean a motion around an exterior point. Thus the earth *rotates* on its polar axis, but *revolves* about the sun.)

Angular momentum is a measure of a body's speed and distance from the center of motion; if one changes, so must the other. Thus if a planet moves closer to the sun, it must speed up; if it moves farther away, it must slow down — exactly as Kepler's second law says it does. A complex rotating object, such as the solar nebula or an ice skater, is made up of many component parts; if, on average, those parts come closer to the axis of rotation, the rotating body speeds up in its spin, as does the skater when she pulls in her arms, and as did the solar nebula when it contracted. Moving the parts out again, of course, slows down the rotation.

The fact that planets obey Kepler's second law proves that the angular momentum of each of them is conserved. This in turn proves that they are subject either to no external force or to a force directed toward or away from the sun. Since they are moving on curved paths, clearly the planets are being accelerated, so there must be a force. And since their orbits curve in toward, not out away from, the sun, the force must be exactly toward the sun—just as Kepler had speculated.

That force is *gravitation*. It makes the planets fall around the sun, just as it makes apples fall to the ground. Newton formulated his theory of universal gravitation also in the *Principia*, and explained how he tested the idea. We describe it in the next scene.

The foregoing discussion is not intended as an outline of the history of science. That subject has become highly sophisticated, and I refer the reader to works by qualified scholars for a more competent account. I simply wanted to share a little of my fascination for how man came to shift the earth from the center of a crystalline universe to a moving planet. But the real message here is not that the earth moves; it is the *nature* of man's understanding of nature.

Science, you see, is the act of building models to represent the universe. Once we have found a model that works, which gives us rules that enable us to interpret phenomena, we can apply those rules to new situations and can see if they work. We are, of course, continually seeking better models that work over even wider realms. Without thoroughly understanding this nature of scientific inquiry, we cannot hope to understand the true significance of neutron stars, black holes, or the "big bang" any more than the reader of Dickens' *Christmas Carol* can hope to understand what was marvellous about the events of that tale without first understanding that Marley was dead.

SCENE 4

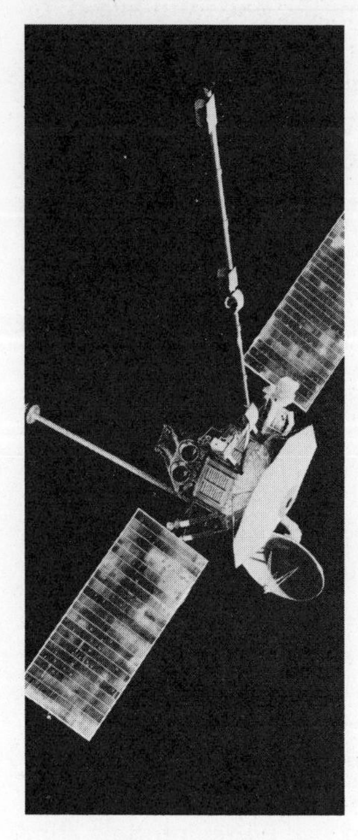

Universal Gravitation

The planets would all move off into space in straight lines were it not for some force, directed exactly toward the sun, diverting them from those linear paths. In this scene, we explore the nature of that force.

The Force of Gravitation

The velocity of each planet is constantly changing as it revolves about the sun and changes its direction of motion; thus the planet is always accelerating. That acceleration toward the sun (or toward the center of revolution of any revolving body) is called *centripetal acceleration*. Newton showed how we can calculate the centripetal acceleration of a planet at each position in its orbit. The computation is complicated, however, by the fact that the distance from the sun and the orbital speed of the planet are continually changing. To solve the problem, Newton had to invent a new branch of mathematics (he did so independently of other mathematicians). He called it *fluxions;* we call it *differential calculus.* There happens to be a simple solution for a planet that has that very critical speed and direction needed to stay in an exactly circular orbit; then the acceleration is proportional to the square of the planet's speed divided by its distance from the sun. In any event, having found how to calculate the centripetal acceleration of a planet, Newton then addressed himself to the problem of finding the mathematical nature of the force required to produce that acceleration.

The clue to the answer was Kepler's first law. Planets were observed to move in ellipses with the sun at one focus. What kind of force could produce just the right centripetal acceleration to make a planet fall continually around the sun in an elliptical orbit? Again, the new calculus came to the rescue. The force had to be one that was greater near the sun, and less, farther away—in fact, a force that grew weaker with the *square* of the distance of a planet from the sun (that is, if one is twice as far away, the force is one-fourth as much; if one is 10 times as far away, the force is $1/(10 \times 10) = 1/100$ as great; and so on).

The *force* accelerating a body is its mass times that acceleration. Thus the actual force on a planet must depend on its mass. But if forces are mutual (as implied by Newton's third law), a planet should accelerate the sun as well, and *that* acceleration should depend on the sun's mass. Because the sun's mass, compared to that of any planet, is so large, its acceleration should be very small; nevertheless, the sun's mass should be involved. Thus, Newton postulated that the force between a planet and the sun was proportional to the masses of both the planet and the sun, and inversely proportional to the square of the distance between them. Here was a scientific hypothesis to be tested.

The Apple and the Moon

Now the genius of the Newtonian concept is this: All natural laws must be *universal*, and must hold not only for the planets but also for everything else in the universe, including apples and oranges. Therefore Newton reasoned that things must fall to the earth in accord with the same law that makes planets fall continually around the sun.

Now Galileo had found that objects uniformly accelerate toward the earth. But near the earth, an object is attracted by *all* material parts of the earth—nearby trees and buildings, the Rock of Gibralter, Mount Everest, the core of the earth, and the Australia Barrier Reef, to name a few. How do the contributions from all the various parts of the earth add up? To solve this problem, Newton had to invent *inverse fluxions*, which deals with the problem of adding up an infinite number of infinitesimal quantities (this branch of mathematics had been explored by others about the same time, but Newton shares the honor of the invention of what is now called *integral calculus*, as well as of differential calculus).

The result of Newton's solution is beautifully simple: The entire earth, all parts of it pulling simultaneously on every object at or near its surface, acts, gravitationally, as though all its mass were concentrated at a geometrical point at the center of the earth—6400 km away. This means that an object several hundred meters above the ground is not enough farther from the center of the earth, 6400 km away, that the force of attraction toward the latter is appreciably less than that on an identical object at the earth's surface. That is why falling objects accelerate uniformly near the earth's surface. At a great distance above the earth, say, 1000 km, the acceleration would be noticeably less.

FIGURE I.39
Newton. *(Yerkes Observatory)*

Galileo also said that all objects fall at the same rate; this, too, is now understandable, because force is mass times acceleration. The gravitational force of the earth on another object is greater the greater is that object's mass, but the greater its mass, the greater in proportion is the force needed to produce a given acceleration. In other words, a small-mass object, with a small force on it, *needs* a smaller force to produce the same acceleration as does a much heavier object, with a proportionally greater force on it. In fact, if the moon could be concentrated into a point and placed near the earth's surface, it would accelerate toward the ground at the same rate as, say, an apple.

On the basis of laboratory experiments the apple is found, like everything else near the earth's surface, to accelerate, that is, to increase its speed of fall, by 980 cm/s every second. This, then, is the acceleration the moon would have at the earth's surface (if the moon were or could be imagined to be a point mass). But the moon is actually about 384,000 km away from the earth's center—about 60 times the distance of the earth's surface. Thus, according to the inverse-square law of gravitation, its acceleration should be 60×60 times smaller than that of the apple, or $980/3600 = 0.272$ cm/s each second.

We can test the idea, as did Newton, because we know what the actual centripetal acceleration of the moon must be to keep it in a nearly circular orbit about the earth, namely, the square of its speed divided by its distance. Carrying out this calculation, we find that the moon's *observed* acceleration to the earth is, in fact, just 0.272 cm/s per second, in marvellous agreement with the theory of gravitation. Imagine Newton's excitement, in doing the calculation the first time, when he found that it worked! It was not just the correct answer to an arithmetic problem; he had shown that the same law of gravitation that holds us on the earth and makes things fall to the ground keeps the moon in orbit! Since gravitation of the same mathematical nature keeps the planets in orbit about the sun, that force must be the same everywhere, truly universal!

---

1. What would be the acceleration of a body at an altitude above the earth's surface equal to nine times the earth's radius?
2. What if a planet were acted on by two forces, an inward gravitational force (toward the sun) and an equal outward force exactly balancing gravitation. How, then, would the planet move? Sometimes orbital motion is described as resulting from such a balance of forces. Comment on this description.
3. When an astronaut dropped a hammer and a feather on the moon, they were observed to fall at the same rate. In your own living room the same result would probably not be observed. Why?

Kepler's Laws Revisited

Armed with the new law of gravitation, Newton worked backward and showed that each of Kepler's laws *must* follow from it. He found that *any* two objects anywhere in space (according to gravitation) must be mutually attracted in such a way that they move about each other in paths that are some kinds of conic sections—either hyperbolas or ellipses. If the planets moved on hyperbolic orbits, they would long since have left the solar system; thus it is obvious why they are observed to have elliptical paths about the sun.

We have already seen how Kepler's second law is a direct result of the conservation of angular momentum. Because the mutual gravitational pull between a planet and the sun is on a line between them, the conservation of angular momentum, and Kepler's second law must follow.

Newton found something very interesting about Kepler's third law. He found that if *any* two objects are in revolution about each other (for example, the sun and a planet, two stars, a planet and its moon, or a dish and a spoon), their period of mutual revolution should depend both on their average separation – in the same way as predicted by Kepler's third law for the revolution of a planet about the sun – *and* on the sum of the masses of the two objects. In other words, Newton predicted that the period of revolution of a planet about the sun should depend not only on its distance from the sun, but also on the combined mass of it and the sun. Even the most massive planet (Jupiter) is only $^{1}/_{1000}$ the mass of the sun, so the combined mass of the sun and any one planet is almost identical to that of the sun and any other planet; that is why Kepler had not realized the role of the masses of the revolving bodies.

On the other hand, Newton's version of Kepler's third law is far more general. It applies, for example, to the mutual revolution of the earth and moon. If we take into account the far smaller combined mass of the earth and moon than, say, the earth and sun, the moon's period of revolution about us is found to bear the same mathematical relation to its average distance as the earth's revolution about the sun does to its distance from that body. We can also use that same mathematical formula to calculate how massive Jupiter must be to account for the period of revolution of any of its satellites, given the distance of that satellite from Jupiter. In fact, we use the same relationship to find the combined masses of the two stars in a binary-star system.

Kepler, in fact, had noted that Galileo's newly discovered satellites of Jupiter obeyed his third law of planetary motion – that is, the square of their periods of revolution about Jupiter are in proportion to the cube of their distances from that planet. Incidentally, Kepler speculated that the other planets had satellites as well, and predicted that Marhs should have two.

The prediction of Kepler was based on numerology, not sound science. It happens, however, that Mars *does* have two moons; they were discovered in 1877 by Asaph Hall at the United States Naval Observatory. It is pure coincidence that about 150 years earlier, in 1726, Jonathan Swift, in his novel *Gulliver's Travels,* described the discovery of two Martian satellites that had characteristics rather similar to Mars' real moons. In the particular episode of the exploits in question, Gulliver was visiting a strange land, the flying island of Laputa. (Readers knowing Spanish may wonder about the name; in other parts of this episode Swift was satirizing what he regarded to be absurd goings on in the Royal Academy of Sciences.) Anyway, Gulliver reported that the Laputian astronomers had discovered

> ... (two) satellites, which revolve about Mars, whereof the innermost is distant from the centre of the primary planet exactly three of the diameters, and the outermost five; the former revolves in the space of ten hours, and the latter in twenty one and a half; so that the squares of their periodical times are very near in the same proportion with the cubes of their distance from the center of Mars, which evidently shows them to be governed by the same law of gravitation, that influences the other heavenily bodies.

From this quote we see, perhaps, the degree to which Newton's law of gravitation had become widely appreciated among scholars within less than a century of its publication.

---

4. What do you think is the most remarkable thing about Swift's description of the satellites of Mars? Why do you suppose I have reported it here?

Artificial Satellites

Newton also described the possibility of artificial earth satellites. Figure I.40, based on one by Newton himself, shows various trajectories of a bullet fired from an impossibly high mountain, well above the earth's atmosphere so that no air resistance can interfere with the motion of the bullet. If the rifle is fired horizontally with moderate speed, the bullet travels a little distance, but eventually falls to the ground (path 1 in Figure I.41). Without air resistance, however, nothing slows it down in its forward motion. It hits the ground only because of the earth's gravitation pulling on it. On the other hand, if the earth's surface were not in the way, nothing would prevent the bullet from completing an entire elliptical orbit about the center of the earth; the earth's center would be at one focus, with the nearest approach to the focus (perigee) being opposite the rifleman.

FIGURE I.40
A diagram by Newton in his *De mundi systematic*, 1731 Edition. *(Crawford Collection, Royal Observatory Edinburgh)*

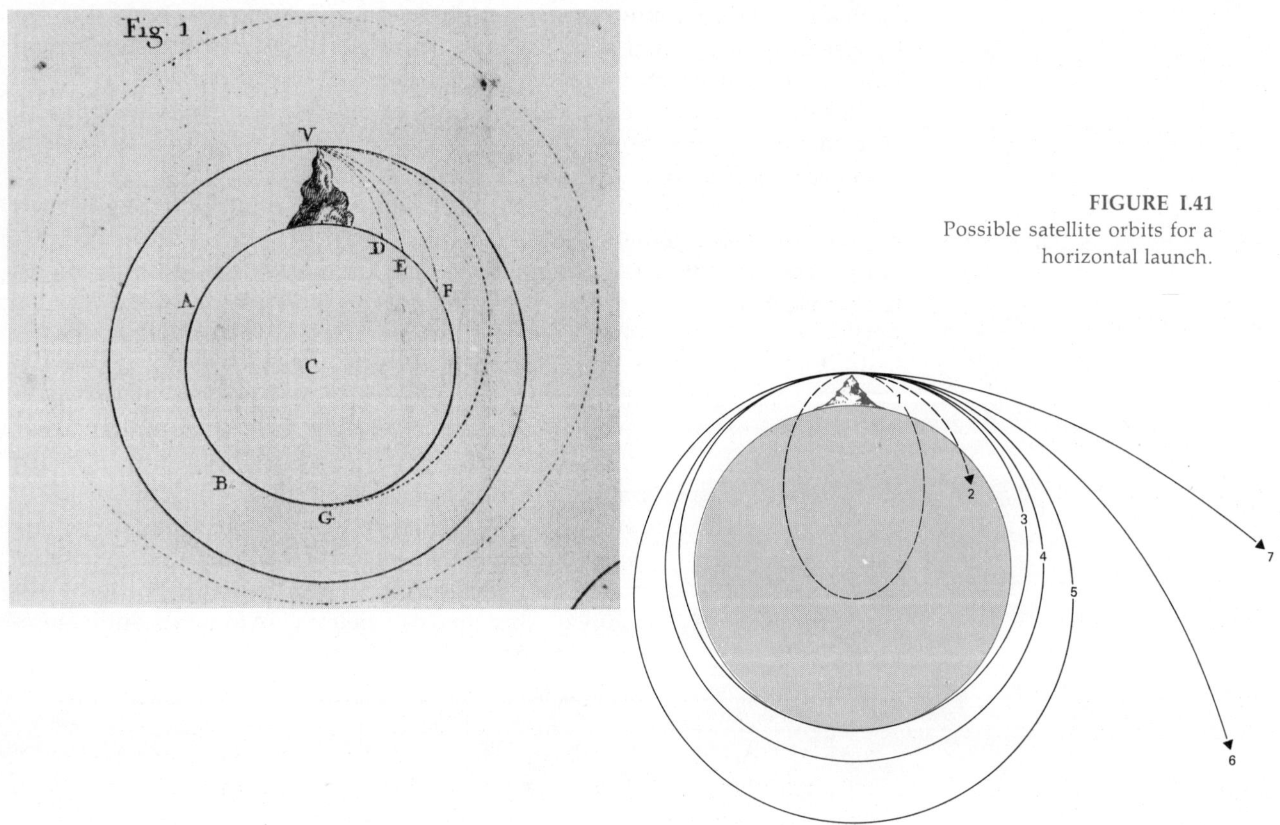

FIGURE I.41
Possible satellite orbits for a horizontal launch.

If the bullet is fired with a greater muzzle velocity (path 2 in Figure I.41), on the other hand, it will travel farther before hitting the ground. If the bullet is fired fast enough, as it falls toward the earth it will have gone far enough that the curved surface of the earth has dropped away beneath it, and it can literally fall *around* the earth, in orbit (path 3). There is a critical muzzle velocity at which the bullet will enter a circular orbit about the earth (path 4); this speed (about 8 km/s) is called the *circular satellite velocity*. A still higher muzzle velocity will put the bullet into path 5, an ellipse, again with the center of the earth at one focus, but with the farthest extent of the path from the earth's center (the *apogee*) on the other side of the earth. Another, higher, critical speed will put the bullet into an orbit that is a parabola (path 6), and the bullet will eventually travel infinitely far away from the earth, escaping it entirely. That critical speed is called the *parabolic speed*, or *speed of escape;* it is about 11 km/s. Still higher muzzle velocities will put the bullet into hyperbolic orbits (for example, curve 7).

Any of the above trajectories for the bullet can be considered to be those of an artificial satellite. Only orbits 3, 4, and 5 in the figure, however, are closed orbits about the earth that clear its surface. And obviously, the object need not be a bullet; it may be a rocket, the Skylab, or a baseball. Of course a baseball or a real bullet has an orbit even smaller than 1 in Figure I.41, and completes only a minuscule portion of its orbit before striking ground.

The first real artificial earth satellite was launched by the Soviet Union on 4 October 1957; the Russians called it *Sputnik I.* I recall that the *Los Angeles Times* telephoned me when it had received word over the news services about the launching, asking my opinion. It happens that I had just been considering the precision of speed and direction needed to launch an earth satellite, and had doubted that any nation yet had that kind of rocket technology. Thus I advanced the opinion that it was probably only a rumor. The next morning from my back yard I saw that rumor, illuminated by the rays of the not-yet risen sun, crossing the sky!

Jules Verne, the novelist, anticipated the Russians by a century. In one story he described how an enemy force had planned to bomb a distant city with a high-speed missile carrying a bomb. Alas, the missile was launched too rapidly, and passed harmlessly over the city in perpetual orbit about the earth.

---

5. Why is "force of forward motion" a nonsense phrase?
6. Suppose a bomb is dropped from an airplane flying with constant horizontal velocity of 1000 km/s due north and at an altitude of 9800 m. When the bomb strikes the ground, where is the plane with respect to the point of impact? (Ignore the effects of the earth's atmosphere.)
7. Describe how a ballistic missile (one which, after initial thrust, coasts to its target) can be considered an artificial satellite. How do the orbits of these objects differ from those of the usual earth satellites?

Several times in the preceding discussion I have said that two bodies in orbit revolve mutually about each other. What this means is that they both revolve with the same period, and in the same shaped orbit, about a point located on a line connnecting their centers. That point is called the *center of mass*, or *barycenter*. If the two bodies have equal mass, the barycenter is halfway between them. Otherwise it is proportionally closer to the more massive body.

The mass of an artificial satellite is so negligible compared to that of the earth that to all intents and purposes the barycenter of the earth-satellite system is at (or extremely near) the center of the earth. However, the moon's mass is not so negligible, but is $1/81$ that of the earth, and the barycenter of the earth-moon system is $1/81$ of the 384,000-km distance from the center of the earth to the center of the moon, which puts it about 1700 km beneath the surface of the earth. It is the barycenter of the earth and moon that revolves annually about the sun. From observations of nearby planets, we can rather easily detect the monthly revolution of the earth about that barycenter, just beneath its surface. In fact, it was by locating the barycenter that we were first able to learn that the earth's mass is 81 times that of the moon.

"Weighing" the Earth

Everything near the earth's surface accelerates as it falls at the rate of 980 cm/s every second, because of the gravitational pull of the earth, which is proportional to its mass. That number which, when multiplied by the mass of the earth (or anything else), tells us how much gravitational force it exerts at a given distance, is called the *constant of gravitation* and is universally symbolized by the letter G. To find the value of G we must actually measure in the laboratory the gravitational force between two objects whose masses are already known in terms of some standard.

Such an experiment is difficult to perform, because it turns out that the gravitational force is exceedingly weak, and is only noticeable when exerted by objects of considerable mass, such as the earth or moon. It was done, however, in 1798 by the English physicist and chemist Henry Cavendish. Cavendish suspended a thin rod 1.8 m long with a 5-cm lead ball at each end by a delicate silvered-copper fiber 1 m long, as shown in Figure I.42. He then placed a 30-cm lead ball about 20-cm from the smaller ball at each end of the rod. The weak gravitational pull between the large and small balls caused the fiber to twist slightly. Cavendish carefully measured that twist, and knowing the force required to twist the fiber, he determined the actual force of gravita-

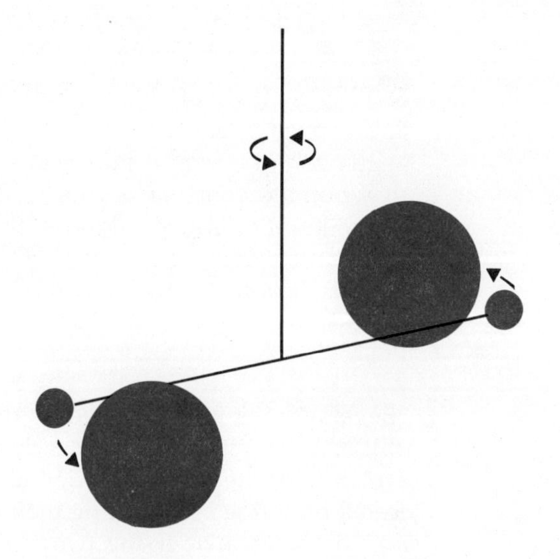

FIGURE I.42
The Cavendish experiment.

tion between the lead balls. Because all the masses and distances in the apparatus were known, Cavendish could determine the constant G. Its numerical value depends on the units used to measure mass, distance, time, and force, but the value in units commonly used by physicists and astronomers is given in Appendix 5.

Since the product of G and the mass of the earth is what gives that acceleration of 980 cm/s per second at its surface, Cavendish had determined the mass of the earth as well as G. (He called it, not quite accurately, "weighing the earth.") Subsequently, many other experimenters have determined these basic quantities more accurately, although they are still known to only a few parts in ten thousand. The earth's mass is about 6×10^{21} (6 with 21 zeros) tons.

Bound and Unbound Systems

Newton's laws tell us that any two objects, moving under the influence of their mutual gravitation, have orbits about each other that are the shapes of conic sections. Now we have also seen that those orbits are either closed, in which case they are ellipses, or open, in which case they are hyperbolas. In the former case, the bodies permanently revolve about each other; we say they comprise a *bound* system. In the latter, they merely pass each other in space, diverting their directions as they do so; they are then said to comprise an *unbound* system. Stars, for example, commonly pass each other in space, deflecting each other as they do so, into hyperbolic orbits about each other. They can never capture each other into closed, elliptical orbits. (We assume here that only the two objects are involved; the matter can be more complicated if there are three or more bodies attracting each other.) Planets in orbit about the sun, satellites about planets, or double stars revolving about each other are in permanently closed orbits, always elliptical in shape, and can never escape each other (again, in the absence of the gravitational pulls of exterior bodies).

A body in motion has a certain amount of energy associated with that motion. That energy of motion is called its *kinetic energy*. There is also a certain amount of energy associated with the gravitational attraction of two bodies; that energy is called their *gravitational potential energy* (or often just *potential energy*). Kinetic energy can be converted to potential energy, and vice versa. For example, if you raise a stone high above your head, you give it a certain amount of potential energy, because the earth is pulling on it (and it on the earth). Now if you release the stone, it falls, picking up speed and also kinetic energy as it falls. But in doing so, it gives up the same amount of potential energy. In an isolated system the sum of the kinetic energy and potential energy is always the same (the principle is called *conservation of energy*). As a planet, moving on its elliptical orbit, falls in somewhat nearer the sun, it releases some potential energy, but speeds up in its orbit, gaining an exactly equal amount of kinetic energy. As the planet moves out again, it slows, converting some of its kinetic back into potential energy.

If the gravitational potential energy of a system is greater than its kinetic energy, the speeds of the objects in it are too small to escape from each other, and the system is bound, as are planets in the solar system. If the kinetic energy is greater than the potential energy, however, the bodies in the system have more than their mutual speeds of escape, and will disperse into space; the system is then *unbound*. Similarly, we have seen that the universe is expanding. If it has a total potential energy greater than its total kinetic energy, all the galaxies will one day fall back together again; in that case, the universe is bound. If, on the other hand, its kinetic energy exceeds its potential energy, the expansion will go on forever, and the universe is unbound.

Universality of Gravitation

All the foregoing concepts are part and parcel of Newtonian mechanics. Indeed, the only assumptions involved are Newton's three laws of motion (defining momentum, force, and mass) and his law of universal gravitation. All else, all the heavenly motions we have investigated, the basis of our entire space program, and, for that matter, much of modern technology, derive directly from those beautifully simple ideas. I say "beautifully simple" first because it really takes no more than simple high school mathematics to understand the basis of the entire Newtonian theory. But isn't it also a beautiful thing that nature conforms to laws so simple?

This does not mean that new things will not be discovered. But the new things should, if the theory is correct, behave in a way compatible with it. For example, in 1781, the German-English astronomer William Herschel discovered a new planet, eventually named Uranus. Uranus is about twice as far from the sun as is Saturn, and has only half Saturn's diameter; thus it is usually invisible to the unaided eye (sometimes, though, it can barely be seen). Uranus fits right into the fold; it revolves about the sun in strict accordance to Newton's and Kepler's laws. In the following scene, we shall see that the strictest application of Newtonian laws to Uranus' motion resulted in yet another triumph for the theory!

Space Vehicles to Mars

Remember that Newton's laws resulted, in part, from Kepler's laws, which in turn came about because of Tycho Brahe's observations of Mars. Mars is an amazing planet from many points of view. Not only did observations of it lead indirectly to our modern technology, but it is the planet most like the earth. It is also the planet that, in the first three-quarters of the twentieth century, was believed to be the most likely of any other in the solar system to have life. That is why the United States launched the Viking space vehicles to Mars in 1975. Those space probes are typical of interplanetary probes insofar as their space journeys are concerned. Let's see how Newton's laws (and Kepler's) are applied in sending Viking to Mars.

First, we must launch the space vehicle with enough speed to escape the earth—about 11 km/s. If we launch it at a somewhat faster speed, it will escape the earth with some speed to spare; when extremely far from earth (say, a million kilometers) the vehicle will still be moving in relation to the earth. But remember that the earth is moving about the sun and at a speed of about 30 km/s. In firing the space probe away from earth, nothing was done to take away from it the speed the earth had to begin with. Thus, while moving, say, 3 km/s with respect to the earth, a million or so kilometers away, its motion with respect to the sun includes, in addition, that 30 km/s the earth has.

Suppose we fire the vehicle away from the ground with a speed greater than 11 km/s in the direction toward which the earth moves in its orbit. Then, when far from the earth, it is moving around the sun in the same direction as the earth, but at a greater speed than the earth. Remember what hap-

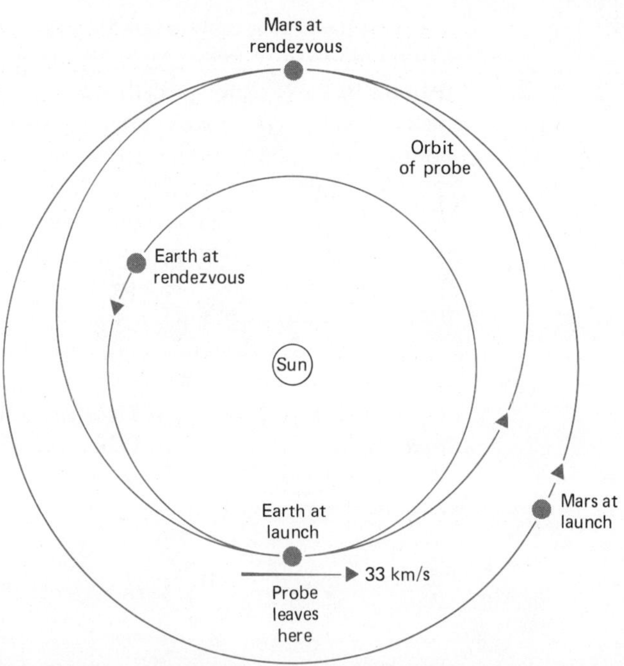

FIGURE I.43
The least energy orbit to Mars.

pened to the bullet fired by the rifleman (Figure I.41) at a speed greater than that for a circular orbit? It moved into an elliptical orbit that carried it farther from the earth. Similarly, the space probe must move out into a larger elliptical orbit. If its speed is just right, that larger orbit will just reach the orbit of Mars (Figure I.43). To do so, it must have a speed of about 33 km/s, or about 3 km/s greater speed than the earth's. This means that if we fire the space probe from the earth in just the right direction and with enough oomph to have a little more than the escape speed of 11 km/s, and, when far away, to have 3 km/s left with respect to the earth, it will be a little planet in its own right, and will travel out as far as the orbit of Mars.

Of course we have to launch the probe at the right time so it will meet Mars in the latter's orbit when it reaches Mars. Kepler's third law tells us the secret of this trick. We know what size orbit the probe must have to reach that of Mars, so we can calculate the period of a hypothetical planet in that orbit. The probe, of course, is such a planet, at least temporarily. That period turns out to be about 17 months, which means that half that time, or $8\frac{1}{2}$ months, is required to traverse half the full orbit and reach Mars. We must time the launching, then, when Mars itself is $8\frac{1}{2}$ months behind that point in its orbit. When the vehicle reaches Mars, additional rockets must be activated in the forward direction *(retrorockets)* to slow it down, so that it can enter an orbit about Mars (otherwise it would pass Mars on a hyperbolic orbit). To land the instrument package of Viking on the surface of Mars, the vehicle containing that package must be disengaged from the orbiting probe and slowed again with an additional rocket, so that it enters a new, smaller orbit that will intercept the planet's surface at the chosen place. Finally, still additional retrorockets are fired to slow the landing package still more, so that it settles softly on the ground, aided by a parachute.

---

8. Why would a space probe to Mars pass Mars by on a hyperbolic orbit if it were not slowed by a retrorocket?

If you have stayed with us this far, you realize that this kind of trip to Mars can be made only when Mars and the earth are in the correct orientation with respect to each other and the sun at the time of launch. Actually, there are many other ways to send a space probe to Mars, but they require a lot more energy, and with available rockets we could not send a payload of instruments along to learn anything. Thus the planetary probes to Mars, and also to Venus, that have so far been launched have orbits of very nearly the type described. Those optimum times for launch to Mars occur about once every two years. There is a period of a few weeks around each optimum date that we can carry out such a mission; that period is called a *launch window*.

In many respects landing the Apollo astronauts on the moon involved the same techniques. The whole thing is very complex in practice, because many details must be taken into account. But the principles are those same simple ones—Newton's laws. And they were discovered because of Brahe's careful observations of Mars; it was Mars, then, that has made such missions as those to Mars possible.

SCENE 5

More Proof of the New Order

In 1766 Titius of Wittenberg discovered an interesting numerical relationship concerning the distances of the planets from the sun. Suppose one measures their mean distances in terms of the distance of the earth from the sun. We call that distance, for convenience, one *astronomical unit,* abbreviated *AU.*

Well, what Titius discovered is that a rather simple progression of numbers can be generated which gives, very roughly, the planets' distances from the sun in astronomical units. We obtain that interesting progression by writing down the numbers 0, 3, 6, 12, ..., and so on, each number being twice the previous one (after the first). Now we add 4 to each number, and then divide by 10. The resulting quotients come close to being the actual semimajor axes of the orbits of the known planets in astronomical units, as we can see from the following table.

Titius Progression	*Planet*	*Actual Distance of the Planet from the Sun in AU's*
$(0 + 4)/10 = 0.4$	Mercury	0.387
$(3 + 4)/10 = 0.7$	Venus	0.723
$(6 + 4)/10 = 1.0$	Earth	1.000
$(12 + 4)/10 = 1.6$	Mars	1.524
$(24 + 4)/10 = 2.8$	–	–
$(48 + 4)/10 = 5.2$	Jupiter	5.203
$(96 + 4)/10 = 10.0$	Saturn	9.539
$(192 + 4)/10 = 19.6$	Uranus	19.18

Even the newly discovered Uranus fit the progression quite well. The sequence of numbers seemed so striking that it was brought into prominence by J. E. Bode, director of the Berlin Observatory. It has since been known as Bode's law (with typical justification).

The Minor Planets

The reader will notice that Bode's law has a hole in it; namely, that there is no planet 2.8 AU away from the sun. Newton's laws do not predict any such planet, but Titius' sequence of numbers does. Thus a search for the "missing planet" was actually organized. While the search was going on, but completely independently of it, the Sicilian astronomer Giuseppe Piazzi was engaged in mapping a region of the sky in Taurus, when on 1 January 1801 he observed an uncharted object. During the next two nights he noticed that the object had moved slightly with respect to the background stars. Thus, he continued to observe it until February 11, after which he fell ill and was forced to interrupt his work. In late March, he reported his discovery to Bode.

Bode suspected that the new object might be the missing planet, but at that time no one knew how to calculate the orbit of an object that had been observed for only 6 weeks. The great German mathematician, physicist, and astronomer Karl Friedrich Gauss (1777–1855) came to the rescue. He invented a new way to calculate the orbit of a body that had been observed for only a short time and which was moving under the influence of Newton's laws. By November (1801) he was able to predict where the object should be. It was rediscovered by von Zach on New Year's Eve, the last night of the year of its discovery.

The new object was named *Ceres,* for the goddess of agriculture and the protecting goddess of Sicily. Ceres is the largest, as well as the first discovered, of the *minor planets,* or *asteroids* (*asteroid* means "starlike"; the International Astronomical Union has adopted "minor planet" as the preferred usage, but the terms are synonymous). Ceres is about 1000 km in diameter—tiny compared to the moon, let alone a planet. Although there are at least 100,000 minor planets, most are only a kilometer or so across, and their combined mass falls far short of making up a typical planet. So Ceres was not exactly the missing planet; yet it does have an orbit of about the right size, as do the majority of the other minor planets.

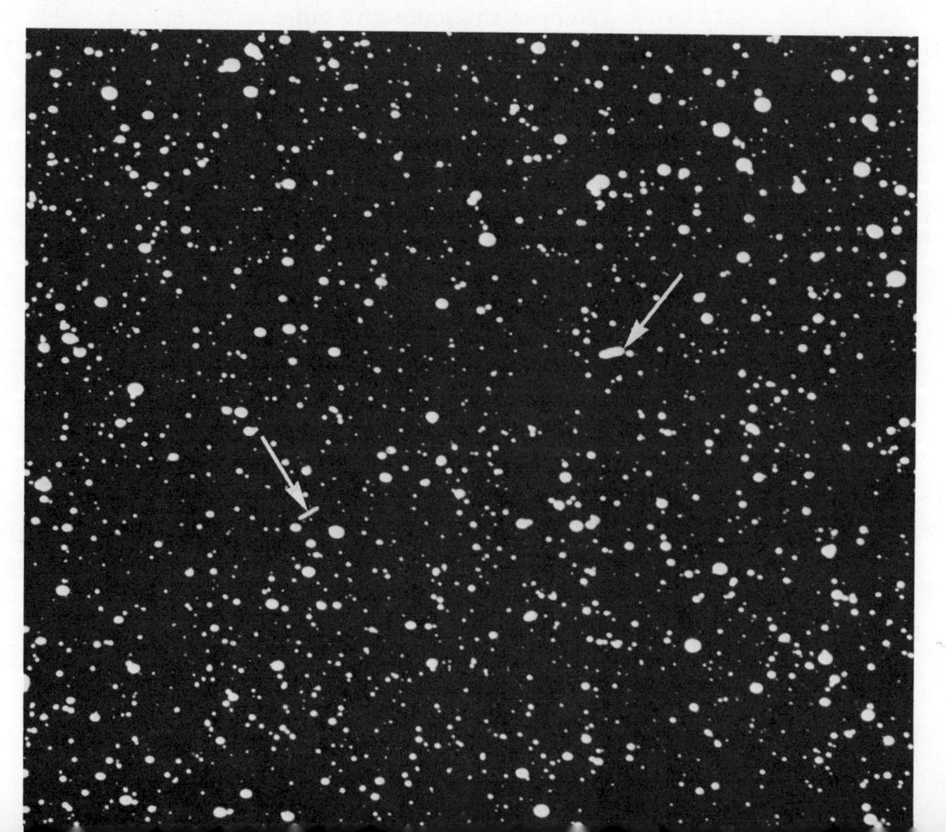

FIGURE I.44
Time exposure showing trails of two minor planets (marked by arrows). *(Yerkes Observatory)*

Then what about Bode's law? Does it carry some wisdom missed by Newton? No. It is simply one of many possible numerical sequences that happen to describe the roughly regular spacing of the planetary orbits. In fact, Bode's law breaks down completely for Neptune and Pluto, discovered much later. There does not happen to be a full-scale planet in the part of the solar system occupied by the minor planets, perhaps because that was a region of the original solar nebula where the gas and dust were of lower-than-average density.

---

1. What distances from the sun, in astronomical units, would Bode's law predict for the planets Neptune and Pluto? How do these distances compare with the actual distances of these planets? (See Appendix 6 for necessary data.)

The most important point is that Gauss, applying Newton's laws to a new circumstance, managed to save the discovery of a new member of the solar system. Ceres was no exception—it was not above the law (of gravitation). Newtonian physics had triumphed again!

The Earth's Shape

Because of its fairly rapid rotation the earth cannot be precisely spherical. Consider an object at the equator. It is moving at about 1600 km/hr in its daily trip about the earth's axis. To conserve its momentum, it would *like* to fly off on a tangent to the earth. But the gravitational attraction of the earth keeps pulling the object around, so that it stays on the ground. It turns out that about 1 percent of the earth's gravitation on that object at the equator is used up just keeping the object on the surface, so the whole brunt of gravity is not available to supply the object's weight, pushing it onto the ground. A person actually weighs a little less at the equator than he does at the poles because of this effect of the earth's rotation.

Well, what do you suppose happens if things weigh down more at the poles than at the equator? Gravity simply squeezes the earth a bit at the poles and forces it to bulge out at the equator. As a consequence, the earth measures about 43 km larger along an equatorial diameter than it does through the poles. The earth is a more perfect sphere than a bowling ball, but is *not* a *perfect* sphere.

---

2. If the earth had its present size and rotation rate, but were far more massive than it actually is, would its shape be more or less oblate? Why?

Precession

It is the oblate shape of the earth that is responsible for its 2600-year precession cycle, discovered by Hipparchus (Scene 2). The moon's orbit is inclined only at 5° to the plane of the ecliptic, so the moon moves through the zodiac, tipped at an average angle of 23½° to the plane of the earth's equator. Thus the moon sees the earth's equatorial bulge as sort of a "spare tire" but

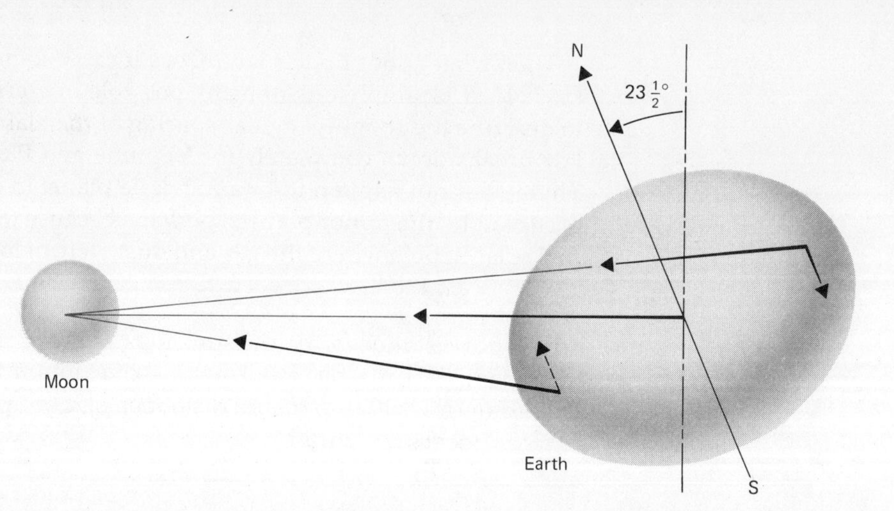

FIGURE I.45
The moon's gravitation pull on the earth's equatorial bulge tries to erect the earth's axis.

tipped at that skewed 23½° angle. The moon attracts that spare tire in such a way as to try to pull it into alignment with the moon's orbit. (Figure I.45). The sun does the same thing, but because of its far greater distance, its effect is less important than the moon's. The earth's axis of rotation yields to that pull, but not the way one might expect. Rather than straighten up with respect to the orbital plane of the moon, it describes the slow conical precession of all tops and gyroscopes under the influence of the earth's gravity. It is the angular momentum of the various parts of the heavy earth, or of the top, that prevents it from toppling immediately in the direction of the force on its axis. Only as friction slows down its spin does a top topple to the floor. But the earth, with no friction to slow it, just keeps on precessing.

The same thing happens to a bicycle wheel if you try to change its axis of rotation. Every cyclist knows that if he leans to one side, the front wheel turns in the direction he leans, instead of the bicycle falling flat to the street; that is how one can steer a bicycle with "no hands."

---

3. How do you suppose precession affects the seasons? After half a precessional cycle (13,000 years) will there still be a reason why there should be summer and winter? Why or why not?

Tides

The cause of tides, like that of precession, is now fully understood as a consequence of gravitation. The moon pulls on the side of the earth nearest it slightly more strongly than it does on the center of the earth, on which, in turn, it pulls slightly more strongly than on the side of the earth farthest from it. The consequence of the moon's different pulls in different sides of the earth is to distort the earth into a slightly elongated shape, like a football, with its long diameter pointing toward the moon (Figure I.46).

To some it seems mysterious that these tidal bulges occur on the sides of the earth both toward the moon and *away* from it. Of course the moon pulls all parts of the earth toward it, and the earth pulls all parts of the moon toward it; as a result, the earth and moon mutually revolve about their barycenter. The tidal distortion we are discussing comes about because of *differences* between the moon's pull on different sides of the earth. Just as the

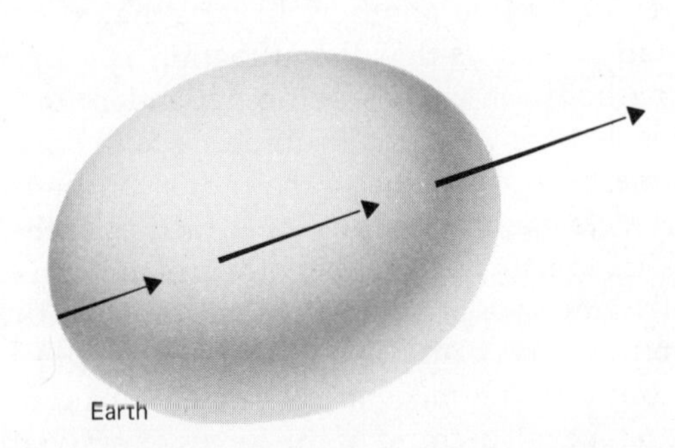

FIGURE I.46
The moon pulls on the side of the earth nearest it more strongly than it does on the side of the earth farthest from it.

side of the earth nearest the moon feels itself pulled away from the center of the earth, so the center feels itself pulled away from the far side. In relation to the earth's center, then, both the near and far sides feel themselves pushed outward.

The effect is actually very small; the solid earth is distorted from its spherical shape only by about 20 cm—this is far less than, and superposed on, the equatorial bulge of the earth due to its rotation. Moreover, as the earth rotates with respect to the moon, this distortion is continually changing. Nevertheless, delicate measurements can detect the rising and falling of the earth's surface by those few centimeters twice each day.

The rigidity of the earth, however, prevents it from deforming quickly enough as it rotates to take on such a shape that the tidal forces are exactly in balance with the earth's gravitation. Consequently the waters of the oceans are still subjected to tiny forces trying to make them flow to the sides of the earth toward and away from the moon. These little forces, changing back and forth in direction very regularly, day after day, eventually set the oceans sloshing back and forth in their basins. As the water at a shore alternately rises and falls, we say the tide is "coming in" or "going out."

The sun also plays a role, but because of its greater distance from the earth, it attracts all parts of the earth with very nearly the same force. The *total* force of the sun on the earth is about 150 times that of the moon's, but the *difference* between the sun's gravitational pull on different sides of the earth is less than half that of the moon's. Nevertheless, when the moon is lined up with the sun—that is, when it is full or new—the solar and lunar tides reinforce each other and are greater than usual. These are *spring tides.* At first- and last-quarter moons, on the other hand, the solar and lunar effects on the tides are working at cross purposes. Then tides are less than average, and are called *neap tides.*

---

4. Draw a diagram showing the earth and moon, but with the earth's axis tipped about 23½° to the earth-moon line. Now consider an observer at a latitude of about 60° North. Show that for him the two high tides each day, at least in an ideal deep ocean, would be unequal in height.

5. Should the interval between high tides be exactly one-half day? If not, why, and what should it be?

Perturbations

Now if Newton's laws are really universal, you may ask, why don't the planets attract one another? In fact, they do, but remember that the law of gravitation tells us that the attraction is proportional to the mass of the attracting body, as well as being dependent on its distance. The planets have masses that are very small compared to the mass of the sun; even Jupiter, the most massive planet, has a mass only 0.1 percent that of the sun. Thus each planet feels mainly the force of attraction of the sun, as if it and the sun were alone in the universe. The other planets also affect the planet, but very slightly, and to a good approximation such tiny effects can be ignored.

But the effects cannot be ignored to the precision with which we routinely carry out celestial mechanical calculations today. Precise tables of planetary motion take full account of the forces of gravitation between the planets. Each planet moves approximately as if it and the sun constituted a two-body system, isolated in space, but not exactly so. The small adjustments to the simple theory, needed to take account of the interplanetary forces, are called *perturbations.* The perturbative effects of the planets are always taken into account in planning the space trajectories of such missions as Viking, Pioneer, and Voyager space probes.

Comets

Comets are an interesting kind of natural space probe. They can occasionally be seen in the sky with the naked eye; then they look like faint diffuse balls—sometimes not much larger than stars—often at the end of a faint whispy tail of luminous matter. At such times comets are mostly very large spheres of very diffuse gas—typically some tens of thousands of kilometers in diameter—set aglow by ultraviolet radiation from the sun. Mixed in with the gas are billions of microscopic dust particles, many of which are pushed away from the comets by the radiation pressure of sunlight. These ejected dust particles, as well as ionized gases that are forced away from a comet by a continual wind of charged atomic particles always blowing away from the sun, make up the comet's tail. Sometimes a comet can put on a pretty good show and can be quite conspicuous in the night sky for several weeks, changing position among the stars night after night as it sweeps in its orbit around the sun.

---

6. Why do comet tails not point toward the sun? On what occasion can a comet tail point in the general direction the comet is moving?

Despite their occasional spectacular appearance, comets are really very tiny objects. All that gas and all those dust particles are extremely thinly dispersed, and have evaporated from the outer few meters of a ball of dirty ice no more than a few kilometers across. We think that there are many hundreds of thousands of millions of comets belonging to our solar system. We suspect that they formed very early in the formation of the solar system and that today most of them travel about the sun in orbits hundreds to thousands of times as large as the orbits of the planets.

According to the favored theory of comets, their orbits are, typically, as large as a third of a distance to our nearest neighboring stars. Over a period

FIGURE I.47
Comet Wilson, photographed 25 July 1961. *(Courtesy, Alan McClure)*

of millions of years, many stars pass the sun in space, and as one star chances to do so, it may move near or through the region of space occupied by the comets. The gravitational attractions of these passing stars perturb the comets in their motions and radically change their orbits. Some comets are ejected permanently from the solar system. Others, slowed down in their orbital motions, enter new orbits that bring them closer to the sun. After several such encounters with passing stars, a comet may chance to enter the inner part of the solar system occupied by the planets.

If a comet comes as close to the sun as Jupiter, the heat of sunlight warms its surface ices and evaporates them, thus creating that vast cloud of tenuous gas and dust around the icy nucleus. It is the gas absorbing and reemitting sunlight and the dust reflecting sunlight that give the comet its light and spectacular appearance. For the most part, we never know in advance when a comet will appear, because we cannot see it far out in space and have no way of knowing it is coming. However, each year a dozen or so comets are usually discovered and observed telescopically. It takes a comet only a few months to pass around the sun and off into space again, during which time it will have evaporated perhaps 3 or so m of its outer layers. When it is far away once again, it settles down to that frozen iceberg, typically 5 to 10 km in diameter.

In the vast majority of cases, we see a particular comet no more than once in our lifetime. The reason, again, is explicable in terms of Newton's laws. Originally, the comet is tens of thousands of astronomical units from the sun, revolving about it in a period of millions of years. Successive perturbations by passing stars have altered its orbit to one where its farthest distance from the sun *(aphelion)* is still out where it began, but its nearest approach *(perihelion)* is only a few astronomical units (perhaps even less than 1 AU). Its orbit is thus a *very* elongated ellipse when we see the comet for the first time

—so elongated that the end of that ellipse we see is virtually indistinguishable from a parabola. In fact, the comet, when coming near the sun, is moving at very nearly its speed of escape from the solar system; just a little more speed would be enough to change its orbit to a hyperbola, making it move out again far beyond its place of origin, into interstellar space.

---

7. Why would a comet moving on a parabolic orbit have a speed equal to its velocity of escape from the solar system?

Well, as the comet comes in, it is mostly affected gravitationally by the two most massive bodies of the solar system, the sun and Jupiter. Of course, the other planets also attract it, but their combined mass is very much less than even Jupiter's, so their attraction on the comet can be ignored to a first approximation. Were it not for Jupiter, the comet would revolve about the sun as though the sun were a point mass, and would obey Kepler's laws. But remember that the comet is moving at very nearly the critical speed it needs to escape the solar system (for it has fallen in from such a great distance). The smallest perturbation in its speed, even a fraction of 1 percent, would be enough to change its orbit drastically.

Although Jupiter's mass is only 0.001 that of the sun, it adds its influence to the sun's just enough to change the comet's speed slightly. The result is that the comet will either speed up, and after it passes the sun, will permanently leave the solar system (in which case it will never be seen again); or it will slow down slightly, depending on which way it is moving with respect to Jupiter. In the latter case the comet will enter a smaller orbit and return to the sun again. Comet Kohoutek, which passed through the inner solar system for the first time in 1973, had its period so shortened, and it will return again in a few tens of thousands of years. A relatively small number of comets, after a few passes of the sun, have their orbits altered enough by Jupiter to enter orbits about the sun with periods of under one hundred years or so. These are called *periodic comets.* The most famous example is Halley's comet, with a period of about seventy-six years. It last appeared in 1910 and is due again in the spring of 1986.

Periodic comets seldom last more than a few hundred passes by the sun (perihelion passes) before their icy nuclei are completely evaporated. As a comet disintegrates, a trail of dusty debris is sometimes left in its orbit. Such cometary debris is responsible for the majority of the meteors (popularly called "shooting stars") seen in the sky at night. The small particles, as they enter the earth's atmosphere moving many kilometers per second, burn up in friction with the air and appear as moving spots of light. Several times each year the earth passes through the orbits of periodic comets that have left such trails of particles, and we experience a greater than average frequency of meteors—phenomena known as meteor *showers.* Two such showers are associated with Halley's comet.

We shall return to the fascinating comets and meteors later; we have mentioned them here to call attention to how Newton's laws account for their motions. In fact, Halley's comet played an important role in the history of gravitational astronomy. Edmond Halley had been influential in getting Newton to publish the *Principia* in the first place (in fact, he *paid* for its publication). Halley had investigated the orbits of a number of comets, checking them out

FIGURE I.48
Halley's comet in 1066, as depicted on the Bayeux Tapestry. *(Yerkes Observatory)*

against Newton's laws, and found that three, those of 1531, 1607, and 1682, were very similar. He correctly surmised that they were the same comet, and predicted its return in 1758. It was sighted on Christmas night of that year; Halley had died by then, but the comet was named for him. Records have now been found for its appearance at every perihelion passage since 239 B.C.

So Newtonian gravitational theory was introduced as a unifying hypothesis in nature. But we see that it did far more than account for terrestrial laboratory experiments and Kepler's laws. It worked for newly discovered planets and minor planets; it accounted for the shape of the earth, for tides, and for precession, and even for the seemingly erratic behavior of comets. The crowning glory of gravitational theory, however, was the prediction and discovery of a new planet.

Discovery of Neptune

Uranus, as we saw in the last scene, was discovered in 1781. By 1790 its orbit had been calculated, and it was found to move according to Kepler's laws. However, on checking old records, it was realized that Uranus had been observed, but not recognized as a planet, on a number of occasions prior to its discovery—in fact back to 1690. To account accurately for its motion for a period of a century, it was necessary to include in the calculations the effects of the perturbations of other planets. However, even after figuring in the perturbations of Jupiter and Saturn, there were small discrepancies. By 1840 the difference between the computed orbit and observed position of Uranus had amounted to 2′—an angle barely discernible to the unaided eye, and certainly a discrepancy that would have been ignored two centuries earlier. Still, by the nineteenth century the discrepancy was unacceptable in terms of the precision of gravitational theory. Evidently, there was another, unknown planet influencing the motion of Uranus.

In 1843, a young English astronomer, John Couch Adams, undertook to calculate where such a planet would have to be. His computations indicated a planet beyond the orbit of Uranus and the position in the sky where it should be found. Adams sent his calculations to Sir George Airy, the Astronomer Royal. Airy, doubting the competence of so young and unknown a person, posed a simple problem to Adams for a test; when Adams did not bother to reply, Airy dropped the matter.

The following year, however, a paper appeared by U. J. Leverrier, a French mathematician, also predicting where the unknown planet should be. On noting that Leverrier's predicted position agreed within 1° with that of Adams, Airy sent the Frenchman the same problem, and received a prompt and correct reply.

Consequently, Airy suggested to J. Challis, director of the Cambridge Observatory, that he look for the new planet in the predicted position. Having no up-to-date star charts of that region of Aquarius, the Cambridge people simply recorded the stars they observed in that part of the sky, planning to compare such records made at intervals of several days to see if they could detect the planet by its motion.

Meanwhile, J. G. Galle, director of the Berlin Observatory, at the suggestion of Leverrier, also looked for the planet. Having good star charts, the German astronomers found it on 23 September 1846, the very night Galle received Leverrier's letter. It was just 52′ from Leverrier's predicted position. It turns out that the Cambridge astronomers had actually plotted the planet previously, but had not recognized it.

---

8. Contrast the discovery of Neptune with that of Ceres. mathematical scientific

---

Appropriately, the honor for the discovery of the eighth planet, Neptune, is shared by Adams and Leverrier. It was, of course, a great triumph for the new physics. But by that time, most astronomers already had complete faith in gravitational theory. On 10 September 1846, Sir John Herschel, son of the discoverer of Uranus, confidently expecting the existence of a new planet, remarked: "We see it as Columbus saw America from the shores of Spain. Its movements have been felt trembling along the far-reaching line of our analysis with a certainty hardly inferior to ocular demonstration."

SCENE 6

How Universal Is Universal Gravitation?

In 1650, less than half a century after Galileo first turned a telescope to the sky, the Italian astronomer Baptiste Riccioli discovered that Mizar, the middle star in the handle of the Big Dipper, appeared through his telescope as two stars. Mizar is the first double star to have been discovered. Subsequently many thousands have been observed and cataloged.

At first it was not known whether the members of such a *binary-star system* were physically associated or were two separate stars lined up in projection. The answer was provided by the bright star Castor, in Gemini, which is seen through a telescope to be two stars separated by about 5″. In 1804 William Herschel, in comparing the pair with earlier charts, found that the fainter component had moved with respect to the brighter. Later observations showed the two stars to be revolving about each other. Here was the first evidence for gravitation beyond the solar system.

Now if Newton's laws apply to double stars, as they do to planets in the solar system, the orbits of the stars in a binary pair should be ellipses, with each star at one focus of the orbit of the other. The first to show that this is indeed the case was Felix Savary in 1827, from his observations of the double star ξ Ursa Majoris, whose members revolve about each other in elliptical paths in a period of sixty years.

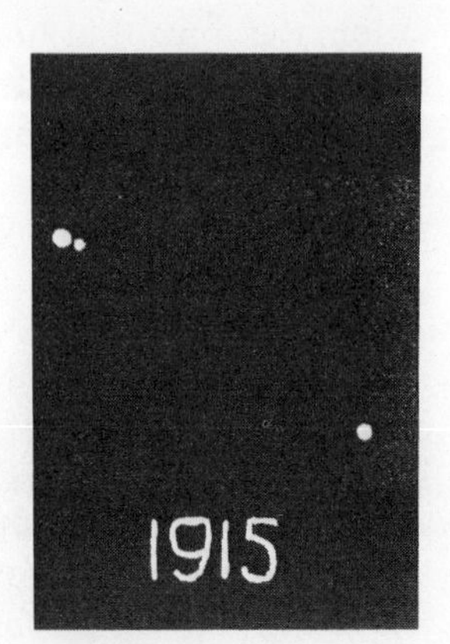

FIGURE I.49
Three photographs of the double star Kruger 60 *(upper left of each)* covering a period of about 12 years, so that the orbital motion can be easily detected. *(Yerkes Observatory)*

FIGURE I.50
The globular star cluster, M92. *(Lick Observatory)*

Among the stars near the sun, at least half, and possibly an overwhelming majority, are members either of binary systems, or of groups of three or more stars, or even of clusters. All binary stars so far analyzed are found to move according to Newton's law of gravitation. Analyses of binary stars are very important, because our only direct way of calculating stellar masses is from the period required for the mutual revolution of two stars, the sizes of whose orbits are known.

There is another interesting example of the success of Newton's gravitational law. Later we shall see how we can probe the internal structures of stars by application of physical principles. A key factor in this study is gravitation, the force that holds the gases of a star together, maintains their high internal pressures, and makes it possible for stars to shine. As we shall see, we have learned a great deal about the structure and evolution of stars, and our theoretical analyses are highly successful in accounting for observations; such success would have been impossible if the Newtonian formulation of gravitation were wrong.

So the generality of Newton's laws has been verified very far beyond the solar system—out among the remote stars. Evidence of gravitational forces extends even further: Some kind of attractive force must keep the stars of a star cluster from dispersing in space, must bind galaxies, and even clusters of galaxies, together. It is indeed a grand and wonderful order that, so far as we know, governs the behavior of matter throughout the universe. *Gravitation is one of the four known forces of nature.*

Now, How about Astrology?

Surely our understanding of the magnificent achievement and universality of Newtonian theory gives us a new perspective of the planet-gods and crystalline spheres of antiquity. As late as the Renaissance virtually all intellectuals believed in astrology. But as the real laws of nature emerged, scientists, with the new realization of the role of the earth as a planet and the grand order that governs the universe, turned away from this ancient religion. Astrology was simply no longer credible in the actual world, any more than are vampires, witchcraft, and the forecasting of events with the entrails of animals. Yet, today hundreds of millions of people in Europe and North America still cling to the ancient belief of astrology.

One reason is probably that astrology claims to tell us something about ourselves—a subject of obvious interest to us all. But so do other superstitions that the modern educated person rejects. I think the main reason that astrology has persisted in our lands is that it is presented (by astrologers) as a science, and most people in our time are just not well enough informed to recognize the difference between science and pseudoscience. As the frontier of knowledge is pushed forward, science has become more and more complex, and scientists themselves have become increasingly specialized. Every new subbranch of science develops its own jargon, and each of these new languages is incomprehensible to those not in it—even to scientists of other disciplines.

---

1. Consider the following terms:
a) deceleration parameter;
b) trine;
c) progression;
d) Robertson-Walker metric;
e) rectification;
f) Hubble constant;
g) periastron;
h) cusp;
i) spicule;
j) refraction;
k) ascendant.
Which do you think are astrological and which astronomical?

Many modern astrologers, in fact, claim a scientific basis for astrology. I have met some in television discussions and debates, who have claimed that planets exert tidal forces on people. In fact, however, even lunar tides on human beings are completely negligible, and because of their greater distances the planets exert tidal forces on earth that are millions of times smaller yet. For example, the tidal force with which Mars, when at its nearest, tries to distort your body (or the fluids in it) is about 50 million times weaker than is the tidal force exerted on you by this very book, placed 2 m away.

---

2. Stars have masses far greater than those of planets. Would you not expect that stars therefore exert far greater tidal forces on people than planets do? If so, explain? If not, how can you prove your answer?

Nor can the light from planets be relevant. Babies are generally born indoors, shielded from the radiation reflected to us and emitted by the planets, and anyway, the light of all the planets combined is minuscule compared to even tiny variations observed in the intensity of sunlight.

One astrologer reminded me of the radio radiation discovered in recent years coming from Jupiter. "Isn't *that* an influence you astronomers didn't know about before?" he asked. It is true; we didn't know of those radio waves from Jupiter until we had built enormous radio telescopes capable of detecting them. Meanwhile, our bodies are continually bathed in radio radiation of the same frequencies from man-made transmitters. Even the radiation we receive from the 100-watt transmitter of a small radio station 100 km away is millions of times stronger than that from Jupiter—no surprise, of course, because we can pick up that station with a pocket transistor radio.

Sometimes astrologers (without understanding the subject) talk about electric and magnetic forces exerted by the planets. Like virtually all bulk matter, however, the planets are electrically neutral. Some do have magnetic fields, but we can detect those fields only with instruments carried to those planets in space probes, not from here on earth. The small permanent magnet in the loudspeaker of a typical portable radio, in comparison, produces an incredibly strong magnetic field in the vicinity of the listener.

To be sure, the sun affects us, and very much so, but in ways understood without invoking ancient gods. And the moon produces tides and reflects sunlight to us. Moonlight can influence the harvester and the hunter, and doubtless can produce psychological effects as well. On the other hand, many of the "well-known facts"—such as the fact that at times of full moon more violent crimes are committed or that more people are admitted to mental hospitals—are not borne out by recent objective investigations. For example, statistical studies of 2497 suicides and 2017 homicides in Texas between 1959 and 1961, of another 2494 homicides in Texas from 1957 to 1970, of 339 suicides in Erie County, New York, and of 4937 mental hospital admissions, all show no correlation with either the phases or the distance of the moon.

---

3. How might you determine if violent crimes are more common at times of full moon? Do you think it would be useful to ask opinions of several policemen? Why or why not?

In short, there is no way in terms of known laws of nature that the planets' directions in the sky can influence human personality and fortune in the manner predicted by astrology. If the planets were to exert an influence on us, it would have to be through an unknown force—and one with very strange properties: that force would have to emanate from some, but not all, celestial bodies; would have to affect some, but not all, things on earth; and would not have to depend on the distances, masses, or other characteristics of those planets giving rise to it. In other words, it would lack the universality, order, and harmony found for every other force and natural law ever discovered that applies in the real universe.

---

4. Can Jupiter affect humans by passing between the earth and sun and shielding us from some of the sunlight? Explain your answer.

What are the properties of such a force? And what evidence is there that it exists? The astrologers answer, "Astrology works." And I must acknowledge that most people who have their horoscopes analyzed by astrologers say that the descriptions they have received of themselves are accurate. However, the descriptions are generally rather vague, sometimes contradictory, and they almost always reveal a good grasp of human psychology on the part of the analyzer. Among the many experiments concerning people's suprise at the success of the astrologer, I shall describe only one especially interesting one.

In a test of the computerized horoscope industry, the French psychologist Michel Gauquelin sent ten sets of birth dates, times, and places to a major advertiser.[1] In order not to reveal himself, he used addresses of various friends. The birth data were genuine, but were not of himself or his friends; they were the birth times and places of the ten most heinous criminals he could find records of. One of these, for example, Dr. Marcel Petiot, was born in Auxerre at 3 A.M. on 17 January 1897. He was executed on 26 May 1946 after a spectacular trial. He had posed as an underground agent promising to help refugees from the Nazis escape then-occupied France. When the unfortunates would arrive at Petiot's home with all their money and most prized possessions, he would murder them and dissolve their bodies in quicklime in a secret chamber of his house. Although indicted for only 27 such murders, Dr. Petiot, cynical to the end, boasted of 63.

What did his horoscope say? In part:

> As he is a Virgo-Jovian, instinctive warmth or power is allied with the resources of the intellect, lucidity, wit . . . . He may appear as someone who submits himself to social norms, fond of property, and endowed with a moral sense which is comforting—that of a worthy right-thinking, middle-class citizen . . . . The subject tends to belong wholeheartedly to the Venusian side. His emotional life is in the forefront—his affection towards others, his family ties, his home, his intimate circle . . . sentiments . . . which usually find their expression in total devotion to others, redeeming love, or altruistic sacrifices . . . a tendency to be more pleasant in one's own home, to love one's house, to enjoy having a charming home. . . .

Next Gauquelin placed an ad in a Paris newspaper offering: "COMPLETELY FREE! Your ULTRA-PERSONAL HOROSCOPE; a ten-page document. Take advantage of this unique opportunity. Send name, address, date and birthplace . . . ." There were about 150 replies. To each correspondent Gauquelin sent the same horoscope—the one he had received for Dr. Petiot. With each he sent a self-addressed envelope and questionnaire asking about the accuracy of the reading. Ninety-four percent of the respondents said they recognized themselves (that is, they said they were accurately portrayed in the horoscope of a man who murdered several dozen people and dissolved their bodies in quicklime), and for 90 percent this positive opinion was shared by their families and friends.

---

5. If you have the opportunity to obtain horoscopic analyses for several different friends (say, by mail order), present copies of all of them to each friend, and see if he can pick out the one that is actually his own.

---

[1] *Science et Vie,* No. 611, p. 80, 1968.

Few astrologers today claim that astrology can predict with perfect precision every event of one's life. "The stars do not compel," they say, "but impel." But if that's the case, how can we test astrology? Only statistically, and unfortunately statistical studies are easy to misapply, easy to misunderstand, and especially easy to fake. Competent reexaminations of studies claiming results that support astrology fail to confirm those results.

One of the most famous astrologers claiming success in statistical tests of astrology was the pro-Nazi occultist Karl E. Krafft. Gauquelin made a detailed study of the results presented in Krafft's major publication *Traite d'Astrobiologie* (1939), and found every one of the allegations to be invalid.[2] Students of mine have collected birth dates of more than 18,000 prominent people, and have found no correlation between their sun signs and various professions. Psychiatrist Carl Jung commissioned a statistical investigation of the sun-moon configurations in the natal horoscopes of 483 married couples to test astrological predictions for marital compatibility. Although Jung himself was interested in astrology and seems to have held some belief in it, his statistical test came out negative. There have been many other such studies, but of them all, I know of none providing statistically significant support for astrological claims.

But why should we expect any such correlations? Astrology, after all, is based on no more than a magical correspondence between the gods of antiquity and the planets that bear the same names. Our ancestors of 2 millennia ago were normal human beings, and among them were brilliant and great thinkers. We have, in fact, built on their knowledge—from Hipparchus to Ptolemy to Copernicus to Kepler to Newton and beyond. But we have come a long way since their time. Science today includes quantum electrodynamics, general relativity, and high-energy astrophysics. The Greeks, who gave us astrology, had not yet learned of algebra. Yet the astrology of today is virtually unchanged from that of Ptolemy.

Knowledge and Certainty

I have dwelt at some length here on astrology. Why so much discussion of an ancient religion in a book on astronomy? Because according to my polls and those of my colleagues, about a third of the people in America (including college students not majoring in the hard sciences) believe in it, and about 90 percent are open-minded about its claims. I think most people really don't know what astrology is, what it is based on, and how it differs from astronomy—thus it has warranted this excursion.

I do not mean to belittle the beliefs of our ancestors. At that time, their beliefs were entirely understandable. We simply know more today; we have been over that ground and have moved on. There is no reason, or no need, to ascribe "ancient wisdom" to Ptolemy and his forebears. Ptolemy was a great man and has earned his place in history without our assuming that he was omniscient. Just because Columbus discovered America is no reason to suppose that he would, at that time, have been able to draw an accurate map of the New York City subway system.

But by the same token, if we have advanced knowledge beyond that of the ancients, it does not mean that we have learned it all. Perhaps someday we

[2] *L'Influence des Astres, etude critique et esperimentale* (Le Dauphin Publications, Paris, 1955, pp. 39–57).

SENATE BILL No. 1280

Introduced by Senator Dills

June 20, 1975

An act to add Chapter 5.5 (commencing with Section 6400) to Division 3 of the Business and Professions Code, relating to the licensing and regulation of astrologers, and making an appropriation therefor.

LEGISLATIVE COUNSEL'S DIGEST

SB 1280, as introduced, Dills. Astrologers: licensing.

There is no provision under existing law relating to the licensing and regulation of astrologers.

This bill would provide for the licensing of astrological schools and organizations and making it a misdemeanor to practice astrology without a certificate issued by a licensed astrology school or organization.

The bill would create within the Department of Consumer Affairs an Astrologer's State License Board composed of five members appointed by the Governor for four-year terms. The bill specifies the powers and duties of the board.

The bill also appropriates an unspecified amount from the General Fund to the board and to the State Controller for reimbursement of local agencies for costs imposed by operation of the act.

Vote: ⅔. Appropriation: yes. Fiscal committee: yes. State-mandated local program: yes.

The people of the State of California do enact as follows:

1 SECTION 1. Chapter 5.5 (commencing with Section
2 6400) is added to Division 3 of the Business and
3 Professions Code, to read:

FIGURE I.51
First page of California State Senate Bill 1280, 1975. The bill was not enacted that year, but similar ones are presented from time to time and have received considerable support and pressure.

may find unexpected ways in which planetary motions correlate with human behavior. I do not anticipate that kind of discovery, but I am certain that there will be many surprises in science. New laws of nature, though, do not negate well-established ones—rather they extend our understanding to new realms where the existing laws do not apply.

Sometimes we forget that science gives us only imperfect models of the real world. Newtonian theory was so successful that by the nineteenth century it was widely believed that Newton had told us fundamentally everything and that only loose ends needed to be sorted out. It was thought that all that happens could, in principle, be understood in terms of cause and effect, all things obeying Newton's laws, right down to the submicroscopic world—even including our thought processes.

Imagine the philosophical impact of this idea—complete determinism. If you read these words, or even commit a murder, it is because the motions of atoms in your body and electrical currents in your brain followed in a precise and predetermined way from their immediately preceding condition—as with a clockwork—as sure as the motions of the planets. There could be no free will or choice; nothing is left to chance.

Twentieth-century science has shown that this determinism is wrong. Newton's laws do not apply to atoms. Ultimately nature is a statistical ensemble. When large quantities are considered, the behavior of matter is predictable, but fundamentally, at the level of atoms—and the initial perturbations that can result in even grand actions—there is an inherent uncertainty and randomness, which gives us the power to select. But we shall return to the behavior of tiny pieces of matter in the next act.

On the other hand, even on the very largest scales, we don't know whether Newton's laws apply either. They work first crack out of the box in the solar system, and seem to be just fine over the range of dimensions of double stars. Star clusters, galaxies, and clusters of galaxies seem to be bound together by gravitation, as we have said. We presume that Newton's inverse-square formulation of gravitation describes the attraction that holds clusters of galaxies together, but we are not sure. It is still an untested model.

---

6. We are confident that there is an attractive force binding galaxies and clusters of galaxies together. We presume it is the same kind of gravitational force that exists locally. We cannot, however, be certain that the forces holding galaxies together and the familiar gravitational force that governs the solar system are described by the same mathematical law. What kind of evidence would we need to convince us that they are, indeed, the same? Is this kind of evidence available for the case of double-star systems? Explain.

But now we must say something even more unsettling, especially after having dramatized the success of Newtonian theory—namely, that it is beset with inherent difficulties.

For example, Newton talked about acceleration and change of momentum. But change with respect to what? He meant a change with respect to a room, laboratory, or whatever, that was itself *not* accelerating (such a system is said to be *inertial*). But everything seems to be in motion with respect to other bodies—the earth and all on it accelerate as we revolve about the sun. Newton postulated that there was such a thing as absolute space, with respect to which all acceleration could be measured. But how is absolute space defined? In terms of the stars? Because of their great distances and consequent low apparent motion, the stars do seem to define something that is approximately an unaccelerated (inertial) reference frame. But the stars are moving, too. This matter of absolute space bothered scientists by the end of the nineteenth century.

Make no mistake about it! Not only is Newtonian theory a magnificent achievement, but it is fantastically useful. It is about the first thing engineers study in college, and is primarily responsible for our modern technology—with its airplanes, bridges, earthquake-resistant buildings, space vehicles, and tremendously more. But it is not the final word, not the *absolute truth*.

In the next act we shall explore the newer science, especially of the twentieth century, and see how it has given us a far richer description of nature.

ACT II

LIGHT, SPACE, AND TIME

SCENE 1

Light and the Telescope

A *straight line*, I learned in high school geometry, is the shortest distance between two points. So defined, a straight line makes possible the logical structure of Euclidean geometry. But I don't think Euclid ever thought about how one might determine which path among the infinitely many paths from one point to another is the shortest. Without a prescription to specify Euclid's straight line, it remains operationally undefined, and has never really existed in the actual world.

But how about using a ruler? you ask; and how do we know the ruler is straight? By *sighting* along it. In other words, in practice we use the path of light to define a straight line. It seems to work pretty well—at least in common experience. We measure the radius, R, of a circle with a ruler—that is, along a light beam and—sure enough!—we find its area to be πR^2. In fact, light gives us pretty good Euclidean straight lines over interstellar distances. On the other hand, there is no assurance whatsoever that the area of a circle whose radius extends over great intergalactic distances is πR^2. We shall see later that other kinds of geometry may be more convenient to the study of the universe as a whole. In particular, we shall see that such *non-Euclidean* geometry is necessary in the vicinity of very dense matter, as in a neutron star. The reason is that the paths of light are themselves affected—in fact determined—by the surrounding matter. For example, stars seen near the limb of the sun (that is, near the apparent edge of its disk as seen in the sky) during solar eclipses are in directions slightly displaced from those they have when the sun is in another part of the sky; the light from them has been deflected in passing near the sun.

We shall return to these interesting complications later in the act. For now we shall accept the approximation, technically inaccurate but pretty good in the laboratory, that light travels in a straight line.

---

1. See if you can think of a method, which makes no use of light paths, to tell if something is really "straight." What, in this case, do you mean by "straight"?

object going in circle let loose, must fly straight

The Speed of Light

Light also travels with the highest possible speed, but that speed is finite. The first demonstration of the finite speed of light was made in 1675 by the Danish astronomer Olaus Roemer. Some of Jupiter's satellites are regularly eclipsed when they enter the Jovian shadow. Roemer was measuring the period of one of the satellites by timing the interval between its successive immersions in the shadow. He found, in effect, that those intervals were longest when the earth is moving in its orbit away from Jupiter, and shortest when the earth is approaching Jupiter.

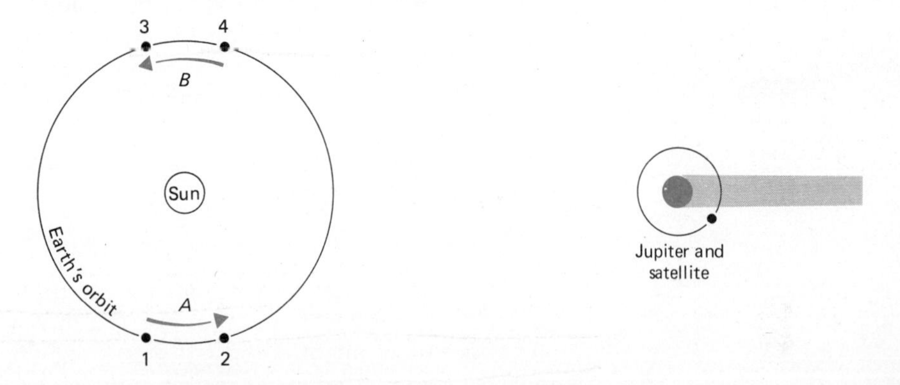

FIGURE II.1 Roemer's method of demonstrating the finite speed of light.

It is easy to understand the reason by referring it to Figure II.1. When the earth is at *A*, approaching Jupiter, it moves from point 1 to point 2 in its orbit during the period between the satellite's eclipses. But light had to travel from Jupiter to point 1 to tell us of the first eclipse, and only the shorter distance from Jupiter to point 2 to tell us of the second one, so we received the news somewhat ahead of the time we would have received it if the earth had been stationary. On the other hand, when the earth is at *B*, moving away from Jupiter, news of the second eclipse, with the earth at point 4, arrives later than it would have if the earth had remained at point 3, where it received the message of the first eclipse. Thus, with the earth at *B*, it seems to us to take longer for a revolution of the satellite than with the earth at *A*.

Such observations show that light requires about 16½ minutes to cross the diameter of the earth's orbit. When the size of that orbit (that is, distance from the earth to the sun) was later determined, Roemer's observations even provided a way of calculating the speed of light.

---

2. Given that the earth is about 150 million km from the sun, what, approximately, is the speed of light in kilometers per second?

Historically the speed of light was determined in many interesting ways. One method involved reflecting light from a rapidly rotating mirror, across a precisely determined distance to a stationary mirror, and back, and timing how much the first mirror had turned during the round-trip light-travel time. In one such experiment the American physicist A. E. Michelson passed light back and forth over the distance between Mount Wilson, near Los Angeles, and Mount San Antonio, 35 km away. Michelson's result gave 299,729 km/s for the speed of light in air.

In modern times, the speed of light is timed electronically, and measured completely in the laboratory, without the necessity of surveying relatively inaccessible distances between mountains. The basic idea, though, is the same: finding the time required for light to cross a given distance. The best modern determination of the speed of light in a vacuum (the speed of light is universally denoted by the symbol c) is $c = 299,792.5$ km/s.

Light is slowed, however, in passing through a transparent substance. In air, its speed is very nearly the same as in a vacuum, but in glass or water it is slowed by about a third. The ratio of the speed of light in a vacuum to that in a transparent substance is called the *index of refraction* of that substance. Air, at sea-level pressure, has an index of refraction of about 1.00029. Water and crown glass have indices of refraction of about 1.3 and 1.5 respectively.

---

3. What is the approximate speed of light in water?

Formation of Images

If light strikes a substance, it either passes through that substance, is absorbed by it, or is reflected by it (or even all three). If light is reflected from a rough surface of an opaque body, it scatters about in a more or less random manner. However, if light strikes a smooth and highly polished surface, such as a mirror, it is reflected back symmetrically with its original direction; in particular (see Figure II.2) the incident and reflected beams lie in the same plane as the line perpendicular to the surface (or *normal* to it) at the point of reflection. Moreover, the incident and reflected beams make the same angle with that normal, as shown in the figure.

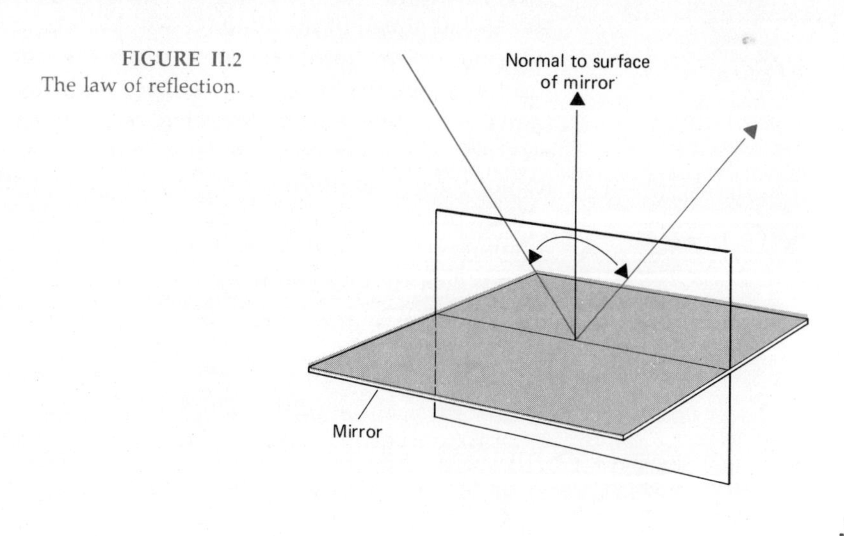

FIGURE II.2
The law of reflection.

---

4. Design a periscope utilizing mirrors for use in a submarine to observe ships on the ocean surface. Indicate carefully the angles at which you must place your mirrors. If it is necessary for the observer in the submarine to scan the sky for airplanes, how would you change your design?

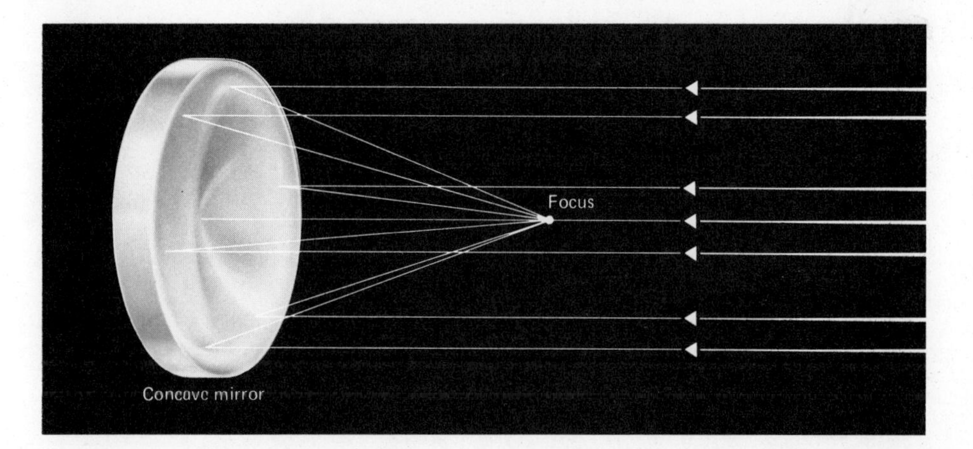

FIGURE II.3
Formation of the image of a point source by a concave mirror.

If a beam of light strikes the curved surface of a concave mirror and if the curvature has just the right shape, all the reflected rays will pass through the same point, called the *focus of the mirror* (Figure II.3). If the light consists of essentially parallel rays coming from a point source very far away (for example, a star), the reflected rays make an *image* of that point source (or star) at the focal point. That image can actually be seen on an opaque card or frosted glass placed at the focus of the mirror.

In the next scene we shall take up what happens to the absorbed light. However, that which passes on into or through a transparent, or partially transparent, substance is *refracted,* or bent, as seen in Figure II.4. The amount by which it is bent or refracted depends on the index of refraction of the substance; the greater its index—or the amount that the speed of light is slowed down in the substance—the greater is the refraction. But it is always bent toward the normal to the surface if it passes from an object of one index into one of greater index, as from air into water or glass. If, on the other hand, it passes from a substance of greater index into one of lesser index, as from glass into air, it is bent away from the normal.

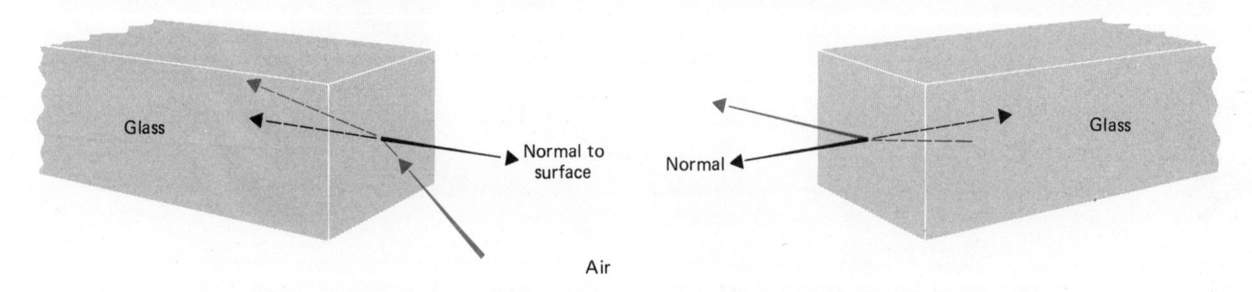

FIGURE II.4
Refraction of light passing from one transparent substance to another.

Now if parallel rays of light from a distant point source (say, a star) pass through a piece of glass whose surfaces have the right shape, such as the lens shown in Figure II.5, they can all be refracted so as to pass through a point on the other side of the glass, known as the focus of the lens. An image of the point source (star) is formed there, just as it is by reflected light in front of a concave mirror.

So both a concave mirror and a lens can produce an image of a distant source of light. This is the principle of the telescope, which uses either a mirror or a lens to gather light from a distant source and concentrate it into an image at its focus. The way an image is formed of an extended source, such as

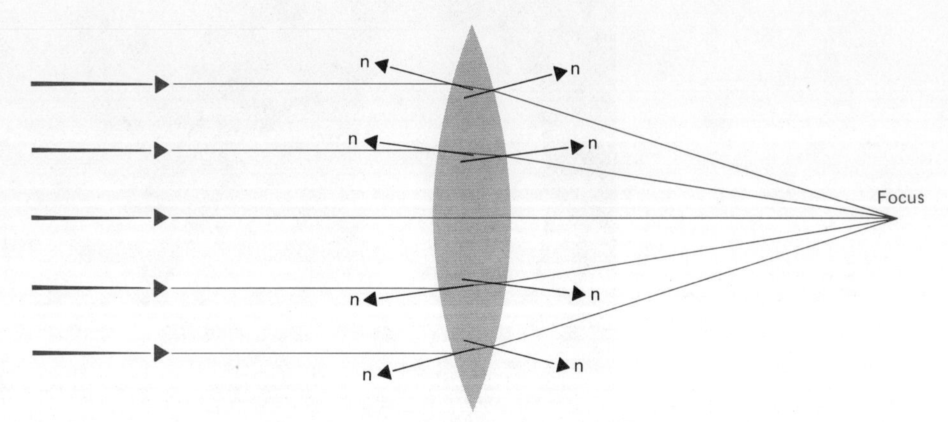

FIGURE II.5
Image of a point source produced by a convex lens.

a group or cluster of stars or the moon, is shown in Figure II.6. The figure shows an image formed by a lens, but the principle is the same for a mirror. Light from each part of the source (the moon) enters all parts of the lens, and is refracted to a focus. However, the location of the focus depends on the direction of the incoming rays, that is, from what part of the source they originate. Thus, in the figure, light from the "bottom" of the moon is focused in one place, and that from the "top" of the moon somewhat below it. Light from intermediate parts is focused in between, so that an entire image of the moon is built up at about the same distance behind the lens. Note that the image is upside down, or reversed.

FIGURE II.6
The image of an extended source produced by a lens.

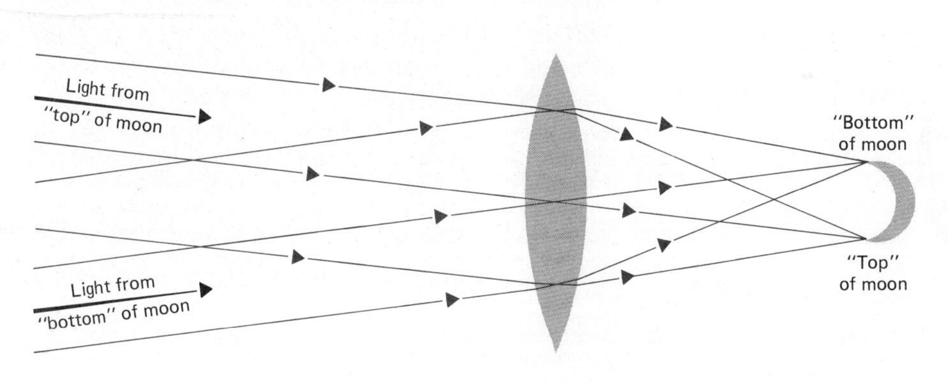

---

5. Design an optical system that produces erect images.

The human eye and a camera are good examples of such image formation by a lens. The formation of the image by an eye is shown on Figure II.7; the location of the upside-down image is at the *retina,* where delicate endings of the optic nerve, sensitive to light, pick up the sensations and carry them to the brain, which interprets the image and turns it around so it appears upright. A camera does the same thing, except that the retina is replaced by the photographic emulsion. After development, the image can be made upright by turning the film around.

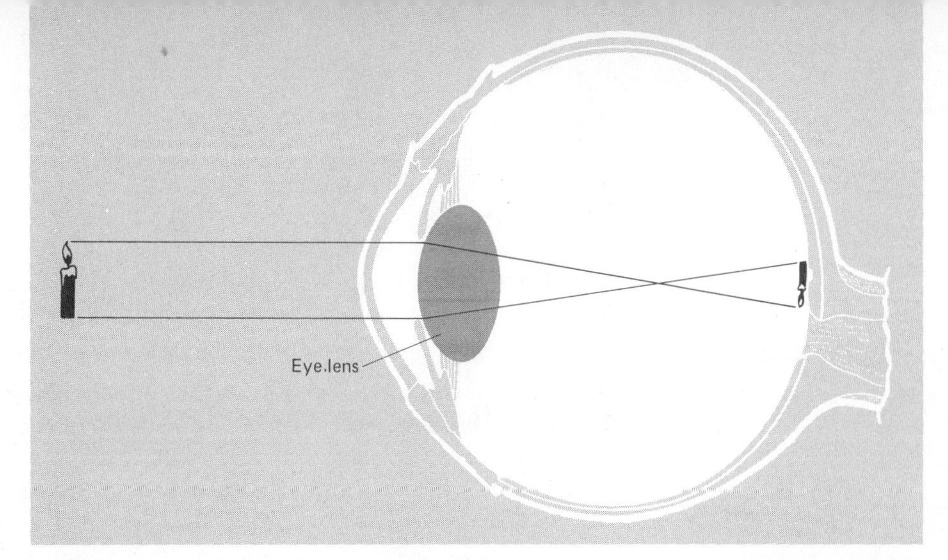

FIGURE II.7
Production of an image on the retina of the eye.

Obviously we have oversimplified things. The images produced are never perfect; there are several kinds of distortions, whose properties depend on the kinds of lenses or mirrors used. To minimize these image defects, telescopes generally use complex lenses of two or more pieces of glass, or mirrors of special shapes. A Schmidt telescope, in fact, utilizes both a lens and a mirror (Figure II.8), and is designed to produce images of good quality over a large field of view.

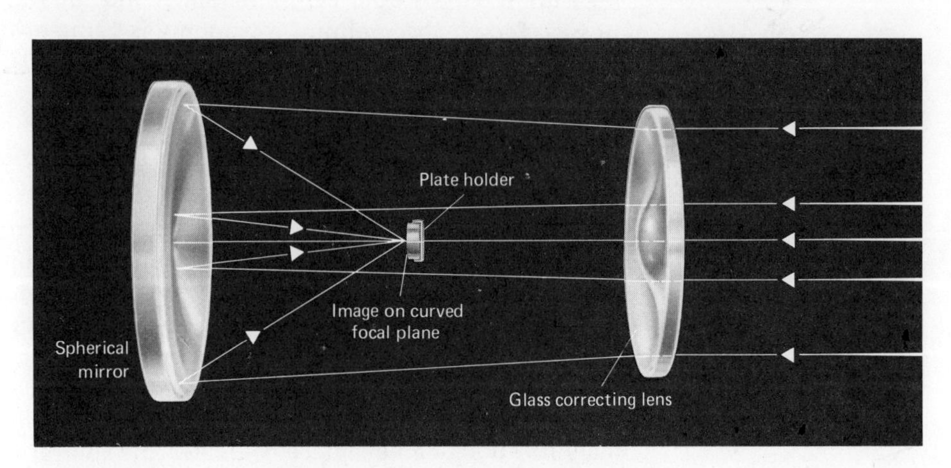

FIGURE II.8
The Schmidt optical system.

Telescopes

Refracting telescopes have lenses that produce images; *reflecting telescopes* make images with mirrors. It is easier to make a large reflecting concave mirror (which has only one optical surface) than it is to make a large refracting lens, which, to produce good-quality images, requires at least four optical surfaces; thus most large astronomical telescopes are of the reflecting type. In either case, however, an image is produced of a celestial object or field of stars at the focus of the telescope's *objective* (its lens or mirror). That image can be measured and analyzed in many different ways.

In the next scene we shall see that one important method of analysis is to observe the *spectrum* of the light in the image (to study the different colors or wavelengths of the light that make it up). We can also simply look at the image, projected on a screen or white card, just as we do the image produced by a slide projector. More often we superpose the image on a photographic emulsion and thus take its picture (which can be analyzed in detail later). One advantage of astronomical photography is that we can make a long-time exposure—up to several hours. The longer the exposure, the more of the faint light in the image is recorded. Indeed, the illustrations of star clusters, nebulae, and galaxies reproduced in this book are all time exposures; direct viewing of these objects with a telescope would never reveal their fainter features recorded on the emulsion, sometimes even after hours of exposure.

---

6. Can you think of a way in which you might determine the relative brightnesses of two remote objects by the use of telescopic photography? Be sure to consider all the steps involved, as if you were setting up a program for actually carrying out such an experiment.

Some substances have the property that, when exposed to light, they emit electrons (subatomic negatively charged particles); these are called *photoemissive substances*. In a vacuum-enclosed tube, these electrons can be attracted to a positive terminal with an electric field, and their flow comprises an electric current. Measurement of such a current is a way of detecting the incident light on the photoemissive substance. A *photomultiplier*, for example, is a highly sensitive tube utilizing a photoemissive surface capable of detecting very low light intensities; photomultipliers are often used at telescopes to measure the light from very faint stars, and thus to determine their brightnesses. Today, observational astronomy is being revolutionized by the use of television techniques and such devices as *image tubes*. With their photoemissive surfaces placed at the foci of large telescopes, image tubes now commonly record in a few minutes pictures that would have required exposures of many hours—in some cases, several successive nights—directly on photographic emulsions.

So we see that telescopes use large lenses or mirrors to make images, and we analyze those images by various techniques in order to record their properties and to make measurements. We've come a long way from Galileo, who just "looked through" his telescope. To "look through" a telescope is simply to inspect with a magnifying glass the image produced by the telescope's lens or mirror. The magnifying glass is called an *eyepiece*, and it is inserted into a sliding tube behind the image to hold it steady during the viewing. Now the magnification of an eyepiece is greater, the shorter is its *focal length*—that is, how far behind it parallel light comes to a focus. Thus it makes no sense to ask an astronomer what the "power" of his telescope is, for he can change the power at will by simply switching from an eyepiece of one focal length to one of another. The number of apparent diameters that a telescope magnifies (called its magnifying power) is the ratio of the focal length of its main lens or mirror to that of the eyepiece being used.

---

7. What is the magnifying power of a telescope whose objective has a focal length of 1000 cm and whose eyepiece has a focal length of 0.5 cm?

8. To the unaided eye the moon subtends an angle of about ½° in the sky. If you have a telescope whose objective has a focal length of 100 cm, what focal-length eyepiece must you use to make an image of the moon that appears 10° in diameter?

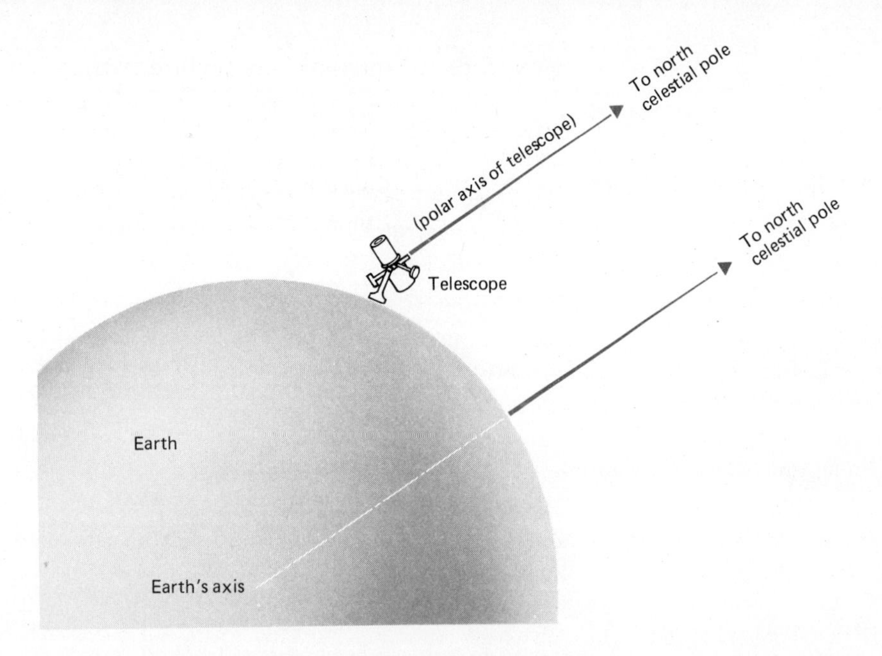

FIGURE II.9
The equatorial mounting of a telescope.

Telescopes are generally arranged in some sort of tube—either enclosed or open (it doesn't matter much which). In a refractor the lens is at the front and the image at the back, as in a common spyglass. In a reflector the concave mirror is in the back, and the image is produced in front of it. Some kind of ladder or platform (in a large telescope) must be provided to allow access to the image. Often, auxiliary mirrors reflect the image (especially in a reflecting telescope) to a more convenient location. The entire tube is mounted on some kind of pivoting device so that it can be pointed toward any direction in the sky. Most often, in an astronomical telescope, that device, or *mount,* consists of two axes at right angles to each other. One axis is parallel to the earth's axis, so that it points along the direction to the north celestial pole. This is called an *equatorial mount* (Figure II.9). The advantage of this system is that a slow automatic rotation of the telescope about that axis can compensate for the earth's rotation and keep the telescope directed toward a particular star or point on the celestial sphere. Professional astronomical telescopes are almost always enclosed inside rooms, usually with a domed ceiling, to protect them from weather. A large slot or oblong window can be opened (by removing a shutter) in the dome, and the dome can be rotated so that the window can be turned to the desired direction in the sky.

---

9. At what rate would an equatorial telescope have to be driven about its polar axis in order to follow the apparent motion of a star as seen from the moon?

Some major astronomical optical telescopes are illustrated in these pages, and some of the more important ones are listed in the accompanying table. By "size" of a telescope is meant the diameter of its objective lens or mirror.

In the early days (including *my* early days) the observing astronomer would typically spend long, cold winter nights on a mountain top, bundled in furs or perhaps in an electrically heated pilot's suit, constantly adjusting the telescope, to insure that it was accurately tracking the object under obser-

Some Major Optical Observatories		
Observatory	*Location*	*Sizes of Largest Telescopes*
Astrophysical Observatory	Caucasus, U.S.S.R.	6-m (236-inch) reflector
Hale Observatories	Palomar Mountain, California	5.1-m (200-inch) reflector 1.5-m (60-inch) reflector 1.2-m (48-inch) Schmidt
	Mount Wilson, California	2.5-m (100-inch) reflector 1.5-m (60-inch) reflector
	Cerro las Campanas, Chile	2.6-m (101-inch) reflector
National Observatory	Kitt Peak, Arizona	4-m (158-inch) reflector 2.1-m (84-inch) reflector
Inter-American Observatory	Cerro Tololo, Chile	4-m (158-inch) reflector 1.5-m (60-inch) reflector
Anglo-Australian Observatory	Siding Spring, N.S.W., Australia	3.9-m (153-inch) reflector 1.2-m (48-inch) Schmidt
European Southern Observatory	Andes, Chile	3.6-m (142-inch) reflector 0.9-m (36-inch) Schmidt
Lick Observatory	Mount Hamilton, California	3.0-m (120-inch) reflector 0.9-m (36-inch) reflector 0.9-m (36-inch) refractor
McDonald Observatory	Davis Mountains, Texas	2.7-m (107-inch) reflector 2.1-m (82-inch) reflector
Burakan Astrophysical Observatory	Armenia, U.S.S.R.	2.6-m (104-inch) reflector
Royal Greenwich Observatory	Herstmonceux, England	2.5-m (98-inch) reflector
Mauna Kea Observatory	Mauna Kea, Hawaii	2.2-m (88-inch) reflector

vation. It was exhilarating, satisfying, and sometimes enjoyable, yet damned hard, cold work. It is still the case at many observatories, but gradually, as astronomy modernizes (more accurately, as we find the funds to update our equipment), things become more automated. At many observatories the guiding of telescopes is now completely automatic. When I observe at the University of California's Lick Observatory these days, I spend my nights in shirtsleeves, in a heated room inside the telescope dome, watching my object on a television screen and recording its characteristics in a computer. We soften in our old age.

SCENE 2

Visible and Invisible Light

Early in the history of the development of refracting telescopes it was found that images formed by them often showed annoying color fringes. At first it was thought that the colors were produced by the glass lenses themselves. It was known, for example, that a glass prism, such as the one shown in Figure II.10, produces a rainbow of color when white light is passed into it. Newton, however, showed that with another prism and some lenses he could recombine those colors back into a beam of white light.

The colors, in other words, were always present in the light; white light, in fact, is simply a mixture of the various colors of the rainbow. But, then, what are the colors? What is it about one color that distinguishes it from the other colors?

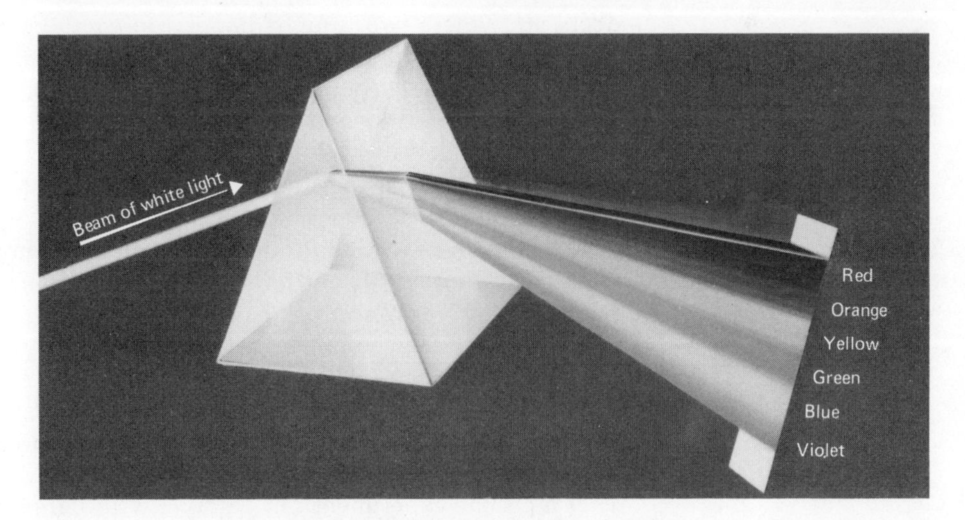

FIGURE II.10 Dispersion of white light by a prism.

Wave Motion and Interference

The answer is supplied by the realization that light is *waves;* it propagates as a train of waves, just as do waves in the ocean. If you stand in the water at the beach, just out beyond the breakers, you can experience the waves passing

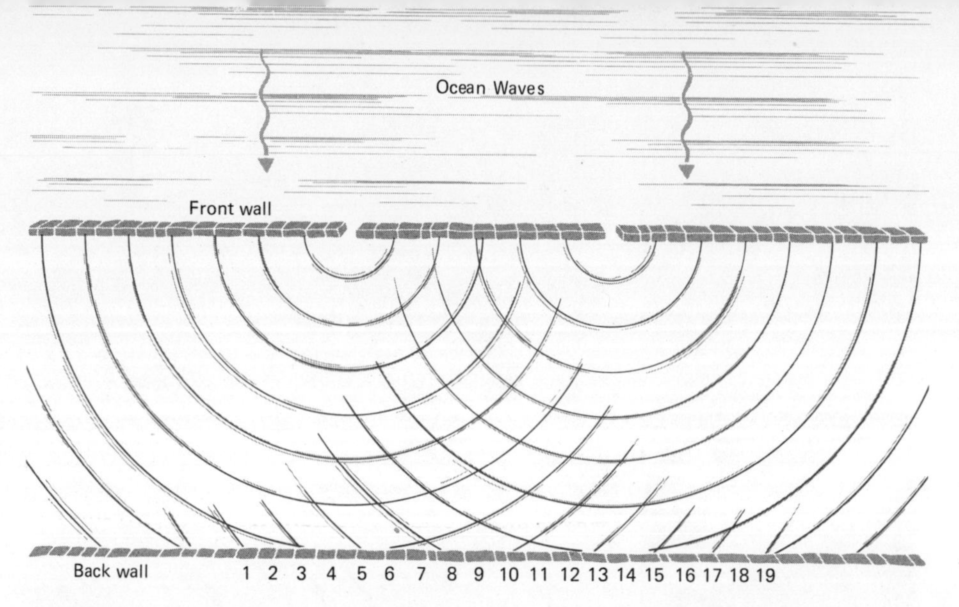

FIGURE II.11
Interference of water waves passing through two openings.

by you, with the water getting deeper as each crest, and shallower as each trough, moves in. Similarly, as light passes a point, it varies in intensity (as the ocean water does in depth); the waves of light pass by so rapidly, however—some thousand million million per second— that you are unaware of the wavelike nature of light.

To explain one of the many experiments that demonstrate the wave properties of light, it will be useful to begin with an analogy. Imagine ocean waves approaching a wall, labeled "front wall" in Figure II.11. Let us suppose there are two openings, or gates, in that front wall. The waves of water are stopped, or reflected back, at every point along that front wall, except where it can flow through the open gates. Now, as the waves of water flow through the gates, they spread out into semicircular waves, radiating from each gate just as the ripples in a pond radiate from the place where a pebble is dropped. But the waves flowing through the two different openings intermingle; where two crests happen to come together, the water is especially deep, and where two troughs happen to combine, it is especially shallow.

Now consider the heights of the water on a second wall, behind the first one, the "back wall" in Figure II.11. At points 1,3,5,7,9, . . . , along the wall, wave crests from the two gates always arrive at the same time. Between the successive arrivals of two crests, wave troughs from the two gates arrive together. Thus at those odd-numbered points the water alternately is very high and very low, as the successive crests and troughs of the waves from the two gates reinforce each other. At intermediate points, however, 2, 4, 6, 8, . . . , the wave crest from one gate always meets with a trough from the other, so the two sets of waves cancel out each other.

If you have followed the preceding discussion, you will realize that if you were standing at the odd-numbered points (in the figure), you would experience large-amplitude waves striking you (*amplitude* simply means the height of the waves compared to the midpoint between crest and trough). In between, at the even-numbered points, the waves would disappear. Well, exactly the same experiment can be done with light and with the same result.

It works best if we select light of a very pure color, and let that light, from a distant source, approach an opaque surface with two narrow slits, exactly analogous to the gates in the front wall in Figure II.11. Now suppose you put a white card or screen some distance behind the slits. Light passing through the two slits spreads out, just as the ocean waves do, and strikes the

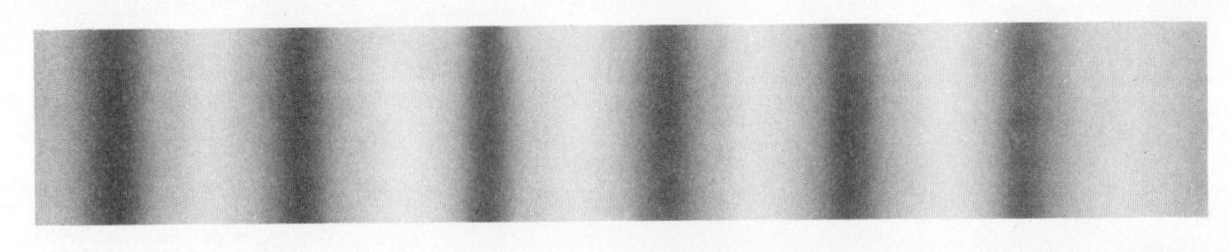

FIGURE II.12
Interference fringes of light.

screen behind the slits. However, you will not see the screen uniformly illuminated by the light, but rather crossed by a series of bright and dark fringes, of the same color as that of the light source (as in Figure II.12). The reason is that light consists of waves, and where the wave crests and troughs of the light passing through the two slits reinforce each other at the screen, you see a bright fringe of light, just as you see large-amplitude waves at the odd-numbered points on the back wall in Figure II.11. In between, though, the waves and crests of the light from the two slits cancel each other, and you see no light; those places are the dark fringes in Figure II.12, and correspond to the even-numbered places on the back wall in Figure II.11.

The exact spacing of the fringes on the screen, or the distance between the odd-numbered points on the back wall in Figure II.11, depends on how far apart the crests of the wave are; that distance between wave crests is called the *wavelength* of the waves. Typically ocean waves have wavelengths of tens of meters, but visible light typically has wavelengths of 0.00005 cm. This phenomenon, whereby light waves reinforce each other or cancel out each other and produce fringes, is called *interference*. The fact that light can display such interference is evidence that it propagates as waves.

---

1. Show by a diagram that the spacing of the interference fringes does, indeed, depend on the wavelength of the light. What would happen if light of a large range of different wavelengths was used in the experiment?

The Spectrum of Light

It is found that light of different colors has different wavelengths. Some colors are complex mixtures of light of many different wavelengths; others are rather pure in that they contain only small ranges of wavelengths. A prism breaks white light (as from the sun) into the continuum of wavelengths that makes it up; a raindrop does the same thing, so we see the same colors produced by a prism that we see in the rainbow. The shortest wavelengths of light give the impression of violet to the eye; they have wavelengths near 0.00004 cm. Light of slightly longer wavelengths appears blue. Successively longer wavelengths appear green, yellow, orange, and red. The longest waves of red light have wavelengths of about 0.00007 cm. This array of color, from violet to red, is called the *spectrum* of visible light.

Light, however—all light—travels with the same speed in a vacuum, c. Therefore, fewer of the long waves of red light can pass a given point per second than of the shorter waves of violet light. The number of waves that pass a point per second is called the *frequency* of the light (or of any wavelike radiation). To restate this more precisely, consider the observer at A in Figure II.13, watching a train of waves pass by him. Now suppose that radiation travels at a speed of c cm/s, as light does. Let us consider a point B on the wave train, a distance c from A. Clearly, a particular wave crest at B will reach

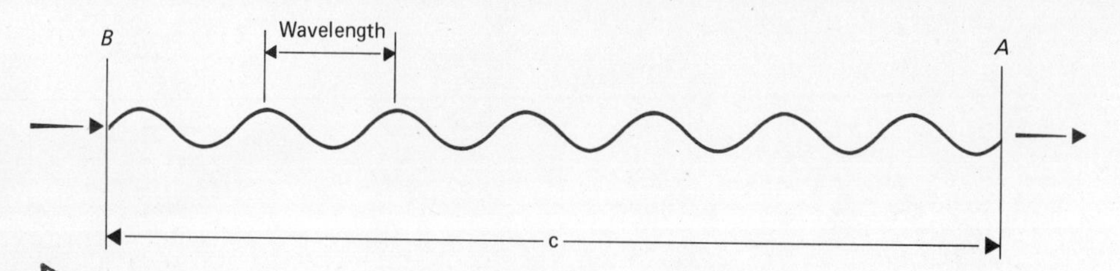

FIGURE II.13
Relation between wavelength, frequency, and speed of wave propagation.

A just one second later (for it travels the c cm to A during one second). The frequency of the radiation, the number of waves passing A per second, is just that number of them lined up between the observer and B. Thus the distance c betweeen A and B is simply the number of waves lined up in that distance (the frequency) times the distance between the crests (or wavelength). The speed of any wave motion is the product of wavelength and frequency. Red light, of relatively long wavelength, has a lower frequency than violet light, of shorter wavelength.

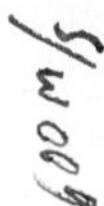

2. Suppose ocean wave crests 30 m apart approach the shore at the rate of one every 20 s. With what speed are the waves moving?

Almost everything we know about the universe is learned by means of light, and it is especially fruitful to analyze the light from stars and other celestial objects by decomposing it into its different wavelengths. A large fraction of the time that a typical astronomical telescope is used is devoted to studying the spectra of stars. An instrument attached to a telescope used to record the spectrum of a celestial object is called a *spectrograph*.

Most spectrographs apply the principle of interference to produce spectra. It is sufficient for our purpose, however, to consider a simpler kind of spectrograph—one that uses a prism; in fact, prism spectrographs are also widely used in astronomy. One is shown, schematically, in Figure II.14. Light is

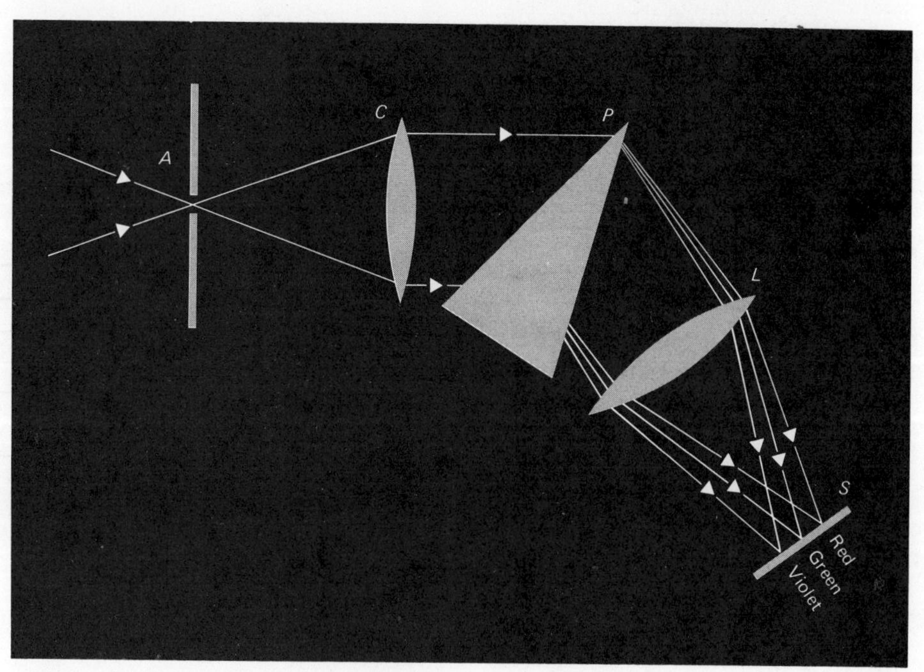

FIGURE II.14
Construction of a simple spectograph.

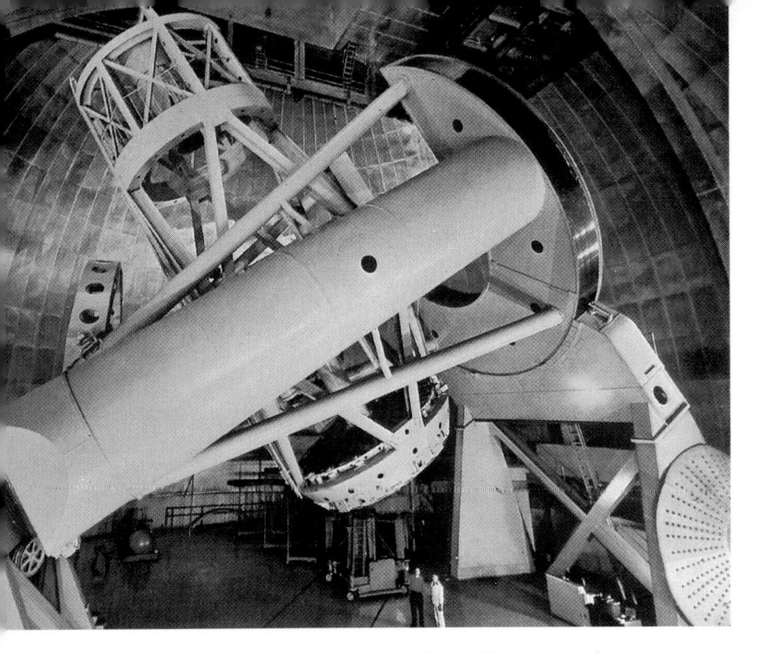

The 5-m (200-inch) telescope. (*Hale Observatories*)

The observer's cage at the prime focus of the 5-m telescope. (*Hale Observatories*)

The 122-cm (48-inch) Schmidt telescope at Palomar. (*Hale Observatories*)

The 3-m (120-inch) telescope of the Lick Observatory. (*Lick Observatory*)

The 1-m (40-inch) refracting telescope. (*Yerkes Observatory*)

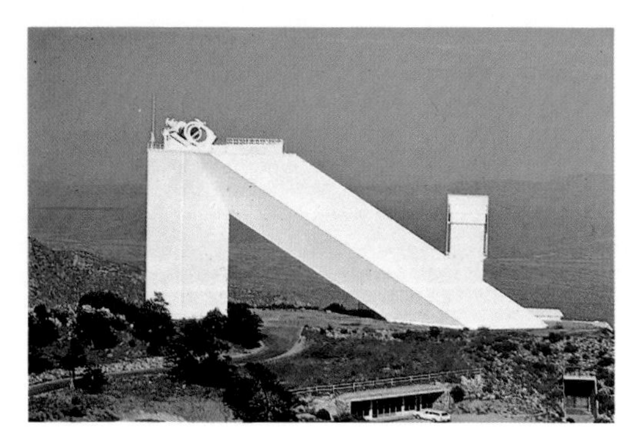

The Robert McMath solar telescope. (*Kitt Peak National Observatory*)

The heliostat mirrors of the McMath solar telescope. (*Kitt Peak National Observatory*)

The Kitt Peak National Observatory. (*Kitt Peak National Observatory*)

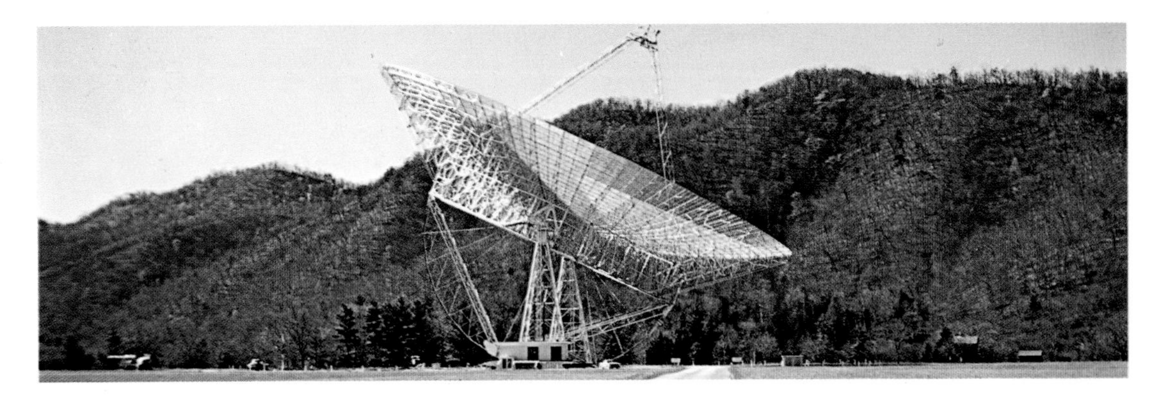

The 90-m (300-foot) radio telescope. (*National Radio Astronomy Observatory*)

Voyager spacecraft en route to Jupiter. (*NASA*)

The 4-m telescope of the Inter-American Observatory at Cerro Tololo, Chile. (© *Association of Universities for Research in Astronomy, Inc. The Cerro Tololo Inter-American Observatory*)

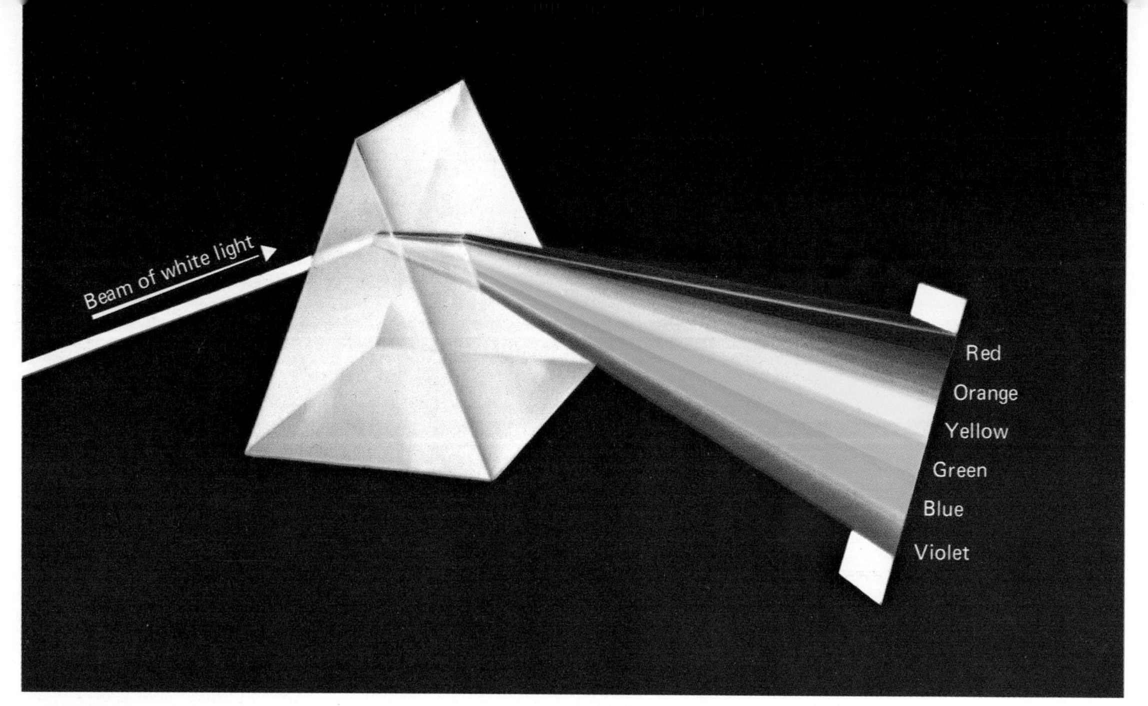

Dispersion of white light into a spectrum by a prism.

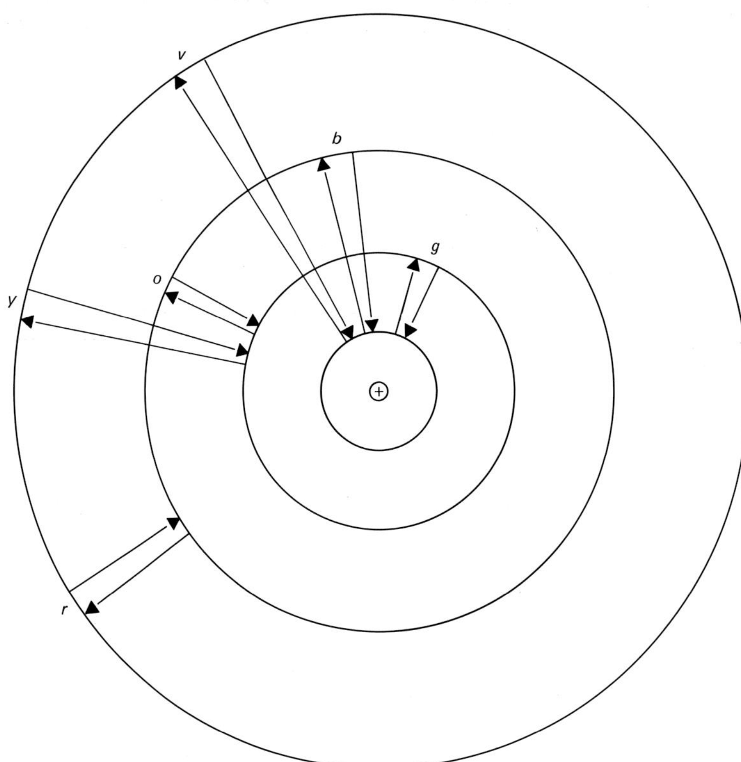

The fictitious atom Abellium and its spectrum.

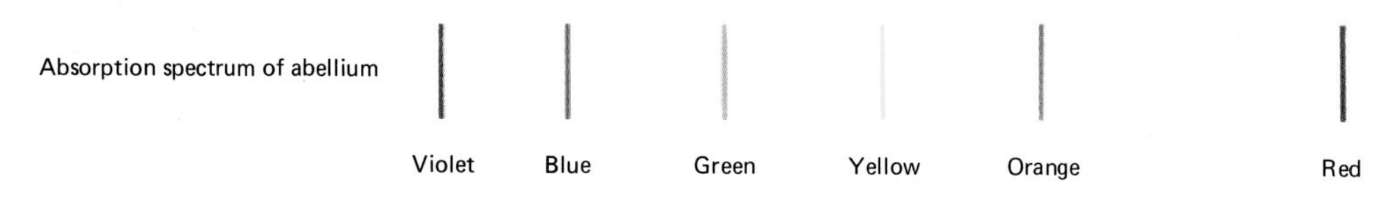

gathered from a star by the telescope lens or mirror and is focused at *A*, where there is an opaque plate of metal or some other material. The light from the star is allowed to pass through a small opening in the plate (usually a narrow slit) at *A*, and as it spreads out behind the plate it is collected by a lens, *C*, which converts it into a beam of parallel light and sends it into a prism, *P*. The prism *disperses* the light into the colors of the rainbow. The reason is that the index of refraction of the glass depends on the wavelength of the radiation, being slightly greater for shorter wavelengths than for longer. Thus the violet light is bent the most, the blue slightly less, and so on, with the red light being bent the least. Finally, another lens projects the image at the spectrum of colored light, leaving the prism at *S*, onto a screen or photographic plate, an image tube, or some other detecting or recording device.

The spectrum of a star (or planet or quasar) carries a wealth of information. We shall see how we can learn a star's temperature, the pressure of its outer gases, whether it has strong magnetic fields, its chemical composition, whether it is ejecting large amounts of matter, how fast it is rotating, how fast it is approaching or receding from us, and often many other things as well. From time to time, we shall return to the study of spectra, but for now we shall tarry only long enough to explain how we can determine a star's radial velocity.

The Doppler Effect

The *radial velocity* of an object is the component of its speed that lies in our line of sight—that is, how fast the object approaches us or recedes from us. It has nothing to do with how fast a star may be moving in other directions, say across our line of sight, but only with the rate at which its distance from us is increasing or decreasing.

Consider Figure II.15. Suppose a star is motionless (with respect to us) and is emitting light steadily. The wave crests simply radiate away from it, like those ripples in a pond where a pebble is dropped. Suppose, however, we are now rushing toward that object from the left of the figure (at *A*). Because we are moving forward to meet the oncoming waves, we see them approaching closer together, apparently at a shorter wavelength (and at a

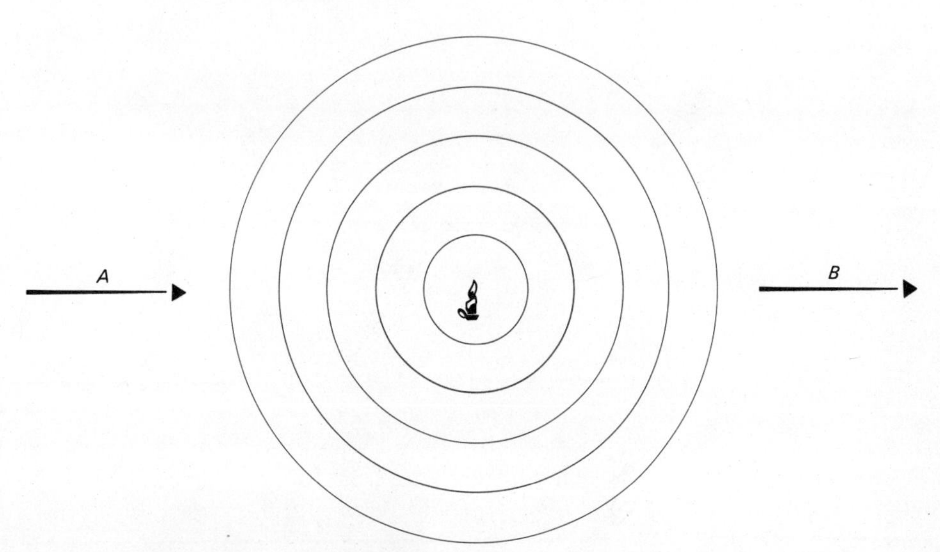

FIGURE II.15
Doppler shift for an observer moving with respect to the source.

higher frequency) than that with which they are emitted. On the other hand, if we are moving away from the source, at *B*, the wave crests have to catch up with us, and so appear spread out as they reach us, at a longer wavelength (and at a lower frequency). In other words, if we approach the source, the wavelengths of its light are shortened by comparison to normal; if we are moving away from it, the wavelengths of its light are lengthened.

Now consider Figure II.16. Here we suppose that we are stationary, but that the source is moving (to the right in the figure). A short time ago, when the source was at position 1, it emitted a crest of light, which has now spread out to *a*, centered on the place where the source was at position 1. But when it emitted its next crest, it was at position 2, and by the time the first crest reaches *a*, the second crest is at *b*. Crest *b* is centered on where the source was when that crest was emitted (namely, at point 2), but is *not* concentric with crest *a*, because the source had moved from 1 to 2 between emission of those crests. Likewise for the successive crests, *c*, *d*, *e*, and so on, emitted when the source was at 3, 4, and 5, respectively. To an observer at *A*, whom the source is approaching, the waves appear crowded together and hence of shorter wavelength. To an observer at *B*, they appear lengthened. Observers at C and *D*, with respect to whom the source is moving across the line of sight, see the light at its normal wavelength.

In practice, both the source and the observer may be moving. In fact, we shall see in Scene 4 that it doesn't even make sense to ask *which* is moving. It doesn't matter, though, because the effect is the same, whether the source, the observer, or both are moving. If we are getting closer together, the waves of light appear shortened; if we are separating, they appear lengthened. This shift in the wavelength of radiation from a source, due to its relative motion in the line of sight, is called the *Doppler effect*, after the physicist Christian Doppler, who explained the phenomenon in 1842.

FIGURE II.16
Doppler shift for an observer seeing a moving source.

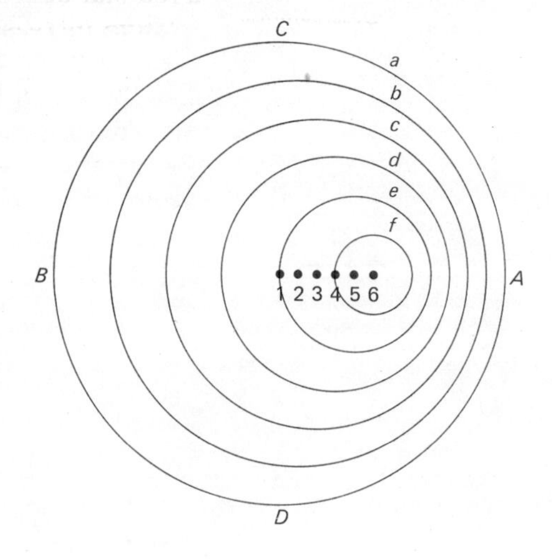

---

3. In the preceding scene we saw how the apparent period of revolution of any of Jupiter's satellites depends on whether the earth is approaching or receding from Jupiter. Explain how this effect is like the Doppler effect, and also how it differs from the Doppler shift in light.

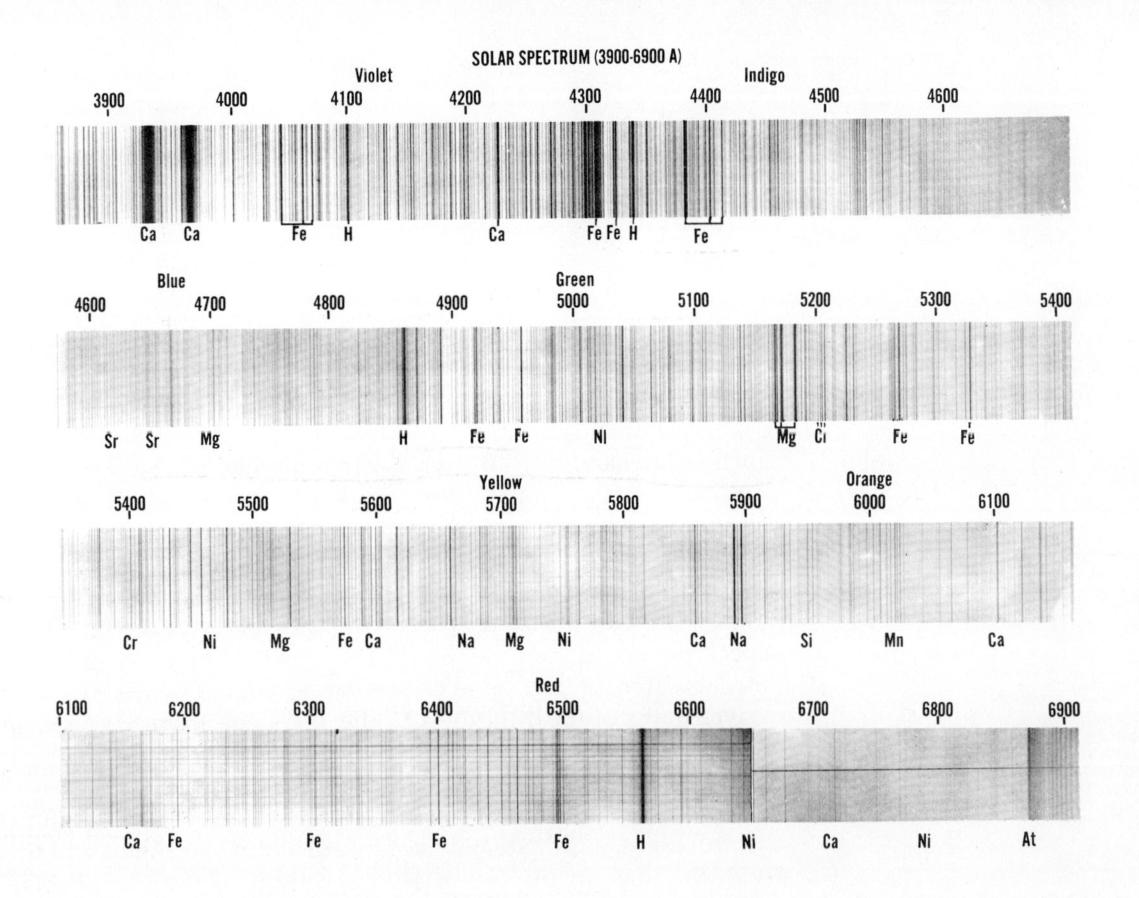

FIGURE II.17
The solar spectrum. *(Hale Observatories)*

Does this mean, then, that a blue star moving away from us looks red, and a red star approaching us looks blue? In principle, yes. However, the actual relative increase or decrease of the wavelength of the radiation is roughly equal to the speed of the source (in the line of sight) divided by the speed of light. Because of the expansion of the universe, very remote galaxies that would appear yellow, if stationary, really do look red. On the other hand, individual stars that are near enough, and hence bright enough, for us to analyze their spectra are almost all in our own Galaxy, and their speeds with respect to us are but a tiny fraction of the speed of light—at most a few hundred kilometers per second, and typically only a few tens of kilometers per second. The shift in wavelength of their light is very slight, and their colors are thus not noticeably changed.

How, then, can we measure the Doppler shifts of stars? Happily, the spectrum of a star is not (with rare exceptions) a continuous band of color, but rather has numerous wavelengths where light is missing or at least greatly diminished in intensity. Thus the spectrum of a star, or of the sun, is usually crossed with many dark lines (see Figure II.17). Later we shall see that those dark lines are due to various kinds of atoms in the outer layers of those stars absorbing some of the light the stars radiate, and absorbing it at certain discrete wavelengths. In fact, it is mainly because of those lines that we can learn so much about stars from studies of their spectra. We can recognize the patterns of dark lines produced by different kinds of atoms, and we know, in most cases, the precise wavelengths at which they occur. All we need do is see if the particular set of lines in the spectrum of a star (that are due to a particular kind of atom) are at their normal wavelengths, or whether their wave-

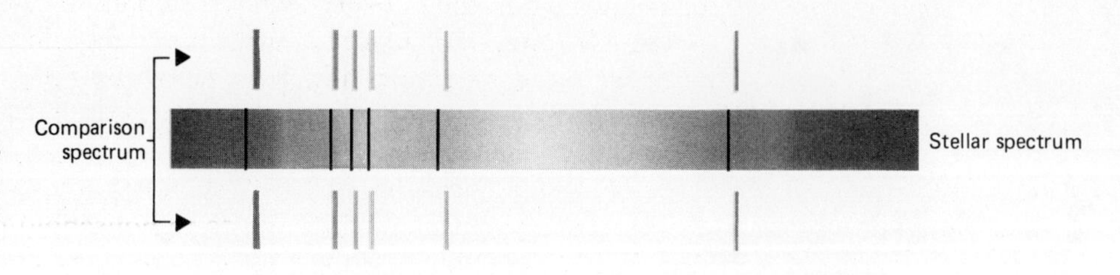

FIGURE II.18
The appearance of spectral lines shifted by the Doppler effect.

lengths are slightly greater or less than normal, which would indicate a motion of that star away from or toward us, respectively.

Whenever we observe the spectrum of a star or other object with a telescope, we nearly always *also* observe the spectrum of some laboratory source, such as the glow of light from gas in a discharge tube. That light consists of bright lines at certain specific (and known) wavelengths arising from atoms emitting light of the same discrete wavelengths they can absorb. We shall see later how this comes about. When the spectrum of a star is photographed in the most usual manner, the spectrum of the laboratory source is photographed with the same spectrograph, and is recorded on either side of the spectrum of the star (Figure II.18). The bright lines in the spectrum of the laboratory source serve as standards, against which we can measure the wavelengths of the dark lines in the star's spectrum. For example, suppose we use a helium discharge tube for a laboratory source. Then we obtain a *comparison spectrum* consisting of bright lines at various wavelengths, which make up the spectrum of the helium. If we see lines of helium in the spectrum of the star, they should be at those same wavelengths and should line up with the lines in the comparison spectrum. If they are displaced, we can measure that displacement and thus calculate the star's motion in the line of sight. That motion toward or away from us (the star's radial velocity) is considered positive if the star is receding, and negative if it is approaching.

There is a reason for this somewhat lengthy discourse on the Doppler effect! Much of our understanding of the universe comes from it; it helps us find the masses of stars, of star clusters, of galaxies, and of clusters of galaxies; it enables us to find the motion of the sun among its neighboring stars; it helps us learn the structure and general properties of our Galaxy; and —most profound of all—it tells us of the expanding universe.

The Electromagnetic Spectrum

IMPORTANT

You may ask, if light of wavelengths in the range 0.000065 cm to 0.00007 cm looks deep red to us, then what would light look like if it existed at wavelengths *longer* than 0.00007 cm? As a matter of fact, it does exist at longer wavelengths, but it cannot be detected by the eye, so we don't call it light; we call it *infrared radiation*. Similarly, radiation of still shorter wavelengths exists which cannot be seen; it is called *ultraviolet radiation*. So there is a lot more to the spectrum than meets the eye.

Infrared extends over a very large range of wavelength—all the way up to about 1 mm, at about which point it is called *microwave radiation*, or very shortwave *radio* radiation. Still longer waves comprise radio waves, and they go all the way up to kilometers in length. Infrared radiation of wavelength less than about 0.00015 cm can be photographed with special emulsions; at longer wavelengths other types of detectors must be used. Many of these detectors utilize semiconductors or other substances that change electrical resistance when they absorb infrared radiation. The detector, with a current flowing through it, is placed at the focus of the telescope where infrared radiation falls on the semiconductor and is absorbed. The amount of that incident radiation is measured by monitoring the change in the current passing through the device.

Even these devices do not "see" the long-wave radio radiation. We can detect it with antennas, much as we do in household radio and television sets. One way to *generate* radio waves is to pass an alternating current back and forth through a wire; radio waves are radiated from the wire with the same frequency as that with which the current alternates. Just as a current going back and forth through a wire can generate radio waves, so radio waves striking a wire can induce a current in it. In commercial broadcasting, powerful alternating currents in the broadcasting antennas produce the radio radiation, and feeble currents are generated in all the antennas on all the rooftops in the area.

---

4. (Project) Consult an appropriate reference (if necessary) and explain briefly how the mere broadcast and detection of radio waves can be used to transmit sound or picture information, as in a television set.

5. With what familiar device can visible light be utilized to carry sound information? How does it work?

Now let's talk about the other extreme: the ultraviolet and other radiation of even shorter wavelengths. Centimeters get a bit unwieldy for measuring such waves. A common unit of length is the *Ångstrom,* abbreviated Å. There are 100 million (100,000,000, or 10^8) Ångstroms in a centimeter. So green light, of wavelength 0.00005 cm, also has a wavelength of 5000 Å. Ultraviolet radiation begins at a wavelength of 4000 Å and goes on down to a wavelength of about 200 Å. Still shorter wavelength radiation is called X rays. (The dividing line between ultraviolet and X rays, however, is not precisely defined—it depends on whether you are talking to a man who considers himself an investigator of ultraviolet radiation or of X rays.)

X rays shorter than about 0.1 Å in wavelength (0.000000001 cm) are no longer called X rays, but *gamma radiation* or *gamma rays.* Gamma rays have the shortest wavelengths of all (and thus, the highest frequencies), and there is, theoretically, no lower limit to their lengths, any more than there is an upper limit to the lengths of radio waves. Ultraviolet radiation, X rays, and gamma rays can be photographed or detected with photoelectric cells.

All these kinds of radiation are the same kind of energy as light, differing from each other fundamentally only in wavelength or frequency. It is as if they were invisible colors of light. They all make up what is called *electromagnetic radiation* or *energy,* and the array of them is the *electromagnetic spectrum.* The name derives from the fact that they are propagated by the build-up and decay of electric and magnetic fields, as we shall see in the next scene. They all travel (in a vacuum) with that same speed, *c.*

Today we observe celestial objects in the wavelengths of all kinds of electromagnetic radiation. Until the mid-twentieth century, however, our view of the universe was limited to what we could observe by means of light and just a small range of wavelengths of ultraviolet and infrared radiation on either side of the visible spectrum. The reason, in part, is that detectors sensitive to the very short and very long wavelengths had not yet been invented.

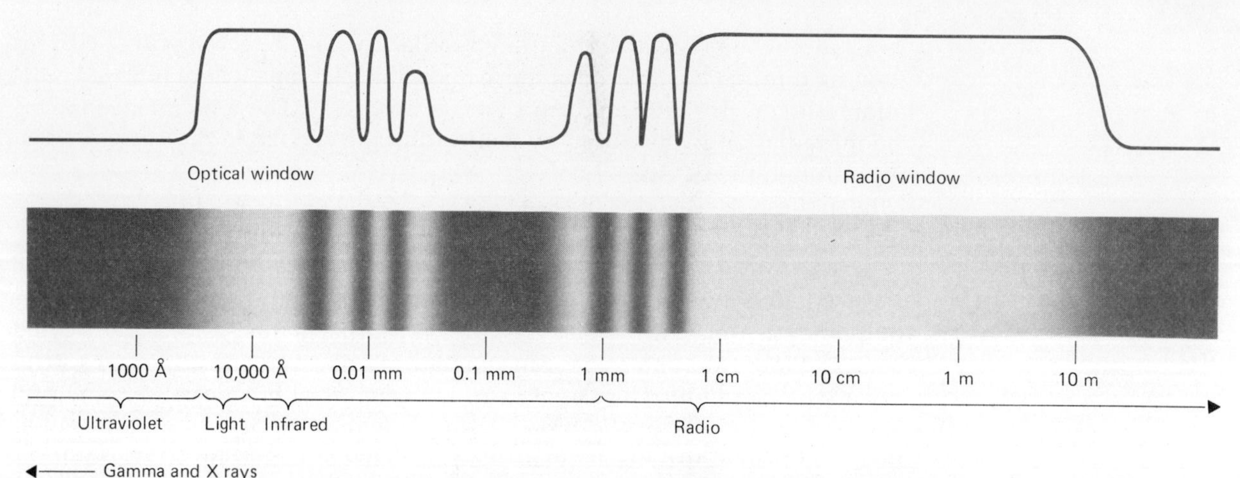

FIGURE II.19 The relative transmission of electromagnetic energy of various wavelengths through the earth's atmosphere.

Energy of many of these wavelengths, however, is permanently hidden to us on the ground by the atmosphere of the earth.

The situation is illustrated schematically in Figure II.19, which shows most of the electromagnetic spectrum. The graph above the spectrum is a plot of the transparency of the atmosphere. Those wavelength ranges at which the atmosphere is opaque (and blocks incoming radiation) are shown as dark regions on the spectrum below. We notice two atmospheric "windows"—wavelength ranges within which radiation can penetrate to the ground. One of these windows is in the visible spectrum and near infrared ("near" meaning near visible light in wavelength). The other is in radio wavelengths from a few millimeters to several meters. Short of the visible spectrum the ultraviolet, X ray, and gamma radiation is absorbed in ionizing atoms high in the atmosphere. In most of the infrared, molecules in the air absorb the radiation. The longest radio waves are reflected back into space by the ionized layers of the atmosphere (ionosphere).

Radio Telescopes

The first peep out the radio window was in 1931. K. G. Jansky, of the Bell Telephone Laboratories, was experimenting with antennas for long-range radio communication when he encountered interference in the form of radio radiation coming from an unknown source. He soon discovered that the radiation was received at regular times each day, and correctly deduced that the interference occurred when the Milky Way was passing overhead. Today, we know that he was receiving radio waves emitted by the clouds of interstellar gas in our Galaxy.

In 1936 the American amateur astronomer and radio ham Grote Reber built the first radio telescope, a special antenna designed to pick up radio waves from space. Reber remained an active radio astronomer for more than 30 years. In 1942, radio radiation was first received from the sun at radar stations in England. The big push in radio astronomy, however, came after World War II, when the technique of detecting radio waves from space developed rapidly. Large radio telescopes were built, especially in Australia, the Netherlands, England, and later in the United States.

The most common kind of radio telescope uses a large concave dish or "mirror"—usually of wire mesh—to reflect and focus the radio waves. It is an

FIGURE II.20
Grote Reber's original radio telescope. *(National Radio Astronomy Observatory)*

almost complete analogue of large reflecting optical telescopes. Just as we need a detector at the focus of an optical telescope to pick up the radiation, so do we need one in a radio telescope. But in place of a photographic plate, or image tube, a radio telescope uses an antenna or a wave guide, or some device in which the radio waves can induce a current, which in turn can be conducted to a receiver that can amplify it and record it, to provide a record of the radio energy received.

The reflecting dish of a radio telescope typically measures tens of meters across. Usually these large radio reflectors are steerable, so that they can be pointed to any direction and can follow the motion of an object as the earth's rotation carries it around the sky. Such is the 76-m (250-foot) radio telescope at Jodrell Bank, England. The 91-m (300-foot) telescope of the National Radio Astronomy Observatory at Greenbank, West Virginia, is partly steerable. Sometimes, however, the dish of wire mesh is stretched into the correct curvature over a natural depression in the ground that is conveniently close to the correct shape. The largest of these is the 305-m (1000-foot) telescope at Arecibo, Puerto Rico, operated by Cornell University and the National Science Foundation. This kind of telescope can observe only objects that pass approximately through the zenith; however, some flexibility is permitted by moving the detector around in the neighborhood of the focus of the bowl. The director of the Arecibo installation, Dr. Frank Drake, points out that the volume of the reflecting dish is roughly equal to that of the world's annual beer consumption.

Radio telescopes can also be used to transmit radio waves into space. We can beam into space radio radiation with enough intensity for us to be able to observe its reflection from planets as far away as Saturn. This field of *radar* astronomy has been extremely useful in giving information about the precise distances and motions of planets and about their surfaces, as we shall see later. We also use such ''broadcasting'' telescopes to transmit radio instructions to space probes and satellites.

FIGURE II.21
The 1000-foot (300-m) fixed radio dish at the Arecibo Observatory, Puerto Rico. *(Cornell University)*

Since the 1950s we have been hard at work exploring the universe through that atmospheric radio window, now as wide open as the optical window, known for centuries. But the rest of the electromagnetic spectrum is observable only from space. With instruments carried in high balloons we have observed a bit more of the hidden infrared, but most of our observations of the hitherto hidden radiation have been made with instruments carried in rockets, satellites, space probes, and by astronauts. We shall have occasion frequently to refer to some of the truly exciting developments in modern astronomy made possible by new access to wavelengths that we were previously shielded from.

Thermodynamics

A warm or hot solid is composed of molecules and atoms that are in continuous vibration. A gas consists of molecules that are flying about freely at high speed, continually bumping into each other, and bombarding against the surrounding matter. That energy of motion is called *heat.* The hotter the solid or gas, the more rapid is the motion of those molecules, and its *temperature* is just a measure of the average energy of those particles. One of the principles of *thermodynamics* (the *second law*) is that heat always tends to transfer from a hot object to a cooler one. Thus a solid or gas that is at a higher temperature than its surroundings radiates some of its heat energy into those surroundings, thereby cooling.

A century ago physicists were interested in the properties of this emitted radiation — that is, how much of which kind of radiation is radiated by warm and hot bodies? The situation is quite complicated, because bodies absorb some radiation, reflect some, and transmit some. A blue sweater, for example, reflects more of the relatively short-wave blue light than it does other colors; that is why it looks blue. A piece of black coal reflects relatively little of any visible light, and a window pane transmits most light through without either absorbing or reflecting it.

All these bodies, though, absorb some radiation, especially when the entire electromagnetic spectrum is considered; moreover, all bodies will eventually come into equilibrium with their surroundings, until they reemit energy at a rate which, averaged over time, is exactly the rate at which they absorb it.

When confronted with a complex problem, as this one is, the scientist usually tries to get a start in his analysis by finding a circumstance in which the problem is simplified. To this end, physicists invented the *ideal* or *perfect radiator*—a hypothetical body that completely absorbs every kind of electromagnetic radiation incident on it, reaches some equilibrium temperature, and then reradiates that energy as rapidly as it absorbs it. Because the perfect radiator absorbs everything and reflects and transmits nothing, it is also called a *black body*.

A black body, however, is not so called because it necessarily *looks* black; since it is radiating energy, it might be very bright indeed. It happens that a piece of coal, which is a crude approximation to a black body, does look black, but that is because (at least at room temperature) its reradiated energy is mostly in the invisible infrared; if we could see infrared radiation, we would find that coal to be glowing brightly. Stars happen to be good black bodies, because the gases they are made of are very opaque to virtually all electromagnetic radiation, and only from a star's outermost layers can energy escape into space. A star like the sun, of course, is very bright indeed.

Black-Body Radiation

Laboratory devices that are close approximations to perfect radiators were invented long ago. It is found that the distribution of energy at different wavelengths emitted by a given unit area of such a black body depends only on the body's temperature; two different black bodies of the same temperature always radiate in exactly the same way. Figure II.22 shows how much electromagnetic energy at various wavelengths black bodies of several different temperatures radiate from each square centimeter of their surfaces.

We note three interesting things about the graph in the figure. First, every black body emits some radiation at *every* wavelength. Second, at *every* wavelength, a hotter black body emits *more* energy than a cooler one. Third, for a black body at each temperature there is a certain wavelength at which it radiates a maximum amount of energy, and the *higher* the temperature, the shorter that wavelength of peak radiation.

The third point is expressed precisely by *Wien's law*, which states that the wavelength of peak emission is inversely proportional to the temperature (or, equivalently, the product of its temperature and that wavelength of maximum radiation is always the same number; that number is given in Appendix 5.) The sun, for example, has a temperature of about 6000°K and emits its maximum light at about 5000 Å. Wien's law means that a star with twice the sun's temperature emits its maximum light at 2500 Å. Such a star appears blue to us, because most of its radiation is at short wavelengths. Some stars

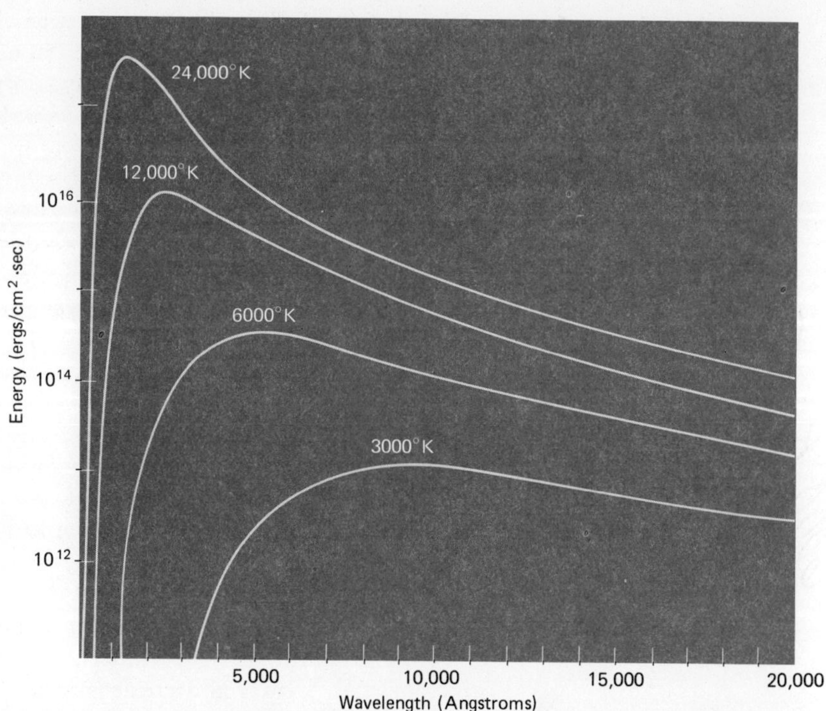

FIGURE II.22
Intensity of the radiation of electromagnetic energy as a function of wavelength, from each square centimeter of black bodies of several different temperatures.

are so hot that most of their energy is emitted in the invisible ultraviolet. On the other hand, a star with a temperature of only 3000°K, half the sun's temperature, emits its maximum light at 12,000 Å – in the infrared. Note that since all black bodies emit some energy at all wavelengths, even the cool stars emit a little ultraviolet light. Moreover, not only do the very hot stars emit some infrared and visible light, but they emit *more* of it (per unit area) than the cool stars do (remember the second point in the foregoing paragraph).

So a circumstance of Wien's law is that we can calculate the temperature of a star from observing the wavelength at which it emits the most radiation. Since that also is what determines the color of a star, its color tells us its temperature.

Now, the sun's wavelength of maximum radiation is that of green light; yet the sun appears *white,* not green. The reason is that white light is the particular admixture of different wavelengths we receive from the sun (or, approximately, from a black body with a temperature of 6000°K). A piece of white paper looks white because it reflects all wavelengths about equally to our eye, so the reflected light from a white paper has the same distribution of wavelengths as sunlight—or any other source of white light. Stars that have higher temperatures than the sun look blue or violet to the eye, and those of lower temperature look yellow, orange, or red, because the peak of their light is at those shorter or longer wavelengths, respectively; a star you might expect to look green has about the same temperature as the sun and is white, more or less by definition. We would have to filter out most of the other colors to make sunlight look green.

---

6. If a star emitted radiation of equal intensity at all wavelengths, what do you suppose its color would be, and why? [Hint: Consider how its proportions of red, blue, and green light compare with the proportions in sunlight.]

Since a hot black body emits more at all wavelengths than a cool one, the combined amount of electromagnetic radiation over all wavelengths emitted by a black body is extremely sensitive to its temperature; it is proportional to that temperature multiplied by itself three times (that is, to the fourth power of the temperature). This fact was known a century ago, and is known as the Stefan-Boltzmann law. If you actually want to calculate the amount of energy emitted by each square centimeter of a black body, you raise its temperature (in °K) to the fourth power, and multiply that number by the *Stefan-Boltzmann constant,* whose value is shown in Appendix 5.

Now imagine two stars of the same size and distance, but one of which is blue and the other red. The blue star, being hotter, radiates very much more efficiently, and so appears far, far brighter than the red one. But what if a red and a blue star, both the same distance, appeared *equally* bright? Knowing that the red star is an inefficient radiator (per unit area), you realize that it must be very much larger to appear as bright as the blue one.

So at once you see how we can use the Stefan-Boltzmann law to figure out approximately how big a star is. First, from its color, we find the wavelength of the star's maximum radiation; that tells us the temperature. The temperature, with the Stefan-Boltzmann law, tells us how much energy the star radiates from every square centimeter of its surface. Now if we know the star's distance (we come to how *that's* done in the next act), we can observe how much energy it emits altogether from its entire surface. Knowing how much it radiates from each square centimeter and knowing how much it radiates altogether tells us how many square centimeters there are in its surface—that is, its size.

---

7. What can you say about the size of a star whose color tells us that it has twice the surface temperature of the sun, yet whose total rate of radiation is only one-hundreth that of the sun?

The earth, moon, and planets are not very good black bodies. Typically, they reflect about half of the sunlight incident on them, although it's not the same for all of them; Venus, for example, reflects more than half, and the moon less than 10 percent. But they all absorb some of the sunlight hitting them, and they must reradiate this energy. We see the planets in the sky by the sunlight they reflect. It's easy to calculate how much sunlight strikes them, so from their brightnesses we find out how much solar radiation each planet must absorb, which is, of course, how much it also emits. Now the Stefan-Boltzmann law enables us to calculate the temperature of a black body that emits the same amount of energy a particular planet does. Although a planet is not really a perfect radiator, its surface temperature will be in the general neighborhood of that black body's temperature.

Thus we find that the earth's temperature should average somewhere around 300°K—about room temperature. (Actually, 293°K is 20°C, or 68°F.) Mars, being farther from the sun and thus receiving less energy from it than we do, is a little cooler. Mercury, being closer than we, is quite a bit hotter.

Wien's law tells us the wavelength at which a black body of the same temperature as a planet would emit its maximum radiation; roughly, this is the wavelength of the peak of the energy radiated by the planet, too. The earth's temperature of 300°K is about $1/20$ the sun's 6000°K; if the sun radiates most

FIGURE II.23
The "Mills' Cross" radio telescope in New South Wales, Australia. Each branch of the cross is a mile-long radio reflecting trough that can be pointed to various elevation angles and can reflect radio waves to a running antenna along the focus of the trough. *(Courtesy, B. Y. Mills, University of Sydney, N.S.W.)*

strongly in green light of wavelength 5000 Å, the earth must radiate most strongly at 100,000 Å (or about 0.001 cm)—way out in the infrared. Everything about us, therefore, including ourselves, is radiating infrared energy. If we had infrared vision, we would easily see each other in the dark—in fact, we would find that our world had continuous illumination. "Night" is night only to visible sunlight; there is no night on earth to its own radiation.

Similarly the other planets radiate in the infrared. We can measure the radiation they emit with infrared detectors attached to our telescopes. By the way, by now you can appreciate one of the problems of the astronomer who specializes in infrared observations. The telescope dome and everything in it, including the telescope itself, and the astronomer and his assistants are all shining brightly in the very wavelengths he is trying to detect from celestial objects. It's rather like trying to watch a motion picture show with the theater lights on. The astronomer must take very special pains to shield his detector from all the radiation except that reflected from the telescope mirror.

---

8. For which planet in the solar system might you expect the peak of its radiated energy to lie at the longest wavelength, and why?

9. Can you suggest a way of determining the size of a minor planet by comparing its brightness in visible light to the intensity of the infrared radiation it emits? Outline the steps clearly. (If you have difficulty, see my Exploration of the Universe, Third Edition.)

The Quantization of Light; Photons

We have seen that quite a bit was known a century ago about the way things radiate. At that time, however, there was not yet a physical theory that accounted for the properties of radiation from black bodies—that is, those particular graphs shown in Figure II.22. We did not understand radiation in the same sense that we understand the motions of the planets in terms of the law of gravitation. Now it was known that currents moving in a wire emit electromagnetic waves (for example, radio waves), so it was thought that charged particles (electrons) within the atoms of the radiating substances were oscillating back and forth to emit all kinds of electromagnetic energy. But if those electrons could oscillate in every way, there was no reason why a plot of the energy emitted at various wavelengths should look like the graphs in Figure II.22.

The answer to that riddle was part of the start of the quantum theory—one of the great new branches of physics of the twentieth century. The quantum theory has profound implications for all of us, and we shall have more to say about it in the next scene. But for now, how about that black-body radiation? In 1900 the German physicist Max Planck succeeded in deriving a theoretical formula for those graphs in Figure II.22. To do so, he had only to assume that the tiny atomic oscillators in the radiating bodies can emit energy at any given wavelength or frequency only in certain discrete amounts. In other words, at each frequency the energy leaving a radiating particle is always some multiple of a minute unit of energy. That unit of energy is the frequency of the radiation multiplied by a tiny number—always the *same* number, symbolized h. That number, h, is now called *Planck's constant.* Its value, along with other important numbers, is given in Appendix 5.

Now comes something really interesting. We have already mentioned photoemissive substances—those which emit electrons when struck by electromagnetic energy. Actually, since about 1888 it has been known that any metal will emit electrons when struck by electromagnetic radiation if that radiation is of short enough wavelength (or high enough frequency). In other words, the radiation must have a certain critical or threshold frequency. Radiation of *just* the threshold frequency barely dislodges electrons from the metal, but that of *higher* frequency dislodges them and gives them some kinetic energy to boot!

The subject was clarified by Albert Einstein in 1905. Einstein proposed that electromagnetic radiation itself travels in units of energy, each unit having energy equal to h (Planck's constant) times the frequency. Moreover, each electron emitted results from the metal absorbing just *one* of those units of radiant energy. Now it takes a certain minimum energy to remove an electron from the metal. Thus, of the units of electromagnetic energy that are absorbed, the only ones that can cause the ejection of electrons are those which have high enough frequencies to possess that requisite energy. On the other hand, any radiation unit of *higher* energy (that is, higher frequency) can be absorbed, and the excess energy absorbed goes into kinetic energy (energy of motion) of the dislodged electron.

In short, Einstein showed that the phenomenon of radiation causing the ejection of electrons—called the *photoelectric effect*—is evidence that electromagnetic energy is itself *quantized,* consisting of little packets of energy. These packets of radiant energy are now called *photons.* Einstein received the Nobel prize for this contribution to our understanding of the nature of radiation—not for his better known relativity theory!

Particles and Waves

The photoelectric effect, and many other experiments, show that electromagnetic energy acts as though it were made up of little energetic particles—photons. But we have already seen that electromagnetic energy also propagates as *waves.* So these photons themselves must be waves, but always containing discrete amounts of energy. Thus photons of the shortest wavelength and highest frequency—gamma rays—have the highest energy, and those of radio waves have the lowest energy; visible light is intermediate.

How can we reconcile our concept of photons of light and other electromagnetic radiation as having the properties of both particles and waves? I suggest two ways: First, we can think of each photon itself as waves propagating in all directions through space, like ripples in a pond. The photon can do all the things waves can do—it can pass through two slits at once and interfere with itself, for example. But as soon as we *observe* it, say by its absorption in the retina of the eye, in a photographic emulsion, or by means of the photoelectric effect, those waves immediately disappear, and the photon itself is known to have been at the place where it was observed.

Second, we can think of those waves as just a mathematical formula, which tells us the probability of the photon—itself a localized particle—being in any particular place. Thus the wave crests are the places where the photon is more likely to be found, and the wave troughs places where it is less likely to be. We never, of course, know where it *really* is until we observe it.

"Ah," you say, "the second way of thinking of photons makes physical sense—one that is conceptually credible." But actually, both ways are exactly equivalent as far as predicting results of experiments are concerned, and science, after all, doesn't concern itself with what a photon "really" is, but only *how it behaves.* Unless we can devise an experiment to distinguish between these viewpoints, it does not even make sense—within the realm of science, anyway—to ask which is *correct.*

In fact, it turns out that material particles—like protons and electrons, and yes, even billiard balls—can similarly be thought of as waves. We shall see that we cannot say just where an electron (for example) is located at any given instant. The best we can do is write down an equation that gives us the probability of its having various locations. The equation turns out to be the equation for a system of waves; we are, in other words, exactly where we are with the photon, and may as well think of the electron itself as a wave.

These concepts may seem less satisfying than the simple, mechanical, predictable system of Newtonian theory, but it is just as successful in accounting for the behavior of matter on the atomic level as Newton's laws are on the scale of the solar system. And the two theories are not contradictory, either. If you flip an honest coin in a random way one million times, you can safely predict that very close to 50 percent of the flips will come up heads. The gentlemen who run the casinos at Las Vegas make their fortunes by knowing exactly how statistics of large numbers work out. Similarly, planets and billiard balls have absolutely enormous numbers of atomic particles. Just as we cannot predict whether an individual gambler will win a particular hand of blackjack, so the behavior of each particle in a billiard ball is uncertain. But the behavior of the total ensemble is highly predictable. As the casino operator knows that on the average the customers will lose (and for that matter, so will each individual if he plays long enough), so the physicist knows how the billiard ball will carom, behaving according to the average of the motions of its tremendous numbers of atoms.

On the other hand, submicroscopic chance events can lead to chains of actions with macroscopic consequences. An individual gambler may win some money at roulette and purchase a new car, thus resulting in the automobile salesman's being promoted, and building a new home, and . . . —you can continue the tale. Analogously, the chance absorption of a photon from a remote galaxy in a photomultiplier on a telescope can eject an individual electron, which can be detected, and the signal could be used, if we wished, to turn on the lights of a city. I wonder how ideas originate in the human brain. Anyway, this innate uncertainty in nature provides an alternative to the philosophy of determinism of the preceding century.

And if you find *these* ideas exciting, hold on!

SCENE 3

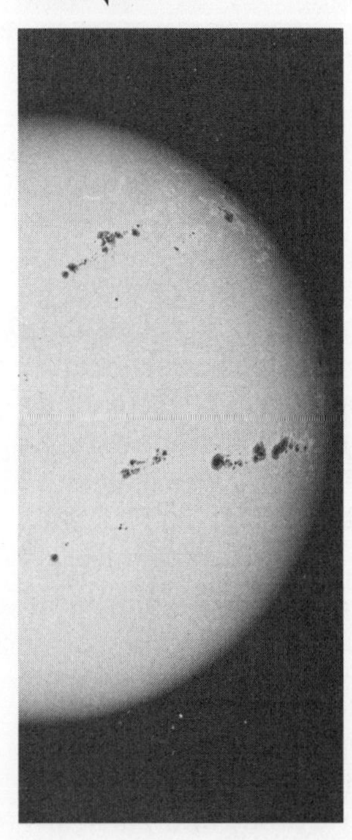

Atoms and Radiation

I learned in high school that there are 92 different kinds of atoms, which are the building blocks of the 92 different elements. Today, however, more than 100 different elements have been studied, although those made of the largest and most complicated kinds of atoms don't last very long; after a while, their atoms spontaneously *decay*, or split up into simpler atoms of other elements. Moreover, atoms, as we shall see, are far from indivisible. Nevertheless, each of the pure chemical elements is composed of a different kind of atom.

Atoms tend to combine into *molecules*. Molecules consisting only of atoms of the same kind are molecules of an element; for example, each molecule of hydrogen gas consists of two atoms of hydrogen, and each molecule of oxygen gas consists of two atoms of oxygen. More often, however, molecules consist of atoms of two or more kinds, and make *compounds*, like water (whose molecule has two atoms of hydrogen and one of oxygen), sodium chloride, and formaldehyde. There are, of course, many millions of compounds, each with a molecule consisting of a different combination of atoms. The atoms in a molecule are held together by the electric forces between their charged parts.

Atomic Structure

At the center of an atom is its *nucleus*, which contains the bulk of the mass of the atom, and a positive electrical charge. Moving around the nucleus are the electrons of the atom, each electron carrying a negative charge. All electrons are identical; the charge is exactly the same on each of them. And in an ordinary atom, the positive charge on its nucleus is exactly equal, but opposite in sign, to the sum of the negative charges carried by the electrons; thus the atom is electrically neutral.

Under some circumstances, on the other hand, an atom can lose one or more of its electrons, leaving it with a net positive charge. It is then said to be *ionized*, and it is called an *ion*. Or it may be able temporarily to cling onto an *extra* electron or two, and have a net negative charge; it is then a *negative ion*.

The positive charge on the nucleus is contributed by *protons*. A proton has exactly the same positive charge as the electron does a negative one; so there are as many protons in the nucleus of a neutral atom as there are electrons outside the nucleus. Different kinds of atoms (of the different elements) are distinguished from one another by how many protons they have in their nuclei. The simplest atom, hydrogen, has 1 proton. The next simplest, helium, has 2. An atom of oxygen has 8 protons, and one of uranium has 92. The number of protons in each atom of an element is called the *atomic number;* oxygen has an atomic number of 8.

---

1. How many electrons are there altogether in three different atoms whose atomic numbers add up to 132?

The mass of a proton is nearly 2000 times as great as that of an electron. In the ordinary hydrogen atom, the nucleus is simply that one proton, with the electron moving about it at an average distance of about 1 Å. Nuclei of heavier atoms, however, generally contain another kind of particle as well, the *neutron.* The neutron has about the same mass as the proton, but is electrically neutral. There are usually about as many neutrons as protons, all the protons and neutrons being bound tightly together in the nucleus. Helium atoms usually have 2 neutrons as well as 2 protons, and most oxygen atoms have 8 neutrons in their nuclei in addition to the 8 protons. The number need not be exactly equal, however; the most common kind of atom of uranium has 146 neutrons. All the atoms of a given element do not necessarily have exactly the same number of neutrons, either. Atoms of a particular element having different numbers of neutrons in their nuclei are said to comprise different *isotopes* of that element. For example, occasional atoms of hydrogen have nuclei with 1 proton and 1 neutron as well; they make up the isotope of hydrogen called *deuterium.* One interesting isotope of uranium has atomic nuclei with 143 neutrons each; it was used to make the first nuclear bomb.

---

2. Suppose a sample of water consisted of atoms containing oxygen and deuterium, rather than oxygen and hydrogen. How might that water differ from ordinary water? Actually, such water does exist. Try to find out what it is called.

Since my school days physicists have discovered a host of *other* kinds of particles, besides protons, electrons, and neutrons. There are sigma particles, K particles, lambda particles, and dozens of others. They are unstable, however, and decay into the more familiar particles in periods that are usually extremely minute fractions of a second. Even the neutron is not stable outside the nucleus of an atom; in 15 minutes or so it changes into a proton and an electron (and an *antineutrino,* but we'll come back to him in Act IV).

Moreover, for each kind of particle, there is an antiparticle. The antiparticle corresponding to the electron is a *positron*—it is just like an electron except that it has a positive charge, like the proton's. The corresponding particle for the proton is the antiproton, which is heavy, like a proton, but has the negative charge of an electron. In principle, it is possible to have whole antiatoms of these antiparticles. No one has ever made one, but we know they would work. Hypothetical matter made up of such antiatoms is called *antimatter.* There can't be any around here; even antiparticles made in the laboratory exist only for brief times (except for antineutrinos). The reason is that if matter and antimatter come into contact with each other (for example, an electron and a positron), they annihilate each other, turning completely into electromagnetic energy! So far as we know, in some far-off part of the universe there could be entire clusters of galaxies made of antimatter, and to us they would look just like ordinary clusters of galaxies. But whoof! goes the astronaut who tries to visit one and makes contact!

Much of modern physics is concerned with probing the atom and its nucleus. A little later in this tale we shall have much interest in the nucleus, because it is inextricably involved with the evolution of stars themselves. We shall also touch on some modern ideas about all those many particles, and how they may fit into some simpler scheme of nature than at first confronts us. But for now, we are concerned, not with the nucleus of the atom but the atom itself and with how it interacts with radiation.

Electric Charge

I think everyone is familiar with the phenomenon of *charge* or *static electricity*. For example, you can put an electrical charge on a rubber comb by combing your hair vigorously on a dry day, or on a glass rod by rubbing it with cat's fur. You have not, however, *created* a charge; you have simply rubbed some negatively charged electrons off the rod, leaving it with a net positive charge. Those electrons, of course, are present and accounted for; they are on the cat's fur, which means you have given it a negative charge equal to the positive one on the rod.

Charge, in other words, is *conserved.* Let us see how this comes about. Charge comes in *units;* it can be either positive or negative, but the unit of charge is the same in either case. It is the negative charge on the electron or the equal but opposite positive charge on the proton. Some of the other kinds of particles have charge too, but their charges are always equal to or multiples of that on the electron or proton; when they decay, it is always into other particles of the same total charge. Neutrons can decay into protons and electrons, but since their charges are opposite, those charges cancel and the net charge is conserved. Particles and their antiparticles can annihilate each other, but that always destroys equal amounts of positive and negative charge.

Thus, so far as science knows today, the sum of all the positive charges in the universe (protons, positrons, and so on), minus the sum of all the negative charges (electrons, antiprotons, and so on), stays the same. On the earth, at least, and possibly throughout the universe, that total net charge seems to be pretty close to zero. We can transfer some electrons from one place to another—from a glass rod to a cat's fur—but on the whole everything seems to be pretty electrically neutral.

But when we do build up a charge locally, which means making one thing positively charged and something not very far away negatively charged, we find that strong forces are at work. These forces seem to work right on down to the level of atoms. Objects of opposite electrical charges attract each other, and those of like electrical charges repel each other. The magnitude of the force is the same, whether it is of attraction or repulsion; it depends only on the charge and the separation of the charges. And here is something exciting: just like gravitation, it varies inversely as the square of the separation of the charges! Two positively charged balls repel each other four times as strongly

FIGURE II.24
Natural lightening is an equalization of charge between water droplets and the ground by a massive flow of electrons between the ground and cloud of droplets. *(United Press International Photo)*

if their distance apart is halved, and a proton and electron attract each other 16 times as strongly if their separation is quartered.

A useful way to look at a force is in terms of its *force field.* The earth attracts things around it. Thus, there is a *gravitational field* around the earth within which things feel attracted to the earth. In principle the earth's gravitational field goes out forever. Of course it gets weaker and weaker farther and farther away, and eventually the earth's field becomes all confused with the fields of other stars and things in the universe, but technically the farthest galaxy ought to move, in some minute measure, according to the earth's pull on it, added to that of everything else. The same is true of electrical forces, with one exception.

The force of gravitation is always attractive; all matter contributes to the attractive force of gravitation, which means that a very large body, like the sun, attracts more strongly than a smaller one, like the earth. Thus we are all familiar with gravitational forces about us. But with electrical forces, some attract and some repel. Since around most matter, at least, the attractive forces between oppositely charged particles are balanced by the repulsive forces of similarly charged ones (because most matter is electrically neutral), most ordinary matter *does not exert appreciable electric forces on exterior bodies.* The positive and negative charges exert attractive and repulsive forces that cancel each other. But *inside* an atom those electrical forces are very strong indeed! In addition to gravitation, the electric force is the second natural force (of four forces) we have so far discussed. (It is a little inaccurate to describe it as an electrical force only, but we shall come back to that point in a few moments.)

Let's compare the relative strengths of the force of gravitation and the electrical force within an atom—say, the hydrogen atom—where the proton and electron attract each other, both because of their mutual gravitation and because they are of opposite charges. It turns out that the electrical force between that proton and electron is greater than their mutual gravitational attraction by more than two thousand million, million, million, million, million, million (or, to be more exact, 2.27×10^{39}) times! Compared to electricity, gravitation is an *incredibly* weak force. The only reason it seems important to us is that matter everywhere around us is almost completely neutral.

Nevertheless, within a molecule the nucleus of one atom feels itself attracted by the electrons of the adjacent atoms, and the same is true within the latticework structure of a crystalline solid. In fact, for a molecule or solid to be bound as a unit, the atoms must fit together in such a way that the forces between the positive and negative charges of neighboring particles can bind that structure cohesively. It is those extremely strong electrical forces that give solids their strengths, that hold rocks together, and that hold *us* intact. I hope you appreciate the tremendous importance of electric forces, even though most bulk matter is electrically neutral.

---

3. We have said that most bulk matter is electrically neutral. On the other hand, can we rule out the possibility that both the earth and the moon have a surplus of positive over negative charges in the amount one part in a thousand? Explain your answer.

4. If the electrical forces binding together the atoms of a solid are so extremely strong compared to gravitational forces between them, why does the earth pull mountains into conformity with the ground, whereas a table stands rigidly on the floor? What would happen to the table if it could be scaled up in size by 1000 times?

Electricity and Magnetism; Maxwell's Equations

Now another phenomenon, known centuries ago, is magnetism. It was also known pretty long ago that magnetism has something to do with electricity. An electric current (which is just the flow of electric charges—electrons) moving through a wire, for example, produces magnetism—a *magnetic field*—around that wire. Thus we can pass a current through a wire and make an *electromagnet* that has all the same properties of a permanent magnet of iron. On the other hand, if we move a permanent magnet back and forth through a coil of wire (Figure II.25) we *induce* a current in that wire! Equivalently, when a conductor moves through a magnetic field, a current flows in that conductor (that's how a generator generates electricity). So moving magnetism produces electricity, and moving electricity produces magnetism; strange. . . .

In 1873 the Scottish physicist James Clerk Maxwell developed the theory that synthesized electricity and magnetism, and in a sense did for it what Newton did for gravitation. Maxwell was actually a member of the Clerk family, but when his great-grandfather married Dorothea Maxwell, heiress of Middleby, he took on her name in order that the couple could keep the

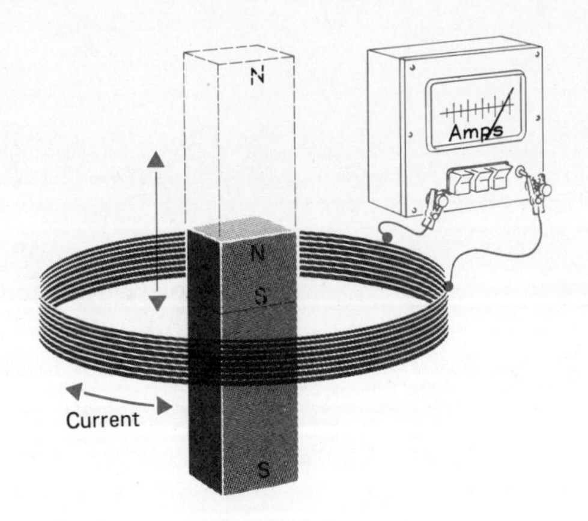

FIGURE II.25
Induction of a current in a coil of wire by the movement of a permanent magnet.

Maxwell fortune. The Clerk family estate, which Clerk Maxwell visited often at holiday time, is still intact at Penicuik, just south of Edinburgh. But whatever's in a name, Maxwell earned his own in physics.

Maxwell's great synthesis of electromagnetism is best expressed in terms of mathematical equations now known as *Maxwell's equations.* Maxwell's equations describe the nature of the electric field around a charge, and the magnetic field around a magnet (or moving charge). They also describe how *changes in one kind of field always produce the other kind.* Suppose, for example, a current is flowing back and forth in a wire. The motion of the charge, which *is* that current, produces a magnetic field around the wire. But every time the current changes direction, the magnetic field collapses and builds up again with the opposite polarity. However, this changing magnetic field, in turn, sets up an electric field. The electric field, though, collapses and re-forms every time the magnetic field does so, and each change in the electric field results in a new magnetic field – and so on. Thus a whole chain of alternating electric and magnetic fields is formed and destroyed, and the energy of those fields propagates through space as electromagnetic waves. These waves, according to Maxwell's theory, travel in all directions with the same speed (in a vacuum) – c.

Radiation and Atoms

Because Maxwell's equations were so successful in accounting for electric and magnetic phenomena, physicists felt that moving charges must be responsible for the emission of electromagnetic radiation from every source – for example, from an atom. We have already seen that individual atoms can be responsible for the emission and absorption of light. For example, suppose a beam of white light is passed through some gas (Figure II.26). The light from the lamp source is from a glowing tungsten filament and has approximately the characteristics of light from a black body. Thus if the spectrum of the lamp is observed, it is a continuous band of the colors of the rainbow. If we analyze the spectrum of the light passing through the gas, however, we find certain wavelengths missing, or at least greatly diminished in intensity; evidently, the atoms that make up the gas have absorbed certain wavelengths, thus producing the particular pattern of dark lines in the spectrum of the light from the lamp. (We have already mentioned that atoms in the outer layers of stars do the same thing to starlight.)

On the other hand, if the gas absorbs some of the lamplight, it must emit it again. Indeed it does; it reemits it in all directions, so that only a negligible

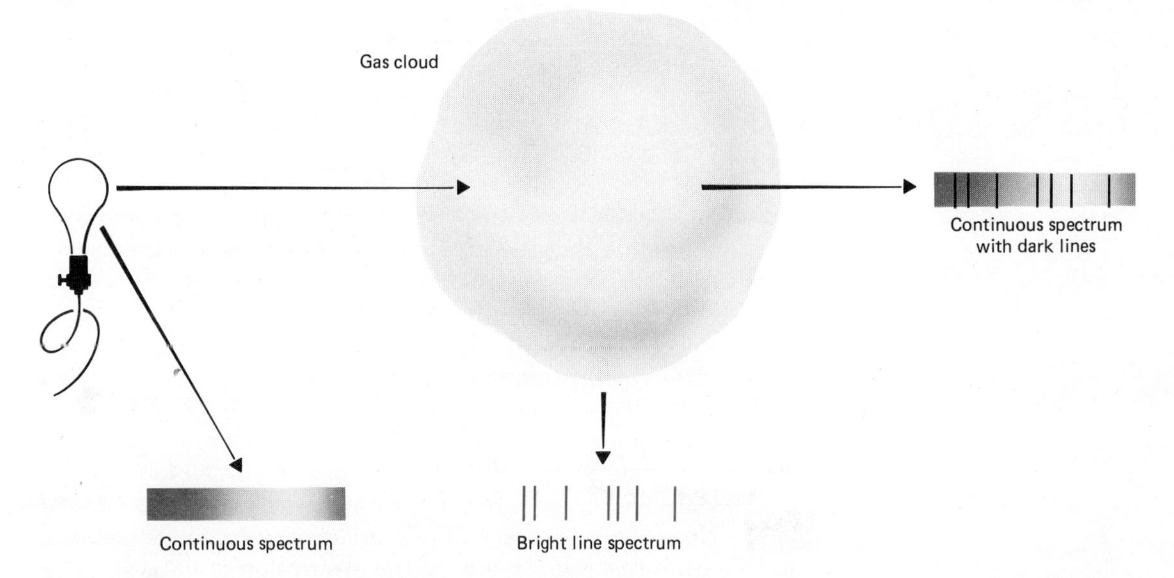

FIGURE II.26
Absorption and emission spectra of a cloud of gas absorbing radiation from a light source.

amount of the reemitted light gets near enough back into the beam to appreciably replace the light the gas abstracted in those dark lines. However, if we observe the gas from such a direction that we do not look at the light of the lamp directly, we can see that the gas is faintly glowing. The spectrum of that reemitted light shows that it exists only at certain discrete wavelengths—those same wavelengths at which the gas absorbs light from the continuous spectrum of the lamp.

It is found that every different kind of gas, consisting of different kinds of atoms, produces its own characteristic pattern of dark lines in the spectrum of light shining through it. Thus each kind of atom evidently has the ability to absorb and reemit radiation at its own peculiar series of wavelengths. The gas cloud shown in Figure II.26 is transparent, so it does not radiate at all like a black body. However, if the gas were very dense or under high pressure, or if it were a hot solid, it would approximate a perfect radiator and it would emit radiation more or less like that of a black body. There must, therefore, be a multitude of ways in which atoms can absorb and radiate electromagnetic radiation under various conditions and at many different wavelengths. How does this all jibe with Maxwell's equations?

Toward the end of the previous century, it was hypothesized that the atoms themselves acted like tiny *local oscillators,* with their charges oscillating back and forth to emit the radiation, just as an alternating current does in a radio broadcasting antenna. It was further supposed that each kind of atom has its own resonant frequencies at which it could oscillate, so that it could emit light only at those frequencies at which its charges could oscillate, or could absorb radiation only at those frequencies which would be able to set the charges oscillating. Thus the local oscillator model accounted for why atoms emit and absorb radiation only at certain discrete wavelengths.

But the local oscillator model fails to explain the quantization of radiation. We recall that Planck was able to account for the distribution of radiation at various wavelengths emitted by a black body only by assuming that the energy came off in little bursts, each burst of energy equal to the frequency times h (Planck's constant). Moreover, the photoelectric effect shows that those bursts of energy are in the form of indivisible photons. To understand

Absorption lines

Emission lines

FIGURE II.27
The Balmer series of spectral lines due to hydrogen, both in absorption from a source with a continuous spectrum (*top*) and in reemission of that energy (*bottom*).

how atoms can emit and absorb light only in quantized amounts, it was necessary to assume that the possible energies of atoms themselves are quantized as well.

A study of the spectrum of hydrogen by the Swiss physicist J. J. Balmer threw new light on the structure of atoms. Balmer found that the wavelengths of light absorbed or emitted by atoms of hydrogen (that is, the lines they produce) in the visible spectrum show a regular spacing (Figure II.27), which he was able to represent by a simple mathematical formula. Later other investigators found that a slight modification of that formula gave the wavelengths of other lines produced by hydrogen in the ultraviolet and infrared. Note that each spectral line occurs at a particular wavelength or frequency, and thus corresponds to the absorption or emission of photons of a particular energy.

The Bohr Atom

Analysis of the mathematical spacing of the lines in the spectrum of hydrogen led the Danish physicist Niels Bohr to the derivation of a model for the hydrogen atom, which seemed to account for the observations. This *Bohr model* has the electron revolving about the nucleus (proton) in one of a number of possible circular orbits. Ordinarily, the electron is in the smallest possible, or innermost, of those orbits (about 1 Å in radius), and the atom is said to be in its *ground state of energy*. However, if the atom absorbs some energy in the form of a photon, the electron moves out to a larger orbit; because of the attraction of the nucleus for the electron, the latter has more energy in the larger orbit, so the atom uses the energy it absorbed from the photon in shoving the electron out a bit. Then, in that higher energy state, the atom is said to be *excited.* It does not remain excited long, however; in a brief time (typically, a hundred millionth of a second) the electron jumps back to its innermost orbit again, returning the atom to its ground state. The energy released in this *transition* goes off as a photon.

So we see how the Bohr picture allows the atom to absorb and reemit energy. But how do we make it absorb or reemit photons of only certain energies? Bohr did that by postulating that only certain circular orbits were allowed for the atom; that is, it is allowed only certain discrete energy states, and can absorb only those photons possessing the requisite energy to raise it from one allowed state to another higher one. Similarly, the only photons emitted are those with energy corresponding to that given up by the atom in jumping back to a lower state.

Perhaps a picture will help. The problem, however, is that even though the Bohr model of the hydrogen atom allows electron orbits of just certain discrete sizes, still there are an infinite number of possible orbits, getting ever larger at higher and higher energies. Thus I want to draw an atom that is even simpler than hydrogen; it will have one proton and one electron, as does hydrogen, but only *four* allowed energy states, or electron orbits. I did this once in a large freshman class in astronomy. But I wanted to emphasize that there is no such atom as one with only four energy states—that it was fictitious and ultrasimple—so I called it an "atom of Abellium." A month or so later a distraught and frustrated coed appeared at my office; in trying to understand the subject she had consulted every encyclopedia and chemistry book she could lay her hands on, but could find no mention of the element "Abellium"! Notwithstanding, Figure II.28 shows an atom of Abellium.

There are twelve possible transitions between the four energy levels of the Abellium atom, and they are all shown in the figure. Those with the arrows pointing *away* from the nucleus correspond to absorption of photons, and those with the arrows pointing *toward* it correspond to emission of photons. The photon absorbed when the electron goes outward from one orbit to a larger one has the same wavelength as the one emitted when the electron goes inward between those same two orbits. I have identified those photons by letters, *v*, *b*, *g*, *y*, *o*, *r*, corresponding to the colors of the parts of the spec-

FIGURE II.28
Possible transitions in an atom of the fictitious and ultrasimple element of "Abellium."

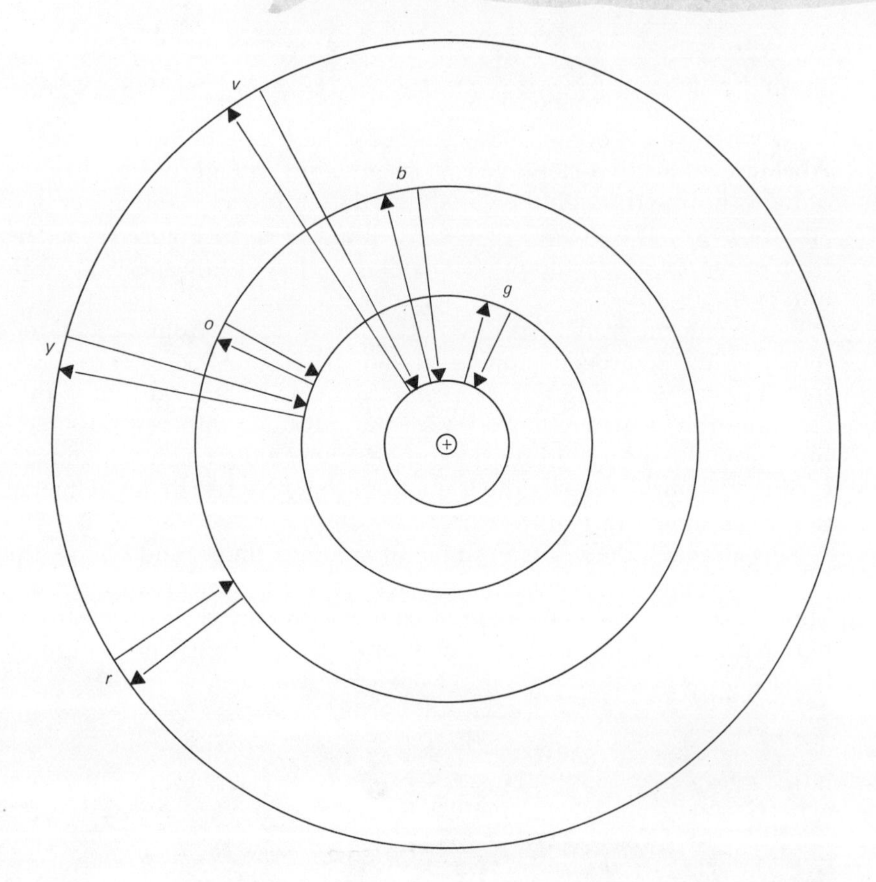

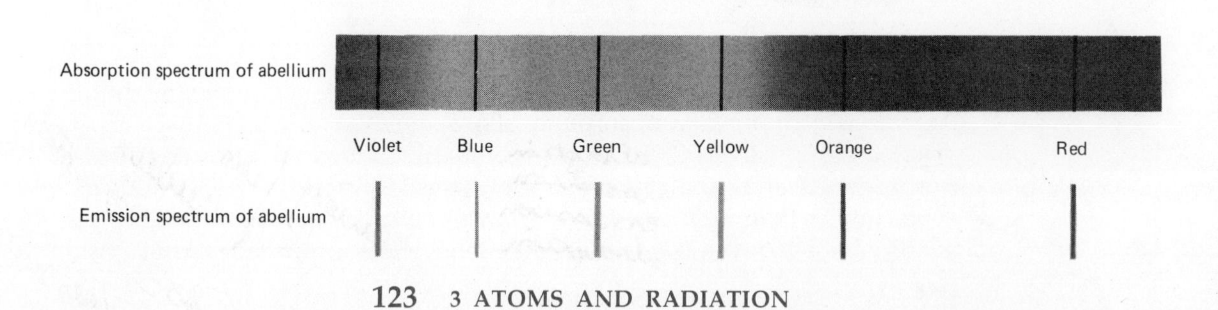

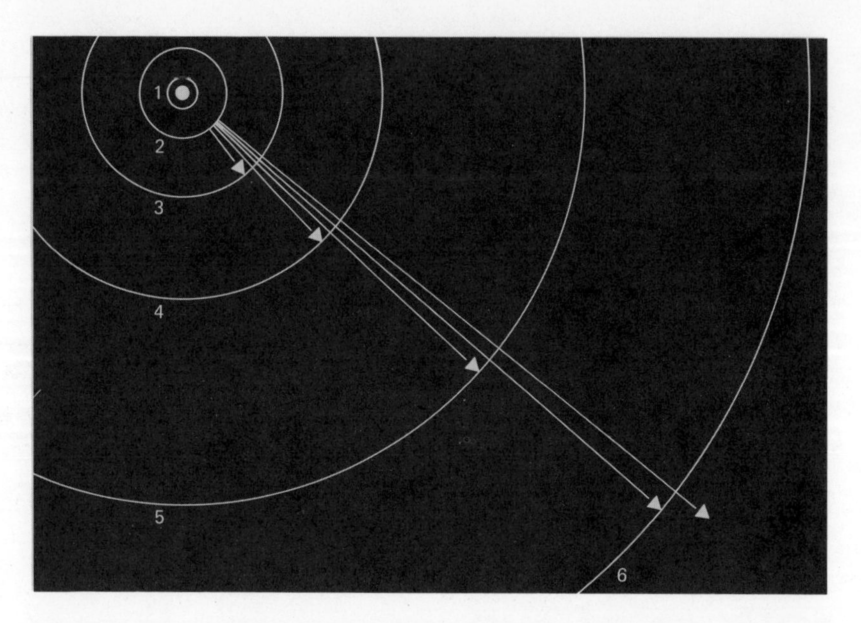

FIGURE II.29
Some of the possible transitions of the hydrogen atom according to the Bohr model.

trum where they occur. Now suppose there are millions of millions of Abellium atoms in a cloud of gas, like the one in Figure II.26. The spectrum of light shining through the gas shows dark lines, as in the upper of the two spectra below the picture of the Abellium atom in Figure II.28. The spectrum of the gas itself, though, shows the emission-line spectrum at the bottom of the figure.

Please remember that Abellium is fictitious. Part of Bohr's actual representation of the hydrogen atom is shown in Figure II.29. Also shown are a bunch of arrows pointing away from the nucleus, and starting at the *second* electron orbit. They are just a few of the infinite number of possible transitions from the second orbit up to higher ones in the real hydrogen atom (according to Bohr); these are the transitions that give rise to the Balmer absorption lines shown in Figure II.27.

Now although the orbits in the Bohr atom get bigger and bigger, they also get so far from the nucleus that its force on the electron becomes very small. Thus it takes quite a bit of energy (relatively) to raise the electron from the innermost orbit to, say, the twentieth, but only a very trivial amount of additional energy to raise it from the twentieth to the twenty-first. In other words, the energies of the higher and higher orbits crowd closer and closer together, and so the spectral lines at the end of the series crowd together in wavelength, as can be seen in Figure II.27. Moreover, an electron in a very large orbit is so far from its atomic nucleus that it is very likely to be bumped off entirely in a collision with another atom. Hence those very big possible orbits are not often occupied by any electrons unless the gas has a very low density, as in interstellar space.

---

5. Of the following transitions in the hydrogen atom, which involve emission, and which absorption, of energy?
a) level 2 to level 20,
b) level 3 to level 4,
c) level 6 to level 1,
d) level 1 to level 5.

6. The following are transitions between levels in the hydrogen atom. Each involves the emission or absorption of a proton. Place these transitions in order of increasing wavelength of the protons involved.
   a) 20–21,
   b) 1–22,
   c) 3–5,
   d) 1–5.

Ionization

If an atom absorbs enough energy—that is, a photon of high enough frequency—its electron can be ejected entirely, leaving the atom ionized. A certain critical *ionization energy* is required to remove the electron from the atom, and that energy is different for each energy level the atom is already in. In other words, it takes a relatively large amount of energy to remove the electron from the atom if that electron is in the innermost orbit, but much less energy if the electron is in one of its larger orbits. The atom can absorb *any* photon whose frequency is higher than that corresponding to the ionization energy. This is what accounts for the continuous absorption of light (over a continuous range of wavelengths) at wavelengths just less than that where the series of Balmer lines of hydrogen crowd together in the spectrum shown in Figure II.27. The energy absorbed over and above that needed to ionize the atom goes into energy of motion (kinetic energy) of the freed electron. If ionized atoms are present in a gas, every now and then one of the ions will meet a free electron and capture it in one of those allowed orbits. In that case, it emits a photon of energy equal to the ionization energy from that level, plus the kinetic energy the electron happened to have at the moment of capture.

Ions can emit radiation in another way, too. Because the positive ion (proton in the case of hydrogen) and electron are of opposite sign, they attract each other, just as two stars do that pass in space. The passing stars go by each other in hyperbolic orbits, and so do the ion and electron. When the two are close together, however, it is possible to emit a photon and change to less energetic hyperbolic orbits, or to absorb one and switch to faster-passing orbits. Because the electron remains free of the ion before and after the encounter, this type of change is called a *free-free transition.*

When a hydrogen atom is ionized, it obviously cannot absorb any more photons in the process of moving its electrons about, because it has no electron left. However, other atoms, having more than one electron, also have energy levels and absorb and emit photons at discrete frequencies, but the theory is more complicated. Helium, with two electrons, has its own pattern of spectral lines, different from those of hydrogen. Ionized helium still has an electron left, but now the energy levels (allowed orbits) are all different, and the pattern of lines for ionized helium is not the same as for neutral helium. If an atom loses two or three electrons, it is *doubly* or *triply* ionized; oxygen, for example, can be ionized eight times. Each kind of atom, in each stage of ionization, produces its own peculiar pattern of spectral lines. For the simpler atoms we can calculate those allowed energies from theory, but in the more complex case it is easier to determine them experimentally. Anyway, we can determine the levels for each kind of atom and ion, and can recognize its spectral lines from their characteristic pattern. Even molecules produce spectral lines, but they have more ways yet to do it, because they have various allowed states of rotation and vibration, as well as of electron orbits.

---

7. How many times can each of the following atoms be ionized?
a) hydrogen,
b) helium,
c) carbon,
d) iron,
e) uranium.
(See Appendix 16 for data, if you need it.)

Excitation and Ionization by Collision

Now we have already said that ordinarily atoms are in their lowest state of energy, with the electrons in their innermost orbits. However, as we have seen, atoms can raise to higher energies, and become excited, by absorbing photons of just the right energy. They can also absorb energy and become excited, or lose it and become *de-excited,* through collisions with other atoms or ions or electrons, or whatever is bumping around. The higher the temperature of a gas, the faster the molecules move, and the harder they bump. Thus more atoms are excited, or excited to higher levels, in a hot gas than in a cooler one. If the temperature is high enough, such collisions result in most or all of the atoms becoming ionized. For example, at a temperature of 10,000°K most hydrogen atoms are ionized, and at 30,000°K most helium atoms have lost both their electrons. So the pattern of spectral lines a particular kind of atom can produce, which depends on the state of ionization of the atom, depends indirectly on temperature as well. Given the temperature and density of a gas, we can calculate the rate of collisions and the state of excitation and ionization of its atoms.

Consider a hot, dense gas. Its atoms are absorbing and emitting radiation all the time by many or all of the mechanisms we have been discussing. It would, of course, be beyond the ability of our biggest computers ever to calculate in detail what every atom is doing. However, the theory worked out by Planck does tell us that if the gas is opaque enough, all those detailed emissions and absorptions will combine to give the distribution of energy we observe radiated from a black body. We don't have to know every detail to know the ultimate result. We accomplish this by recognizing that radiation is quantized.

The Modern Quantum Theory

Now we have also seen how by quantizing the allowed energy levels, as in the Bohr model, we can understand the patterns of spectral lines produced by gases that are not dense and opaque. Some people point out that the motions of the electrons in orbits about the nucleus of an atom are an analogy to a tiny solar system, with the sun representing the nucleus and the planets electrons. But the analogy is not a good one.

One place it breaks down is that the solar system is too cluttered to resemble an atom. The sun's distance is about 150 million km, which is only about 100 times the diameter of the sun. The typical distance of the electron from the nucleus of the hydrogen atom is only about 1 Å, but this is 100,000 times the size of the proton! In terms of volume, the atom is a thousand million times emptier than the solar system. All the solids we know about on earth are fantastically empty space!

---

8. (Project for those who like computations) Compare the mean (average, smoothed-out) density of the hydrogen atom and of that part of the solar system no farther from the sun then the earth's orbit. Assume the radius of the hydrogen atom to be 1 Å (10^{-8} cm). Other necessary data are in the appendices.

The analogy of atoms and solar systems also breaks down in a very fundamental way. The electrons do not move in circular or elliptical orbits. We don't know how they move; neither do they. We can talk about the most likely distances of electrons from the nuclei of their atoms, and hence the most likely energies those atoms have, but not where the individual electrons are or exactly what they are doing. Yet if Bohr's simple picture is wrong, at least the idea of energy states of atoms is correct; it's just that we must not think of electrons moving about between orbits in that nice, neat way. Even the energy states or levels of an atom are not precisely sharp. A particular energy state is one of the more likely energies an atom can have, but there is almost as good a chance that it will have a very slightly different energy. So the energy levels themselves are fuzzy, and therefore so are the spectral lines. Those lines are not *absolutely* sharp, but just pretty sharp.

The Uncertainty Principle

The vagueness of where an electron is in an atom arises from the inherent uncertainty and statistical nature of fundamental units of matter touched on in the foregoing scene. The German physicist Werner Heisenberg considered an idealized experiment in which one tries to measure with perfect precision the simultaneous position and momentum of a particle. He showed that even with theoretically optimum circumstances, there are certain inherent limitations (such as the finite speed of light) to the experimental procedures that prevent those two quantities from ever being known better than to a certain limiting accuracy. There is always an uncertainty in the position, and one in the momentum of the particle; the more accurately one of the two is known, the less accurately is the other. The product of those two uncertainties is ultimately never less than a certain small number (Planck's constant, h, divided by 2). The result is called the *Heisenberg uncertainty principle.*

The uncertainty principle applies to everything, large and small. However, a large object like a bowling ball has enough mass that even a tiny uncertainty in its velocity results in the momentum being uncertain enough that its position (so far as this quantum-mechanical limitation is concerned) can be specified to a far higher accuracy than we could ever hope to measure. Suppose, for example, the speed of the bowling ball is known to within 0.1 mm/s—in other words, darned accurately. Then, the uncertainty principle says that theoretically the ball's position can never be known to better than 0.000000000000000000000000000015 cm (more easily written 1.5×10^{-29} cm)—not a very interesting limitation to the average bowler.

But an electron has a tiny mass, and if its velocity is uncertain by that same amount, its position is uncertain by more than one meter!

"But," you say, "you're just talking about our inability to make a precise measurement. Obviously the real momentum and real position *exist* to perfect precision." But how can we define something like perfect precision if there is no possible experiment, even theoretically, by which we can determine it? According to the rules of science something that cannot be defined by means of a measurement or experiment (that is, operationally) does not exist—at least not in the realm of science. Since the limitation is theoretical, we must regard the uncertainty principle as *inherent* to nature. The *electron* cannot simultaneously know its precise position and momentum—those precise quantities do not exist at all! There is, in fact, very ample evidence to prove that such is the case. Indeed, we shall see in Act IV that our existence on the earth would not be possible if perfect precision existed in nature.

Uncertainty and Probability

Remember how it is with the photon (previous scene)? We can never say where it "is" until we observe it. Prior to that, all we can do is write an equation (the equation of a wave) describing the probability of its being in various different places. It's the same with all matter, and the effect is very important with small particles like electrons. Think back to that experiment with ocean waves approaching the front wall in Figure II.11. Now consider the similar apparatus shown in Figure II.30. An electron gun shoots electrons at random toward the "front wall," which, in this case, is a metal plate with two holes in it. If you were to block up either of the two holes, the electrons hitting the back wall would be bunched around a point behind the other hole, as you would expect. You could detect them with many kinds of devices (such as a phosphor screen or photographic film), and would find that they behave like any other self-respecting particle. But if both holes are open, you find the electrons bunched up in many evenly spaced positions, as

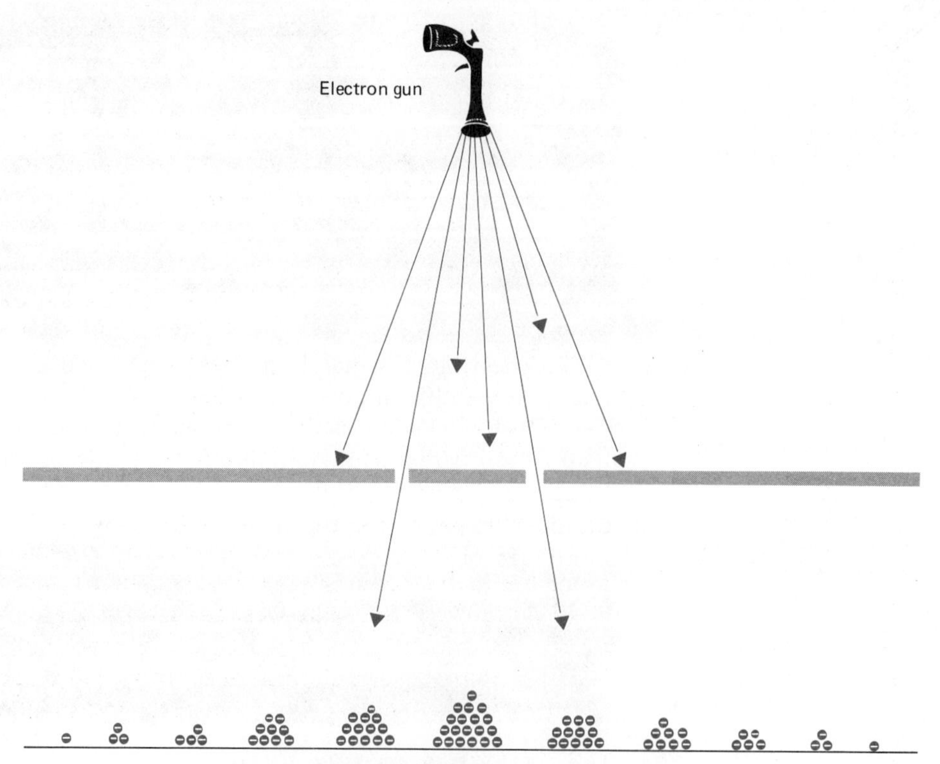

FIGURE II.30
Electrons fired through two apertures interfere with each other just as light waves do.

shown in Figure II.30—just as the high waves bunched in certain spots against the back wall in the ocean wave experiment or light waves made fringes behind the slits. In other words, the electrons *interfere with each other,* just as if they are waves.

"But," you argue, "if an electron is an indivisible particle (and it is!), how can it pass through both holes like a wave? Surely each electron went through one hole or the other!" But if you devise an experiment to tell which hole each electron goes through, the fringes disappear, as they did when you closed one of the holes. With one hole closed, you already know that each electron passing through the plate went through the other hole. Suppose you shine light upward from the back wall through the holes, and detect the electrons as they come through by the flashes of light (photons) they deflect on the way. Then the fringes disappear, and the electrons act like particles again. Only when *we don't know where the electron is* does it behave like a wave.

Is the wave, then, just a kind of formula that tells us the probability of finding the electron in various places? Yes, but it's deeper. The electron itself doesn't know where it is in the wave until it is detected somewhere—that is to say, its precise position does not exist. Its location is an innate uncertainty built into nature. Nature, in its most fundamental way, is described by probabilities of what will happen; you can never tell the future for an individual particle, such as an electron. Of course, with large numbers of particles, like those casino operators mentioned earlier, you can predict statistically what will happen—for example, that you'll get those fringes behind the two holes.

Suppose you toss up two dice, and when they fall the number 7 comes up. Now it is meaningless to calculate the probability that you got 7—that probability is unity, because you already know 7 turned up. You cannot calculate the probability of a single event occurring that is already known to have occurred. But when the dice are still in the air, the number 7 does not yet exist on the table; there is only a certain probability that it will exist in a moment. Maybe it's sort of that way with particles like electrons. After one particle has been detected (observed), it is known to have existed as a particle, but before then it exists only as a probability formula spread about in space—a wave.

This doesn't seem to tell us what an electron is, but only how it behaves. Perhaps the "real nature" of the electron seems to defy logic—common sense. Yet, maybe the nonscientist has the advantage here over those of us who have studied some mathematics and physics. As a young student I found geometry and algebra beautiful and rational systems of logic, and highly satisfying. Then, when I studied mechanics, it was particularly satisfying to be able to understand the behavior of nature in terms of sensible, logical, concrete models. Perhaps some of us have learned, or have been trained, to need those concrete ideas to be really happy with a description of something.

On the other hand, maybe the person not trained in science would not experience such discomfort with our less than concrete description of the electron. Take my mother, for example. Mind, she was a marvelous person, and bright, witty, and charming. She was an expert at literature and poetry, and could make up limericks you wouldn't believe. But when it came to mathematical or physical things, she was a disaster—confused, befuddled. Once I asked her to draw an angle for me, and, so help me, it had hairs on it! I doubt if she would have been the least disturbed about our description of the electron—as at the same time a particle and a wave. It would probably never have occurred to her that it should be anything more specific.

But then, she would never have *discovered* the quantum theory, either.

SCENE 4

Special Relativity—Meaning of Space and Time

Consider a baby just old enough to hold a rubber ball in his hand. He releases it, and it stays where he left it, floating in midair. "Ah," thinks the baby in his baby-think language, "the rubber ball stays put; good." Then he grabs it again and once again releases it; this time it floats upward and stops at the top of his carriage. "Ah," he thinks again, "it's gone up; good." He grabs it a third time, but this time, when he releases it, it floats off to one side. "Ah," he thinks, as he begins to perceive the world around him, "sometimes it stays, sometimes it goes up, and sometimes it goes off to one side."

None of this is strange to the baby. Why should it be? He has had no previous experience to base his notions on about how matter ought to behave; he is still learning from experience.

"Now," he thinks, "how about this wooden block . . . ."

Unlike the baby we are disadvantaged by having too much previous experience. But it's a limited experience, which has shaped our notions of what is natural. If we were living in an environment where most things about us were moving at speeds near that of light, our gut feelings would be very different about the natural way things should behave. Ours is a very special and rather extreme environment, in that we encounter only speeds that are very slow compared to that of light.

Relative Motion

Let us imagine an experiment. I am on a train moving past you at 50 km/hr. Just before I reach you, I throw a ball in the same direction toward which the train is moving at a speed of 20 km/hr with respect to the moving train. Because the train is moving 50 km/hr and the ball is moving another 20 km/hr, you and I will have no problem agreeing that the ball approaches you at a speed of 70 km/hr. On the other hand, if, after I and the train had passed you, I were to throw the ball *back* to you, we would agree again, I suspect, that the ball approaches you at only 30 km/hr.

1. Suppose you are standing at the front of an airplane, and throw a baseball to a friend down the aisle toward the rear of the plane at a speed of 50 km/hr. The plane, meanwhile, is moving at a speed of 1000 km/hr. How fast does the ball approach your friend? How fast would an observer on the ground say the ball is moving? Roughly, how fast would a hypothetical observer on the sun say it is moving?

You and I represent two different coordinate systems—yours on the ground and mine on the moving train. Yet, the speed of the ball I throw can easily be transformed from one of our coordinate systems to the other in a perfectly logical way. Galileo would have followed this discussion and have agreed with our logic. In short, the velocity of the ball includes the *velocity of the source* from which it is thrown.

Next let us consider another, very different experiment. I am sitting on a raft in the middle of a lake. I am creating waves in the lake by pushing an oar or something else back and forth. The waves travel out from my raft, like ripples on a pond, and strike the lake shore, moving, say, 10 km/hr. Now you get in a boat and row out toward me at a speed of 5 km/hr, while I go on creating those waves. Surely, we agree that you, moving forward to meet the waves, find them passing your boat at 15 km/hr. And then, if you turn around and row back to shore, you are moving away from the waves, and they have to catch up to you, so they are moving only 5 km/hr with respect to you. Everything is perfect common sense, and Galileo would surely have agreed with all of it. In other words, the speed of the waves with respect to you depends on your own speed—the *speed of the observer.*

Now let us imagine some strange, nonsensical, nightmarish world where the ball I throw from the train approaches you at 20 km/hr whether or not the train was approaching you or receding from you when I threw it. (We assume that I always throw the ball with the same force.) And, similarly, what if those ripples in the pond approached your boat at 10 km/hr whether or not you were approaching them or rowing in the opposite direction. You would certainly conclude that certainly something was wrong—maybe this is a trick or some magic. But you try everything out, over and over, and no magic! Everything checks out—completely crazy. That is *exactly* what happens with light; its speed depends on neither the speed of the source nor the speed of the observer.

The Constant Speed of Light

Consider the following hypothetical experiment: You enter a spaceship and travel away from earth at a speed of 200,000 km/s. Now I send you a light signal. The light waves leave me at 300,000 km/s, but they have to catch up to your ship, moving away at 200,000 km/s, so with respect to you, they must be moving 100,000 km/s. If you measure the speed of the light shining through your back window toward the front of your ship, surely you will find it going only 100,000 km/s; right?

Wrong! 300,000 km/s.

But then, obviously something must be wrong; Right? Wrong! obviously right!

Well, "obviously," after many, many experiments are performed.

One such experiment was that of Michelson and Morley in 1887. They hypothesized that the speed of light was constant in *space*, just as the ripples in the lake move with constant speed in the water. They reasoned that as the earth approached a lamp, the light from it should appear to approach us at a higher speed than when the earth is moving away from the lamp, exactly analogous to your rowing the boat toward or away from me in the lake and observing the ripples. I shall not take the space to describe the details of the Michelson-Morley experiment (it is described in very many books) but will simply report that they found absolutely no difference in the apparent speed of light, whether the earth (carrying the observer) was moving toward the source or away from it. Many other similar experiments were performed, with the same result. It is an *experimental fact* that the speed of light does not depend on the motion of the observer.

However, the Michelson-Morley experiment, and others like it, did *not* rule out that the speed of light might depend on the speed of the source (like bullets from a gun or the ball I threw at you from the moving train). Other experiments, however, both astronomical and in the nuclear physics laboratory, eliminate that possibility. As Gertrude Stein might have put it, "c equals c equals c."

---

2. One such experiment involves the observation of double stars. We find many examples of pairs of stars revolving about each other at speeds of tens of kilometers per second. We always observe them moving in orderly ways in elliptical orbits. Remember that stars have such distances that many years are required for light to reach us from them. Can you see why the observed orderly orbital motion of double stars about each other rule out the hypothesis that the speed of light depends on that of its source, as does the ball thrown from a moving train? Explain.

But how in heaven's name can different observers, moving with respect to each other, always measure the same speed for light from the same source, in each of their comoving coordinate systems? Einstein showed how, in 1905. The different observers, moving uniformly with respect to each other, perceive space and time in different ways from each other.

Postulates of Special Relativity

Einstein made two postulates. The first is that the speed of light is an absolute constant of nature; that all observers, in spite of their relative motion, will measure the same speed for electromagnetic radiation. This point is hardly arguable; it is an experimental fact, borne out by experiment after experiment after experiment.

Einstein's other postulate was the *principle of relativity*, which states that one can never detect the rate of his own *uniform motion*. This principle *does* appear to be common sense, and is also borne out by experiment. Everyone who has ridden a train has probably had the experience that the train on the next track appears to move, when in reality it is his train that has moved (or vice versa). If one train jolts or starts up too fast, you can tell, because *acceleration* is not the same as uniform motion; you can tell if you are accelerating.

FIGURE II.31
Albert Einstein. (*The Bettman Archive*)

But uniform motion, according to the principle of relativity, cannot be determined. You can ride on an airplane and feel motionless. You can play catch on the same airplane or perform experiments with billiard balls, pendulums, and anything else, so long as the plane does not lurch, turn, or otherwise accelerate. All your experiments work just the same as in the laboratory. (Try some on your way home tonight, if you are a passenger in a smoothly riding car, and you will see what I mean.) All experiments you can perform give the same results, irrespective of your *uniform motion*. "Oh," you say, "I can look out the window of the car or airplane and see that I am moving." But how can you prove the earth is not moving and the plane is stationary? In fact, for centuries, men thought the earth was stationary and the sun was moving. We don't feel uniform motion, and no laboratory experiment can detect it; that, too, is an experimental fact.

---

3. Do you think that an experimenter on a merry-go-round and one standing on the ground outside it would find that the same laws of physics apply in their respective systems? Explain. yes,

---

Spacetime

Now let's look at coordinate systems. It takes three coordinates to locate something in space. Suppose you pick the southwest, bottom corner of the room for an origin. Then you can locate the bulb in the hanging lamp in the middle of the room by specifying its distance north, east, and above (or higher than) that corner. But that is not all. What if someone removes that lamp? So you must also say *when* that lamp was there to fully specify its location in space and time.

The time part might seem to be a trivial complication. But not really; bear with me. First, imagine yourself seated in the rear tier in the Hollywood Bowl listening to a concert. During the concert you notice that the orchestra is always a fraction of a beat behind the conductor, because it takes sound a large fraction of a second to reach you from the podium. Then, when the concert is over, you hear people in the rear applaud first, those in the middle next, those at the front last, because it takes sound longer to reach you from the front. You do not hear things over the entire theater at the same time.

Nor do you see them simultaneously; however, light travels so rapidly that the delay in light travel time from the front to the rear of the Hollywood Bowl is unobservable. But not so when you look at the sky. You see the moon not where it is now, but where it was just over a second ago; light has required that long to reach you from the moon. You see the sun (or did, befor it set) as it was 8 minutes ago. You see the planets in the sky as they were anywhere from several minutes to a few hours ago (depending on the planet in question). The nearest star you could see from the Hollywood Bowl with the unaided eye is Sirius, and you see it as it was eight years ago. Most of the visible stars in the sky are so remote that light has taken hundreds or even thousands of years to come to us from them. At the same moment you are looking at the sky, telescopes may be observing remote galaxies as they were hundreds of millions of years ago, and feeble radio waves from some distant part of the universe that left their source thousands of millions of years ago. In other words, you never see a snapshot of the sky as it is "now." You see different events in different places in time in the past. A snapshot of the universe (or for that matter, even a scene on earth) as it appears exactly at a given moment in time does not exist, and has never existed. Time is a dimension just as surely as space is; we see events in space and time (or *spacetime*).

But this does not mean that time is just an "add-on" to three-dimensional space to make the picture more complete. Time is a coordinate inseparable from space. We cannot transform the appearance of an event as seen by one observer to how it would be seen by another observer moving with respect to him without involving time, because observers moving in space with respect to each other also have their pasts and futures all mixed up. Your time is a very personal thing, and how it relates to someone else's time depends on where you are in space and how you are moving with respect to him. Let's see why.

Imagine the following setup. You and I are in fast trains on parallel tracks. Let us suppose that as seen by an observer on the ground, your train moves west and mine east (Figure II.32). However, it is dark, and whereas we each can see the other's train by its lights, neither of us realizes that his own train is moving. Now, just as we are abreast of each other, the ground observer sets off two flares on the ground between the sets of tracks, one flare 5 km east of us and the other 5 km west of us. The flares are carefully timed to go off at precisely the same instant, as seen from the ground at the point where we are passing each other. I am moving toward flare *B*, while you are moving toward flare *A*. By the time the light from flare *B* reaches me, I am as shown in Figure II.32. You, meanwhile, are some distance in the opposite direction from where we passed, and are just now receiving the light from flare *A*. Sometime later, I receive the light from flare *A* and you from flare *B*.

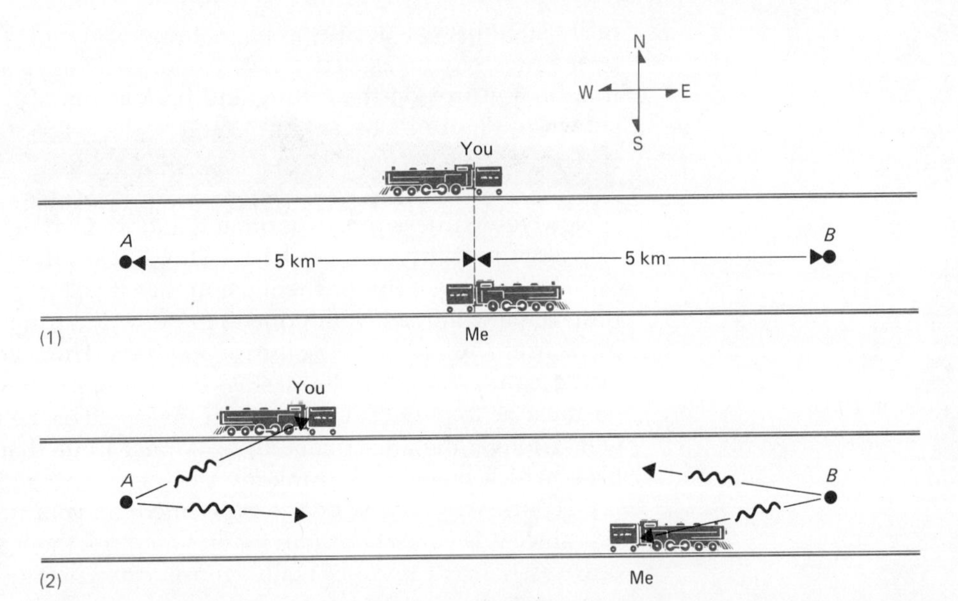

FIGURE II.32
An experiment demonstrating the nonabsoluteness of simultaneity.

Now, remember that we are moving so smoothly that neither of us can tell he is in motion. We can see each other and agree on our relative motion, but each of us insists he is stationary relative to the ground. Thus I (over our intercom) will argue that flare *B* went off before flare *A*, and you will argue the other way around. Because there is no way to settle on who is moving, one of us is as "right" as the other. We simply disagree on which event occurred first.

Thus different observers in relative uniform motion can never agree on the simultaneity of events; the past of one can be the future of the other. Time, we see, is all locked up with one's location and relative motion.

---

4. Rediscuss the experiment described above and illustrated in Figure II.32, but for the circumstances in which I, moving eastward, observe the flares to go off simultaneously. Then what do you and the observer on the ground see?

Special Relativity

So how is it then that two observers in relative uniform motion perceive space and time differently? This is what *special relativity* is all about. Einstein published the special theory in 1905. The *general* theory (the other theory of relativity) came in 1916; we discuss it in the next scene. One nice thing about special relativity is that once you are prepared to accept those experimentally verified postulates on which it is based (the principles of relativity and the universality of the speed of light), the main bones of the theory can be understood with only elementary algebra—it's as easy as Newton's law of gravitation.

Let's see how two observers in relative motion—say you and I—disagree on time. Again, we are in two trains on parallel tracks passing each other. Again each of us insists that he is stationary and that the other is moving. Since I am telling this story, let's assume for the moment that it is I who is really stationary. I decide to regulate my clock by timing the path of light over a precisely measured distance, namely, from a source on the floor of the train to a mirror on the ceiling and back again. My experimental setup is as shown in Figure II.33a. So I shine the light to the mirror and time its return with a delicate mechanism, and find that it takes t seconds (t is, of course, much less than 1, unless we have awfully tall trains).

Now you witness this experiment, and, in fact, perform it as well yourself. You *also* time with your clock how long it takes light to make the round trip. You, however, are under the illusion that it is I who is moving, and that my train actually moved from point A to point B during the round trip the light made from the source to the mirror and back. Thus, you claim, the actual path of the light was along the dashed line in Figure II.33b. On the other hand, we must be in perfect agreement on the speed of the light. Since you think it went a longer distance than I do, you must argue that it took longer to reflect back, which means you think my clock is running slow.

Of course, if *you* now do the experiment on *your* train, I correctly point out that since it is you who is moving, if you think the light went straight up and down (whereas I know it really went farther at the same speed), it must be *your* clock that is running slow.

In fact, if we stop our trains and get together and compare our clocks, we find that they both run precisely together. We disagree only when we are in relative motion. Each of us always thinks the other's clock is running at a slower rate. Since (according to the principle of relativity) there is no way of settling the argument, we must agree that identical clocks disagree when they are in relative motion—with the "other" one always going slower.

We also disagree on lengths. A convenient way to measure length is to time how long it takes light to traverse it. But if we disagree on our timing, then we must disagree on our lengths as well. Each of us finds that the other's clock runs too slowly, and that his second is therefore too long by a certain factor. Each also claims that the other's units of length, as measured in the direction of relative motion, are too short, and should be lengthened by that same factor. With elementary algebra one can calculate by how much two observers in relative motion disagree about length and time.[1]

"But then who is *really* right?" you ask; "whose ruler is *really* shorter and whose clock *really* runs slower?" The answer is that both are right. Each person's clock is *really* slow compared to the other's. We simply perceive space and time differently, but your perception is as good as mine, because there is no experiment by which we can tell which of us is *really* in motion. Length and time are not innate quantities of nature; they depend on the observer and his relative state of motion; they are *relative*.

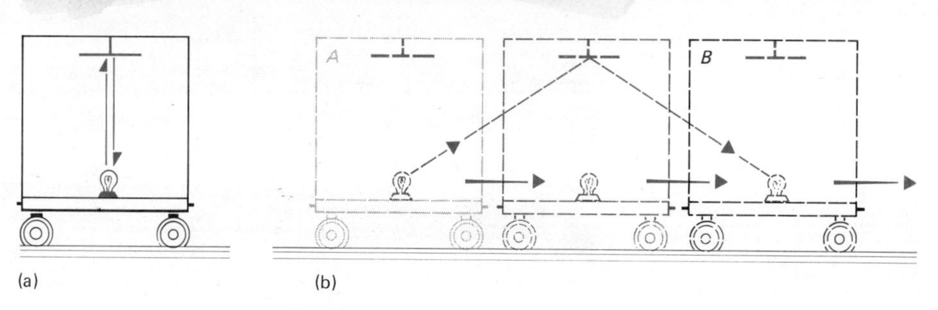

FIGURE II.33
Measuring time intervals in two comoving coordinate systems.

[1] If their relative velocity is v, that factor is $1/\sqrt{1 - v^2/c^2}$.

As we said earlier, we live in a world where most perceptible motions (at least most of those noticed prior to 1905) are slow compared to the speed of light. At low relative speeds observers differ in their measurements of length and time by a negligible amount. What if you and I were moving 1000 km/s (about 3500 times as fast as the fastest passenger jet) with respect to each other? Even so, we would differ on our measures of length and time by only 0.00055 percent. In modern times, however, there are many occasions in which we encounter relative speeds great enough that the relativistic effects are very noticeable indeed.

Time Dilatation

Please realize that if you are moving with respect to me, not only does your clock run slower, but your *time* runs slower. That means you age less than I do; everything involved with the passing of time occurs at a slower rate to you, who are moving, than it does to me. Of course you think just the reverse is true, that *my* time runs slower and that yours is perfectly normal.

A dramatic verification of the reality of this effect is provided by the *muon*. A muon is one of those subatomic particles that can exist for very brief periods. One way muons are formed in nature is by the bombardment of the upper atmosphere of the earth by cosmic rays from space (cosmic rays are the nuclei of atoms moving at extremely high speeds; we shall return to them in Act IV). Collisions of cosmic ray particles and molecules of the air more than 10 km above the ground break up those molecules and produce muons in large numbers. But muons spontaneously decay into electrons and other particles in an average time of only 2 microsecond (0.000002 s). Even if they were moving at the speed of light, they could hardly survive long enough to travel a kilometer; yet they are observed at the ground in very great numbers. The reason is that they are moving at a speed so close to that of light that even though they *do* break up in 2 microseconds according to their own clocks, as *we* see them their clocks are running some 50 times too slow, so they make it all the way to the ground. You see, this time dilatation is *real*.

If a spaceship can travel to some remote star fast enough and return to earth, astronauts aboard it will age only slightly, even though the ship may seem to require hundreds or thousands of years for the round-trip voyage, according to us at home. The experiment has never been tested on people (the Apollo astronauts neither went fast enough nor were gone long enough to tell if they had aged less than we on earth), but it has been done with sensitive instruments in orbit, and works out as predicted. It is, in principle, a means by which interstellar travel should be possible in a human lifetime. The rub is the energy requirement; we shall return to this fascinating subject in Act V.

---

5. Suppose an astronaut visited a remote star and returned to earth, moving all the way at a speed so close to that of light that he aged only slightly compared to people on earth. What if he claimed that his ship was actually stationary, and that it was the earth that did the moving?—then he should appear aged and people on earth younger. How can this paradox (a famous one) be resolved? That is, how can the astronaut realize that it was, in fact, he that did the space traveling, and hence that he must have aged less?

Relativistic Momentum and Mass

If you and I, in relative uniform motion, disagree on length and time, we must also disagree on velocity—which is distance covered per unit time. Suppose we both witness the collision of two billiard balls. We will, by virtue of our relative motion, disagree on the speeds and directions of the balls before and after impact. Yet, according to the principle of relativity, we must agree that the laws of mechanics work in your system as well as in mine. In particular, each of us must agree that momentum is conserved in the impact of the two balls.

We can stand side by side, and inspect those balls and find them to be identical. Nevertheless, when we are in relative motion, in order that we both find momentum to be conserved in the impact, we must disagree on how much momentum each ball has. In particular, an object in motion with respect to us but of course at rest in its *own* coordinate system acts as though it has a different mass from that which it has when it is stationary. This effect is commonly observed in the laboratory.

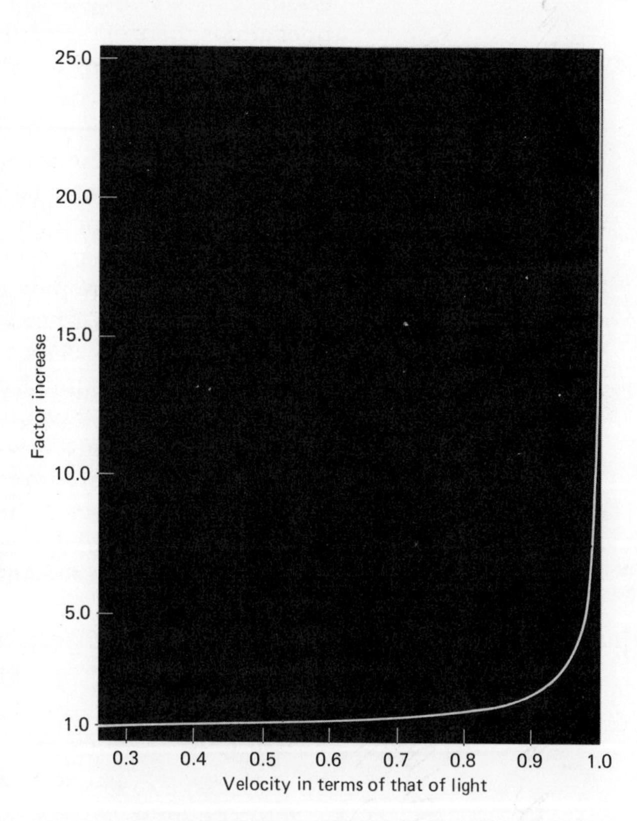

FIGURE II.34
The factor by which mass seems to be increased, by which seconds are lengthened, and by which lengths are shortened in a system moving at various speeds with respect to that of light.

The mass of an object is thus effectively increased, by virtue of its motion, over that which it has when it is at rest (called its *rest mass*) and by that same factor by which length and time are affected. Figure II.34 shows the amount by which the effective mass of a moving object is greater than its rest mass, as a function of its speed compared to that of light. The same graph also shows how much faster time flows for us than in a moving system, and how much greater the length of a standard ruler is to us.

Equivalence of Mass and Energy

Thus we have seen that the conservation of momentum requires that the effective mass of a body depends on its speed. Now, a moving body has energy of motion (kinetic energy). When account is taken of how the body's mass depends on its speed, it is found that the energy of motion is that mass times the square of the speed of light. Einstein postulated that even when the body is at rest relative to the observer, it still has energy corresponding to its rest mass, called its *rest energy*, which is equal to its rest mass times c^2.

The idea of rest energy describes an equivalence between mass and energy, expressed by the famous equation

$$\boxed{E = mc^2}$$

where E is the energy and m the rest mass. Thus special relativity predicted that mass can be converted into energy and vice versa. The energy obtained from the conversion of some matter is the mass of that matter times the square of the speed of light. Now the speed of light is a very large number, and its square is absolutely enormous; thus the conversion (or destruction) of even a tiny amount of mass creates an extraordinary amount of energy. The equivalence of mass and energy is, of course, the source of nuclear power, and, as we shall see in Act IV, the major source of the energy of the sun and stars.

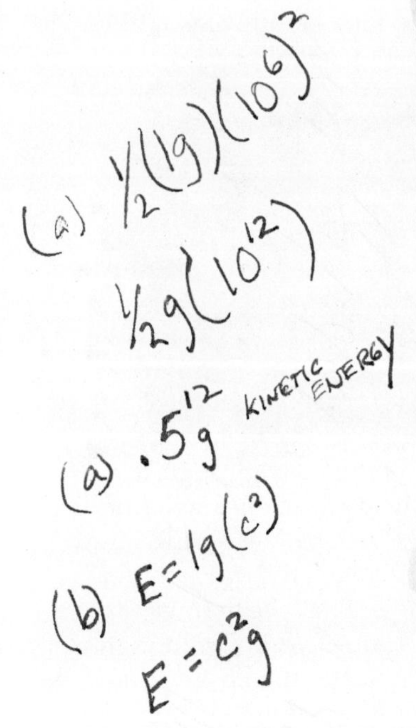

6. According to Newton's laws the ordinary kinetic energy of a body of mass m, moving at a speed v, is ½ mv^2. Calculate the kinetic energy of a body of mass 1 g moving with a speed of 10^6 cm/s (about one-third the orbital speed of the earth). Now calculate the energy associated with the rest mass of the body. How do the two energies compare?

The conversion of energy back into mass is observed on a small scale in the laboratory; it must be on a small scale, because so much energy is required to make a little bit of mass. We think that creation of mass from energy occurs on a pretty big scale, though, in certain stellar explosions (for example, in *supernovae*), and that it is probably responsible for the existence of the elements with atomic numbers greater than that of iron. We shall return to this matter, too, in Act IV.

It may seem that the principle of conservation of mass is therefore violated. The total energy of a system, however, *is* always conserved (at least so far as is known today), if we include its rest energy. We can express by the above equation all the mass in a system in terms of its equivalent energy, and add that to the other energy in the system. Then that total energy does not change unless we somehow pump energy into the system from outside or draw energy away from it. The conservation could also be demonstrated, of course, by expressing all the energy of the system in terms of its equivalent mass.

The energy of a body with a small speed, such as an airplane or a rifle bullet, is its rest energy plus its kinetic energy, but because c^2 is so big, virtually all that energy is the rest energy. If the speed of a particle is near that of light, however, its total energy becomes incredibly great. This is the problem with astronauts traveling at nearly the speed of light. The spaceship would have to have almost unbelievable energy; yet energy has to be conserved, which means we would have to supply that energy to the spaceship somehow. We shall consider this matter more specifically in Act V; meanwhile, remember that we cannot create energy out of nothing, even with relativity!

Relativistic Velocity

We have already mentioned that if you and I are in relative uniform motion and if we both observe a third moving body, we will disagree on how fast *it* is moving. The elementary mathematics of special relativity tells us how to calculate the velocity I will observe for the body, given the velocity you measure for it in your moving system, and vice versa. But here's an important point. Neither of us, and no one else, will *ever* see it moving as fast as light. No matter how close it is to the speed of light in *your* system, and no matter how fast you are moving with respect to me (always less than *c*, of course), I will nevertheless *always* find that its speed is less than that of light. Here's an example: You are in a spaceship that takes off in one direction from earth at 90 percent of *c*. I, remaining on earth, now fire a missile in the opposite direction from the way you went, also at a speed of 90 percent of *c*. Common sense would tell us that you should see the missile moving away from you at a speed of 180 percent that of light. But you have read too far in this scene to believe in old-fashioned common sense any more! In fact, if you work it out, you find that you will see the missile moving at 99.45 percent of *c*.

In order to move with the speed of light, the momentum of a material body (and its effective mass) would have to be infinite, as would its energy; it would take more energy than there is in the universe to get it moving that fast! The only thing that can move with the speed of light is something with *no* rest mass. Two things that have energy but no mass are photons and neutrinos (as promised before, we shall come back to neutrinos later). In other words, light itself and other electromagnetic energy (photons) *can* travel with the speed of light. In fact, they *must,* for anything with no rest mass, if traveling less than the speed of light, could have no energy. To repeat: No material body (one with nonzero rest mass) can *ever* travel with the speed of light. Photons and neutrinos, with *no* rest mass, *must* travel with the speed of light.

No material body can ever go as fast as light, let alone pass through it to higher speeds. Yet, time and again people tell me that surely science will find a way to do it someday. They read science fiction stories by the wrong authors who talk about breaking the light barrier. Dear readers, it is *not* a matter of technology; it is *not* that we do not, today, know how. I believe that we can travel to Mars if we want to, and suspect that it is even possible (although difficult) for man to travel to the stars. But man going faster than light is *not* possible. It is as fundamental a principle in physics as that 2 and 2 are 4. Even though special relativity may someday be superseded by a more accurate theory of motions, it will not be proved fundamentally wrong in areas where it has been well tested and known to apply, any more than general relativity (next scene) has thrown out Newton!

But do not let this dishearten you. Although, as we shall see later, the energy requirements for travel near the speed of light are prohibitive, in *principle* one can travel at a speed as close to that of light as he wishes, but just not quite *at c*. The closer he is to the speed of light, the more his time slows down. He can make any trip to any place in the universe as short as he wishes by merely approaching closely enough to the speed of light. If the trip to the most remote quasar could be made in one second, by going *almost* as fast as light—a speed theoretically (if not practically) possible—why would you want to go faster than light, which is impossible and which would mess up the trip anyway? Indeed, if an imaginary, massless astronaut could travel *at* the speed of light (say, riding on a photon), his time would stand still, and he would be everywhere in the universe at the same instant by his own clock.

Some of these concepts are probably new to you and may seem mysterious, but only because of our lifetimes of experience at low speeds. The baby with a rubber ball, placed in an environment of high velocity, would doubtless adjust to the concepts of special relativity without difficulty. Rest assured, though, that the special theory is not controversial today. It is extremely well tested, and no respectable physicist questions its validity. It is fundamental and central, and absolutely vital in our modern technology.

SCENE 5

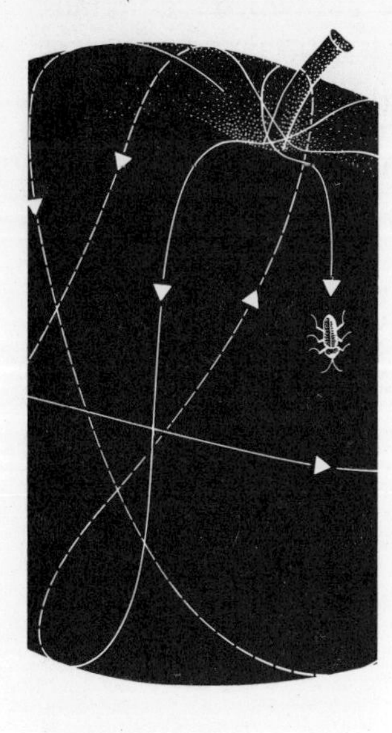

General Relativity; A New Meaning of Gravitation

A young couple on the beach are playing catch with a beach ball; they are shown in Figure II.35. Everyone knows that if the boy throws the ball straight toward his friend (along path *A*) it will not reach her, because gravity will make it fall to the ground somewhere in front of her. The boy knows it, too, so he aims the ball somewhat above her head; as it falls away from its straight line path, it moves along route *B* and reaches her arms.

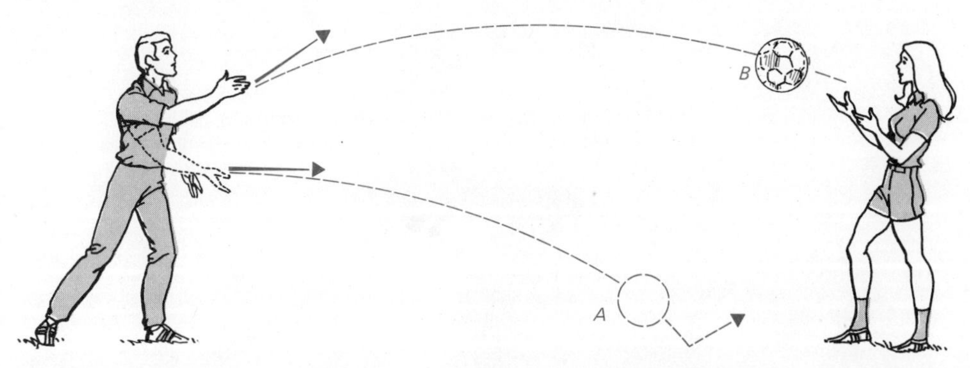

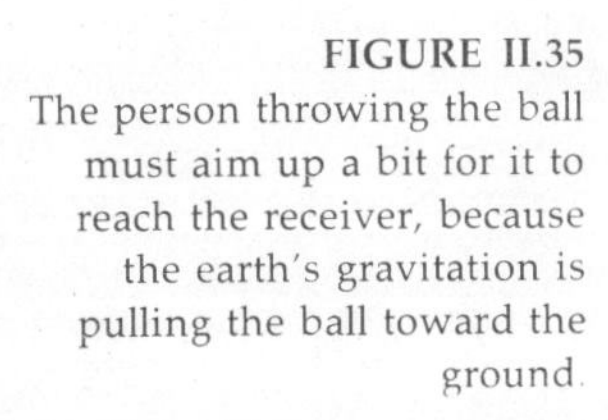

FIGURE II.35
The person throwing the ball must aim up a bit for it to reach the receiver, because the earth's gravitation is pulling the ball toward the ground.

Suppose our couple walk over to an enormously deep chasm, and they stand on the opposite cliffs of the canyon facing each other, as in Figure II.36. We now make two unreal assumptions: first, that there is no air inside the chasm to impede motions, and second, that our friends are devoted enough to science to carry out the following experiment: At the same instant, they both jump off their cliffs, into the chasm, with the boy carrying the ball. When at position *A*, he throws it, but this time aiming it directly at her. Now, he, she, and the ball all fall at the same rate (remember Galileo?), so when she has reached point *B*, the ball is there at the same time and meets her. She immediately throws it back, aiming straight at her friend, and it leaves her hand when she is at *C*. All three are accelerating, so they are falling faster now, but still together, and the ball returns to the boy at *D*.

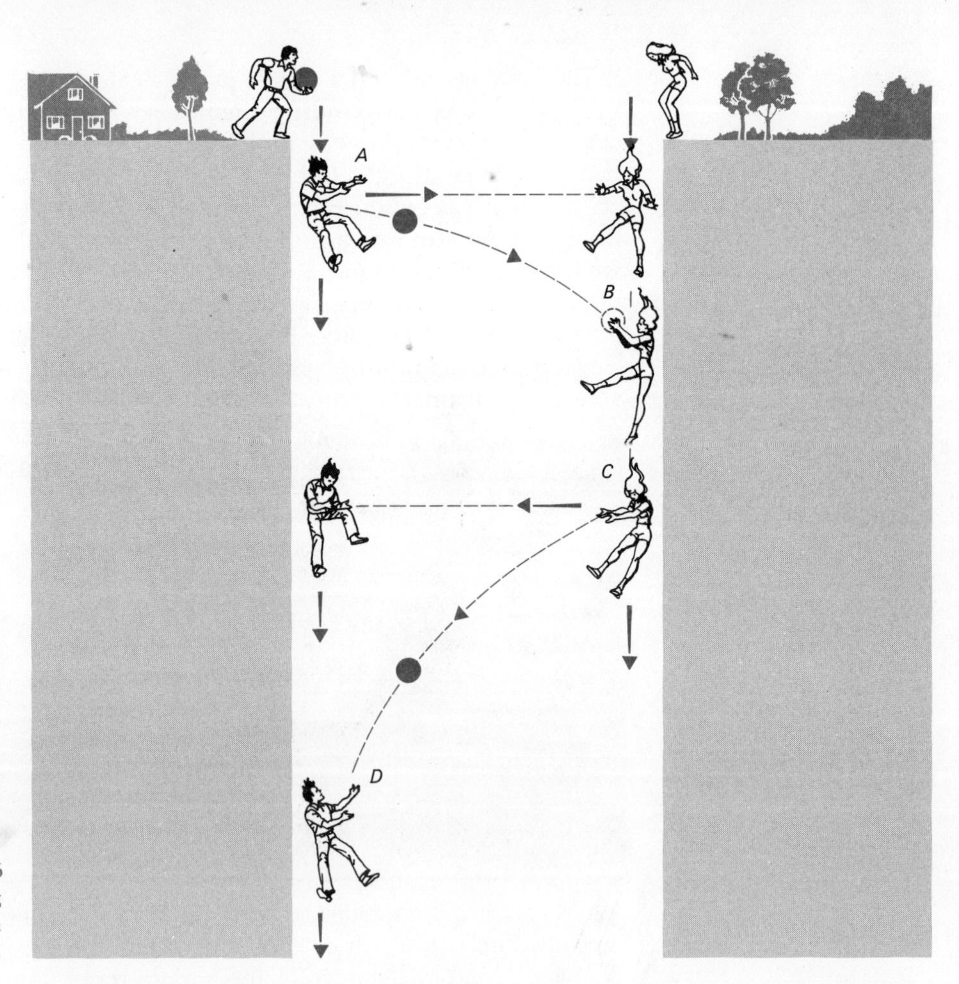

FIGURE II.36
The brave couple playing catch as they descend into a bottomless abyss.

---

1. A very intelligent monkey, hanging from the limb of a tree, sees a hunter aiming a rifle directly at him. The monkey knows that light travels faster than any bullet. Thus he waits until he sees the smoke from the gun, indicating that it has been fired, and at that instant lets go of the limb and drops, expecting the bullet to pass harmlessly over his head. Is the monkey smart enough to save himself with this idea? Explain.

Accelerating Coordinate Systems

When you are standing or sitting, you feel your own weight. The various organs of your body are pulling on their respective membranes, and your muscles are straining to keep you upright (straining more with some of us than with others). But when we are in free fall (as are the boy and girl now somewhere deep in that chasm), all parts of us are falling together, and we no longer feel that tugging and straining. We *feel* weightless. We are not really weightless, for the earth is still pulling on us; but since we are accelerating downward with that pull, the internal effects within our body are no longer present. We can think of the body as defining a coordinate system for those internal organs; by suitably accelerating that coordinate system we have removed the effects of gravitation.

The same is true of the couple playing catch in the chasm. We could imagine a huge box surrounding them, also in free fall. Inside that box (the couple's coordinate system, in this case) they feel no gravitational effects. The ball goes in a straight line at a constant speed with respect to the floor, ceiling, and walls of the box; it is the same as if it were far off in space, remote from all gravitating bodies. Because the box and all in it are falling together, we have, in effect, found a way to accelerate a coordinate system to remove the force of gravitation—at least within it.

You can also create forces by accelerating your coordinate system. When an elevator ascends rapidly, you feel an extra weight. At the amusement park, you can lie against the inside vertical wall of a rotating cylindrical room; its rotation creates a *centrifugal force* that pushes you outward. The centrifrugal force does not exist outside the rotating room; an outside observer would say that you are trying to fly off in a straight line, tangent to the cylinder, but that the wall keeps pushing you inward toward the center, making you describe a circular path. But inside, the rotation (hence acceleration) of the coordinate system has created an outward force.

If we can remove the effects of a force by suitably accelerating, or can create one by a different acceleration, a force must be equivalent to an acceleration of the coordinate system. This concept is called the *principle of equivalence.* It is the basis of the *general theory of relativity*. Einstein showed how to choose a proper coordinate system to remove the force of gravitation. All motions within that system are then unaccelerated; bodies all move in a straight line in accord with Newton's first law of motion (or more precisely, if speeds are high, according to the rules of special relativity).

---

2. A cork is connected by a spring to the inside of the bottom of a bucket of water. The cork tries to float upward, and stretches the spring somewhat, but not enough for the cork to reach the surface of the water. Now, suppose the bucket is dropped vertically down an elevator shaft. What happens to the cork and spring, and why?

An interesting *real* example of how an acceleration can remove the local effects of gravitation is provided by the Skylab. Astronauts spent months at a time in this orbiting laboratory in 1973 and 1974. The Skylab, they, and everything else inside it were in free fall about the earth; thus the astronauts felt weightless, and everything in the laboratory behaved as if there were no gravitation. (The tidal forces of the earth on different parts of the vehicle, as well as the mutual attractions of the small items on each other and on the walls of the lab, were too small to be noticeable.) An article sent in motion moved in a straight line at a constant speed, until it hit a wall. An item placed in midair remained there, where it was put.

The wife of one of the astronauts told of an amusing incident soon after her husband had returned from space. After completing his morning shave, he was applying some lotion to his face. When he was done, he laid the bottle down midair in the bathroom, whereupon it crashed to the floor. A wife would notice that.

---

3. Some of the Skylab astronauts exercised by running around the inside wall of their cylindrical vehicle. How could they stay against the wall while running, rather than float aimlessly inside the Skylab?

FIGURE II.37
In the Skylab everything stays put or moves uniformly because there is no apparent gravitation acting inside the laboratory. *(NASA)*

In the example of Skylab, gravitation has not really been removed, because it and its contents were still revolving about the earth. The problem is to find how to transform to a coordinate system where all gravitational forces are absent, so that everything travels in a straight line within that system. By a "straight line," of course, we mean along a path traveled by light. But note that if, in order to remove the effects of gravitational forces, we transform one coordinate system to another, the path of light must be different in the two. Thus light itself (or electromagnetic radiation in general) must be affected by gravitation. This is something *not* predicted in the Newtonian theory.

Geodesics in Spacetime

Now, how do we find a system of coordinates in which the path of a beach ball, and the path of the earth around the sun, are both the same as the path of light? Einstein found that it cannot be done in space alone, but it can be done in *spacetime*. The path an unaccelerated object follows in spacetime is called its *geodesic*. Let us look a little more at the nature of time as a coordinate in four-dimensional spacetime.

When a draftsman wants to represent a three-dimensional object on a plane, he projects it. For example, he can show the floor plan of a house, or, on a separate drawing, an elevation—that is, a side or end view of the house. We do the same sort of thing in Figure II.38, which is a two-dimensional projection of four-dimensional spacetime. On a plane, we could show any two dimensions of space, or one of space and one of time. We do the latter. Along the horizontal axis is one of three coordinates of distance (or space); I have labeled the distance "forward" to the right, and "back" to the left. There are also right and left and up and down coordinates, but I cannot show them on this diagram. Time is along the vertical axis.

The units we use are arbitrary. For a unit of distance I have chosen the distance that light goes in one unit of time. Thus if time is in seconds, distance is in units of 300,000 km. With these particular units the diagram has convenient proportions, for light travels at 45° lines (in one unit of time, it goes one unit of distance; in five time units, it goes five distance units, and so on).

The origin, *O*, of the plot (where the axes intersect) is here and now. The future is upward along the vertical axis. Thus, if we do not move in space, we are still moving in spacetime; our geodesic is just up along the increasing

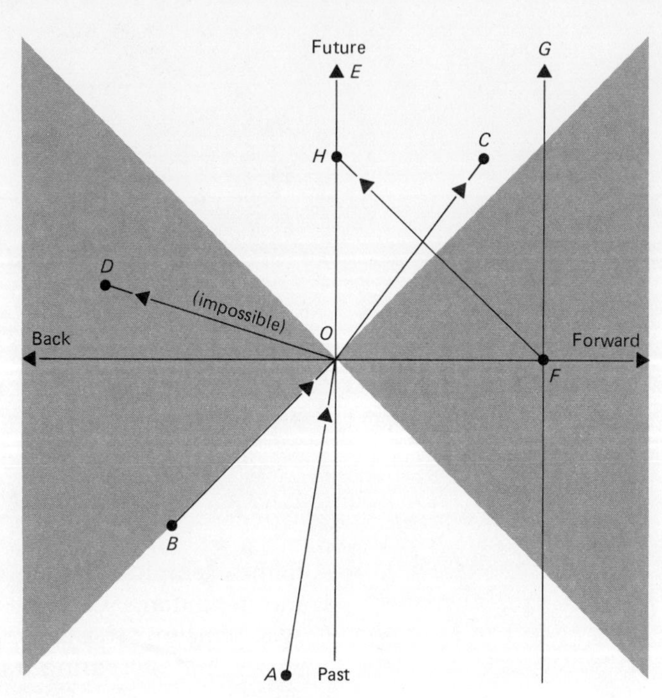

FIGURE II.38
A spacetime diagram.

time axis—path E in the figure. Another person may be standing far in front of us, and at this instant be at F. His geodesic is along path G. The dashed lines starting from O and going into the future (upper right and upper left) are two of the many possible geodesics followed by light rays that we now shine into space. Because light must go in 45° on the diagram (with our units), the dashed lines are the only two of those light paths that I can fit on this picture. The dashed lines coming up to O from the past (lower right and lower left) are two of the many possible geodesics for light rays that are just now reaching us—say from stars we now observe. We see stars, of course, in all directions in space, but because I can't show more than the straight forward or straight back directions in this projection, those are the only stars whose approaching light rays fit on this picture.

For example, suppose a star emitted some light from point B in spacetime. The light from it is just now reaching us along the geodesic from B to O. Or suppose a spaceship from an alien civilization started toward us from point A in spacetime. It just now gets here along the geodesic from A to O. In principle, it is possible for us to launch a spaceship now, which could travel along the geodesic from O to C, reaching a point in space some distance in front of us at some time in the future. However, we could *not* launch another ship (in this case in the opposite direction) along the path from O to D; to reach that point so near in the future, it would have to travel to a greater distance than light could travel in the same time—that is, it would have to have a speed greater than that of light. In other words, we cannot communicate with any of the points in spacetime that lie in the shaded regions of the figure. In particular, we cannot know what the person at F is doing right now; if he sends us a radio message, we will not receive it until we are at point H in the future, by which time his radio waves (at the speed of light) have traveled along the path from F to H.

I hope the above remarks take a little mystery out of spacetime. For a more earthly example, consider the part of a baseball game in spacetime shown in Figure II.39. It is an attempt to show three of the four dimensions in a

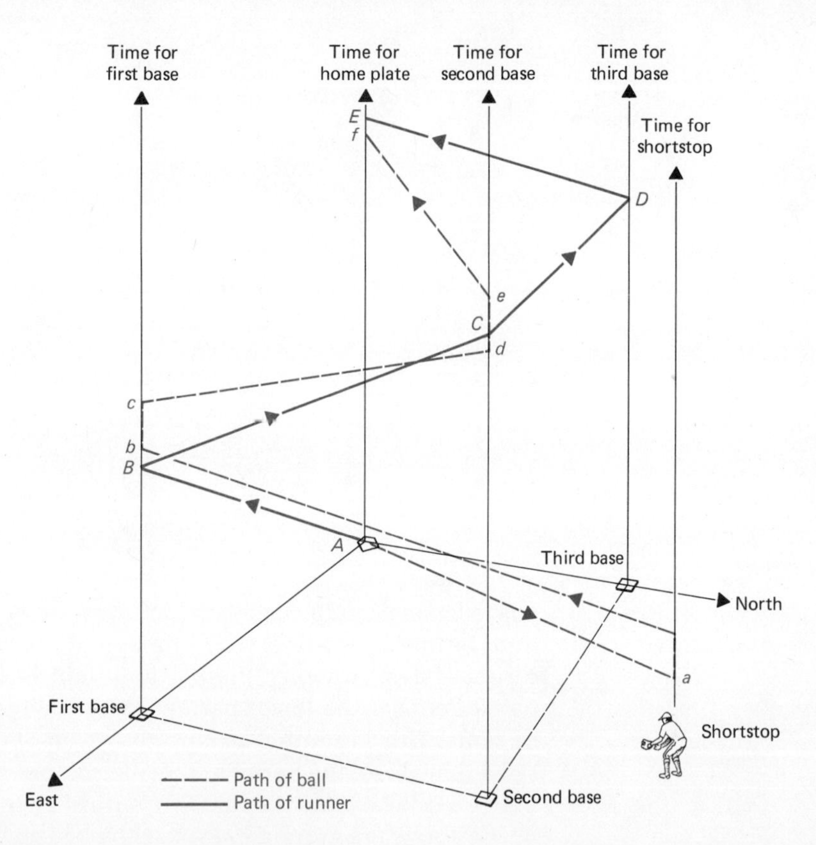

FIGURE II.39
A baseball play in spacetime.

projection. The diamond is laid out with first base east of home plate and third base to the north. Time increases upward in the figure. Geodesics for the bases and the shortstop, all of which do not move in space (in this case), are shown as vertical lines.

The batter, at *A*, hits a grounder to the shortstop, and it reaches him at *a*. The batter runs to first base, getting there at *B*, just ahead of the shortstop's throw, which reaches there at *b*. But the first baseman bobbles the ball, so the runner moves on to second, getting there at *C*. The first baseman threw the ball, at *c*, to second, and it actually beat the runner, arriving at *d*. However, the second baseman drops the ball, so the runner goes on to third, at *D*, is waved on home, and arrives at *E*. But the second baseman recovers, and at *e* throws the ball to home, where it arrives at *f*—just ahead of the runner, who is tagged out. You see how spacetime can help you appreciate the game!

---

4. Represent the following trip by a geodesic spacetime diagram: A salesman starts at Palmburg and drives due north on a straight highway. For the first hour he drives 50 km/hr. He then drives 180 km during the next two hours. Finally, he drives 160 km at 80 km/hr.

5. Make up a new example of a geodesic in spacetime and show it on a spacetime diagram.

Let us reconsider the couple playing catch. Figure II.35 is partly reproduced in Figure II.40. Note that the ball, to arrive at the girl, must follow quite a curved path. Now even though light may fall to the earth with the same acceleration as the ball, it travels so fast that it reaches the girl before it

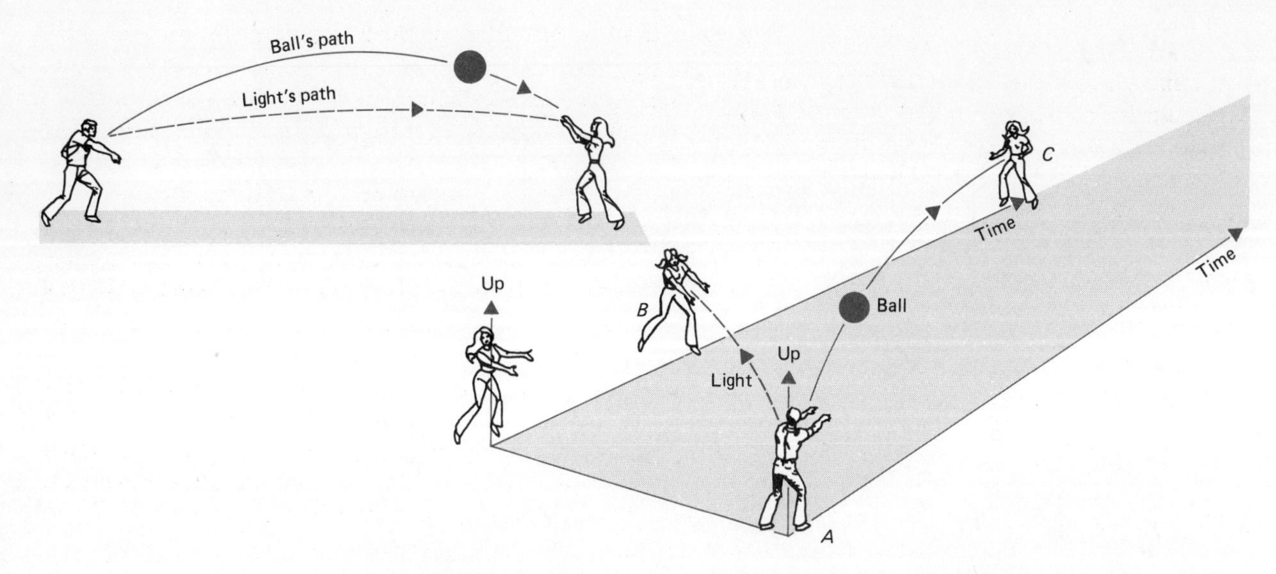

FIGURE II.40
In spacetime the ball and light follow paths of the same curvature, although of different lengths.

has fallen measurably at all; in fact, we grossly exaggerate the curvature of the light beam (dotted line) from the boy to the girl. How can we define a coordinate system in which the ball and light beam both have the same shaped path? Perhaps the lower part of Figure II.40 can give us an idea how it can be done. This is another perspective view of three of the four dimensions of spacetime. Here the horizontal plane contains the time axis, and the distance between the boy and girl and the vertical axis is the height of the ball, or light beam, above the ground. Suppose the ball and light beam both leave the boy at *A*. The light beam reaches the girl very quickly; in that time she has moved, along the time line, only as far as to *B*. The ball, going much slower, does not reach her until she is at *C*, so it follows a very long path in spacetime. Now the curvature of the light path, from *A* to *B*, is the same as that of the much longer path of the ball, from *A* to *C*. Because of their different speeds, the light and ball travel along different geodesics, but they are equally "straight."

6. The earth moves in its nearly circular orbit of 1 AU radius at a speed of only 1/10,000 that of light. What is the radius of the circle that most nearly matches the path of a beam of light passing the sun at the earth's distance from it?

The Warping of Spacetime by Matter

In his *general theory of relativity*, published in 1916, Einstein simply replaced the effect of gravitational forces by choosing the right coordinate system, namely, spacetime warped by the presence of matter. Near a massive dense object, where the Newtonian gravitational force would be very strong, spacetime is strongly warped. Far from large masses, the warping is very slight. Within this matter-distorted spacetime, all objects move in the most natural possible way, unaccelerated and in paths as straight as light paths. In a sufficiently local region of spacetime (inside the Skylab, for example) those paths of particles describe straght lines for which Euclid's geometry works. But in general, in the presence of matter, spacetime is non-Euclidean, and over large distances parallel lines may meet or may diverge, and areas of circles may be greater than or less than pi times the squares of their radii.

Einstein derived the so-called *field equations,* which tell how matter warps spacetime, and the *geodesic equations,* which tell how to calculate the paths of particles (geodesics) within the warped spacetime. The mathematics of the general theory is somewhat difficult, and is best handled with the use of *tensor analysis.* The results, though, are beautiful and, in a way, simple.

There is a very nice analogy at the beginning of the book *Gravitation,* by Misner, Thorne, and Wheeler. I hope those authors will not mind if I steal their idea and show it in Figure II.41. An ant is crawling about on the surface of an apple. The ant is blind and has no idea what the shape of the apple is, but it has an excellent sense of direction, and at all points on the surface of the apple it walks in an absolutely straight line. If the apple were perfectly spherical, the ant's path would be a closed great circle. But the apple is a deformed sphere, with depressions at either end of its core. Thus the actual path of the ant, to an exterior observer, is a complicated one that crosses on itself many times. The ant's spacetime—the surface of the apple—is warped in a complex way, but his path is as natural and unaccelerated as it can be.

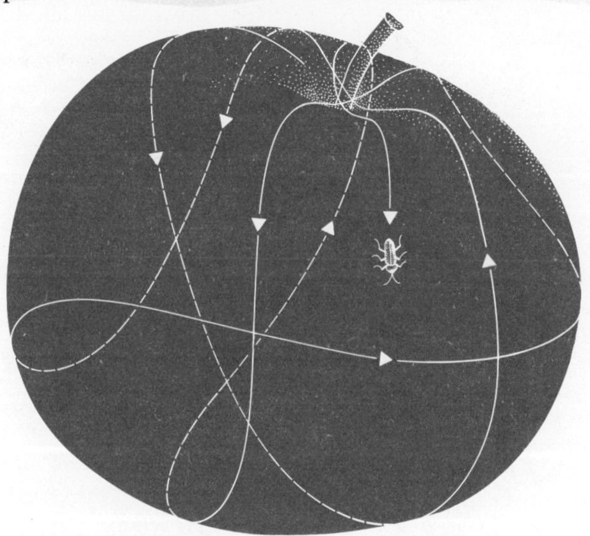

FIGURE II.41
An ant following a geodesic on the apple.

Relativity replaces complex motions, under the actions of forces in a Euclidean world, by simple motions, with no forces, but in a more complicated coordinate system. But, you may well be wondering, does relativity tell us anything new or different, or is it just an equivalent way of looking at Newtonian gravitation? In most familiar applications, such as the orbits of planets, the two theories are essentially equivalent, and in most cases Newton's version gives the same results to measured accuracy. There are easily noticeable differences, however, where the Newtonian gravitational field is very strong, or, in the relativistic language, where spacetime is strongly warped. For example, light itself is affected by the presence of mass, according to relativity. Also, gravitational signals travel at the finite speed of light.

Gravitational Waves

What do we mean by gravitational signals? If bodies move unaccelerated through spacetime warped by matter, what signal is there? The signal occurs when the arrangement of matter changes, for any change in the mass distribution must result in a *rewarping* of spacetime.[1] But it does not rewarp instantaneously over the entire universe; the rearrangement of spacetime occurs in a wave that spreads out from the disturbance (the rearrangement of the matter) with the speed of light. Such waves are called *gravitational waves*.

If you stamp your foot, wave your hand, or throw a ball, you are, technically, generating gravitational waves, because you are moving matter around. Recall, however, how incredibly weak the gravitational force actually is; it is correspondingly very difficult to detect gravitational waves—they involve very small energies if generated by small masses being shoved around. On the other hand, the sudden gravitational collapse somewhere in the Galaxy of a great amount of matter, corresponding to that of many stars, should generate detectable gravitational waves. Several investigators are currently trying to detect gravitational waves.

Tests of General Relativity

Einstein himself suggested three astronomical tests of general relativity. These are the so-called "classical" tests. One prediction of relativity is that rays of light passing near the sun's surface should be deflected (by the relatively strong warping of spacetime there) by 1″.75. Thus a star seen near the limb of the sun in the sky should be displaced by this angle from its direction at other times of the year when the sun is elsewhere on the ecliptic (Figure II.42). From the earth's surface, stars can be observed near the sun only during total solar eclipses. Photographs of the sky around the sun were taken at the solar eclipse of 1919 and at almost every eclipse since. Positions of the stars on the photographs are carefully measured and compared with their positions on other photographs taken at a different time of the year. The expected displacements are very small, and thus are hard to measure, but nevertheless they are found, and to the accuracy of the measurements, the predictions of relativity are confirmed.

A second prediction of Einstein's regards the orbit of Mercury. According to Newtonian theory, if the sun and Mercury were the only two bodies in the universe, Mercury's orbit would be a permanent and stationary ellipse in space. However, the mutual perturbative gravitational forces of the planets on each other cause their orbits to change slowly. Recall that it was the analysis of just such changes in the orbit of Uranus that led to the discovery of Neptune. One kind of change is a slow rotation in space of the longest diam-

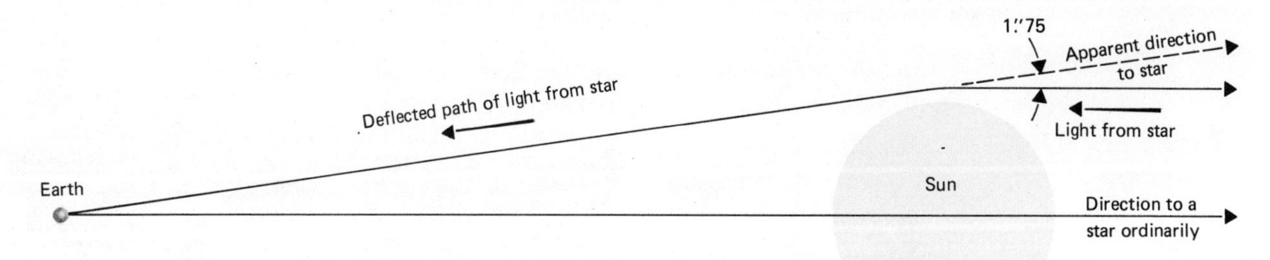

FIGURE II.42 Deflection of starlight passing near the sun.

[1] Unless the mass distribution changes perfectly symmetrically about a point.

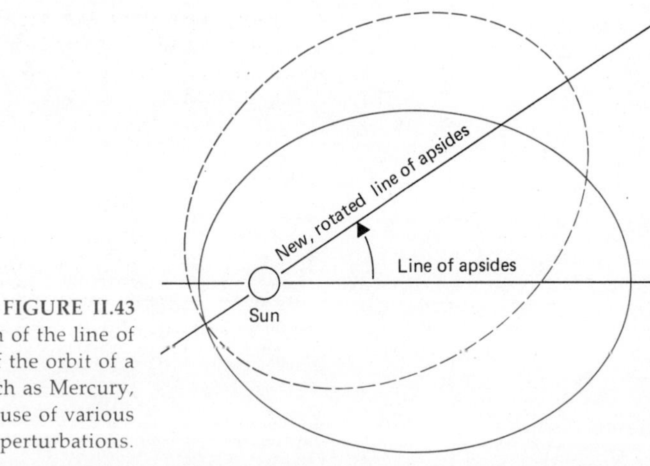

FIGURE II.43
Rotation of the line of apsides of the orbit of a planet, such as Mercury, because of various perturbations.

eter of a planet's orbit—its *line of apsides* (see Figure II.43). Now the line of apsides of the orbit of Mercury rotates 574″ per century. Computation of the gravitational effects of the other planets, however, showed that it should rotate only 531″ per century—a discrepancy of 43″. Relativity theory, on the other hand, predicts a slightly different rotation rate for Mercury's line of apsides than Newton's laws do, and completely removes the 43″-per-century discrepancy.

Einstein's third astronomical test is that there should be a *gravitational redshift.* If we shoot a rocket into the sky, as it rises against gravity it loses energy and slows down. Now if photons are similarly pulled on by gravitation, they too must lose energy as they leave a star. They cannot, on the other hand, slow down, and to lose energy they must decrease in frequency or increase in wavelength. Thus the spectral lines of a star should show shifts to longer wavelength—the gravitational redshift. The effect is too small to observe for ordinary stars like the sun. There exist, however, small dense *white dwarf* stars with very much stronger surface gravitational fields (we shall discuss them in Act IV). A few white dwarfs are members of binary-star systems in which their companion stars are more or less like the sun, and are expected to have negligible gravitational redshifts. In such a case, we can tell, from the motion of the other star, what the radial velocity of the white dwarf must be, and any additional redshift must be due to gravitation. The first significant confirmation that the gravitational redshift is really present was made by D. M. Popper in 1954, from his analysis of the white-dwarf companion of the star 40 Eridani.

The gravitational redshift can now be observed completely in the laboratory. Modern techniques allow us to measure very accurately the frequency of gamma rays emitted from the nuclei of certain radioactive elements. Today, comparison of the emitted frequency (or wavelength) of such gamma rays with the frequency they have after traveling upward only a few tens of meters against the earth's gravity can be made with sufficient precision to verify the gravitational shift.

So all of Einstein's predictions for general relativity have now been confirmed. The theory, however, is not confirmed to the same extent that the special theory of relativity has been. In particular, there are some more recent alternate relativity theories that predict very nearly the same results. The best known of these competing theories is the *scalar-tensor field theory* of Brans

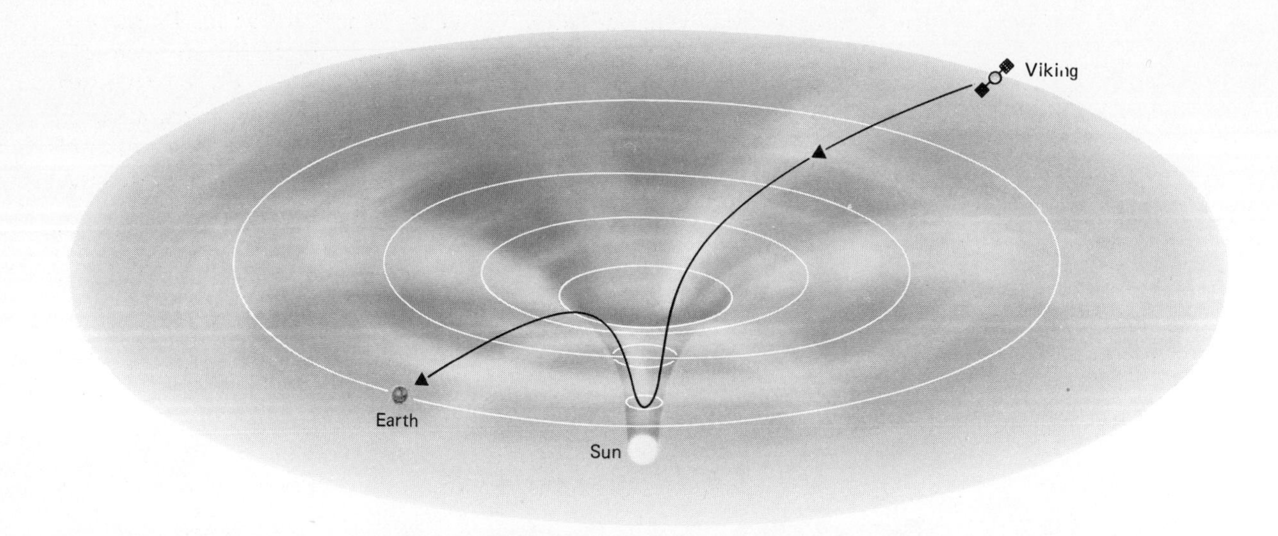

FIGURE II.44
Radio signals from Viking are delayed because they have to pass near the sun where spacetime is relatively strongly warped.

and Dicke. Extremely sensitive experiments are required to discriminate between the different predictions of these alternate relativity theories.

One such test is shown schematically in Figure II.44. In 1976 Vikings 1 and 2 space laboratories were landed on Mars, while their parent ships remained in orbit about the planet. While those space probes transmitted much valuable data back to earth, Mars, in its orbit, passed nearly behind the sun as seen from earth. Thus, for us to receive the radio signals, those waves had to pass rather close to the sun, where it rather significantly warps spacetime. In the figure, that warping is represented by a depression, in the neighborhood of the sun, in the plane containing Mars and the earth. It is somewhat like a heavy stone placed in the middle of a rubber sheet stretched on a circular frame. The path of the radio waves, traveling in warped spacetime, is analogous to that along the depression in the stretched rubber surface; the waves had farther to go than they would have if the sun were not there.

Consequently, the radio signals from the Viking probes began arriving at earth at slightly delayed times, compared to the times predicted if spacetime were flat and Newtonian. The delay actually observed was as great as 200 microseconds, and was within 0.1 percent of that predicted by the Einstein theory.

So far, the weight of the experimental and astronomical evidence favors the Einstein theory of relativity over its rivals. Although modern physicists would not care to commit themselves publicly until the case is airtight, I think most regard Einstein's general theory of relativity as the best description yet discovered of the behavior of gravitation.

Today, there is a relatively new field of astronomy, that of *relativistic astrophysics,* which deals with problems in which the general theory of relativity makes a difference. And those are some of the most exciting problems at the very forefront of modern astronomy! They include the theory of *black holes,* which we discuss in Act IV, and of the evolution of the entire universe, which we take up in Act VI.

Almost always many men play key roles in the advance of new ideas. Still, here and there are the names of people who by any standard must be regarded as great, especially in their impact in their field. It's a good party game to decide on one's favorites—Hipparchus, Copernicus, Bach, Shakespeare, Newton, Maxwell, da Vinci, Napoleon, Gauss, Schubert, Franklin—we could carry on worthy debates for hours on the subject. But in twentieth century physics, one outstanding name, I think, would be on everybody's list, or anyway almost everybody's, even including those who know absolutely nothing about what he did. That name is Albert Einstein. He was, and is, not only respected by scientists, but loved by his friends and associates, and he never lost his humility right up to his death in Princeton in 1955.

ACT III

THE DEPTHS OF SPACE

SCENE 1

Mapping Our Own Neighborhood

To the unaided eye the stars and planets appear as tiny points of light in the sky. Yet we know much about their distances, sizes, masses, temperatures, ages, what they are made of, how they are moving, and even a good deal about their interiors. It is a grand example of detective work to see how we analyze the often subtle clues Nature has given us to so chart the universe.

Position on Earth

We must start our mapping at home, because each step out builds on the previous one. Today we are all used to detailed maps of the earth, but those maps did not always exist. The ancients, as we have seen, knew that the earth is spherical, and even its size was rather well known by the end of the third century B.C. But to draw those first maps, somebody had to figure out where on earth he was.

Finding your latitude is pretty straightforward. In Scene 1 of Act I we saw that the whole aspect of the sky depends on your latitude. For example, if you can locate the north or south celestial pole in the sky at night and measure its altitude above the horizon, you know your latitude at once. In the daytime you need merely see how far north or south of the zenith the sun crosses the meridian about noon. In fact, navigators commonly use noon sun sightings to find their latitudes at sea.

---

1. If the sun is observed to pass through the zenith at noon on March 21, what is the latitude?

2. What is the latitude if the sun passes through the zenith at noon on June 22?

But longitude is another matter. It requires knowing the time at some particular place on earth. Because the earth is spherical, it is obvious that time is different at different places around the world. One side has day and the other night. If it is noon where you are, it must be midnight on the opposite side of the world. We normally base our time on where the sun is, but actually time could be based on *any* celestial object or on just a point in the sky. Indeed, astronomers use a kind of time based on the vernal equinox, called *sidereal time.*

Time of Day

Time—any kind of time—is measured by the *hour angle* of the object it is based on. Sidereal time, for example, is the hour angle of the vernal equinox. Hour angle simply tells us how far around the sky the object is from our local meridian—that imaginary circle in the sky running north and south through the zenith. The sun is east of the meridian in the morning and west of it in the afternoon. When something is *on* the meridian, its hour angle is zero. At noon, as indicated by a sundial, the sun's hour angle is zero.

When something is exactly halfway around the sky from the meridian, where the sun is at midnight, its hour angle is 180°. By convention, however, we measure hour angle in hours, minutes, and seconds of time. In this notation, 1 hour is the same as 15°, 6 hours is 90°, 18 hours is 270°, and 24 hours is a complete circle of 360°. A minute of time is, of course, 1/60 of an hour, and a second of time 1/60 of a minute. So if, say, the vernal equinox is setting due west, its hour angle is 6 hours.

---

3. What is the sidereal time when the vernal equinox rises?
4. If it is 3:00 P.M. local apparent time, what is the hour angle of the sun?

Time of day, of course, is based, not on the vernal equinox but on the sun. For obvious reasons of convenience, however, time by the sun is not the hour angle of the sun, but its hour angle plus 12 hours. That makes the day start at midnight, rather than at noon. The hour angle of the sun would give us the P.M. time. But this kind of time—the hour angle of the sun plus 12 hours—is not the time your watch keeps; it is the kind of time a sundial keeps, and is called *local apparent time.* We don't use local apparent time anymore because it's rather irregular.

The reason apparent solar time is irregular is that solar days (each consisting of 24 apparent solar hours) are not all of exactly the same length. To be sure, the earth rotates at a very nearly uniform rate. A solar day, however, is longer than a rotation period of the earth, because while the earth rotates it also revolves about the sun. Suppose we started counting a day when the earth is at *A* in Figure III.1, with the sun on the meridian of the observer shown. By the time the earth has completed one true rotation (say, with re-

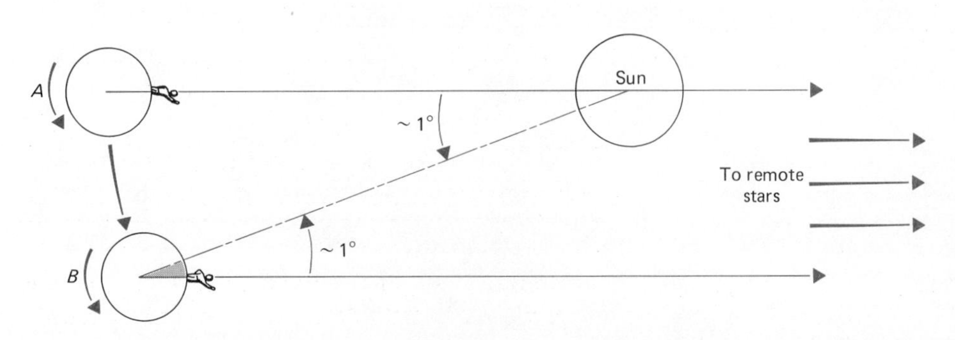

FIGURE III.1
A solar day is longer than a true rotation period of the earth.

spect to the remote stars), it has moved about 1° in its orbit to position *B*, and the sun is not yet on the meridian of our observer. The earth must rotate a bit more (about 1°) to bring the sun back to the meridian and complete a day. It takes the earth about 4 min to rotate through 1°, so a solar day is about 4 min longer than one rotation period of the earth.

But the exact interval by which a day is longer than one rotation period of the earth (it averages 3 min 56 s) varies from day to day. One reason is that the earth's orbital speed varies (in accord with Kepler's second law), so that its angular motion, with respect to the sun, is not uniform. Another reason is that whereas the earth *rotates* from west to east (by definition), the direction of its orbital motion varies during the year by up to 23½° away from due east, because of that tilt of its axis to the perpendicular to its orbital plane (that is, the obliquity). Because of these complications, the precise lengths of individual days can vary from the mean by up to half a minute. Moreover, for weeks at a time, days can be longer or shorter than average, so the difference between a uniformly running time and apparent solar time can accumulate to about a quarter of an hour.

Consequently, we base our time, not on the real sun but on a fictitious one, called the *mean sun*, which moves eastward around the celestial equator at an absolutely constant rate. We define *mean solar time* as the hour angle of the mean sun plus 12 hours.

Longitude

We have said already that time is different all around the globe. Now we shall be more specific. The difference between the longitudes of two places is precisely the difference between the times at those places. This statement is true for *any* kind of time—mean solar, apparent solar, sidereal—if that time is defined as the hour angle of some object. Look at Figure III.2. The difference in longitude between the two places *A* and *B* is just the arc along the equator between them (I have selected places on the equator just for simplicity, but that's not essential to the discussion). Now, again for simplicity, suppose the date is such that the sun is on the celestial equator, and let the sun have the direction shown. The directions of the zeniths of *A* and *B* are also

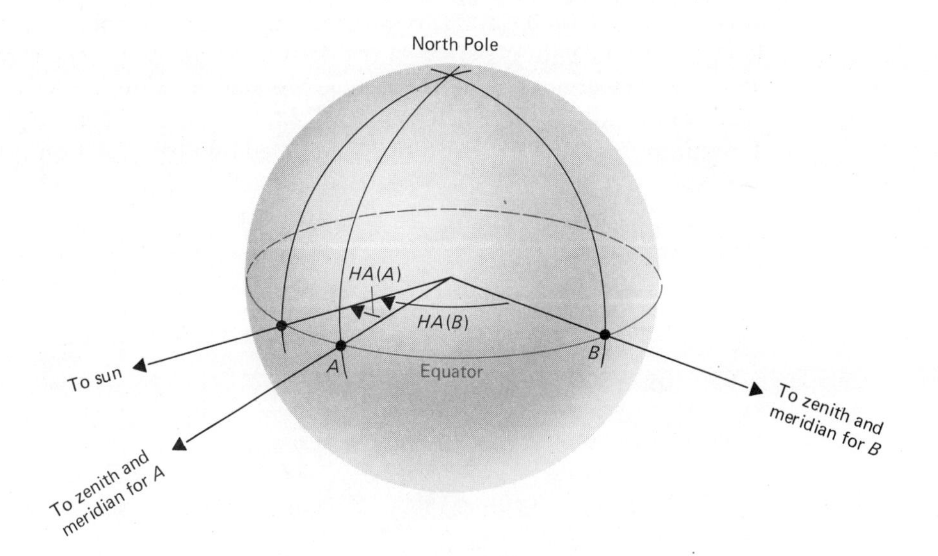

FIGURE III.2 The difference in the longitude of two places is equal to the difference between their times.

shown, and of course the meridian of each goes through its zenith. The hour angle of the sun as seen from A is shown as $HA(A)$ on the picture, and that as seen from B as $HA(B)$. You see immediately that the difference between those hour angles is just the difference between the longitudes of A and B.

So the trick to finding your longitude is to have a good clock that keeps the time of someplace you know—say, Greenwich, England. You can always find your local time by observing the hour angle of the sun and then correcting your local apparent time to mean time by looking up in a standard reference what the difference between the two kinds of time is on that date. You can also find local time by noting what time various stars rise and set cross the meridian or reach certain altitudes in the sky. (The navigator knows all these tricks.) The difference between your local time and the time at Greenwich is your longitude.

In modern times you can always find Greenwich time from shortwave radio broadcasts; in the old days ships had to carry very accurate clocks, called *chronometers.* You can see why it was so important to navigation in the seventeenth through nineteenth centuries to improve the technology of making clocks. You have probably been amused at the funny maps of the New World made in the 1700s. The reason those maps are so funny is that good clocks were not available in those days, and the explorers just didn't know their longitudes very well.

We're not quite yet done with time, though. Not only do our watches not keep apparent solar time; they don't keep mean solar time either. Until near the end of the previous century we did use mean solar time; this meant that every city had its own private time, which went with its own longitude. What if you live in Oyster Bay and commute daily to New York City? As you ride through the East River tunnel, you must set your watch back slightly, because time in Oyster Bay is 1.6 minutes more advanced than that in Manhattan.

Such details as this bothered the railroads, too, until 1883, when the United States divided itself into four time zones. Now everybody in the same zone keeps the time of the central meridian of longitude in that zone, and travelers need reset their watches only when crossing from one time zone to another; then the adjustment is an even hour. In 1884, at an international meeting of 26 nations, the same thing was done worldwide, so now the whole world is divided into standard time zones. Over land the zone boundaries are irregular, because they jog about to follow international and state boundaries. In the open sea, though, the zone boundaries are along meridians of longitude; at sea we call it *zone time,* on land we call it *standard time.*

To summarize: We keep time by the sun, but the sun is an irregular timekeeper; so we average sun time, and call it mean time, but mean time varies from place to place, so we standardize it.

---

5. San Francisco keeps the local time of the meridian of longitude 120°W. By how many hours is the time at Greenwich more advanced than at San Francisco?

6. If you observe the hour angle of Sirius to be 3 hours, but if Sirius is known, at that instant, to be on the meridian at Greenwich, what is your longitude?

7. You wish to telephone a friend in London, but do not want to bother him during the night hours when he would normally be sleeping. During what local times (for you) should you avoid placing the call?

Time Standards

We have now discussed time of day pretty thoroughly, but we have not indicated what we use as standards for the rate of passage of time. It used to be the rotation of the earth (with respect to the remote stars, of course, not the sun). The earth's rotation is pretty regular, but not perfectly so, and changes in it have been discovered in the twentieth century. These are of two types. First there are *periodic* changes, whereby the earth slows down for a while in its rotation and then speeds up for a while. These changes come about because of small expansions or contractions of the earth's crust or because of seasonal shifting of ice and snow deposits. They are very tiny changes, but they are measurable and can accumulate to clock errors of a few seconds one way or the other.

Moreover, superposed on these periodic changes is a very gradual slowing of the earth's rotation rate that has been going on for millions of years. The length of the day is actually increasing by about 0.0016 second per century. This small change was discovered because the clock error accumulates over many centuries, and after some 2000 years it amounts to a few hours. Now from gravitational theory we can calculate how the moon and planets are moving with great precision, which means we can also calculate exactly when and where eclipses of the sun should have occurred during antiquity. But comparison of calculations and ancient records showed that the timing was off by those few hours, and hence that the earth must be slowing down.

The reason for this lengthening of the day is that the earth is losing rotational energy in friction in the tides. The earth's angular momentum due to rotation is very gradually being transferred to the moon, with the result that the moon is slowly spiraling farther from the earth. It takes thousands of millions of years for the changes to be large, but nevertheless we can find better standards than the rotation of the earth.

One standard is gravitational theory itself. We simply keep track of the positions of the moon and planets, for they tell extremely accurately how much time has passed. Modern astronomical tables are based on this time found by keeping tabs on the motions of the planets. We call it *ephemeris time*. Ephemeris time is used as a constant check, but in everyday application we keep track of the passage of time with atomic clocks, and we call that *atomic time*. Atomic clocks operate on the principle that atoms emit and absorb radiation at certain discrete wavelengths or frequencies. Since 1967 the standard atomic clock has been the *cesium clock*. An *atomic second* is defined as the time it takes the isotope of cesium, whose atomic nuclei contain 55 protons and 78 neutrons, to emit 9,192,631,770 cycles of radio radiation. The accuracy of the cesium clock is claimed to be about 1 second in 6000 years.

Since 1 January 1972, international time has been coordinated by the *Bureau International de l'Heure (BIH)* in Paris. Seven member institutions, including the United States Naval Observatory and the National Bureau of Standards, contribute time data to the BIH, which in turn provides *Coordinated Universal Time (UTC)*. Coordinated Universal Time is the time at Greenwich, England, but it is regulated by atomic clocks and planetary observations. It is broadcast all over the world on shortwave radio. You may have picked it up yourself on station WWV or WWVH. When you dial for time on the phone, the time you are given, if the telephone company is on the ball, has also been checked against UTC.

FIGURE III.3
A modern cesium clock maintained by the National Bureau of Standards in Boulder, Colorado. *(National Bureau of Standards)*

But, alas, the earth keeps changing its rotation rate. What happens, I am sure you are wondering, when the difference between UTC and earth time accumulates to several hours, so that the sun crosses the meridian, not at noon (roughly) but, say, at 6 in the morning, or 5:30 in the afternoon? Well, the time people thought of that matter, too. The difference between time of day (at Greenwich) and the smoothly flowing atomic time is allowed to accumulate until it amounts to a full second. Then UTC is simply adjusted by that second to keep it in tune with where the sun is. By convention, these adjustments are made only at midnight in Greenwich on June 30 or December 31. The first such adjustment was made on 30 June 1972, when an extra atomic second was added, so that that day had 86,401 seconds, rather than the usual 86,400. Such "leap seconds" usually need to be added about once a year.

The Calendar

By the way, there are not an integral (whole) number of mean solar days in a year. In antiquity it was known that one year contains about 365¼ days. That is why Julius Caesar (on the advice of the astronomer Sosigenes) introduced, in 45 B.C., the convention of leap year. In this *Julian calendar* three years of 365 days each are followed by a leap year of 366 days, giving the year an average length of 365.25 days. In fact, however, the year of the seasons contains 365.242199 mean solar days, a bit less than 365.25. By the time of Pope Gregory XIII in the sixteenth century the error in the Julian calendar had accumulated to 10 days since the Council of Nicaea in 325 A.D., when the dates of Easter and certain other religious holidays were established. Gregory thus introduced a further calendar reform, which, over the next several centuries, was adopted throughout the world. The *Gregorian calendar* eliminates three leap years from the Julian calendar every 400 years. Under the Gregorian calendar rule, century years (like 1700, 1800, and 1900), all divisible by 4 and thus leap years in the Julian calendar, are *not* leap years unless they are also divisible by 400 (as are 2000 and 2400).

---

8. The American Colonies remained on the Julian calendar until 1752, when England finally adopted the calendar reform initiated by Pope Gregory in 1582. Moreover, at that time England and the Colonies had to adjust the date to bring their calendar into agreement with the new Gregorian calendar by eliminating eleven days. To make matters worse, until 1752 England still had followed the ancient practice of beginning the year on March 25 rather than on January 1. Now by the Gregorian calendar (in use today) George Washington was born on February 22, 1732. On the other hand, what date would have been indicated by a calendar on the wall of Washington's home at the time he was born?

Finding the Moon's Distance

We saw in the foregoing act how time is all mixed up with space; now we have seen that it is even involved in so mundane a thing as mapping the face of the earth. But having accomplished that, how do we go about mapping the rest of the universe?

The moon, our nearest natural neighbor, comes first. Its distance had been determined pretty accurately even by Hipparchus, and later also by Ptolemy. Figure III.4 illustrates a simple way of doing the job; it is similar, in principle, to the method used by Ptolemy. Suppose you saw the moon at your zenith. This means you must be on the line from the center of the earth *(C)* to the moon, namely, at *A* in the figure. A few hours later the rotation of the earth carries you to, say, position *O*. Now, you know the angle at the center of the earth between *A* (where you were) and *O* (where you are), because you know how much the earth has turned in those few hours. Moreover, you can observe the angle, z, between the moon's direction and the zenith, so you also know the angle at your location in the triangle made by the center of the earth, you, and the moon. Finally, you know the radius of the earth—the distance from C to *O*. This means you know two angles and an included side in the triangle, and you recall, no doubt, from high school geometry that everything else in the triangle is therefore determined, including the distance from

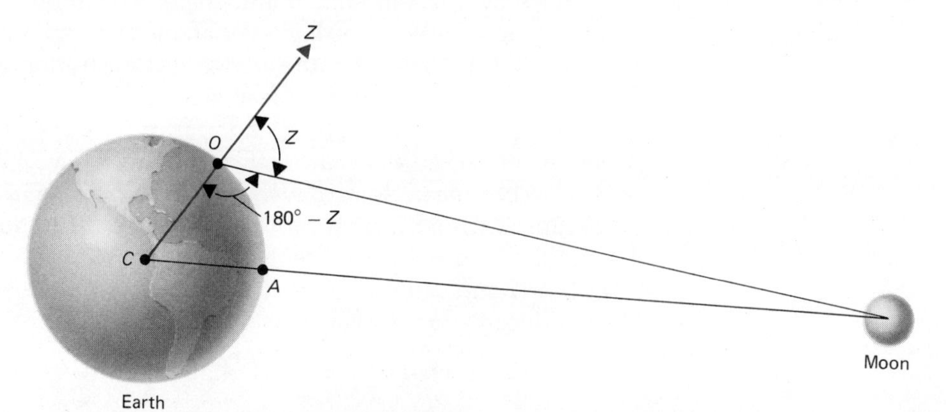

FIGURE III.4
Surveying the distance to the moon.

the center of the earth to the moon. There is one complication: The moon moved in its orbit while the rotation of the earth carried you from A to O. No problem though; you know how fast the moon moves, so you can correct for that motion.

The foregoing discussion has oversimplified and idealized the problem somewhat; for example, from most places on earth you will never see the moon at your zenith. Nevertheless, those of you who have studied trigonometry will see immediately how you could measure the moon's distance from your own back yard. The basic trick is that we are *surveying* the distance to the moon, and are using a large portion of the earth for a baseline.

Surveying the Solar System

The same can be done for the nearer planets, but only with a telescope. The planets are so much farther away than the moon is that even the full diameter of the earth provides too small a baseline to observe a shift in the direction of a planet (or parallax) with the unaided eye. For the planets we must use a portion of the *orbit* of the earth.

---

9. On the assumption that the earth and Venus have perfectly circular orbits centered on the sun, can you suggest a way of finding the radius of the orbit of Venus in terms of that of the earth? [Hint: Consider the situation when Venus appears at the largest possible angular separation from the sun.]

Kepler had to do just this to discover that the planets have elliptical orbits. We show the idea of Kepler's approach in Figure III.5. For simplicity, we assume the earth's orbit to be circular (Kepler used the actual shape, which he had determined) and we ignore the fact that the earth's and Mars' orbits are not quite in the same plane (Kepler took the different inclinations into account). Remember, Kepler had in his possession tables, compiled by Tycho Brahe and based on 20 years of observations, that gave for each date the angle between the sun and Mars, as seen from earth.

Suppose on a particular date the earth in its orbit is at the known position 1 in the figure ("known" because the earth's direction from the sun is just

FIGURE III.5
Kepler's method of determining the orbit of Mars.

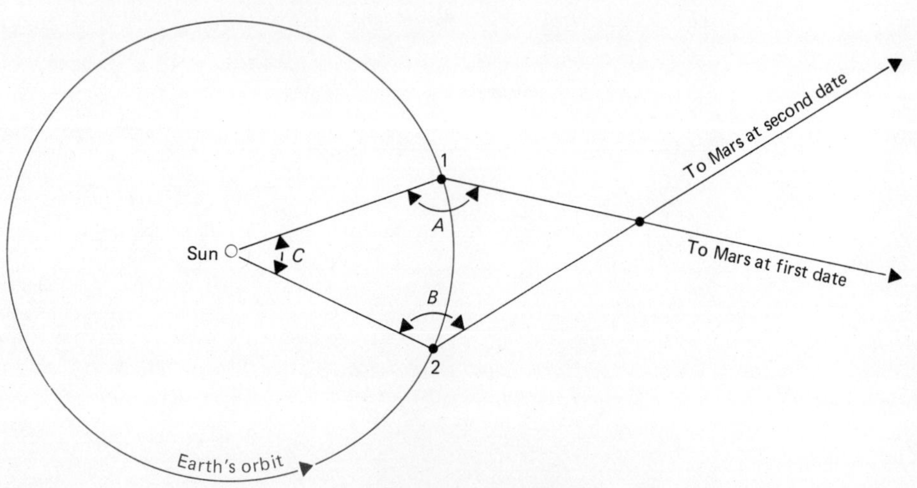

opposite that of the sun from the earth). Now on that date, Brahe's tables show Mars to be at an angle *A* away from the sun in the sky, and hence in the direction shown. Now Mars is known to revolve about the sun in 687 days, so exactly 687 days later Mars will be at the very spot in space that it was when the earth was at place 1. During that time the earth, going around the sun in 365¼ days, will be short only 43½ days of having completed two full revolutions. This will put the earth at position 2 in the figure, with the angle *C* being about 43° (the earth moves about 1° per day). On that second date, Brahe's tables show Mars to be at the angle *B* from the sun, and hence in the new direction shown on the figure. All we need do is construct the figure accurately, drawing in the earth's orbit and the directions to Mars, and where the lines indicating those directions intersect is where Mars has to have been at those two times. (Of course you can also solve the problem with trigonometry, avoiding the need of drafting ability.)

Selection of two other dates, also separated by 687 days, will locate Mars at some other point in its orbit. By using a large number of pairs of dates separated by 687 days, Kepler plotted out the whole orbit of Mars. Note, though, that here too it is found only in terms of the size of the earth's orbit. It is the earth's orbit that gives us the baseline for this surveying problem.

Kepler's method required lots of observations spaced over years. We have already described how Gauss invented a method of calculating the entire description of the orbit of a planet from observations covering only a short time, and how he used his method to save the discovery of the minor planet Ceres that Piazzi observed for only a few weeks. Other mathematical techniques in addition to Gauss' are used today, but they are all more or less equivalent. In most cases it is possible to calculate the orbit of a body from only three observations of its direction from the moving earth—the observations spaced about a week or so apart. The methods are too complicated to go into here, but they are commonly studied by students taking courses in celestial mechanics. In fact, I have often worked during summers with interested high school students who learn to make three observations of a minor planet and calculate its orbit. Of course, having calculated its orbit, we know just where the planet is at any time, as well as its distance from earth.

Gauss' method, like those of Copernicus and Kepler, still is one of surveying distances to planets (or other solar system objects) by using a portion of the earth's orbit as a baseline. All these distances to the planets, therefore, are

FIGURE III.6
The semimajor axis of an ellipse.

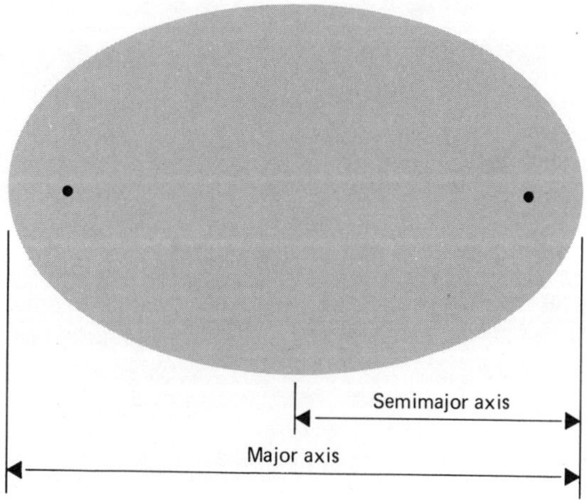

found by these procedures only in terms of the size of the earth's orbit. The shape of the earth's elliptical orbit was known to Kepler, and its semimajor axis is defined as one astronomical unit (AU), so we have the distances to all the planets to very high precision in terms of astronomical units. But just how big *is* the AU? It's like having a very accurate scale map of Europe, without being provided a scale of kilometers. Paris may be 14.37 cm from Munich on the map, but how many kilometers on the highway does a centimeter on the map correspond to?

The Length of the Astronomical Unit

Thus we need to find the length of the AU. To be sure, we know the distance to the moon in kilometers, but that doesn't help, for we cannot survey the moon's distance in AUs from different places in our orbit; the moon comes right along with us. To set the scale, we must find the distance to a planet not only in AUs, which is known, but also in kilometers.

As I said above, planets, unlike the moon, are too far away to display to the unaided eye daily parallaxes—that is, shiftings back and forth in direction as the rotation of the earth carries us from one side of it to the other. With telescopes, however, we can and do observe these parallaxes, especially for the nearer planets, such as Mars. That makes it possible, for example, to calculate the distance to Mars in kilometers, and so determine the scale of the solar system. The results are not very accurate though, because the parallax of a planet is small and hard to measure. It is better to observe the parallax of a close-approaching minor planet, because it can be measured more accurately. In 1932, for example, the minor planet Eros passed only 22 million km from earth—less than half Mars' nearest possible approach. A worldwide coordinated effort was made to observe the parallax of Eros, and pin down the value of the AU. Even better than observing its parallax was the observation of the earth's gravitational perturbations on the motion of Eros, because the earth's gravitational influence depends more sensitively on distance.

Once the length of the AU is known, one need only multiply it by a planet's distance in AUs to find its distance in kilometers. Over the past century, better and better determinations of the length of the AU were obtained, which led to a continual refinement of planetary distances. But by the beginning of the space age, the length of the AU was still known too inaccurately for us to hope to put a space probe in the vicinity of Mars. Fortunately, the space age provided its own solution.

To communicate with space vehicles by radio requires large radio-telescope type antennas, equipped not only to receive radio signals but to send them as well. Thus it became possible to use these new large antennas, like the Jet Propulsion Laboratory's installation at Goldstone, California, to transmit radio waves to the moon and planets and observe the reflected signal. That is, we can now observe the planets by *radar*. Since radio waves travel with the known speed of light, the round-trip travel time of the signal tells us the distances to the objects observed. We can also observe the planets' orbital speeds in km/s from the Doppler shifts of the reflected waves; since we also know the planetary speeds in AUs per second, this is another way of getting the number of kilometers in the AU.

Thus radar observations have determined the astronomical unit accurately enough for us to send space vehicles to the planets and to land Viking on Mars. Those vehicles themselves transmit radio waves to us, and by monitor-

FIGURE III.7
The Goldstone tracking station of the Jet Propulsion Laboratory. *(NASA/JPL)*

ing them, we gain still further information about the scale of the solar system, as well as about the masses of the planets and many other things, which we shall be discussing in Act V.

So we now know the length of the AU pretty well, and hence the distances to the planets with considerable precision—about a part in a million. In fact, one of the uncertainties today in the scale of the solar system is the speed of light itself. It's known to be absolutely constant, but there is a limit to how precisely that constant speed has yet been measured. "But enough of this!" you are thinking. "What are the answers?"

The semimajor axis of the orbit of the moon (best determined, of course, from radar observations of the moon) is 384,404 km, with an uncertainty of about 0.5 km. The best determination of the length of the astronomical unit (in 1975) gives 1 AU = 149,597,892 km, with an uncertainty of about 5 km (this value, a determination by the Jet Propulsion Laboratory, is based on the assumption that the speed of light is exactly 299,792.5 km/s). Thus the sun's distance is about 400 times the moon's.

The sizes of the orbits of the planets are given in Appendix 6. But for our appreciation of the structure of the universe it will help to have some kind of general "feel" for what those distances are, as opposed to precise numbers. Thus, very crudely, the distances of the planets from the sun, in astronomical units, are given in the following table.

Planet	*Distance (AU)*
Mercury	$2/5$
Venus	$3/4$
Earth	1
Mars	$1\frac{1}{2}$
Jupiter	5
Saturn	10
Uranus	20
Neptune	30
Pluto	40

Venus, at its closest, is about 40 million km away, and Mars, at its closest, is about 60 million km away.

Now that we have sketched the procedures by which we survey the solar system, let us tackle the incredibly more difficult problem of surveying distances to the stars.

SCENE 2

The Search for Stellar Parallaxes

Scientists and philosophers from Aristotle to Brahe had argued that the earth could not be in motion about the sun, or else we would see parallaxes of stars. But by the late seventeenth century, when the evidence favoring the heliocentric hypothesis was quite convincing, it was clear that those stellar parallaxes must be there, whether or not they were observable.

Stellar Parallax

Figure III.8 shows what was expected. A neighboring star is shown very close to the earth's orbit, to exaggerate the effect for clarity. As we see the star

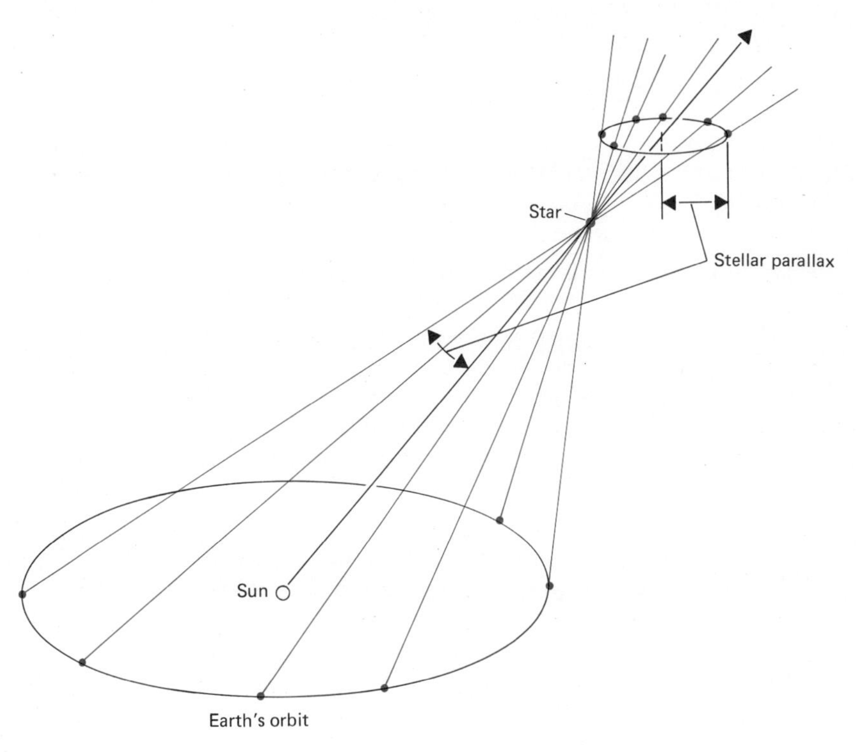

FIGURE III.8 The paraliactic ellipse.

from different places in our orbit, it appears to shift about in direction, and describes a small ellipse in the sky—the *parallactic ellipse.* As can be seen, the star always appears to move in the opposite direction from the earth's motion. The *shape* of the parallactic ellipse depends on the star's direction with respect to the ecliptic plane. If the star were 90° from the ecliptic in the sky, that is, in a direction perpendicular to the plane of the earth's orbit, its parallactic ellipse would be nearly a circle, because the earth's orbit is nearly circular. The ellipse would appear more and more eccentric (flatter) the closer the star is to the ecliptic. If the star were on the ecliptic, it would simply appear to shift back and forth in a line, for then the earth's orbit would be edge-on to it.

The *size* of the parallactic ellipse, on the other hand, depends on the star's distance; the more remote the star, the smaller the ellipse. The semimajor axis of the parallactic ellipse is given the specific name *stellar parallax.* The stellar parallax is also the angle at the star subtended by (that is, made by) the radius of the earth's orbit. Thus the total angle through which a star appears to shift because of the earth's orbital motion is twice the stellar parallax.

So everyone knew what effect to look for; it was just that the stars were evidently so distant that their parallaxes were too small to be observed. It took more than a century of diligent effort before the first stellar parallax was detected. Meanwhile, however, some rather interesting other discoveries were made serendipitously.

---

1. Most ancient astronomers believed that all stars were actually on the inner surface of a literal celestial sphere. Suppose the sun were at the center of the sphere, so that all stars were the same distance from the sun. Would the orbital motion of the earth, in this case, produce any changes in the directions of the stars? Explain.

Aberration of Starlight

One astronomer who made an ingenious attempt to observe stellar parallaxes was James Bradley. He stuck a telescope up his chimney in 1729. I presume he didn't light the fire. His idea was to hold the telescope as rigidly stationary as possible over several months, and to observe those stars that passed through the telescopic field of view as the rotation of the earth carried them overhead. With an illuminated reticle or scale in the telescope, he could measure rather accurately the north-south positions of the stars as they crossed the meridian. He had hoped that relatively nearby stars, because of their parallaxes, would cross his scale in slightly different places at different times during the year.

What he found, instead, was that *every* star shifted position, and by the same amount—20″.5. Moreover, instead of shifting in a direction opposite to that of the earth's motion, as expected for parallax, the stars always were displaced in the *direction* of the earth's motion. What Bradley had discovered is the phenomenon of *aberration of starlight.*

Aberration of starlight is a consequence of the finite speed of light. To understand the effect, consider the analogy of falling raindrops. If there is no wind, the drops fall vertically. Yet if you walk in the rain, they appear to be coming from in front of you, and hit you in the face. Driving through the

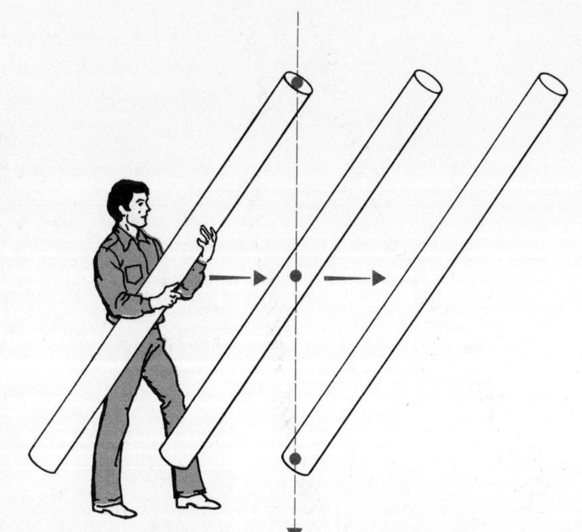

FIGURE III.9
A hollow tube, being carried through the rain, must be tipped forward in order that raindrops will fall through it.

rain, you can see the apparently slanted path of the raindrops very nicely on the side window of a car. You are simply moving forward to meet them, so the drops appear to have a component of motion toward you. You can imagine what happens if you try to carry a drainpipe vertically in the rain; the drops are all swept up by the rear inside wall, and none falls cleanly through. If you want the drops to fall *through* the pipe, you must tilt it forward, as shown in Figure III.9.

---

2. Suggest a way to calculate the speed of fall of raindrops from their apparent angle of fall along the side window of a car that is moving through the rain with a known speed.

It is the same with light. Its speed of approach is not affected by the earth's motion, but its apparent direction *is*. Just like the drainpipe in the rain, for a star to appear centered in a telescope, the telescope must be tipped slightly forward, in the direction toward which the earth is carrying it. The amount of tilt depends on the earth's speed in ratio to that of light. It happens that the earth moves with an orbital speed of about 1/10,000 that of light, which is just right to displace the stars by the 20″.5. Bradley's discovery demonstrated both the motion of the earth and the finite speed of light. He did not, however, discover stellar parallax.

Discovery of Binary Stars

William Herschel had another idea. As we have already mentioned, in 1650, Riccioli discovered the first double star (Mizar in the Big Dipper). It was a long time, though, before it was realized that most double stars are physically associated. Usually one is brighter than the other, and William Herschel assumed that the fainter star was simply more distant. He thought, therefore, that a convenient way to detect parallax would be to measure the shift of the brighter, presumably nearer, star compared to the fainter. Consequently, he began keeping track of double stars as potential candidates for measuring parallax. Between 1782 and 1821 he published three catalogs listing more than 800 double stars.

In 1804 Herschel did, in fact, observe that one star in the double system, Castor, had moved with respect to the other! Subsequent investigation, though, showed that the stars were in mutual revolution. We have already described how important this discovery was for showing that Newton's laws work beyond the solar system. Moreover, analysis of binary stars enables us to calculate stellar masses. Herschel had failed in his attempt to detect parallax, but had found something even more important. It was like Saul, he said, who had gone to seek his father's asses and had found a kingdom. Herschel's son John, by the way, continued the search for double stars, and cataloged more than 10,000 systems.

Allow me to digress for a moment from our discussion of parallax to describe how stellar masses are determined from observations of binary-star systems. Recall that Newton reformulated Kepler's third law, and found that the period of revolution of two mutually revolving bodies depends on their masses as well as on their separation. As a result, the mass of a double-star system can be found if the size of the orbit and period of revolution are known. Double stars provide our only direct means of learning the masses of stars other than the sun.

Visual Binaries

The binary systems noted by Herschel are called *visual binaries,* because we can see both stars in such a system through the telescope. More than 64,000 visual binaries are now cataloged. Only in a small fraction, however, can we actually determine the orbit of one star about the other; in most, the stars are so far apart that it would take thousands of years to see their orbital motion. Figure III.10 shows how the orbit of one star about the other would look in a typical system where the complete motion can be observed. Although both stars are really moving about each other, we show the orbit of one (for example, the fainter) with respect to the other (the brighter).

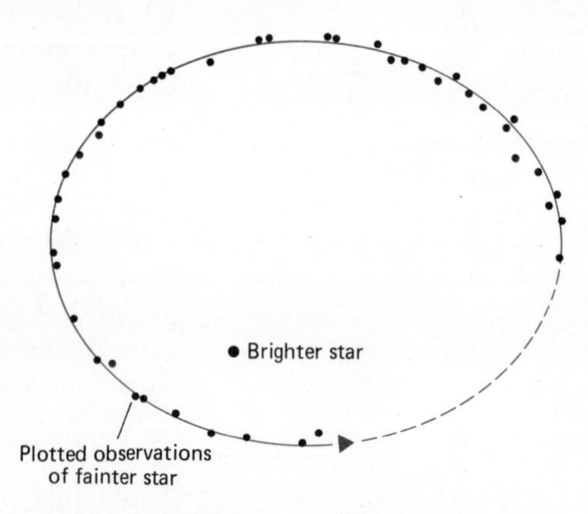

FIGURE III.10
The apparent orbit of one star about another in a hypothetical visual binary system. The plotted points on the ellipse represent the observed positions of the fainter star, relative to the brighter, at many different times, spread over a large number of years.

We can directly observe what the period of revolution is. Moreover, we can calculate the true size of the orbit from its angular size, once the distance to the stars is known. Thus we can calculate the combined mass of the stars in such a system. Now, the two stars actually revolve about their center of mass, or barycenter, and that center of mass, faithfully obeying Newton's laws, moves in a straight line through space. If we have enough good observational data, we can locate that point in space between the stars that is moving in a straight line compared to the background of remote stars, and then see how far each of the two stars in the system is from that barycenter, thereby learning how much of the total mass of the system belongs to each star.

---

3. Suppose a 100-kg man and his 25-kg son try to balance on a seesaw. How much farther from the fulcrum (point of support) must the boy sit than does his father? What has this to do with visual binaries?

Among the tens of thousands of known visual binary stars, however, there are only a few dozen that are so disposed to observation that the analyses have been carried out in sufficient detail to obtain the masses of their individual stars to a precision of, say, 10 percent. Fortunately, we can roughly double this body of data by the analysis of other types of binary systems, which, since we have digressed this far, we may as well mention now.

Spectroscopic Binaries

It turns out that very often the stars in a binary system have separations of only a few astronomical units or less, and periods of from a few days to a few months. Under these circumstances, unless the system is very close to us (as stars go), we will not be able to see the two stars as separate points of light in the telescope. Often, however, we are nevertheless able to know that the "star" we see is a system of two stars.

Unless the orbit of mutual revolution happens to lie exactly face-on to our line of sight (which is unlikely), each star is periodically changing its distance from us as it revolves about the center of mass of its system, and thus has a periodically changing radial velocity. More often than not, one star is enough brighter than the other that its light dominates the observed spectrum. Its spectral lines can be seen, over the days or weeks, to be shifting back and forth, exhibiting periodically changing Doppler shifts as that star moves alternately toward and away from us during its revolution about its companion. Sometimes the two stars are of nearly enough the same luminosity that we can see the spectral lines of both, shifting back and forth, out of phase with each other. A graph that displays the changing Doppler shifts of the lines of one or both stars, shifting back and forth, is called a *radial velocity curve*. Figure III.12 shows how the radial velocity curve for such a *spectroscopic binary* can appear.

It is rather interesting that the first spectroscopic binary to be discovered is one of the two components of the star Mizar—the first visual binary to be found. The discovery was made at Harvard in 1889 by E. C. Pickering. Then, in 1908, it was discovered that the other component of Mizar is a spectroscopic binary as well! Thus Mizar is a double double—four stars, all in mutual revolution.

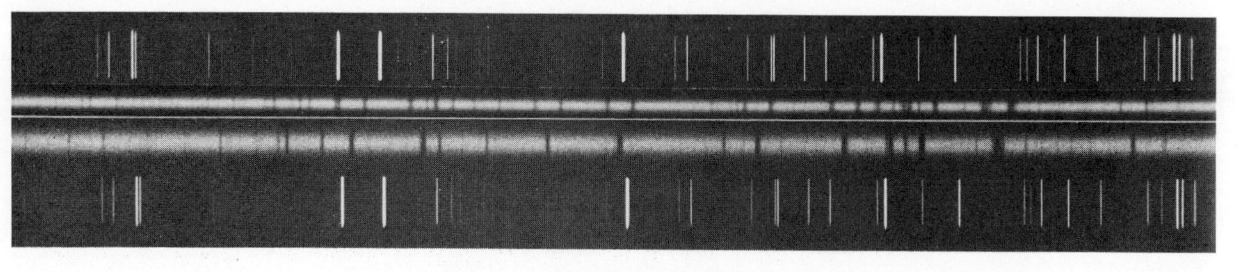

FIGURE III.11
Two spectra of the spectroscopic binary κ Arietis. When the two stars in the system are moving at right angles to our line of sight *(bottom)*, the lines are single. When one star is approaching us and the other receding *(top)*, the spectral lines of the two stars are separated by the Doppler effect. *(Lick Observatory)*

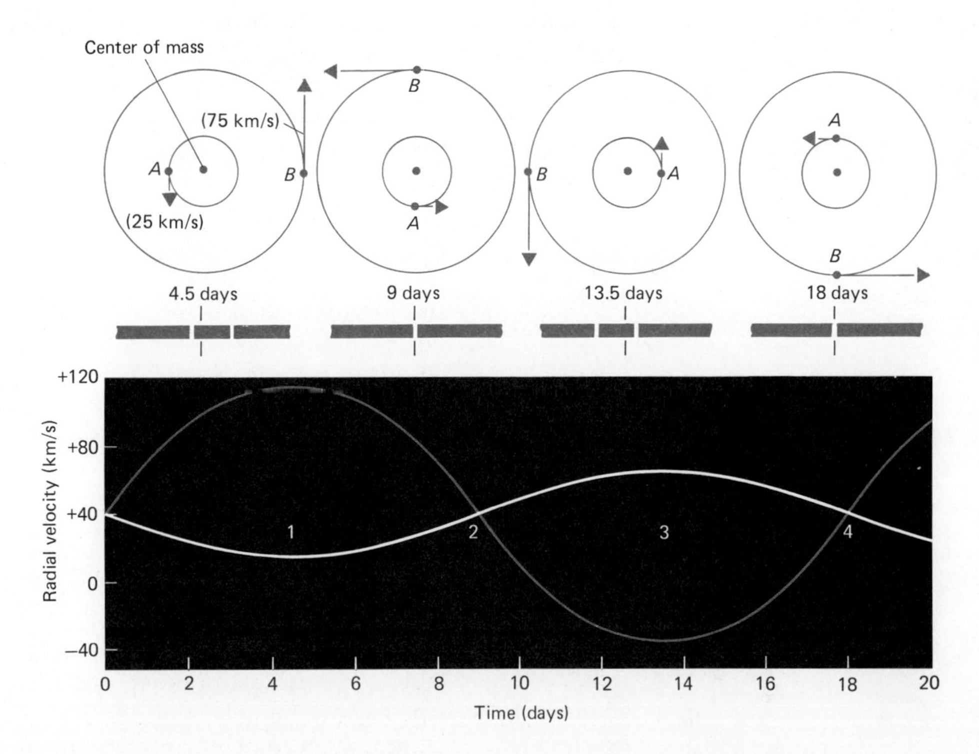

FIGURE III.12
The radial velocity curves of the two stars in a hypothetical spectroscopic binary in which the orbits are circular. The orientations of the stars in their orbits are shown above, and a graph showing their radial velocities as a function of time is below.

The total mass of a spectroscopic binary could be determined from the period of mutual revolution, which is found at once from the radial velocity curve and from the separation of the stars—that is, the size of their relative orbit. Now it might seem that since the radial velocity of each star tells us how fast it is moving, and the period tells us how long it takes to go around, we could easily calculate how big the orbits are, and thus the size of the relative orbit of revolution. But there is a catch: the orbit is, in general, inclined at some unknown angle to our line of sight, and we observe only a part (a component) of that relative velocity.

Analyses of spectroscopic binaries, in other words, do not in themselves tell us enough to derive stellar masses. We do not know the inclination of the orbit of revolution of a particular system to our line of sight, and hence how much greater the real orbital speeds are than are the observed motions toward or away from us.

---

4. Draw a diagram to show why we observe only a portion of the orbital speed of a star in a binary system from observation of the Doppler shifts in its spectrum.

Eclipsing Binaries

Sometimes, though, nature provides us another clue. Some binary-star systems revolve in orbits that are nearly edge-on to our line of sight. Then the stars can totally or partially eclipse each other with each revolution. Such are called *eclipsing binaries*. One of the most famous of eclipsing binaries is the star Algol, in Perseus. Normally, Algol appears as a fairly bright star in the sky, but every 2.867 days it fades to a third of its regular brightness; then after a

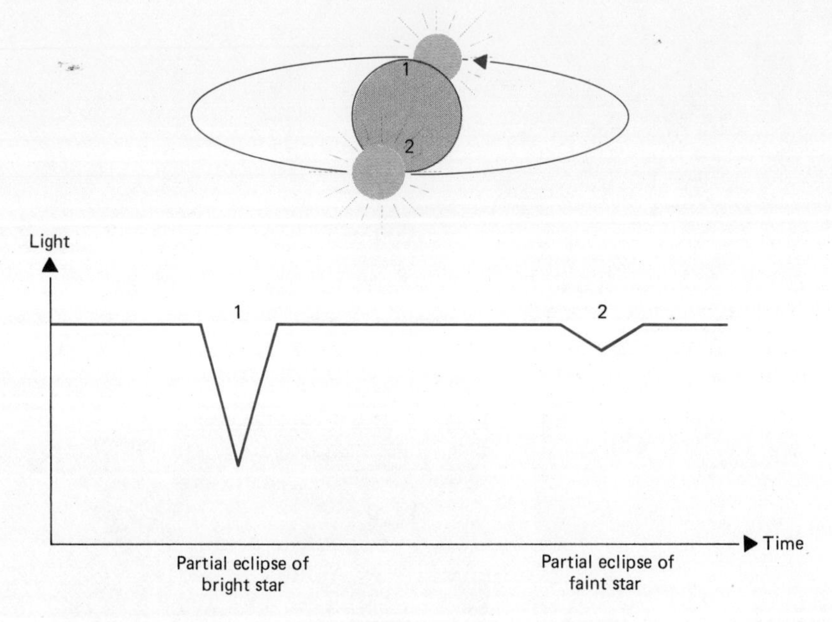

FIGURE III.13
A schematic diagram of the light curve of the eclipsing binary star *Algol*. The partial eclipses of the bright and faint stars are illustrated above.

few hours it returns to normal again (see Figure III.13). It was John Goodricke, a remarkable amateur astronomer and deaf mute, who, in 1783, suggested the explanation, namely, that the periodic dimming is due to a brighter star passing with each orbital revolution about a fainter companion. Today more than 4000 eclipsing binary systems are cataloged.

In principle if we observe a spectroscopic binary-star system in which we can see the spectral lines of both stars, we can find everything we need to calculate the masses of the individual stars except for the unknown angle of inclination of the orbits of the stars. However, in the happy circumstance in which the stars also eclipse each other, we then know that the orbit has to be almost edge-on. In fact, from the detailed way in which the light varies, we can calculate the actual orbital tilt to our line of sight. There are a few dozen examples of eclipsing binary systems where the complete analysis has been carried out with some fair precision, and for which we have found the masses of the individual stars.

The mass of an object, like a star, is one of the most fundamental things that describe it. Analyses of binary-star systems enable us to learn something about masses of some typical stars and, as we shall see, to make some tentative, but highly significant, conclusions about stars, how they are formed, how they evolve, and how they die; the sun (with its system of planets) is just a typical example.

---

5. Suppose a binary-star system has an orbit exactly edge-on to our line of sight, and that the two stars have equal luminosities. One star, however, has ten times the radius of the other. Sketch a graph showing how the total light from the system changes with time.

Observation of Stellar Parallaxes

But back to our story of parallax. Yes, it was finally detected, but not until 1838, and then by three people almost at the same time. Friedrich Bessel, in

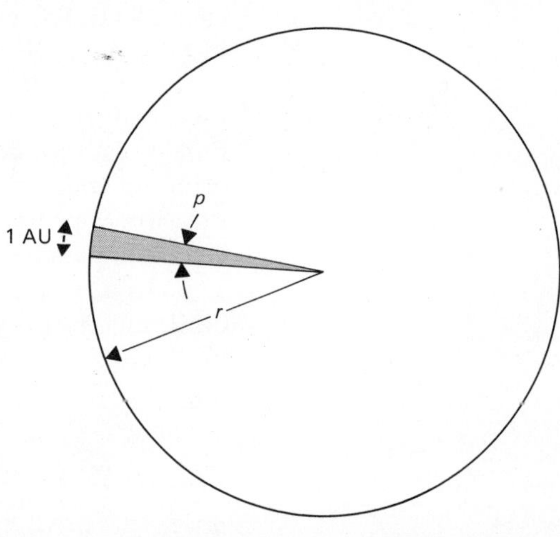

FIGURE III.14
The relation between parallax and distance.

Germany, observed the parallax of the star 61 Cygni; Thomas Henderson, in South Africa, of Alpha Centauri; and Friedrich Struve, in Russia, of Vega.

It is easy to see why a knowledge of the stellar parallax determines a star's distance. Remember that the stellar parallax is the angle subtended by one astronomical unit (the radius of the earth's orbit) at the distance of the star. Now see Figure III.14. The star is shown at the center of an extremely large circle, of radius *(r)* equal to the star's distance from earth. Along the circumference of the circle is laid out the distance 1 AU. (The AU is so small compared to the size of the circle that the tiny piece of arc it spans is indistinguishable from a straight line.) The stellar parallax, p, is thus that angle at the star subtended by that tiny 1-AU arc or segment. Now it is obvious that the 1-AU arc bears the same relation to the entire circumference of the circle that the angle p does to 360°. If we express that relation as a simple equation and solve for r, we find that its value, in astronomical units, is 206265 divided by the parallax, if p is expressed in seconds of arc.

Expressing the distance of a star in AUs, however, is rather like expressing the distance from New York to London in meters. So astronomers have invented a convenient unit of distance, the *parsec* (abbreviated pc), which is equal to 206265 AUs. Then the distance to a star in parsecs is just the reciprocal of its parallax in seconds.

6. What is the distance of a star with a parallax of 0″.04?
7. What is the parallax of a star with a distance of 100 pc? What is the distance to that star in AUs?

If a star had a parallax of 1 second, its distance would be 1 pc. Thus the origin of the name: one *par'sec* is the distance to a star with a *par*allax of one *sec*ond. No star is that close. The nearest (beyond the sun, of course), is Proxima Centauri, with a parallax of 0″.763, and hence a distance of $1/0.763 = 1.31$ pc. Proxima is an outlying member of a system of three stars; the other two form the double star that to the unaided eye appears as the single bright star Alpha Centauri, with a distance of 1.35 pc. Proxima gets its name from being the nearest of the three stars in that system, but it is about 100 times too faint to be seen without optical aid. For that matter, Alpha Centauri is too near the

South Celestial Pole to be seen from most parts of the United States and Europe.

The next nearest naked-eye star is Sirius, the brightest appearing star in the sky. Sirius has a stellar parallax of 0.″377, and hence a distance of 2.65 pc. Parallaxes have now been measured for thousands of stars, but for most the errors of measurement are as big as the parallaxes themselves. In order to measure the parallax of a star, and hence its distance, to a precision of 10 percent, that parallax has to be at least 0.″05, which means the star must be within 20 pc. That's pretty nearby, as stars go. Still, many stars are that close. A standard reference is Gliese's catalog of nearby stars; the 1969 edition lists 1049, but including faint undiscovered ones, the total number of stars within 20 pc may be around 4000.

Proper Motion

But, once again, I have glossed over a complication. It would be nice if the stars would just sit there and let us measure their parallaxes while the earth revolves about the sun. Stars, however, have independent motions of their own, and these are usually many times as great as the parallaxes we are trying to measure. A star's annual change in direction in the sky, as it would be observed from the sun, is called its *proper motion,* and is usually expressed in seconds of arc per year. Fortunately, the proper motion of a star just keeps adding up, year after year, until after several years the star has often changed direction enough to that the change can be measured rather easily. A common technique for measuring proper motions is to measure precise positions of star images on two photographs taken a number of years apart. If a star were observed, for example, to shift its direction by 40″ between its measurements on photographs separated by 20 years, its proper motion would be 1/20 the total change, or 2″.

The star of largest known proper motion is Barnard's star. It is too faint to be seen with the unaided eye, even though it is the nearest star beyond the Alpha Centauri and Proxima Centauri system. Its proper motion is 10.″25, which means that in 2000 years it would shift in the sky by more than 5°; even in a human lifetime its motion would be apparent to the unaided eye, if it could be seen. If a Greek astronomer of antiquity who was very familiar with the sky—say, Hipparchus—could return to life today, he would notice that some of the naked-eye stars appear in slightly the wrong places. Sirius, for example, with a proper motion of 1.″33 would have shifted by about one and one-half times the moon's angular diameter—an easily discernible amount. We see, then, that the "fixed" stars aren't really fixed at all. A few hundred have proper motions exceeding 1″. On the other hand, the average

FIGURE III.15 Two photographs of Barnard's star taken 22 years apart, showing its proper motion. *(Yerkes Observatory)*

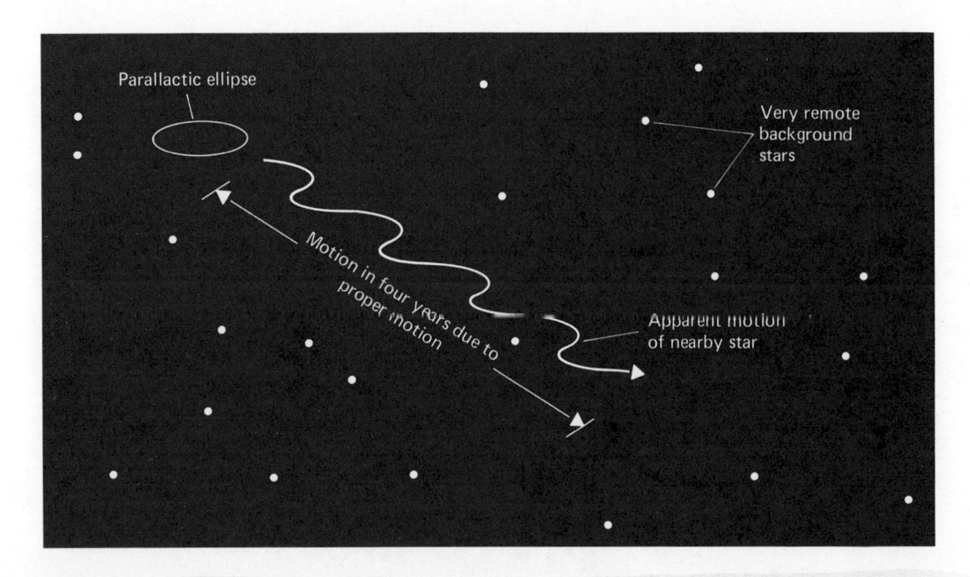

FIGURE III.16
The apparent motion of a nearby star is a combination of its parallax and its proper motion.

proper motion of all naked-eye stars is less than 0.″1, so Hipparchus would have thought most stars had stayed put. A million years from now, on the other hand, the sky will bear no resemblance at all to its present pattern of familiar constellations.

To make a long story longer, the observed proper motions and parallaxes of stars are mixed together. The result is that a relatively nearby star might appear to move against the background of extremely distant stars along a wavy line, as sketched in Figure III.16. The general motion of the star to the lower right (in the figure) is due to its proper motion, whereas the wiggles are the superposed reflection of the earth's orbital motion, that is, the superposed parallactic ellipse. It thus requires many observations over several years to disentangle the two effects.

---

8. What is the smallest possible number of observations of a star that are required to determine its parallax? Explain.

What if a star were in a binary-star system, but its companion were too faint to be visible? How can we be sure that wiggles superposed on proper motion, like those in Figure III.16, aren't due to the revolution of the star about its unseen partner? As a matter of fact there are many such examples, and when the duplicity of a star is ascertained by regular variations in its proper motion, it is called an *astrometric binary*. The star Sirius was in this class until 1862, when Alvan G. Clark was able to observe telescopically the faint companion of Sirius. It turned out to be an example of the class of stars known as white dwarfs, which we have mentioned before. Sirius and its white-dwarf companion mutually revolve in a period of 50 years, and 50-year wiggles could not be due to parallax. Variations due to the parallactic ellipse must have a period of exactly one year, and must always be in a direction opposite to that of the earth's orbital motion; that's how they can be recognized.

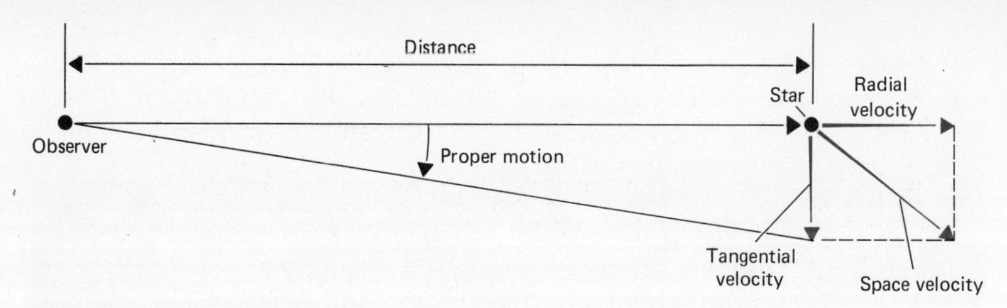

FIGURE III.17 Relation between the proper motion, radial velocity, distance, and space velocity of a star.

FIGURE III.18 The space motions of surrounding stars, *(a)* show a systematic drift away from the direction of *Lyra*. The reason is that the sun is actually moving toward *Lyra*, *(b)*. If we correct for this solar motion, then the stars around us appear to move in random directions.

Motions of Stars in Space

Proper motion alone does not tell us how fast a star really moves in space. A star could be nearby and moving only moderately fast and still have a sizable proper motion, or it could be very remote and really be tearing through space without showing its motion. However, if we know both the proper motion and the distance of a star, we can then calculate its actual speed, in km/s, across our line of sight. This is called its *tangential*, or *transverse, velocity*. If we also know its radial velocity (from its Doppler shift), that tells us how fast,

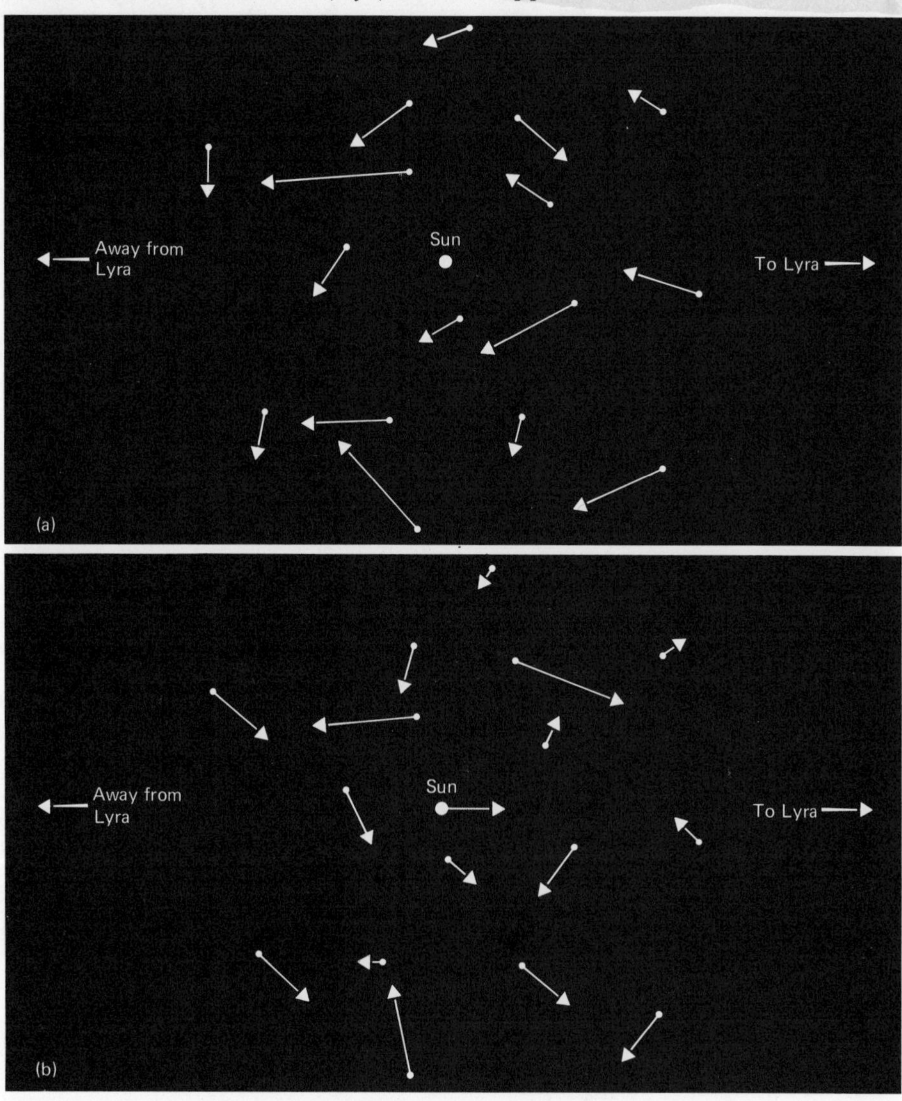

at the same time, the star approaches or recedes from us. Then, by application of the theorem of Pythagoras, we can calculate the actual velocity of the star in space, relative to the sun; that quantity is called its *space velocity*, (Figure III.17).

Most neighboring stars have space velocities of a few tens of kilometers per second. A few, though, are moving more than 100 km/s. If we analyze the space velocities of stars about us in the sky, we find that they are not random. Each way we look, we find stars moving in all possible directions, but on the average those in the general direction of the bright star Vega in the constellation of Lyra seem to be approaching us, and those in the opposite part of the sky, again on the average, seem to be receding. Halfway between, again we find some stars going one way and some the other, but most of them seem to have proper motions carrying them away from the direction of Vega. The reason, as you doubtless have guessed, is because the sun itself is in motion with respect to the average of its neighbors, carrying us, of course, along with it. The direction toward which the sun is moving is called the *solar apex;* it lies about 10° southwest of Vega (see Figure III.18). The speed of the sun, among its neighbors, is about 20 km/s. We shall see in Scene 4, however, that the solar motion and the other stars' space velocities are really only small differences between their general streaming about the center of our Galaxy at a speed many times greater.

A good way to discover nearby stars is to search for stars of large proper motion. Such stars are usually relatively nearby, whereas those of small proper motion are usually relatively distant. We can't tell for sure if this is true or not for an individual star, of course, but statistically, proper motions tell us something about stellar distances. There are a couple of standard techniques for analyzing the proper motions of many stars to find, in the mean, how far away the stars in that sample are. We shall not go into these details here, but I think you can appreciate that some useful data about stars' distances can be found from studies of their proper motions.

---

9. Discuss the analogy between the distances and proper motions of stars and the distances and apparent motions of airplanes seen in the sky. Consider fast and slow planes, and those moving in all possible directions. In each case, discuss the analogy with stellar motions.

The Milky Way Galaxy

When we look at the night sky, we find that some stars appear quite bright and others exceedingly faint. Actually, stars differ enormously from one to another in their total intrinsic output of radiant energy, or *luminosity*. But before stellar parallaxes had been detected there was no way to learn the luminosity of a star. Thus William Herschel, in the late eighteenth century, assumed simply that all stars are intrinsically the same, and that their apparent brightnesses differ because they are at different distances from us. He reasoned that in regions of the sky where there are many faint stars, it is because they extend in those directions to very great distances, while in regions where the stars are sparse, it is because they thin out at relatively small distances. With this idea, Herschel tried to determine the distribution of stars in space.

FIGURE III.19
William Herschel. *(Yerkes Observatory)*

Herschel observed many selected areas of the sky with his telescope, and counted the numbers of faint stars he could see in those directions. He found the stars to extend to great depths in directions along the great circle in the sky where we see the Milky Way, but not in other directions. He concluded that the sun is one star in and near the center of a great system of stars, and that the system is shaped like an enormous "grindstone." If we look edge-on through the grindstone, we see the light from so many stars projected in our line of sight that we have the illusion of the Milky Way. In other directions we are looking out either of the flat faces of the grindstone and see only relatively nearby scattered stars. Herschel's grindstone is shown edge-on in a schematic representation in Figure III.20. In 1750, Thomas Wright, in the *Theory of the Universe,* had speculated that the sun is in such a great system of stars; Herschel demonstrated it. Today we call that system our *Galaxy*. Herschel's model is only partly correct, though, as we shall see in Scene 4; in particular, we are *not* at the center.

FIGURE III.20
A copy of a diagram by Herschel showing a cross-section of the grindstone-shape he derived for our Galaxy.

10. What would Herschel have concluded about our location in the Galaxy if the Milky Way had been observed to extend only halfway around the sky?

Inverse-Square Law of Light

Herschel could have done very much better if he, as we do, had known parallaxes, and thus fairly reliable distances, to over a thousand stars. He could then have calculated the luminosities of those stars from a knowledge of their apparent brightnesses and distances. The calculation involves the very simple idea we call the *inverse-square law* of light, which states that the apparent brightness of a point source of light is in inverse proportion to the square of its distance.

Let's see why. Figure III.21 shows a light source, with its light rays spreading out in all directions. All the light emitted at one instant will, one second later, be crossing the surface of an imaginary sphere of radius 300,000 km (the distance light goes in a second), centered on the source. In two seconds, that same light will have spread out over a sphere of radius 600,000 km, and hence of four times the area of the first. In three seconds it will have spread itself over nine times the area, and so on. How bright the source appears to be depends on how much of that light enters your eye or whatever other detecting device you are using. Since an eye pupil or a telescope aperture has a certain area, the amount of light it gets depends on how far from the source the detector is.

In Figure III.21 the area of the detecting aperture (say, your eye pupil) is represented by the small black square picking up a part (albeit a small part) of the radiation crossing one of those imaginary surfaces centered on the star. As you can see from the figure, if you move your eye farther from the source, it picks up a smaller fraction of the total light the source emits, and that fraction goes down with the square of your distance from it; go twice as far away, you get one-fourth as much, go ten times as far away, you get $1/100$ as much, and so on.

To find the total luminosity of the source of a star, all we need do is measure how much electromagnetic energy our detector picks up, and calculate what fraction this is of the total amount emitted by the source. Let's see how it works out for the sun. Careful measurements show that just outside the earth's atmosphere, every square centimeter receives 1.37×10^6 erg/s from

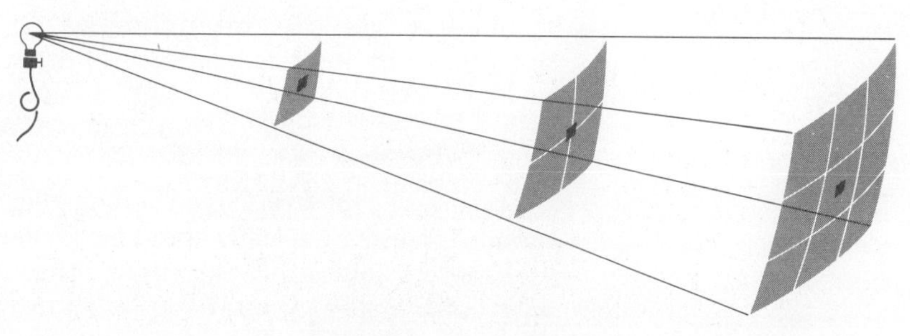

FIGURE III.21
The inverse-square law of light propagation.

the sun (an erg is a small unit of energy). The distance of the sun is about 150,000,000 km, or 1.5×10^{13} cm, which means that the energy emitted by the sun, or its luminosity, is spread over a sphere of radius 1.5×10^{13} cm by the time it reaches the earth. We simply multiply the 1.37×10^{6} erg/cm^{2}s by the number of square centimeters in that sphere (the area of a sphere is 4π times the square of its radius), and find the sun's luminosity to be 3.90×10^{33} erg/s. Now 10^{7} erg/s is one watt of power, so the sun's luminosity is about 4×10^{26} watts. Those ergs from the sun make up one whopping omelet!

But how do we measure the light coming to us from a star? Today, as we described in Act II, we use a light-sensitive device such as a photomultiplier placed at the focus of a telescope. Until recently, however, it was not possible to measure very accurately the light flux from a star. Nevertheless, the human eye is a remarkably sensitive detector of light. Even though it may not be possible for the unaided eye to measure light flux quantitatitvely, the eye is still excellent for comparing the brightnesses of different sources.

The Magnitude System

In the second century B.C., Hipparchus prepared a catalogue of about 850 stars. For each star he listed its *magnitude,* which was an estimate of its relative brightness. Hipparchus placed the brightest stars in the first magnitude, and the faintest ones in the sixth magnitude. He assigned intermediate numbers to the stars of intermediate brightnesses. Later observers preserved Hipparchus' magnitude system.

Eventually, astronomers learned how to make quantitative measures of light flux, and by 1856, enough was known about the relative brightnesses of stars of different magnitude that Norman R. Pogson suggested a scheme for putting the old rule-of-thumb magnitude system on a quantitative basis. It turns out that those stars Hipparchus called first magnitude deliver to us about 100 times as much light as those he called sixth magnitude. Thus Pogson suggested defining a difference of five magnitudes as a ratio of light flux of exactly 100. Measurements showed further that stars with a constant magnitude difference always have about the same ratio of brightness. Thus Pogson's scale was defined such that a difference of one magnitude corresponds to a ratio of light equal to the fifth root of 100, which is about 2.512. Thus a fifth-magnitude star is 2.512 times as bright as a sixth-magnitude star, and a fourth-magnitude star is 2.512 times as bright as a fifth-magnitude star, or 2.512 *times* 2.512 times as bright as a sixth-magnitude star, and so on. This makes a first-magnitude star 2.512 to the fifth power, or 100 times as bright as a sixth-magnitude star.

This is the magnitude scale we use today. The modern scale, based on precise photoelectric measures of light flux, is adjusted to match the old system started by Hipparchus, so the 20 brightest stars, which Hipparchus would have called first magnitude, have an average magnitude of about 1.0. But the 20 brightest stars are not all the same. Sirius, the brightest of them, turns out to be ten times as bright as the average first-magnitude star. A factor of 10 in the brightness means 2.5 magnitudes; thus Sirius actually has a magnitude of about $-1\frac{1}{2}$.

The brighter planets have even smaller magnitudes. Mars and Jupiter, at their brightest, are usually about at -2; Venus at its brightest is at magnitude -4 (it is actually visible in daylight). The full moon, 250,000 times as bright as an average first-magnitude star, has a magnitude of -12.5. The sun's magnitude is −26.5; it appears brighter than Sirius, the brightest appearing star, by 10 thousand million times.

Stars too faint to be seen with the unaided eye have larger magnitudes than do stars classified by Hipparchus. Good binoculars can make stars of 10th magnitude visible, and a 15-cm telescope can reveal stars of 13th magnitude. The faintest stars that can be photographed (with a long-time exposure) with the world's largest telescope are a bit fainter than magnitude 24, and that's as much fainter than Sirius as Sirius is fainter than the sun.

---

11. In each of the following pairs, which has the larger magnitude;

a) the sun and the moon;
b) a candle and a 100-watt bulb, both at a distance of one kilometer;
c) a 100-watt bulb at a distance of 1 km and a typical candle at a distance of 1 m;
d) a firefly at a distance of 100 m and a 60-watt bulb at a distance of 3 m?

Absolute Magnitudes

Now we know that a star that is two magnitudes brighter than another appears about 6.3 times as bright. So if we assign magnitudes to all the stars (easy to do), we can compare their relative brightnesses, at least with a bit of arithmetic. But that doesn't tell us anything about the stars' real luminosities, because they are at all different distances. However, once we know how far away a star is, we can easily enough calculate how much brighter or fainter it would appear at some other distance—all we need do is use that inverse-square law of brightness.

To facilitate comparison of stellar luminosities, astronomers define the *absolute magnitude* of a star as the magnitude it would have if it were 10 parsecs away. Example: Suppose a star were 20 pc distant and had a magnitude of 4.5. Now if it were only 10 pc away, it would be two times as close, and hence four times as bright. It works out that a factor of four in light corresponds to 1.5 magnitudes; thus, if the star were only 10 pc away, its magnitude would be $4.5 - 1.5 = 3.0$. This would be its absolute magnitude. (Note that bringing it closer would make it brighter, and hence make its magnitude *less,* not more).

On the other hand, 10 pc is about 2 million astronomical units (remember that 1 pc is 206,265 AUs). Thus if the *sun* were 10 pc away, it would be two million times farther off, and would appear four million million times as faint, which corresponds to 31.5 magnitudes. Thus the sun's absolute magnitude is $-26.5 + 31.5 = 5$. Put out at that relatively nearby distance of 10 pc, the sun would appear as a faint star, not much brighter than the naked-eye limit and probably not visible at all in the lights of a typical big city.

---

12. Arrange the following in order of decreasing absolute magnitude (those of smallest absolute magnitude at the end):

a) the sun,
b) our Galaxy,
c) a firefly,
d) a standard flashlight,
e) a 100-watt bulb.

Colors of Stars

Up to now the magnitudes we have been talking about refer to the light in the visual part of the electromagnetic spectrum. These are technically called *visual magnitudes,* or *absolute visual magnitudes.* We can also measure the light from a star through various color filters, either with a photomultiplier or photographically, and obtain magnitudes corresponding to other spectral regions. Comparison of the magnitudes of a star in different spectral regions gives us information about its color, and hence its temperature. Suppose, for example, we take two photographs of the same region of the sky, one on an emulsion sensitive to blue light and one on a red-sensitive emulsion. Such a pair of photographs is shown in Figure III.22. The hot blue and violet stars appear brighter on the blue photograph, and the cool red stars brighter on the red photograph. By measuring the exposures of the images and comparing the same star's brightness in blue and red, we can rather accurately determine its temperature. You can see what a wealth of information is obtained on just two such photographs containing images of thousands of stars.

---

13. Prepare a sketch of part of Figure III.22, indicating on your sketch ten relatively hot and ten relatively cool stars. Can you say anything about their relative distances? Why?

FIGURE III.22
A blue *(a)* and red *(b)* exposure of the same region of the sky. Hot, blue and cool, red stars can easily be distinguished by comparing the two photographs. *(Courtesy of the UK Schmidt Telescope Unit. Royal Observatory, Edinburgh)*

Summary

Let's now see where we stand. We know distances in the solar system very well—to one part in a million or so. But the nearest star (beyond the sun) is 1.3 pc away, nearly 300,000 times the sun's distance and nearly 7000 times the distance of Pluto, the most distant planet. About 4000 stars are estimated to be within a distance 15 times as great as that of the nearest star.

Although the parsec is a very convenient unit of distance for the purposes of calculation, you might have a better feeling for the distances if we express them in terms of how long it takes light to traverse them. Light comes from the moon in about 1.3 s. It takes light 8 minutes to come from the sun and a few hours to come from the most distant planets. Light takes about 4 years to make the journey from Alpha Centauri and Proxima Centauri; we say those stars are about 4 *light years* (LY) away. Thus a light year is the distance light travels in a year; 1 pc equals 3.26 LY.

Suppose you draw a scale map of the solar system showing the planets, out to Pluto. Suppose, further, that the map would barely fit on a page of this book. On the same scale the nearest star would be about 0.6 km away; small wonder we know the distances to even the nearest stars only to about 1 or 2 percent precision. Again, on that same scale, the most distant stars for which we can measure parallaxes, and obtain distances, to about 10 percent accuracy would be some 8 km away. Even these, however, are only our very nearest stellar neighbors. Even our own tiny little Galaxy extends more than a thousand times farther!

Obviously we can't measure parallaxes to most stars in the universe, or even most stars visible to the naked eye. We must use indirect methods to survey them. But with our discussion in this scene, we are now armed with enough knowledge to see how it can be done. Sit tight!

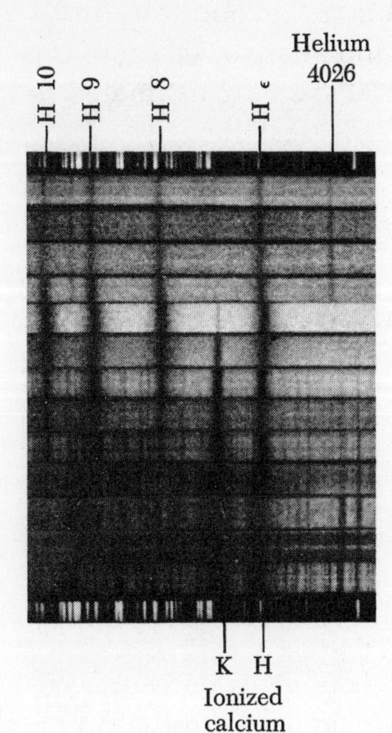

Analyzing Starlight

In Scene 2, we saw that if we know the distance to a star, we can measure its apparent brightness, or magnitude, and then calculate how much brighter or fainter it would appear at some other distance. For example, we can calculate its absolute magnitude—how bright it would appear at the standard distance of 10 pc.

Stellar Distances from Absolute Magnitudes

But now let's suppose we did not know the star's distance, but did observe its magnitude, and by some hook or crook learned what its absolute magnitude was. Do you see how we have just turned the problem backward? Knowing the star's absolute magnitude is knowing how bright it would appear at 10 pc distance, which is knowing how bright it really is. Now, knowing how bright the star really is, we can calculate how far away it has to be to look as bright or as faint as it actually does.

I'll give an example. Let's suppose we observe a star of the tenth magnitude. Now suppose a reliable source tells us that the star has an absolute magnitude of five. That means, if it were 10 pc away it would have an apparent magnitude of five, so would then appear 100 times as bright as it does at its actual distance.[1] To appear 100 times as bright, it would have to be ten times as close. If 10 pc is ten times as close as the star actually is, the star must be 100 pc away. In other words, if we know both the apparent magnitude and the absolute magnitude of a star—that is, its apparent brightness and how bright it really is—we can always calculate its distance with the inverse-square law of light.

---

1. If a star has a distance of 100 pc and an absolute magnitude of 3, what is its apparent magnitude?

[1] We often say "apparent magnitude" for "magnitude" to avoid confusion with absolute magnitude.

2. How distant is a star whose apparent magnitude and absolute magnitude are the same?
3. A star has an apparent magnitude of 12 and an absolute magnitude of 2. What is its distance?
4. How distant can a star of absolute magnitude 6 be, and still be visible to the unaided eye?

Everyone who drives a car at night does the same thing, subjectively, all the time. If you see a red stoplight ahead down the road, you judge its distance from its apparent faintness. You know how bright the stoplight really is, for you have seen many such lights up close. Perhaps you estimate it to be more than a mile away, and think to yourself, "No reason to slow down now; the light will most certainly be green before I get there."

Now, to be sure, if the night is foggy you might be in for trouble; in that case the light could be a lot closer than it looks. Astronomers have to worry about the same thing—not fog, of course, but interstellar dust, microscopic solid grains strewn about much of space. Sometimes stars are dimmed greatly because they are trying to shine through much of this dust, and hence they look a lot farther away than they really are. In the next scene we'll see that before we knew about that dust, it sometimes led us pretty far astray. Now, though, we usually can tell when it's present and correct for its effect. For the time being, we shall assume that space is clear and transparent.

So now the secret is out! For most stars for which we cannot measure parallaxes, we calculate distances by comparing their magnitudes and absolute magnitudes. The only minor detail we haven't touched on yet is how we know their absolute magnitudes. It would never have been possible were it not for that healthy sample of a thousand or so stars whose parallaxes *are* observable. We *know* the absolute magnitudes of *those* stars, because we know their distances. For the stars in that sample, it was learned that most stars fall nicely into rather specific types, and that each star of a given type has a characteristic absolute magnitude. These types are based on features seen in stars' spectra, so now we must say something about stellar spectra.

We saw in Act II how we can observe the spectrum of the sun or a star. In 1802 William Wollaston first observed the dark lines in the solar spectrum. By 1815, the German physicist Joseph Fraunhofer had cataloged about 600 of those lines. Thus the absorption lines in the spectrum of sunlight are still called *Fraunhofer lines.* By 1823, Fraunhofer had observed that the spectra of stars have similar dark lines crossing the continuous bands of color. Some of the lines in stellar spectra were first identified with terrestrial elements by Sir William Huggins in 1864.

About the same time it became apparent that stellar spectra fall into different types. In 1863 the Jesuit astronomer Angelo Secchi classified stars into four groups according to the appearance of their spectra. Secchi's scheme has since been modified and augmented until today we recognize seven principal spectral types that include most stars. They are given letter designations that seem to have no sensible order, and indeed they do not, because it was early this century before it was even realized why stars differ so much in their spectral characteristics.

For example, those stars classified as spectral type A show very strong lines due to hydrogen absorbing light in the Balmer series of wavelengths (described in Act II), those of type K show strong lines of many different metals, and those of type M show lines (actually bands of many closely spaced lines) due to absorption of light by molecules of titanium oxide. Does this mean the A-type stars are made mostly of hydrogen, K-type mostly of metals, and M-type of titanium oxide?

At one time it was thought that the spectral differences were due to different chemical compositions of stars, but now we know that most stars are made of very nearly the same distribution of the various chemical elements. The differences arise because the atoms in the outer layers of different kinds of stars find themselves in very different physical environments. The most important effect is temperature; to a much lesser extent the density matters as well. As we shall see later, stars are so hot throughout that all their elements are in the gaseous state, even those such as iron and tungsten, which are solids on the relatively cool earth. But different stars have different temperatures in their outer layers. Now, you recall from the foregoing act that at high temperatures collisions between the atoms raise them to excited states, or ionize them. The higher the temperature, the greater the amount of ionization. On the other hand, a high density tends to inhibit ionization, because the ions find more nearby electrons to capture and become neutral again.

Well, in very hot stars, hydrogen is mostly ionized and cannot absorb light at discrete wavelengths to produce the dark lines. At relatively low temperatures the hydrogen is neutral, but practically all the atoms are in their lowest energy state, where they can absorb at certain wavelengths but not at those that are seen in the visible part of the spectrum. Similarly, other elements present in stars of different temperatures are in different states of ionization and excitation and produce different patterns of dark lines. In the coolest stars, certain molecules, such as titanium oxide, can exist.

When the role of temperature was fully appreciated, the various spectral types that had been assigned to different stars could be put into order of temperature. Some classes turned out to be redundant and were eliminated, so that seven remain. They are listed in the following table, along with the colors of the stars of each class and some examples. There are some additional

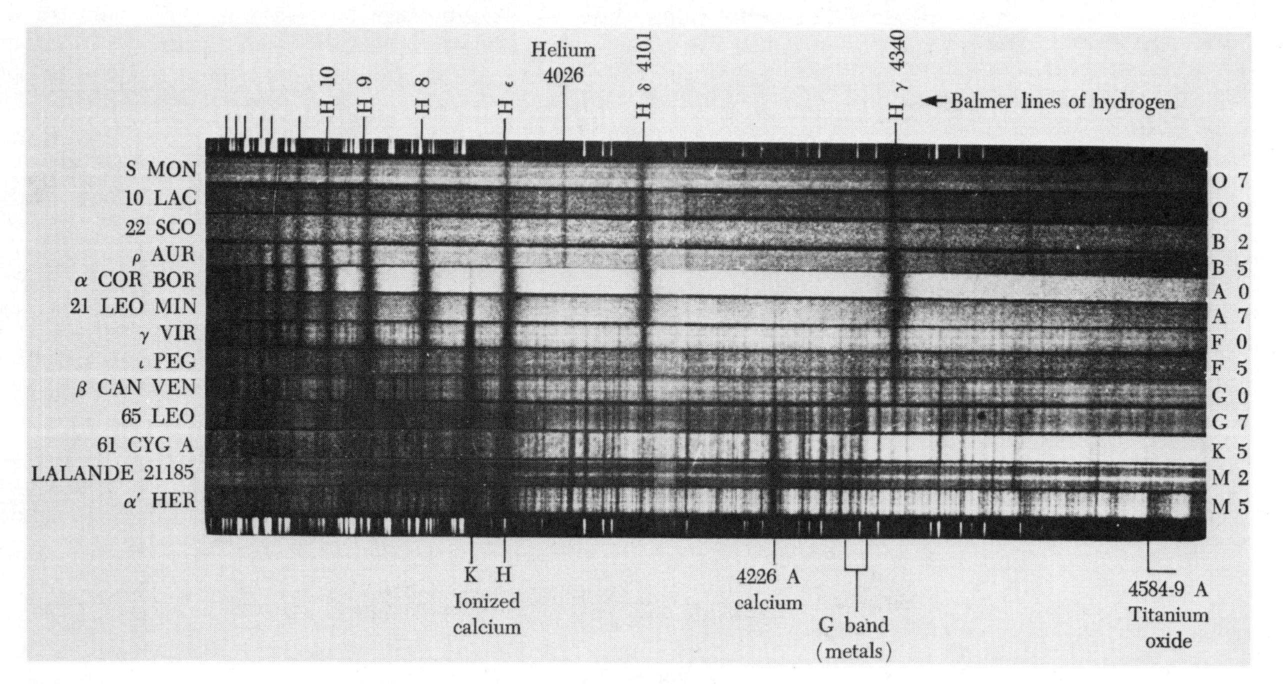

FIGURE III.23 Spectra of stars of the seven principal spectral types. *(UCLA)*

Spectral Class	*Temperature Range (°K)*	*Color*	*Examples*
O	more than 25,000	blue	10 Lacertae
B	11,000 to 25,000	blue	Rigel, Spica
A	7,500 to 11,000	blue	Sirius, Vega
F	6,000 to 7,500	blue to white	Canopus, Procyon
G	5,000 to 6,000	white to yellow	Sun, Capella
K	3,500 to 5,000	orange to red	Arcturus, Aldebaran
M	less than 3000	red	Betelgeuse, Antares

spectral types designated for certain unusual kinds of stars, but they need not concern us here.

The list of spectral types, in order from hotter to cooler, is called the *spectral sequence.* Since it is a temperature sequence and stars exist of temperatures over a continuous range, each class really is a range of slightly different kinds of stars. Thus astronomers now subdivide each spectral class into smaller divisions. For example, the sun is of type G2, which means it is two-tenths of the way from the hottest to the coolest spectral-type G stars. But these details are of interest mainly to the professional.

The Hertzsprung-Russell Diagram

In 1913, the American astronomer Henry Norris Russell got the idea of comparing different things we know about stars to see if there are any relations between them. For a number of stars of known distance, and thus of known absolute magnitude, he compared those absolute magnitudes with the stars' spectral types. The result is a diagram similar to that in Figure III.24. Actually, in preparing the figure, I have used more modern data than Russell had, but the idea is exactly the same. Each point on the figure represents a star of known spectral type and known absolute magnitude. Some of the more famous stars, and some I have made a point of mentioning, are labeled on the plot. Already *you* should notice, as did Russell, that there is a pattern to the picture.

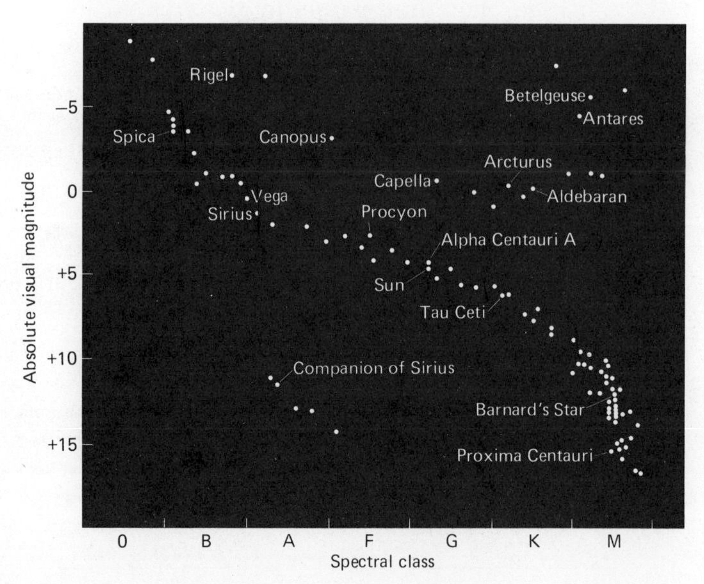

FIGURE III.24 The Hertzsprung-Russell (H-R) diagram plotted from a sample of nearby and bright stars.

The same pattern was noticed independently in 1911 by the Danish astronomer E. Hertzsprung in the course of his study of clusters of stars. All the stars in a particular cluster are at about the same distance from us, so their relative apparent brightnesses are really measures of their relative absolute magnitudes. Hertzsprung did not observe the spectra of the stars, but their colors. As we now know, however, both the color of a star and its spectrum depend mainly on its temperature, so the two really show the same thing. Anyway, Hertzsprung found that the real or absolute brightnesses are correlated with their surface temperatures, whether the latter is measured by spectral type or color. Thus a diagram of the type shown in Figure III.24 is now universally called *Hertzsprung-Russell,* or *H-R, diagram.*

The most conspicuous feature of the H-R diagram is that most stars lie in a diagonal swath running from the upper left (hot, luminous) corner to the lower right (cool, faint) corner. This sequence of stars is called the *main sequence.*

There are also some stars in the middle upper right, including Capella, Arcturus, and Aldebaran, that lie above the main sequence. These stars have about the same temperatures as main-sequence stars like the sun and Tau Ceti, but are more luminous. Now, according to black-body radiation laws (Act II Scene 2), the emission of energy per unit area depends only on temperature. Therefore, the only way those stars can be more luminous than main-sequence stars of the same temperature is for them to have larger radiating surfaces—that is, to be bigger. Thus, they are called *giants,* or sometimes *red giants,* for many are cool and red. There is also a scattering of stars at the top of the H-R diagram, including Betelgeuse and Antares, that are more luminous even than giants of the same temperatures; astronomers, with characteristic imagination, call them *supergiants.*

There are also a few stars lying below the main sequence. Most of these are quite hot—hotter than the sun—and thus are efficient radiators. But despite the fact that one of these stars emits far more energy per unit area than the sun does, they are typically hundreds of times fainter than the sun. These, then, must be very small stars. They are the *white dwarfs;* as you can see, the companion of Sirius, already mentioned, is among them. They are called *white* dwarfs, partly because some are white or blue-white in color (actually, most are blue), but mainly to avoid confusion with just plain *dwarfs.* Astronomers often call main-sequence stars "dwarfs" to distinguish them from giants and supergiants, even though some main-sequence stars are considerably larger than the sun. The term *red dwarf* thus refers to the common main-sequence stars of spectral-type M, like Barnard's star.

---

5. From the data in Appendix 10, plot a Hertzsprung-Russell diagram for the nearest stars. Now on the same diagram, but with different colored points, plot the brightest appearing stars from the data in Appendix 11. What is the striking difference between the sample of nearest and the sample of brightest appearing stars? Can you suggest an explanation?

6. What is wrong with the science fiction movie I once saw that dealt with a rocket ship on its way to the main sequence to visit a star there? (It's not as bad as the student who once wrote on an exam paper that a Hertzsprung-Russell diagram is a plot of Hertzsprung against Russell!)

Absolute Magnitudes from Stellar Spectra

In a few moments we shall see that the densities of the gases at the surfaces of white dwarfs are extremely much higher than they are of main-sequence stars of the same temperature, and that the densities of the surface gases of giants are very much lower. Supergiants have surface densities even lower yet. Remember that the density of a gas, as well as the temperature (although to a lesser extent than temperature) affects the amount of ionization. Consequently there are subtle differences between the spectra of supergiants, giants, and main-sequence stars and rather obvious differences for white dwarfs. Thus Capella and the sun have similar-appearing spectra—both are spectral-type G stars—but their spectra are not exactly alike; the experienced astrophysicist can tell at a glance whether he is looking at the spectrum of a supergiant, giant, main-sequence star, or white dwarf.

In other words, even without knowing the absolute magnitude of a star, from a careful examination of its spectrum the astronomer can locate the star pretty uniquely on the H-R diagram, which of course determines its absolute magnitude. To be sure, we learned how to do this from those stars whose distances were known in advance, but having calibrated stellar spectra with stars of known distance, we can now use their spectra to tell us the absolute magnitudes of stars whose distances cannot be measured directly. Therefore we can indirectly find the distance to any normal star bright enough for us to analyze its spectrum.

I said "normal" star because a relatively few stars do not fall into the regular pattern. Some are of rare types, perhaps with anomalous compositon. Others are pulsating stars, or stars that have recently undergone major evolutionary changes. We shall return later to these unusual stars. At least we can tell from its spectrum whether a star is of one of the usual types, and then find its distance from its spectrum, although not with absolute precision. There is a fair amount of scatter in the distribution of stars around the main sequence and giant sequences, so there is a little uncertainty in the absolute magnitude of a star as determined from its spectrum. However, we usually can find distances to stars this way to an accuracy of 20 percent or so.

---

7. Four main-sequence stars all appear equally bright through a telescope. Their spectral types are G, B, M, and K, respectively. Which is most distant? Which is nearest?

8. A main-sequence star of spectral type G2 has an apparent magnitude of 5. What is its distance?

9. Arrange the following stars in order of increasing size:
a) main-sequence spectral type G;
b) white dwarf;
c) supergiant of spectral type M;
d) giant of spectral type K.

Properties of Stars

As we have seen, the H-R diagram gives us a way of finding a star's approximate distance. It also helps us understand and appreciate a lot of other interesting properties of stars. For example, we saw in the previous scene how we find masses of stars from studying binary systems. Well, it turns out that the farther up the main sequence a star is, the greater is its mass. This correlation between stellar masses and absolute magnitudes is known as the *mass-luminosity relation*. We understand the reason for this relation today, but will come to it in the following act. It turns out that a star's luminosity depends very sensitively on its mass. Stars don't differ from one another very much in mass (most stars fall in the range from 1/10 to 10 times the sun's mass), but they range enormously in luminosity. The least massive stars, main-sequence M-type stars of around 0.08 solar masses, are more than 200,000 times fainter than the sun (but even they deliver 2×10^{21} watts of power!), whereas the most luminous supergiants known—perhaps of 50 solar masses—are a million times as bright as the sun.

The location of a star on the H-R diagram also allows us to determine its size, because stars radiate pretty much as black bodies do. Remember that the energy a black body radiates from each square centimeter is proportional to the fourth power of its temperature. A couple of examples may astound you.

Locate Betelgeuse on Figure III.24. It is an M-type star with a temperature less than 3000°, a little under half the sun's; let's round off and say it is just half. That means every square centimeter of Betelgeuse radiates only $(1/2)^4$, or 1/16 as much radiation as the sun. Yet Betelgeuse is 10 magnitudes brighter than the sun (in absolute magnitude), which corresponds to 10,000 times in light. The surface area of Betelgeuse must thus be $16 \times 10{,}000$, or 160,000 times the sun's. The surface area of a sphere is proportional to the square of its diameter (or radius), so Betelgeuse has a radius about 400 times that of the sun. (With the correct numbers, it comes out nearer 500.) If the sun could be placed at the center of Betelgeuse, the surface of that star would extend out beyond the orbit of Mars. You can see why we call it a supergiant. And there are even larger stars.

At the opposite extreme, consider a white dwarf like the companion of Sirius. Its temperature is roughly twice the sun's, so each unit area of its surface radiates 16 times as much light as a unit area of the sun's. Yet, being seven magnitudes fainter than the sun, its total luminosity is about 630 times

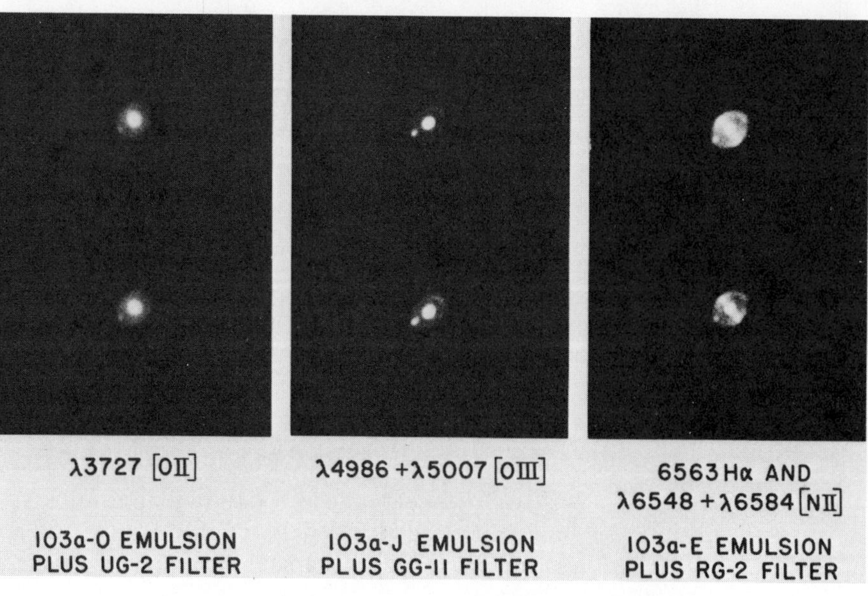

FIGURE III.25
One kind of star that has undergone a recent evolutionary change is a *nova*, a star that violently ejects some of its outer gases. Here are three photographs, in ultraviolet, green, and red light, of Nova Hercules, photographed with the 5-m telescope on Palomar Mountain in 1951, 17 years after its outburst. *(Hale Observatories)*

less, so its surface area must be less than the sun's by 630 × 16, or 10,000 times. This puts its radius at 0.01 times the sun's, or about the radius of the earth. And we shall see later that there are very much smaller things that we generally consider to be stars.

Now let us consider how much we would weigh on the surfaces of various stars compared to what we weigh on the earth. I don't mean to imply that we could actually weigh ourselves there; even stars that we call "cool" are so hot that all elements are in the gaseous state. However, the gases at the surfaces of those stars have weight, just as does the earth's atmosphere, so the calculation is not nonsense. Recall that your weight is the pull of the earth's gravitation on you. It is proportional to the earth's mass, and inversely proportional to the square of the earth's radius (the radius being our distance from its center). Now the sun has a mass of about 330,000 times that of the earth, and a radius 109 times as great. A simple calculation shows that your weight would thus be about 28 times greater at the surface of the sun than it is on earth.

How about Betelgeuse? I don't know its mass, but let's say it's at most 10 times the sun's. But its radius is 500 times as great as the sun's, so your weight on Betelgeuse would be 25,000 times less than on the sun or about 1000 times less than on earth. On the other hand, on the surface of the companion of Sirius you would weigh 10,000 times as much as on the sun. At the surface of that star a standard bowling ball would weigh about 2 million kilograms. That might cramp even Don Carter's style.

Now, as we've said, the atmospheric gases in the giant and white-dwarf stars have very low and very high weights as well. Yet, they are still at high temperatures. Thus in the giant stars the gases expand greatly and distend to very low densities; in supergiant stars the surface gases are at even lower densities; in white dwarfs they are extraordinarily dense. This is precisely why there are those spectral differences between main-sequence stars and non-main-sequence stars of the same temperatures.

Further Study of Stellar Spectra

The study of stellar spectra themselves tells us yet a great deal more about stars. Once we know a star's temperature and the density of the gases in its outer layers, for example, we can calculate the ionization and excitation states for the various kinds of atoms. When those physical conditions are taken into account, from the strengths of the various spectral lines (that is, how much they subtract from the light coming from within the star) we can calculate the relative numbers of various kinds of atoms and learn the chemical composition of those surface layers.

Magnetic fields affect spectral lines, too, so we can tell if a particular star has strong magnetic fields from its spectrum. We have already seen that we can find a star's line-of-sight velocity from the Doppler shift of its spectral lines. In more subtle ways, Doppler shifts give us additional information about the star, for example, the state of turbulent motions of its outer gases, and its rotation.

In Figure III.26 is the sun, with the sense of its rotation shown. Below is a small portion of a spectrogram of the light coming from the east (approaching) limb (*A* in the figure). A dark line is shifted to the left—to shorter wavelengths—with respect to a comparison line on either side of the spectrum. However, light from the center of the sun's disc (*B*) is not so shifted, because

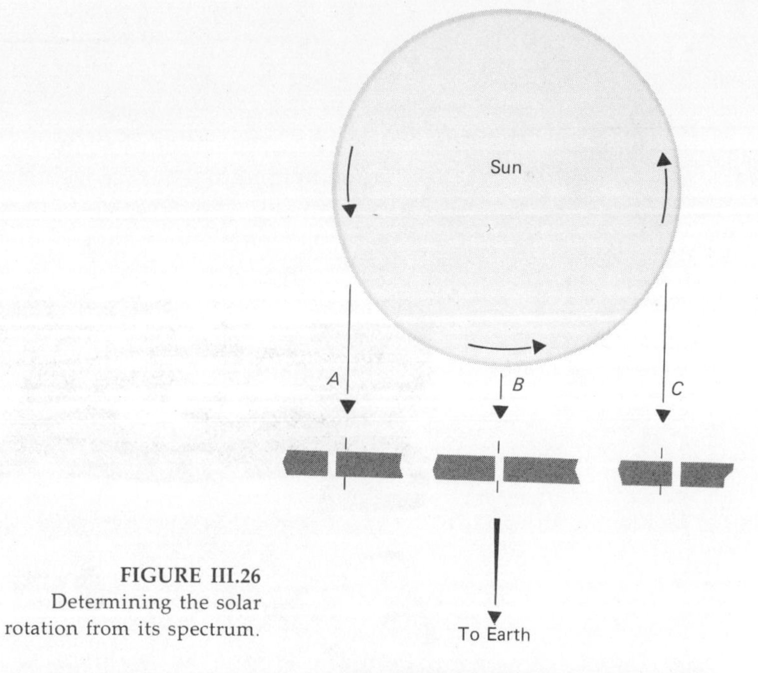

FIGURE III.26
Determining the solar rotation from its spectrum.

that part of the sun moves across our line of sight, and light from the receding limb (C) is shifted to longer wavelengths. Measurement of these Doppler shifts tells us the speed with which the surface is moving, so we can immediately calculate the sun's rotation rate.

Similarly, if a star is rotating, that part of the light from its approaching limb is blue shifted and that part from its receding limb is red shifted, just as in the sun. However, we are obliged to see the spectrum of the star's light coming from all parts of its disk at once. Thus each spectral line, arising from absorption by atoms in both approaching and receding parts of the star, is smeared out into a broad line that has a very characteristic appearance. The sun rotates slowly, taking about a month for a complete turn, and it would be very difficult to detect the sun's rotation from its spectrum if it were as remote as the other stars; the broadening of its lines would be too little. Many stars, however, are very rapid rotators, spinning in fractions of a day, and their rotations can be measured, except for those few we happen to observe pole-on. Most stars that are rapid rotators are much hotter than the sun—those of spectral types O, B, and A.

---

10. Show by a diagram how a star's spectral lines can be broadened if it is rapidly rotating.

Large-scale turbulent motions in the outer layers of a star also broaden its spectral lines, but in a different-appearing way that we can recognize. We should also mention stars that reveal by their spectra that they are ejecting matter into space. Suppose a shell of gas is ejected by a star. That gas in front of the star (between it and us) is approaching us, and produces an absorption line shifted to shorter wavelengths than the stellar lines. The gas in the shell, however, is reemitting the light it absorbed, and we sometimes see superimposed on the star's spectrum bright emission lines coming from this gas. In the next act we shall discuss stars that eject matter into space.

Still other stars vary in light with regular periods; although such stars are very rare compared to normal stars, over 13,000 in our Galaxy are cataloged. Each kind of variable star has a characteristic way in which its light changes in time. The period of a complete cycle of changes ranges, among the different types of variables, from a few hours up to more than a year. Now if we observe the spectrum of one of these stars at different times in its period, we find that the Doppler shift of its spectral lines varies as well and with the same period. The reason is that these stars are physically pulsating, and the changing Doppler shift occurs as the stellar surface alternately rises, moving toward us, and falls, moving away from us. Most pulsating stars are giants and supergiants which have reached stages of evolution at which they become temporarily unstable. We shall have more to say about them later.

The Stellar Population

Now let's take a look at the stellar population in general. The sun happens to lie pretty nearly in midrange of stellar masses, luminosities, temperatures, and sizes. Note in Figure III.24 that it is right about in the middle of the main sequence. It can thus be described as a pretty garden-variety star. It is *not* however, an *average* star.

If we were to judge the stars by the sample of brightest appearing stars, such as those in Appendix 11, we would arrive at very biased conclusions. Most of those brightest stars are relatively distant and appear bright because they are extremely luminous. Of the 20 brightest stars in Appendix 13, only six are within 10 pc. Even among the brightest 3000 stars, only 60 are within 10 pc. The overwhelming majority of naked-eye stars are more luminous than the sun. We might think, then, that the sun is not average because it is too faint. Just the opposite, however, is the case.

Most stars are faint main-sequence stars that can be seen only if they are quite close. We get a more nearly unbiased sample in Appendix 10, listing the known stars within 5 pc. Because 11 of the 39 stars listed are really binary

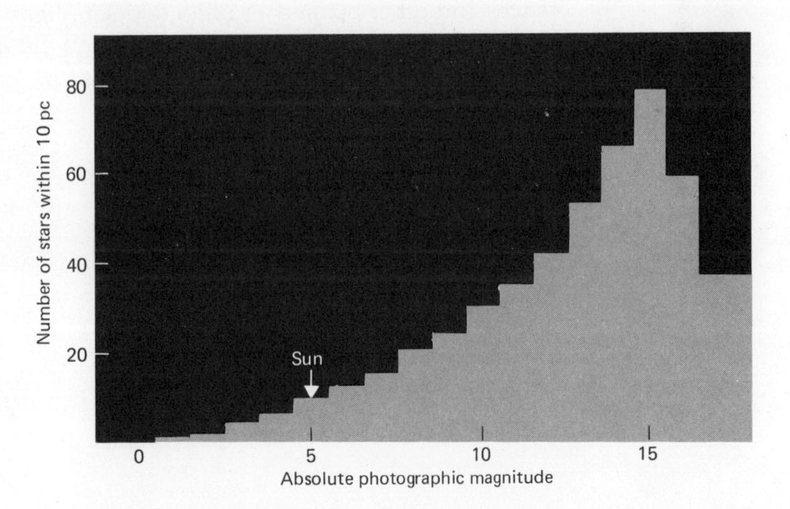

FIGURE III.27
Relative numbers of stars of various absolute magnitudes. *(Adapted from data published by W. J. Luyten)*

or multiple, the table includes a total of 51 stars. Only eight of these 51 stars are visible to the unaided eye, and only three are as intrinsically luminous as the sun. Figure III.27, adapted from work reported by the astronomer W. J. Luyten, shows the relative numbers of stars of various luminosities in the neighborhood of the sun. We see that the most common stars have absolute magnitudes near 15; they are only 1/10,000 as luminous as the sun. The brightest main-sequence stars, giants, and supergiants are relatively too rare to even show up on the plot. The sun, in other words, is bigger, more massive, hotter, and more luminous than 95 percent of the stars.

Those nearest stars also tell us something about the average density of stars in space. Within 5 pc there are at least 52 stars (including the sun). A sphere of radius 5 pc has a volume of 524 cubic parsecs. Thus there is about one star per 10 cubic parsecs. The mean separation between stars is about 2.1 pc (about 7 light years).

Summary

So *now* where do we stand? Up to the end of Scene 2, we saw how to find the distances to the very nearest stars. Now we see how we can find the distances to almost any star bright enough for us to photograph its spectrum. We have the means, therefore, to learn distances to a large number of stars as far away as 1000 parsecs or more to an accuracy of 20 percent or so.

Moreover, we have learned a lot about the ranges of masses, sizes, luminosities, and other properties of stars. We find that the sun lies pretty much in the middle of the range, but that most stars are faint and dwarfish compared to it. We also have a pretty good idea about how close together stars are in our part of the universe.

In Scene 2 we said that if the solar system could be plotted on a page of this book, the stars whose parallaxes could be measured would extend to about 8 km. Now let's suppose we plot all those neighboring stars such that that 8-km radius sphere would fit on this page. We now have the means to learn distances and other properties of a good sample of stars that, on the same scale, would extend out to more than 8 m—well outside my office. Not bad. Now we are in good enough shape to see how we can chart our entire Milky Way Galaxy.

SCENE 4

The Milky Way Galaxy–The Old Universe

The idea that the sun is part of a large system of stars is not new. We have already seen that Thomas Wright speculated on the possibility in 1750. Immanuel Kant had a similar idea in 1755. After all, in 1610 Galileo had shown that the Milky Way is made up of individual stars, too faint to be seen with the unaided eye. And William Herschel, in 1785, showed that we are surrounded in space with a great disk of stars. It appeared, indeed, that Herschel had discovered the flattened shape of the universe.

Herschel's Universe

The only problem is that Herschel's universe had the sun roughly in the center, and the stars thinning out with a distance of 1000 pc or so. We had come a long way from antiquity. No longer was the universe a system of crystalline spheres centered on the earth. The earth was only one of several planets in revolution about the sun. And the sun was only one of an uncountable number of stars in the sky, and not even too atypical among them. But then, why should it just turn out that the sun was in the very center of the universe, after all? The question may not be scientific, but it certainly bothered people. It bothered Harlow Shapley, for example.

By the beginning of the twentieth century, the reason for the sun-centered universe was no longer the pomposity of man but a trick of Nature. Maybe I should not say "trick," for we owe our very existence to that trick, as we shall see in Act V. But at least we allowed ourselves to be fooled. The point is, as I explained near the beginning of the last scene, that the space between the stars is not empty. There is scattered about extremely sparse interstellar gas, and even sparser dust. The dust consists of microscopic particles that, like an interstellar fog, scatter about light. Stars shining through the dust look fainter than they would in its absence, and hence farther away. The interstellar dust makes the stars appear to thin out in the distance, seeming to mark an end to their distribution before that end is actually reached. People who live in large cities, as I do, will know what I mean, for they have too often seen the rela-

FIGURE III.28
The Milky Way in *Cygnus*.
(Hale Observatories)

tively nearby landscape fade away into nothingness in that all pervading smog that plagues us.

So the stars seemed to thin out in the distance, long before the edge of the Galaxy is reached. Here and there this interstellar smog even appears to hang as opaque curtains in the sky, hiding much of the Milky Way beyond. Especially conspicuous among these *dark nebulae* is the great Cygnus rift, seemingly dividing the summer Milky Way into two parts. At one time (for example in Herschel's time, but also much more recently), such clouds of opaque matter were thought to be holes through which we could see out beyond the edges of the Galaxy. But why should so many "tunnels" be arranged along our line of sight?

Shapley's Galaxy

It was Harlow Shapley who awakened us to the more nearly true picture of the Galaxy in which we live.

When I first met Shapley, he had already retired. He was nevertheless extraordinarily alert and had far broader interests than I would have imagined. In fact, he seemed as interested in talking about philosophy, sociology, history, and politics as he did about astronomy. He impressed me as a grand man, and one full of ideas.

Shapley was a young staff member at the Mount Wilson Observatory, just north of Pasadena, California. Among other things, he had made a study of certain kinds of variable stars, and had, from an analysis of their proper motions, derived average distances, and thus average absolute magnitudes for them. Among them are the *RR Lyrae stars.* These variables have periods of less than 1 day. Shapley found that when they are halfway between their maximum and minimum light, they are about 100 times as luminous as the sun. There are more than 4400 of these variables in our Galaxy, and they are often found in certain star clusters.

FIGURE III.29
Harlow Shapley. *(Harvard University News Office)*

There are two kinds of star clusters in the Galaxy. The most common are the so-called *open* or *galactic cluster*. Over a thousand are cataloged; tens of thousands or more must exist. An example is the Pleiades, in Taurus; the brighter members are visible to the unaided eye. Open clusters typically contain 100 to 1000 stars, and are a few parsecs in diameter. They are loose, irregular groupings of stars (which is why they are called "open") and are generally found in the main plane or disk of our Galaxy (which is why they are called "galactic").

But we are concerned at this point with the other kind of cluster, the rarer kind. Only 125 are cataloged as belonging to our own Galaxy. We believe they are among the oldest star systems around. They usually contain at least 100,000 member stars, arranged with high spherical symmetry into systems 100 pc or so in diameter. They are not found preferentially in the plane of the Galaxy, but outside it, like bees swarming about a flower. They are called *globular clusters*. Nearly all of them contain RR Lyrae stars, and some contain more than 100.

Having learned the intrinsic luminosities, or absolute magnitudes, of RR Lyrae stars, Shapley was able to determine the distances to the globular clusters that contained these stars—at least those globular clusters near enough that he could identify the variable stars in them. Many of the globular clusters, however, are so remote that those RR Lyrae stars they contain are too faint to observe with the telescopes in use before 1920.

However, Shapley found the distances to a good number of globular clusters in which he could identify and study the RR Lyrae stars. He was thus able to learn the actual sizes (diameters) of those clusters from their known

FIGURE III.30
The globular cluster M13 in *Hercules. (Hale Observatories)*

distances and angular sizes. The idea is shown in Figure III.31. Three clusters are included in the figure, all having the same angular size as seen by the astronomer with the eye. Suppose a cluster is known to have the distance r. Then it must have the linear (actual) diameter D. If its diameter were less, say D_l, its distance would have to be less, r_l, and if its diameter were greater, say D_g, its distance would have to be greater, r_g.

Shapley found that the range of sizes of globular clusters is relatively small. Thus, he assumed that those more distant clusters have the same average size as do the nearby ones, and he estimated their distances from their observed angular diameters. It's Figure III.31 over again; there is only one distance that will fit with a given real linear size and an observed angular diameter.

---

1. Suppose we observe two globular clusters. Cluster A has exactly twice the angular diameter of cluster B. On the assumption that the clusters are intrinsically identical, how does the total amount of light we receive from cluster A compare with that we receive from cluster B?

FIGURE III.31
Relation between angular diameter , linear diameter, and distance.

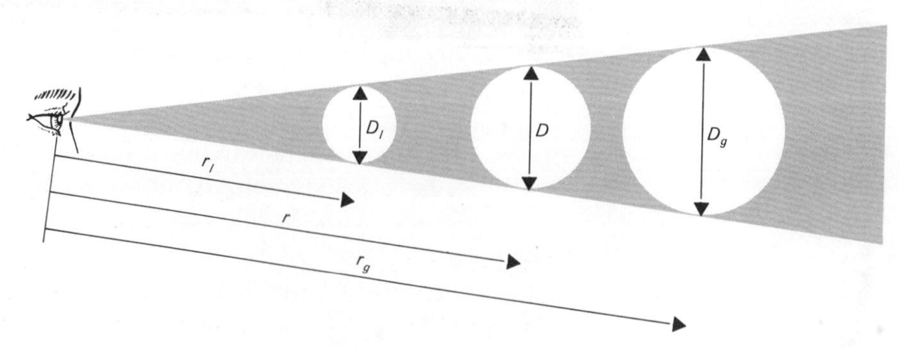

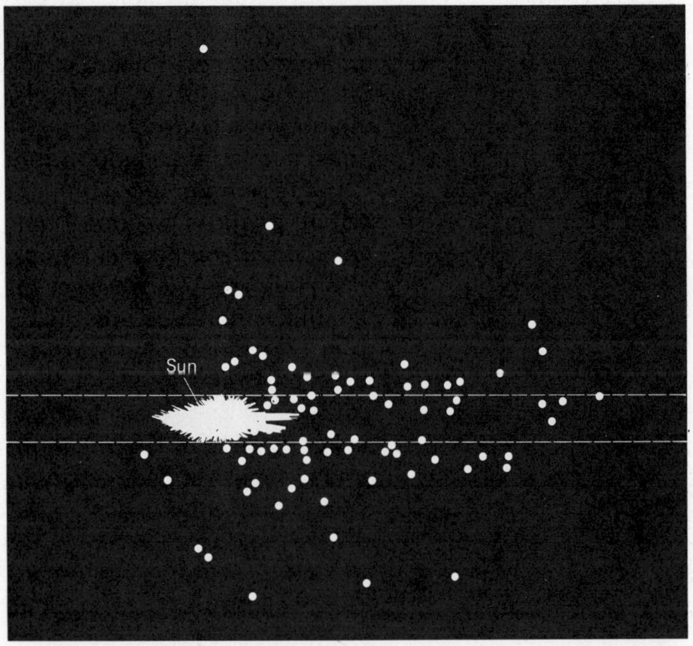

FIGURE III.32 Copy of a diagram by Shapley showing the distribution of globular clusters in a plane perpendicular to the Milky Way, and containing the sun and the center of the Galaxy.

Thus Shapley, from 1917 to 1919, derived distances, or at least approximate distances, to the 86 globular clusters then known. From their distances and directions, he could map out their three-dimensional distribution in space. What he found is illustrated schematically in Figure III.32. The globular clusters seemed to define a more or less spherical distribution in space. But the sun is *not* at the center of that distribution; there are many more clusters, and they extend farther, in the general direction of Sagittarius (in the Milky Way), than in the opposite direction. The whole cloud of clusters seemed to extend over a diameter of 100,000 pc or so. The sun seemed to be more than halfway from the center to one edge. The plane of Herschel's "grindstone" includes the sun, of course, and if extended, that plane would seem to pass through the center of the system of globular clusters, as shown by the dashed lines in Figure III.32.

Shapley then made the bold assumption that the globular clusters outlined the "bony frame" (as he put it) of our Galaxy. He suggested that the center of the Galaxy is at the center of the system of globular clusters, with the sun way off to one side.

---

2. Sketch a view of the distribution of globular clusters, like that in Figure III.32, but shown from the direction of the galactic pole, looking down on the plane of the Galaxy. Show the sun's position on the plot.

The Rotation and Mass of the Galaxy

Subsequent observations showed the sun to be moving with respect to the system of globular clusters. From the radial velocities of the globular clusters, we find that most of those in one direction are approaching us while most of those in the opposite direction are receding from us. Analysis shows that the sun, and also most of its neighboring stars along with it, is moving at a speed of from 250 to 300 km/s, in a direction lying in the plane of the Galaxy and roughly at right angles to the direction to the center. It appears that the sun has a roughly circular orbit about the Galaxy, much as the earth does about the sun. From our distance from the galactic center and from our orbital speed, we find that the period of the sun's revolution—the *galactic year*—is about 200 million earth years.

From the speed with which the sun moves in its orbit about the galactic center, we can calculate the approximate mass of the entire Galaxy. It turns out to be about 2×10^{11} solar masses, that is, 200 thousand million suns. If the average star has a mass of half that of the sun, that means the Galaxy may contain about 4×10^{11} stars.

---

3. Describe how you might estimate the mass of the Galaxy by using Newton's formulation of Kepler's third law. What assumptions and approximations must be made?

Think of all that has happened since the beginning of the Christian era (or, roughly, the time of Ptolemy). There are just about 1000 million minutes in that period of nearly 2000 years. Had Ptolemy started counting the stars in the Galaxy, and counted 400 per minute (about 7 per second), he would just now be nearing completion of the task. We have come a long way since the bible of astrology was written.

The Spiral Structure of the Galaxy

We know that between the stars of our Galaxy there are great clouds of interstellar matter in the form of gas and dust. We shall discuss the evidence for it and its nature in the next act. Suffice it to say here that we find glowing clouds of luminous gas (emission nebulae) and obscuring clouds of dust (dark nebulae) and also evidence of star formation in regions where the interstellar matter is found. Among the newly formed stars are occasional ones of very high luminosity. Those bright stars are markers showing regions of star formation, which is where the interstellar matter of our Galaxy is located.

The same is true in some other galaxies—those which seem to have interstellar clouds of gas and dust. Now most galaxies with interstellar matter are flat, like wheels, and have that material concentrated in arms spiraling out from their centers. The highly luminous young stars in the spiral arms give those galaxies the appearances of Fourth-of-July pinwheels. Since our Galaxy is flat and rotating, and contains interstellar matter and young, bright stars, there is every reason to expect that it, too, should show that spiral pattern. Have we evidence for it?

Yes!

One way of detecting the spiral structure of our own Galaxy is from optical observations of those glowing gas clouds and those high luminosity stars often associated with them. Unfortunately, because of the obscuring dust clouds in the plane of the Galaxy, light does not penetrate very far through it, but we can find glimpses of at least three segments of spiral arms in our local region of the Galaxy. They are shown in Figure III.33.

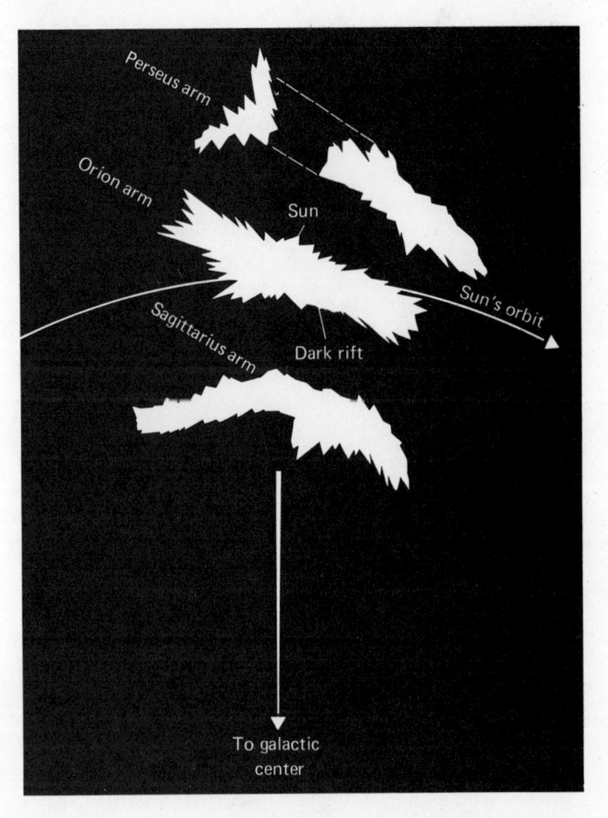

FIGURE III.33
A plot of three small sections of spiral arms as found from optical observations.

---

4. Why is it so much easier to see the spiral structure in external galaxies than in our our Galaxy, where the spiral arms are, in a sense, right here at home?

Our best evidence for spiral arms, though, comes from observations at radio wavelengths. Radio waves are unhindered by those dust particles, and we can receive radio radiation from interstellar gas in all parts of our Galaxy. The only trick is knowing from what distance it originates.

The 21-cm Line

How that trick might be managed was suggested in 1944 by the Dutch astronomer H. C. van de Hulst. The principal gas in interstellar space is hydrogen. Most of the hydrogen is neutral, which means that the atoms consist of protons with electrons moving about them somehow. Now the protons all have a spin, which gives them a certain tiny amount of angular momentum. Also the electron, even though its orbit is undefined (remember the uncertainty principle?) nevertheless has a certain orbital angular momentum. The angular momenta of the proton and electron can be represented by vectors, or arrows, as in Figure III.34. Normally, the two angular momentum vectors are in opposite directions (Figure III.34a), which means that the electron and proton have angular momenta that are oppositely directed.

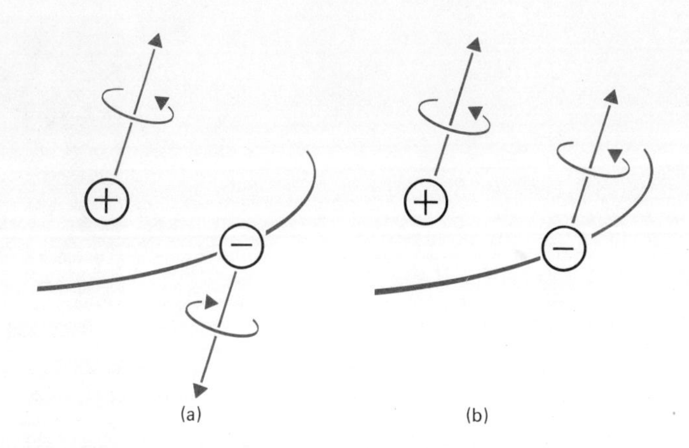

FIGURE III.34
The proton and electron in a neutral hydrogen atom normally have their angular momenta oppositly directed *(a)*; upon excitation, however (say, by collision), the angular momenta are directed in the same direction *(b)*.

However, in that near vacuum of interstellar space, very rare collisions occur between the particles. Sometimes such a collision causes the electron and proton to flip, as it were, with respect to each other, so that their angular momenta are aligned, as in Figure III.34b. In that configuration, the atom of hydrogen has very slightly more energy than normally, and so is very slightly *excited.* After a very long time, on the average about 10 million years, the proton and electron flip back to their normal configuration again, and emit that excess energy as a photon. The energy is so low that the photon is at radio wavelengths—in fact at a wavelength of 21.11 cm.

Hydrogen atoms are very sparse in space, and collisions between them are very uncommon. So in a small volume, for example, a million cubic kilometers, emission of 21-cm photons should be very rare. However, over the vast expanses of interstellar space—tens of thousands of parsecs—there are enough atoms emitting this way that the radiation, van de Hulst thought, should be observable. By 1951, sufficiently sensitive radio telescopes had been built that the interstellar 21-cm line of hydrogen was observed. Since then, a rather major effort in radio astronomy has been to observe the Galaxy in the "light" of 21 cm.

Two circumstances make this radiation important. First, it comes from the hydrogen gas of interstellar space, which, if our Galaxy is a spiral like many others, should be concentrated in spiral arms. Second, that radio emission is in a discrete spectral line, and even though at a radio wavelength it shows a Doppler shift like any other wavelike radiation.

It is inappropriate here to go into the gory details about how the analysis of the spiral structure of our Galaxy is carried out. But I think you can get the idea from a few quick comments. First, like the planets in the solar system, different parts of the Galaxy revolve about its center at different rates. Suppose we knew in advance exactly how fast all parts of the Galaxy turned and exactly where the spiral arms were located. Now, suppose you aim your radio telescope in a particular direction—say along the direction indicated in Figure III.35, where your line of sight intersects two different spiral arms. Knowing in advance where those arms are and how fast those parts of the Galaxy are turning, as well as how fast we (you) are moving, you can easily enough calculate how fast the segment of each spiral arm intersected by the field of view of your radio telescope is moving toward or away from us—that is, in the line of sight. Thus, you could predict exactly the radial velocity that gas would have and at what wavelength you would be receiving the Doppler-shifted 21-cm radiation from each piece of spiral arm.

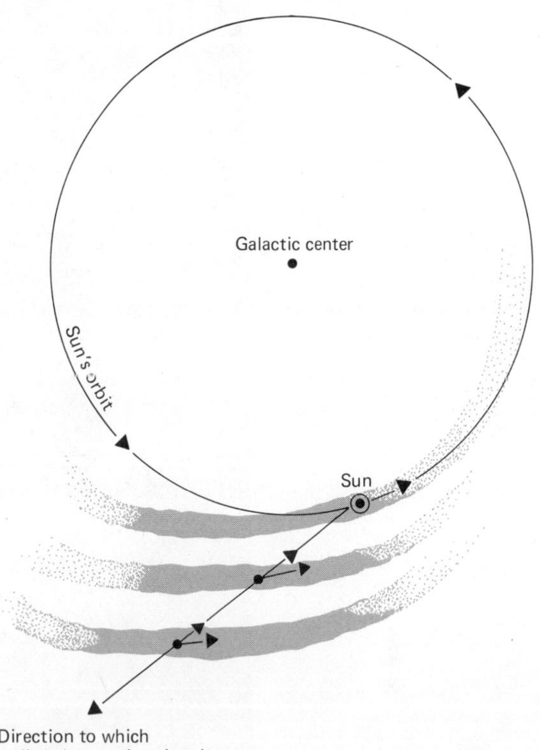

FIGURE III.35
Different segments of spiral arms intercepted by our line of sight should display different Doppler shifts.

Well, obviously we turn the problem around. We observe at what wavelengths we actually receive the radiation, and hence know at what radial velocity each of those other spiral arm segments are moving with respect to us. We then figure out where they have to be to have those relative velocities.

---

5. Most radio telescopes can be tuned to different frequencies or wavelengths, just as we tune a home radio. Suppose a radio telescope is directed along the plane of our Galaxy so that the line of sight intercepts two different spiral arms. Sketch the apperance (schematically) of the output of the receiver that detects and amplifies 21-cm radiation from the hydrogen gas in those arms as the receiver is tuned through frequencies corresponding to wavelengths near 21 cm.

Actually, the problem is very complex and many difficulties are encountered. Moreover, I do not want to give the impression that the spiral structure of our (or any other) Galaxy is thoroughly understood. It is an important and highly mathematical field of modern astronomy to understand the dynamics of our Galaxy. Nevertheless, we have at least shown that we do, indeed, live in a spiral galaxy, and considerable progress has been made in mapping out the spiral structure. Thousands of hours of observations with radio telescopes are combined to produce rough, if incomplete, maps of the spiral arms, such as that shown in Figure III.36.

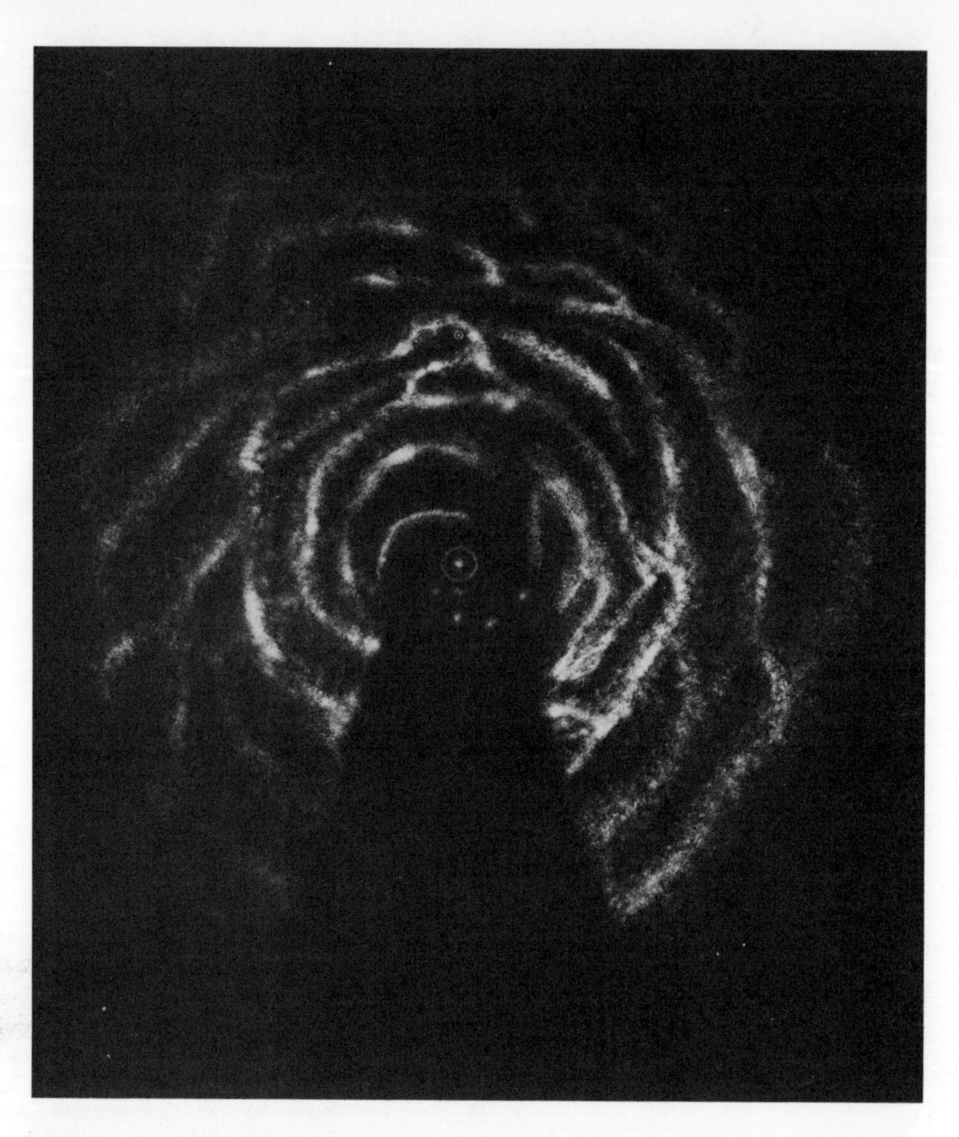

FIGURE III.36
A 21-cm map of the spiral structure of our Galaxy.
(Courtesy, Gart Westerhout)

Modern View of the Galaxy

So, though details are to be worked out, here is the general view of our Galaxy, as we see it today. Refer to Figure III.37. At the center is a large, roughly spherical bulge of stars. A significant fraction of the mass of the Galaxy is in that *nuclear bulge,* but the stars within it are still widely separated—throughout most of it, they are still parsecs apart. There are strange goings on at the central *nucleus* of the Galaxy, in the middle of that bulge, which we shall return to in Act VI.

The nuclear bulge is the central part of the main *disk* of the Galaxy. The disk and bulge together probably, but not certainly, contain most of the mass of the Galaxy. The whole disk is about 30,000 pc in diameter (or about 100,000 LY) with an uncertainty of about 20 percent. The sun is roughly in the central plane of the disk about 10,000 pc from the nucleus (again with an uncertainty of about 20 percent). The whole disk and bulge system is rotating, although not as a solid wheel. At the sun's distance, the rotation period is about 200 million years. The thickness of the disk is a few thousand parsecs, but the edges and surfaces are not sharply defined.

Embedded in the disk are clouds of interstellar material, mostly in the form of hydrogen gas, but with about a 20 percent admixture of helium and a few percent of other elements. About 1 percent of the interstellar matter is in the form of dust particles. The total mass of the interstellar matter is a few percent that of the whole Galaxy. We shall discuss it more in the next act. Most of the interstellar matter is concentrated into spiral arms, probably two of them, winding out from the nucleus, as shown in Figure III.37. Star formation occurs only where the interstellar matter is—that is, in the spiral arms. There, too, are to be found the very luminous stars, which, as we shall see, must be very young.

The sun is in or near one of the arms. Dust in that arm and in neighboring arms hides our view of most of the Galaxy if we try to look edge-on through it—that is, into the Milky Way; in those directions, our optical telescopes penetrate at most a few thousand parsecs, although we observe radio waves from all over the Galaxy. Open star clusters are located throughout the disk, but especially in and near the spiral arms.

FIGURE III.37
Schematic representation of our Galaxy.

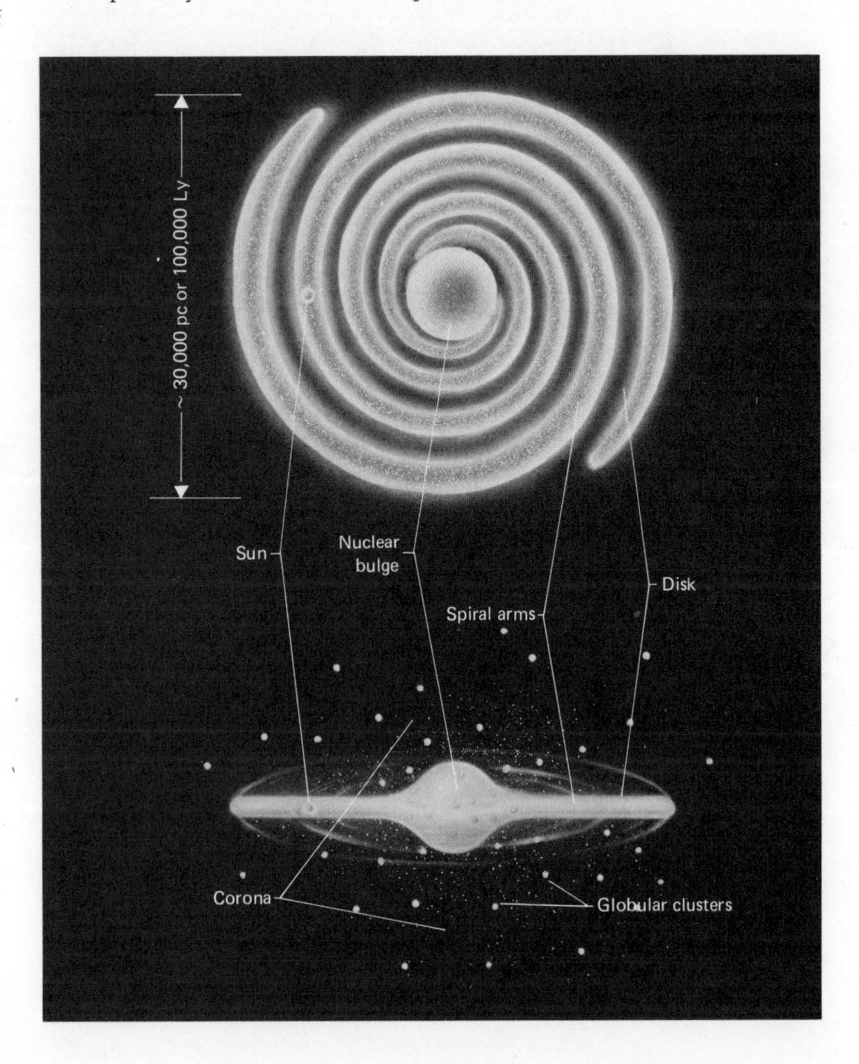

Outside the disk of the Galaxy and surrounding it in a roughly spherical configuration is the *corona*. The corona contains most of the globular clusters, which revolve about the galactic center in highly elongated orbits, but do not partake of the general rotation of the Galaxy. Out among the globular clusters is a sparse distribution of individual stars as well. From another galaxy, those coronal stars would produce a faint haze of light extending far from the center and the plane of the Galaxy. There is some speculation today that the corona of the Galaxy might actually extend very far into space—several times the size of the main disk, in which case it could contain a large fraction of the total Galaxian mass.

Well, there it is, in a few terse words. As you can see, we've learned a lot since Shapley's pioneering investigation. All the same, it was really Shapley who served the role of a twentieth-century Copernicus. By 1900 we had moved to the realm of the Galaxy, with the sun only one of many stars. But the sun was still in the center. It was Shapley who moved the sun away from the center, as Copernicus moved the earth from the center nearly four centuries earlier. Shapley thus played a key role in moving us into a new era of astronomy. After him, the Galaxy became a new and incredibly large universe. Ironically, within a decade Shapley's new universe was to crumble to insignificance, following a debate that Shapley won, but of which he was on the wrong side.

SCENE 5

Beyond the Milky Way–The New Universe

In 1781 the French astronomer Charles Messier prepared a catalog that he felt would be of considerable use to himself and other astronomers. Messier was interested in finding new comets. Now comets, when first discovered telescopically, usually appear as rather fuzzy, indistinct patches of light. Unfortunately, in Messier's view, lots of other things in the sky—"unimportant" things—also appeared fuzzy and indistinct. He could tell that they were not comets, though, because they didn't move in the sky from night to night. He cataloged about 100 of them so other astronomers wouldn't have to waste their time watching them to see if they moved.

It turned out that Messier's catalog (reproduced in Appendix 15), lists some of the most important objects in the sky. Some are conspicuous star clusters, like the globular cluster, Messier 13—or M13, as it is known among friends. Many others are described there as *nebulae*. *Nebula* is Latin for "cloud." Some really are clouds of gas in space, most of them in the Milky Way. About a third of the objects in Messier's catalog, however, are cloudlike objects in directions of the sky far away from the Milky Way. It is those "nebulae," far from the plane of our Galaxy, which are obviously *not* clusters of stars, with which we are concerned in this scene.

Messier was not the first to see some of these objects. In fact the brightest, M31, known until recently as the *Andromeda nebula*, is actually visible to the unaided eye on a clear autumn evening. In 1755, the famous German philosopher Immanuel Kant speculated on the nebulae as follows: The "analogy [of the nebulae] with the system of stars in which we find ourselves . . . is in perfect agreement with the concept that these elliptical objects are just [island] universes—in other words, Milky Ways . . ."[1] Kant had thus suggested that the nebulae are galaxies, like our own, but far beyond it in space. Kant's *island universe* hypothesis intrigued astronomers and philosophers alike, and led to a controversy that was not resolved until just over half a century ago.

[1] *Universal Natural History and Theory of the Heavens.*

FIGURE III.38
The Andromeda galaxy, M31, photographed with the Palomar Schmidt telescope. *(Hale Observatories)*

Catalogs of Nebulae

The first steps in the study of new kinds of objects are discovery, cataloging, and classifying. The age of discovery and cataloging began, roughly, with Messier, and continues today. The towering pioneer in the telescopic survey of the sky was William Herschel (Figure III.19). Herschel, born in Germany in 1738, was a talented musician, and in fact some of his chamber works are performed today (mainly at astronomical conferences). He served for a while in the army during the seven-years' war. He was, however, of delicate health and not suited to a military career. Thus his father got him out of the army and sent him to England, where he took up residence. There he gave concerts, composed, conducted, played at weddings and the like, and gave music lessons to support himself. He was, in fact, quite successful as a musician.

Herschel had a real interest, however, in astronomy. He rented a small telescope of the reflecting type (the type that had been invented much earlier by

Newton), and wanted to have a larger one of his own. Fortunately, he did not have funds to purchase one. I say "fortunately" because, with the help of his brother Alexander and his sister Caroline, both of whom he had arranged to have join him in England, he set about to build his own. He eventually constructed the world's first large reflecting telescope, with light-gathering powers never before realized.

Herschel's reflecting mirrors were of speculum—highly polished metal—rather than of aluminized glass, as we use today. His largest mirror was 4 feet (122 cm) in diameter and was mounted in a giant (at that time) telescope of 40-foot (12-m) length. With one of his telescopes Herschel discovered Uranus in 1781, an accomplishment that won him knighthood.

Herschel's most important work, in my judgment, was carrying out the most extensive systematic survey of the sky ever attempted to that date. Herschel's instruments did not have drives to keep them trained on the moving sky. Rather, he would direct his telescope to a point on the meridian and watch what crossed the field of view from his (probably precarious) observing ladder. Herschel would call out the description of what he saw, while sister Caroline, at the foot of the ladder, would record the information and the time. By such a technique he could survey the objects in a continuous thin east-west strip around the sky. By changing the telescope to a different elevation angle on a different night, he could survey another strip, north or south of the previous one. In this way, he and Caroline eventually surveyed the sky visible from Great Britain.

In the course of these investigations, Herschel discovered large numbers of faint patches of light—nebulae—and he carefully cataloged the objects. Later, his son, John, took his father's instruments to South Africa to extend the survey to the southern sky. In 1864, John Herschel published the *General Catalogue of Nebulae,* containing 5079 objects, of which 4630 had been discovered by him and his father.

The *General Catalogue* was revised and enlarged in 1888 by J. L. E. Dreyer; Dreyer's *New General Catalogue* contains 7840 nebulae and clusters. Even today, most bright galaxies are known by their *New General Catalogue* (*NGC*) numbers. For example, the Andromeda nebula, M31, is also NGC 224. The *NGC* has been supplemented twice (in 1895 and 1908) with the *Index Cata-*

FIGURE III.39
Herschel's 40-foot reflecting telescope. *(Yerkes Observatory)*

logues (*IC*), and has recently been revised with corrections and annotations by Tifft and Sulentic, of the University of Arizona. By 1908 nearly 15,000 nebulae had been cataloged. A few of these are true nebulae (gas clouds) in our Galaxy, and a few are star clusters. The overwhelming majority, however, are those faint, amorphous patches of light of the sort that Kant called "island universes."

---

1. Why is the term "island universe" a misnomer?

The Controversy over the Nebulae

In 1908 the 1.5-m (60-inch) telescope was completed on Mount Wilson, north of Pasadena, California. A decade later the 2.5-m (100-inch) telescope was put into operation at the same location. With these great reflectors it was possible to photograph the nebulae with relatively high resolution. In some of them, individual stars could clearly be seen. If there were some way of knowing the absolute magnitudes of those stars, astronomers could have used the inverse-square law of light to calculate their distances, and hence the distances to the nebulae in which they appeared. If they were stars like the sun, for example, the nearest nebulae would be a thousand or so parsecs away—well inside our own Galaxy. If, on the other hand, those stars were like the most luminous stars in our Galaxy, then, to account for their apparent faintness, they and their nebulae would have to lie far beyond the Milky Way Galaxy, and would have to qualify as "island universes," suggested by Kant.

---

2. What kinds of normal stars would you expect to be the easiest to observe in neighboring galaxies? Why?

Those stars were far too faint for their light to be analyzed spectrographically. On the other hand, in 1917 a few *novae* were found in some of the nebulae, and searches of older photographs revealed other novae that had occurred in previous years. Now, novae are stars that suddenly flare up to great brilliance, and then fade out again over a period of a few months. (We shall discuss their probable origin in the next act.) By 1917 there had been 26 known novae in our Galaxy, and at maximum brightness they reached several thousand times the luminosity of the sun. But the novae observed in the nebulae appeared so faint that if they were like those in our Galaxy, the nebulae would have to be remote island universes. So that would seem to have settled the matter.

But it didn't. For one thing, among the novae in the nebulae, there were two "ringers"—stars that flared up in brightness by amounts many thousands of times that of the other novae. One of the two was S Andromeda, which appeared in the Andromeda nebula in 1885, and was almost bright enough to be seen with the unaided eye. We know today that those bright fellows were *supernovae*, stellar explosions of an entirely different order from ordinary novae. We shall be meeting them also in Act IV. But in the early twentieth century no one knew about supernovae, and thus it was not clear which (if either) of the kinds of novae seen in the nebulae were like those in our Galaxy.

The globular cluster M13. (*U.S. Naval Observatory*)

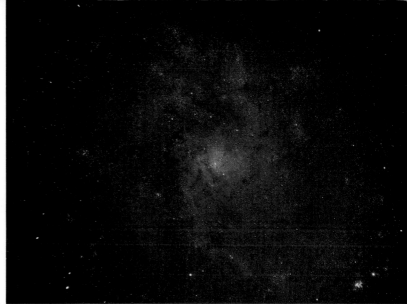

The spiral galaxy M33, photographed with the 5-m telescope. (*Hale Observatories*)

The Milky Way; the light streak is the trail of an artificial earth satellite. (*U.S. Naval Observatory*)

The spiral galaxy M31 in Andromeda, photographed with the Palomar Schmidt telescope. (*Hale Observatories*).

FIGURE III.40
Two photographs of the galaxy NGC5253, both made with the Palomar Schmidt telescope. The second, taken 16 May 1972, shows a supernova; the first, taken 4 June 1959, does not. *(Hale Observatories)*

A further complication was a set of observations that turned out to be just plain wrong. Those observations, spread over more than a decade and spanning the 1920s, were the work of Adriaan van Maanen, an astronomer at the Mount Wilson Observatory. Now, I emphatically do *not* want to give the impression that van Maanen was a bad astronomer because he made a mistake. Quite the contrary. He made many important contributions, and was highly respected. *Every* astronomer makes mistakes, and especially when he is groping at the frontier of knowledge, trying to make exceedingly difficult measurements.

I mention the van Maanen work to illustrate how science progresses: sometimes by trial and error; sometimes by lucky hunches; often with many wrong guesses and blind alleys; by many steps, some sideways, some backward, and some forward. Gradually the foreground gets pretty well trampled over, and the wrong ideas get weeded out, so that foreground becomes pretty well understood. Just beyond, though, is the front line of science, where there is always uncertainty, doubt, and many mistakes. Science does not roll ahead in limousines; it struggles ahead on hands and knees. Only in hindsight do we look back over newly conquered territory and say, "My God! How obvious! Why didn't we see that before?"

But back to van Maanen. He had attempted to see if he could observe proper motions of the nebulae, or of stars in them. His procedure was to compare two photographs of the same nebula taken, say, 10 years apart, and to measure carefully on each photograph the positions of stars or other resolved objects in the nebulae with respect to faint stars in our own Galaxy. Well in those nebulae which looked like spiral pinwheels (we now call them *spiral galaxies,* of course, and we live in one), he found that the stars seemed to be moving and with ordered patterns indicating that the nebulae were rotating. This seemed to prove that the nebulae were not galaxies, for if they were that far away, in order for us to see them rotating, even after 10 years, the individual stars in them would have to be moving faster than light! (In actual fact, the spiral galaxies *do* rotate, but in the *opposite* sense from that measured by van Maanen.)

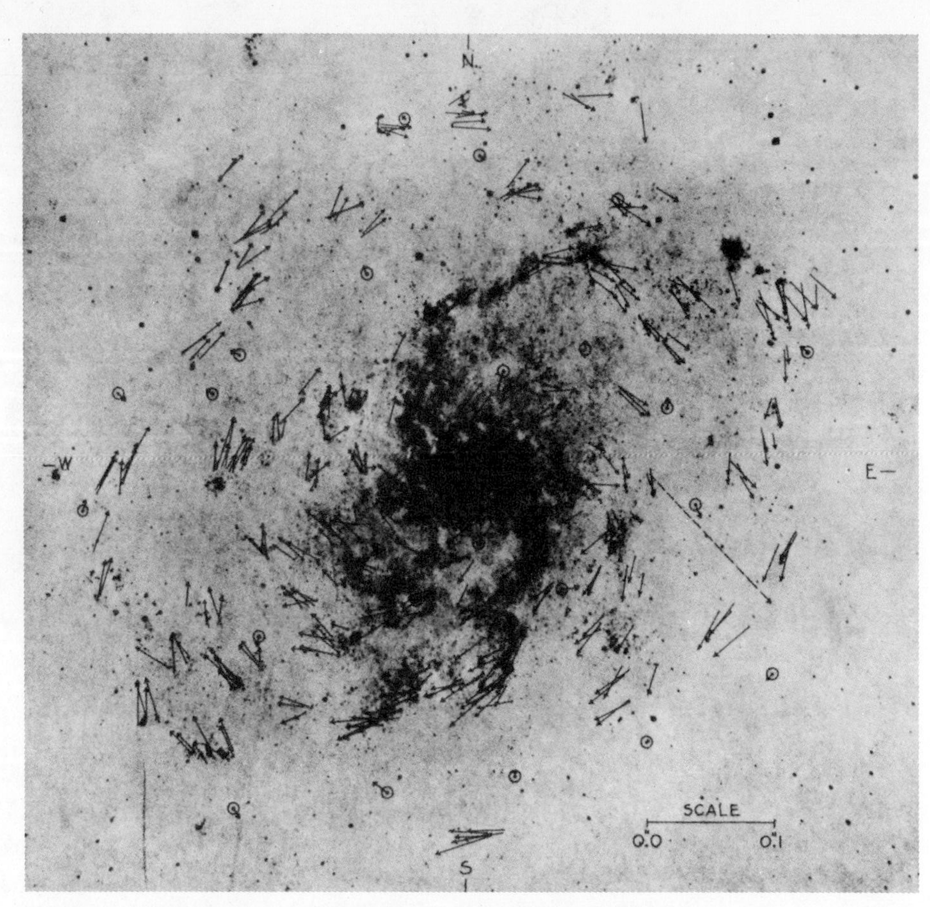

FIGURE III.41
The proper motions of stars and other resolved features indicated on a photograph (negative print) of the spiral galaxy M33. This plot, published by A. van Maanen in 1923, shows measurements now known to be incorrect. *(Yerkes) Plate XIX* Astrophysical Journal, *Vol. 57, 1923*

There has been much discussion on how van Maanen went astray. The motions he measured were exceedingly small, and any number of minor sources of error can give rise to the apparent results. Not the least of these is the very human tendency to be looking very hard for an effect, and think you see it buried in the observational "noise." (Another example of this sort of thing is that, in the first half of this century, many astronomers who observed the planet Mars visually through a telescope thought they saw "canals.") I don't think van Maanen should be faulted for his work; in fact, it was a clever idea to look for rotation in the nebulae. The mistake was in not suspending final judgment until the data were more definitive. It is, I think, a marvelous object lesson that we should all bear in mind (including astronomers); it is especially appropriate to that extremely exciting frontier we shall be probing in Act VI.

The Shapley-Curtis Debate

The controversy over the nature of the nebulae reached its culmination in the famous debate before the National Academy of Sciences on 26 April 1920, between Harlow Shapley, of the Mount Wilson Observatory, and H. D. Curtis, of the Lick Observatory (the Lick Observatory, owned by the University of California, is near San Jose, California). Curtis supported the island-universe hypothesis in the debate, and Shapley opposed it. It is generally agreed that Shapley got the better of the debate, and his views certainly more nearly represented the mainstream of astronomical thought at the time. Allan Sandage put it nicely: "Perhaps the fairest statement that can be made is that Shapley used many of the correct arguments but came to the wrong conclusion. Curtis, whose intuition was better in this case, gave rather weak and sometimes incorrect arguments from the facts, but reached the correct conclusion."[2]

But less than five years after the Shapley-Curtis debate, the astronomical horizon had been lifted to reveal a universe more than a million million times the volume envisaged by that "mainstream of thought" of the early 1920s. The new cosmos was revealed mainly through the work of one young astronomer, near the beginning of his long career at the Mount Wilson Observatory—Edwin Hubble.

The Resolution of the Controversy

In 1922, John C. Duncan, at Mount Wilson, while searching for novae in the spiral nebula, M33, found three variable stars. In 1923, more variables were found in the nebula, NGC 6822. Hubble meanwhile was searching for novae in the Andromeda nebula M31. In the fall of 1923, he found two novae, but also a fainter object (of the 18th magnitude) that looked like a nova, but was visible on older plates. Hubble's subsequent investigation showed that the faint star was a *Cepheid variable.* Now Cepheids are known to exist in our Galaxy. They are giant pulsating stars with periods that fall in the range of from a few days to about 100 days. The greater the period of pulsation of a Cepheid, the more luminous it is at median light (halfway between its brightest and faintest).

The Cepheid Hubble found in M31 had a period of one month, and a periodic variation in light like that illustrated in Figure III.42. At that time such a Cepheid was thought to have an absolute magnitude of −4; to account for its faint apparent magnitude, Hubble found that M31 had to have a distance of about 300,000 pc. The most recent calibration of the luminosities of Cepheids shows that Hubble's variable is even more luminous than he had assumed, which puts M31 even farther away; our modern distance estimate for it is 680,000 pc, or about 2,200,000 light years.

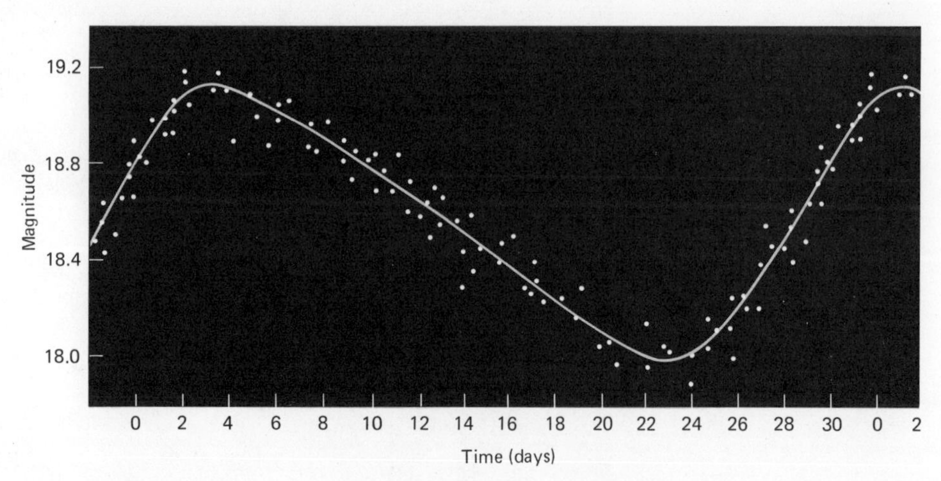

FIGURE III.42
Light curve of a typical Cepheid variable.

[2] A. R. Sandage, *The Hubble Atlas of Galaxies.* Carnegie Institution, Washington, D.C., 1961, p. 3.

3. How might you go about searching for Cepheid variables in a neighboring galaxy?

Wishing to make sure before publishing his result, Hubble studied M31 again the following fall and found 12 more Cepheids, and also 22 Cepheids in M33. His analysis of the newly found Cepheids confirmed that M31 and M33 are at great distances—far beyond our Galaxy. By this time, a number of other astronomers heard of Hubble's discovery. Henry Norris Russell (of the Hertzsprung-Russell diagram) urged Hubble to present his results in a paper at the meeting of the American Astronomical Society to be held in Washington, D.C., from 30 December 1924 to 1 January 1925 (they worked on holidays in those times!). In fact, the meeting was being held in conjunction with one of the American Association for the Advancement of Science, and there was to be a prize for the best paper. Hubble could not attend the meeting himself, but he promised to send a paper which could be read for him.

Hubble's paper, as it happened, was slow in arriving. On the first night of the meeting, Russell remarked to some colleagues at his hotel that he had not yet received it. He and Joel Stebbins, secretary of the Society, thus decided to go to the telegraph office and send a wire to Pasadena to see what had happened. On the way, however, they stopped at the hotel desk to make out the proper telegraph form. There, on the floor behind the counter, where it had fallen unnoticed, Russell saw a crumpled brown envelope with his own name on it. Of course it turned out to be Hubble's paper, and when Russell and Stebbins returned with it to the lobby, their friends were most impressed at the efficiency of the telegraph system.

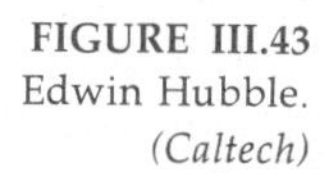

FIGURE III.43 Edwin Hubble. *(Caltech)*

Hubble's paper won him the prize—or at least half of it; he shared it with a scientist in some other field. The summary was published early in 1925, and publication of the full details of his work soon followed. This was just the beginning; Hubble was not only a pioneer, but a leading (probably *the* leading) investigator of galaxies up to his death. And as if he hadn't already made enough of a mark, in 1929 and 1931 he announced, with his friend and colleague Milton Humason, the discovery of what is now known as the *Hubble law:* All remote galaxies are moving away from us, with radial velocities, as indicated by the Doppler shifts in their spectra, that are greater in proportion to the distances of those galaxies. The Hubble law shows the universe to be expanding. We come back to this subject in Act VI.

So Hubble played the key role in opening up the field of extragalactic astronomy. As a graduate student I was fortunate enough to have a course from him, and as part of my student assistantship even worked for him for a while. He had a certain nobility about him; he was a gentleman in the true sense of the word. Some people have described Hubble as aloof, but I found him to be warm and kind, and also found that he always had time to talk with anyone who sought his advice or help. He died in 1953 at the age of 63.

Nature of Galaxies

Most of the so-called nebulae, then, are not nebulae at all, but galaxies—vast stellar systems. Many are spirals, like our Galaxy, and contain stars, clusters, and extensive clouds of gas and dust (the *real* nebulae) within them. Many others, perhaps the majority, are less flattened, being spherical or ellipsoidal; these *elliptical* galaxies also contain stars and clusters, but very little conspicuous interstellar material. A few percent of the galaxies are irregular, and are typified by the Large and Small Clouds of Magellan, the two nearest external

FIGURE III.44
The Clouds of Magellan.
(Sky and Telescope)

galaxies, both about 50,000 pc away. The Clouds of Magellan are conspicuous to the unaided eye, but not from Europe or from most parts of the United States, for they are too near the South Celestial Pole. They look rather like detached pieces of the Milky Way, and were described by the famous circumnavigator for whom they are named.

Now if all galaxies were identical, once we know how far away one is, we would know its absolute magnitude, and thus that of every other galaxy, so we would easily know their distances as well. But galaxies differ enormously from one another. Spirals range in intrinsic brightness by more than a factor of 100, and ellipticals by more than a factor of a million. Our own Galaxy is more or less typical among large spirals, but the largest ellipticals are intrinsically dozens of times brighter than our Milky Way Galaxy, and the smallest are many thousands of times fainter. Thus we must be more clever, and use indirect techniques to measure galaxian distances.

The Extragalactic Distance Ladder

The first step, we have already seen, is provided by the Cepheid variables. When we recognise a Cepheid in a galaxy, we can calculate its distance from the apparent magnitude of that Cepheid at median light. But Cepheids, although giant and supergiant stars, are still bright enough to be seen only in about 30 of the very nearest galaxies. Cepheids, however, are not the brightest stars in other galaxies. For those galaxies in which we *do* see Cepheids, we can obtain distances and thus learn how intrinsically luminous are the very brightest stars in galaxies. Then if a galaxy is too remote for us to see Cepheids in it, but not too remote for us to see its very brightest stars, we can estimate its distance from their apparent faintness. About 1000 galaxies are within the range in which we can see those brightest stars.

FIGURE III.45 A field of stars in the Andromeda galaxy, M31, with two variable stars marked. Photographed with the Palomar 5-m (200-inch) telescope. *(Hale Observatories)*

4. Suppose the brightest stars in a certain galaxy are 36 times as luminous as are the brightest Cepheid variables in that galaxy. Suppose, further, that the brightest Cepheids are just barely detectable with a telescope. How much farther away would the galaxy be for its brightest stars to be just barely detectable?

Elliptical galaxies, on the other hand, do not contain Cepheids, and moreover, for reasons we shall come to in the following act, their brightest stars are far less bright than those in spirals. In some ellipticals we can see globular star clusters, though, and we can estimate their distances from the apparent faintnesses of those clusters. You see, the trick is to recognize some kind of "standard candle" in a galaxy – some kind of object that we can (1) recognize, so we know its intrinsic luminosity, (2) measure its apparent brightness, or magnitude, and (3) by comparing that to the intrinsic luminosity, calculate its distance.

Another trick is to use a standard yardstick – that is, recognize something whose intrinsic *size* is known. Remember (previous scene) how Shapley found distances to remote globular clusters from their angular sizes and known real diameters? One such kind of yardstick that has been used by A. R. Sandage, at the Hale Observatories (the new name of the Mount Wilson and Palomar Observatories), and G. Tammann, at Basel, Switzerland, is the linear diameter of the largest bright nebulae – glowing gas clouds, like the Orion nebula – in other galaxies. They determined those linear diameters by measuring angular sizes of the nebulae in several nearby spiral galaxies of known distance. Then, in somewhat more remote spirals in which they don't see individual stars but can still make out the glowing nebulae, Sandage and Tammann measure the angular sizes of those nebulae and calculate their distances on the assumption that they are of the same intrinsic size as the nebulae in the nearer galaxies.

Sometimes the appearance of a galaxy provides an estimate of its absolute magnitude. About 1960, S. van den Bergh, an astronomer at the David Dunlap Observatory in Toronto, showed that there is a rough correlation between the morphological appearances of nearby spiral galaxies of approximately known distance and their absolute magnitudes. Van den Bergh sorted spirals into what he called *luminosity classes,* and calibrated the absolute magnitude that goes with each class. Now for spirals that are too remote even for their nebulae to be measured, it is sometimes possible to make out their detailed spiral shapes on high-quality photographs, and assign them to their proper luminosity classes, thereby learning their approximate absolute magnitudes and distances.

The overwhelming majority of spiral galaxies, however, are too remote even to classify into the van den Bergh types, and no such morphological classification is possible for the more amorphous-appearing elliptical galaxies. In fact, at present we have no way of knowing the absolute magnitudes, and hence distances, to the individual faint galaxies that show up by the tens of thousands on our telescopic photographs. We shall see, however, that most – if not all – galaxies occur in groups and clusters. Our own Galaxy is part of a small cluster of 21 known galaxies (most of them small ellipticals) with an overall diameter of about a million parsecs. We call it, rather affectionately, our *Local Group*. Small groups like the Local Group are exceedingly

FIGURE III.46
The nearest great cluster of galaxies, that in Coma Berenices, also known as A1656 (from the author's catalog). Photographed with the Palomar Schmidt telescope. *(Hale Observatories)*

common, but scattered about at wide separations we find the occasional rare great clusters of galaxies. A typical great cluster that is not too distant may contain hundreds or thousands of galaxies bright enough to photograph.

A great cluster contains enough member galaxies to provide a good statistical sample of them so that we can pick out the individual members that are intrinsically of the highest luminosity. Approximate distances are known to some spirals that are members of relatively nearby great clusters. Thus we know roughly how far away some clusters are, and, in turn, how intrinsically bright their brightest members are. Now if we make the (uncertain) assumption that the brightest galaxy in each rich cluster is the same as in every other rich cluster, we can tell the relative distances of the other clusters.

---

5. Suppose the brightest galaxies in one cluster of galaxies appear four times as bright as the brightest galaxies in a second cluster. Which cluster is probably more distant, and by how many times?

Now, I don't want to give the impression that it's all as easy as that. I have oversimplified some points and left out other details. Moreover, the measurements are complicated and difficult to make, and are plagued with uncertainties. It has been my intention, rather, to give a rough idea how "it is done," and, in particular, to illustrate how we build up a great, if shaky, extragalactic distance structure, with each layer resting firmly or not so firmly on the one beneath it.

It might have occurred to you that since we know that there is a correlation between the radial velocities and the distances of galaxies, we could simply use the radial velocity of a galaxy to estimate its distance. Well, we do often infer the distance of a galaxy from the Doppler shift in its spectrum, but the velocity-distance relation, or Hubble law, for galaxies must be calibrated; it must be established from galaxies whose distances we already know. Moreover, the relation is not one of exact proportionality when large distances are reached, and it is important to establish just what the relation is between the velocities and distances of remote galaxies in the investigation of the expanding universe.

We shall, of course, be returning to these questions later, and we shall have a lot more to say about galaxies. In this scene, I have wanted to show how we have extended our knowledge of the astronomical distance scale to extremely remote objects in the universe. As you would expect, the distance scale is more and more uncertain at greater and greater depths in space. During the summer of 1976, I was at a meeting on the subject in Paris in which leading experts in the field were disagreeing with one another on the scale of extragalactic distances by a factor of two. But on the other hand, it is quite clear now that early estimates by Hubble of the distances to the most remote galaxies were too small by as much as 10 times. So we have made considerable progress. I think that we probably know distances to our nearest neighboring galaxies (for example, the Clouds of Magellan and M31) to an accuracy of about 20 percent. Our distance determinations to the more remote galaxies are far more uncertain, but in my judgment the correct values are very likely to lie between 60 percent and 160 percent of those of Sandage and Tammann —a far cry better than Herschel's knowledge!

Summary

So *now* where do we stand? Recall that if we put the solar system map (including Pluto's orbit) on a page the size of the one you are reading, the nearest star would be 0.6 km away, and those neighboring stars within 20 pc would extend out to 8 km. On that same scale, our entire Galaxy would occupy a region about 12,000 km across. But now, if you barely fit our entire Galaxy on this page, the other galaxies that we can observe would lie at distances out to 10 km and cover a region comparable to the size of Manhattan Island. About 1000 million galaxies are potentially observable with our present telescopes, and they extend to a distance of at least several thousand million light years.

It is possible that somebody reading this book may one day happen to become an astronomer (although I am not writing it with him in mind). If so, he will spend his life observing cosmic objects as they are revealed by light, radio waves, X rays, or other kinds of radiation. Now if he chooses *not* to study objects in our own solar system, it follows that almost every new thing he learns about the universe during his entire career will be by means of radiation that has already left the objects of his study, and indeed is already almost here. The cosmic events he will be investigating occurred long ago, many of them when the great dinosaurs were roaming the earth.

ACT IV

THE BIRTH, LIFE, AND DEATH OF A STAR

SCENE 1

Anatomy of a Star

As we have seen, galaxies extend throughout the universe. There may be atoms of gas in the space between galaxies; there may even be a great amount of such intergalactic matter. We don't know. It is pretty well agreed, though, that at least a good fraction, and perhaps virtually all, of the matter of the universe is in galaxies.

Galaxies, we have also seen, often have lots of interstellar matter. In our Galaxy, the interstellar gas and dust make up at most only a few percent of the total mass. Some galaxies have a greater percentage than ours; many others (for example, elliptical galaxies) evidently contain extremely little interstellar material. So it is safe to say that most of the matter in galaxies, and hence probably of the entire universe, is concentrated into those little balls we call stars. If we want to understand the universe, and ourselves, therefore, we had better try to understand something about stars. Fortunately, there is one star pretty nearby, and available for study.

Gross Properties of the Sun

The distance to that star—the sun—averages about 150 million km. Since we know how far away it is, and by how much it accelerates the earth, we can easily calculate the mass of the sun with the help of Newton's laws. It is about 2×10^{33}g, or 2×10^{27} metric tons, which is about 330 thousand times the earth's mass (more accurate values for many of these quantities are given in Appendix 5).

We've already explained how we measure the luminosity of the sun; it is about 4×10^{33} erg/s. By the way, note that the sun radiates into space about 2 erg/s of energy for every gram of its mass.

We have also told how we can measure the sun's rotation. Now it is very interesting (and, we shall see, significant) that the outer layers of the sun do *not* rotate all at once—that is, the sun does not spin as a solid body would. At its equator the period of rotation is about 25 of our days. However, at a latitude of 30°, both north and south, the rotation period is *longer* by about $2^1/_2$ days. At 75° latitude it takes 33 days, and near the poles about 35 days, to turn

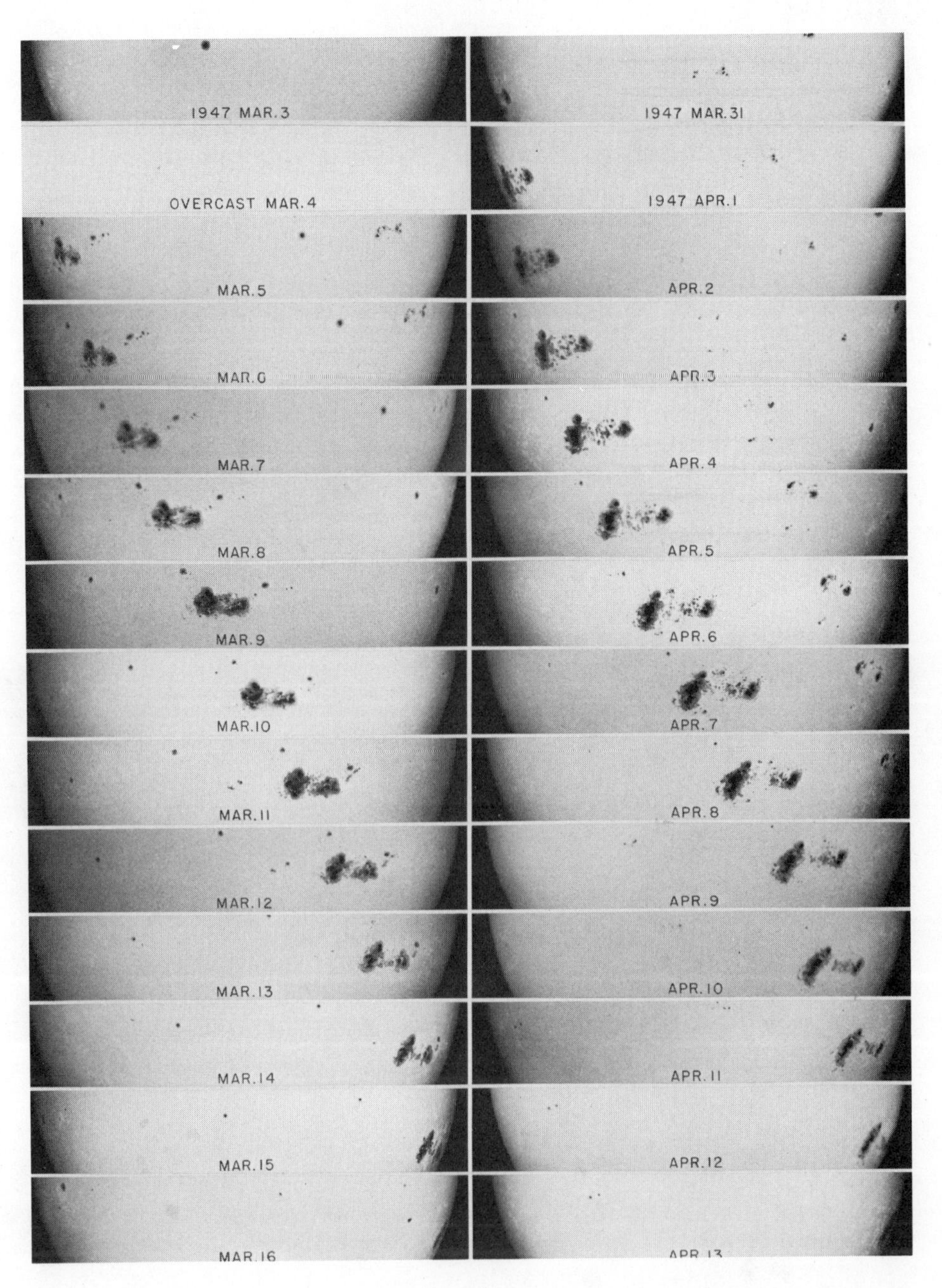

FIGURE IV.1
Series of photographs showing the motions of sunspots from day to day, and indicating the solar rotation. *(Hale Observatories)*

around once. We refer to this variation of period with latitude as the *differential rotation* of the sun

The next question is how *big* is the sun? That's a harder question to answer because it depends on how we observe it. If we look at the sun in the "light" of long-wave radio waves, it has a diameter of a few million kilometers. It looks smaller, though, at shorter wavelengths, until we get down to about 1 cm. In visible light, infrared, and microwave radiation, the sun seems to be about 1.4 million km in diameter, or 109 times the diameter of the earth. The point is there is no actual surface on the sun; it's a big ball of hot gas, and the gas keeps getting thinner and thinner farther and farther out into space. Technically, the outermost parts of the sun surround the earth.

1. By what factor does the volume of the visible sun exceed that of the earth?

The Solar Photosphere

The "surface" we see when we look at the sun is the depth into which our vision can penetrate no deeper as we look toward its center. That depth, in turn, depends on where the gases become opaque to the radiation we are observing. They become opaque to light in the region called the *photosphere* (meaning, literally, "light sphere"); the photosphere is the region in the sun from which visible radiation escapes into space. The very tenuous outer gases of the sun that make up its *corona* are opaque to long-wave radio radiation, but are transparent to light. That's why we normally do not see the corona. Oh, it does emit a little visible radiation, but very faintly—about a millionth as much as comes from the photosphere, which makes the corona appear about half as bright as the full moon. Its feeble light is therefore hidden by the bright glare of the photosphere except on those rare occasions when the photosphere appears covered by the moon during a total solar eclipse.

FIGURE IV.2
The sun, photographed 15 September 1957, and showing many spots. *(Hale Observatories)*

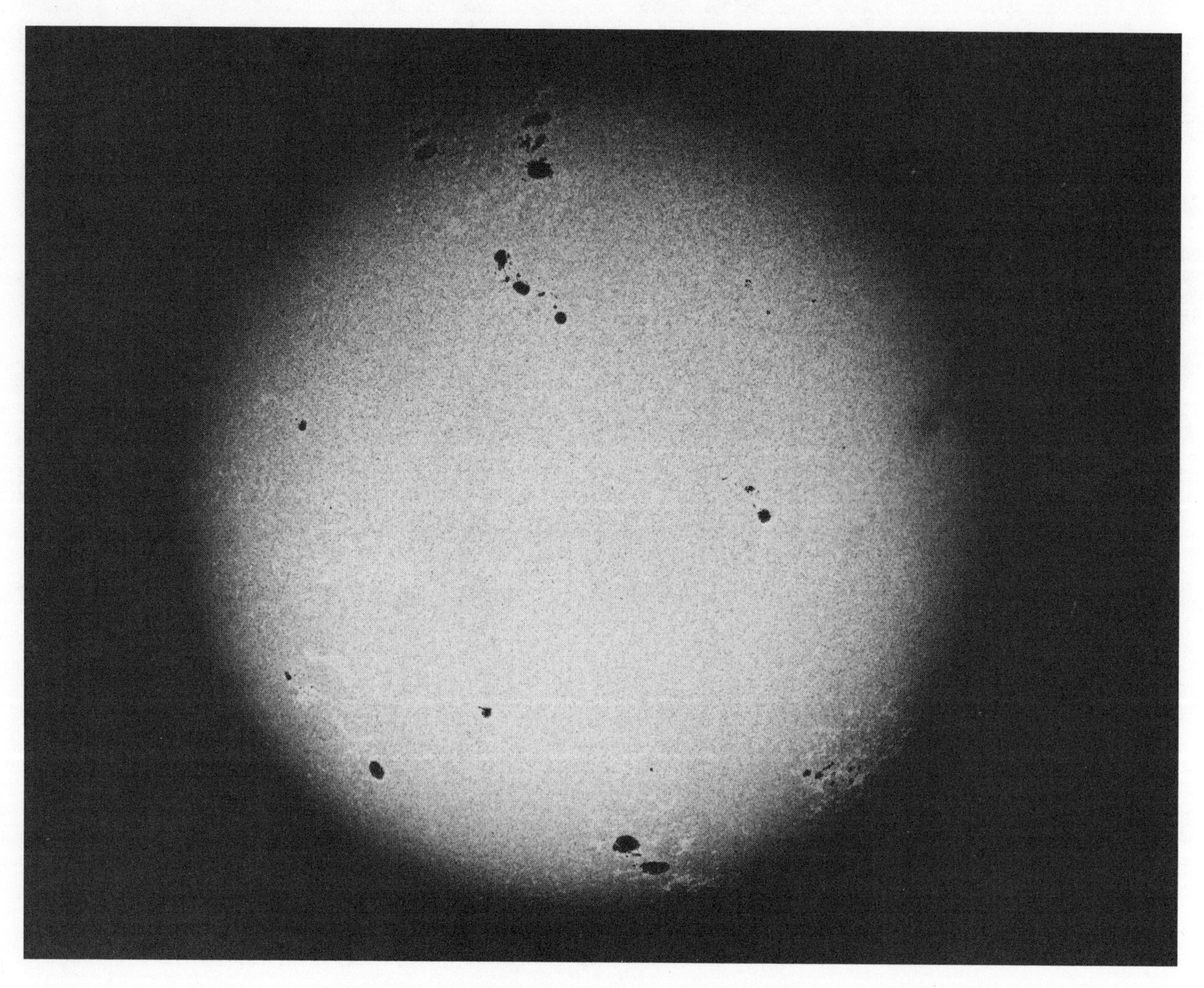

Even the photosphere is not a discrete surface or layer, but rather a range of depths about 300 km thick. The energy of the sun originates deep in its interior, but it cannot flow directly out. Because of the high opacity of the solar material, a photon inside the sun travels, on the average, only about 1 cm before it is absorbed. The energy is, of course, soon reemitted and is thus passed from atom to atom as it zigzags through the solar interior. It takes energy—in this "random walk"—something over a million years to filter from the center to the photosphere. The bottom of the photosphere is where there is a slight chance that an emitted photon can escape directly into space, and the top of the photosphere is where virtually all photons make it out.

Although the photosphere is not an absolutely sharp layer, it is pretty sharp to all intents and purposes, because 300 km is rather insignificant compared to the 1.4 million-km diameter of the sun. It is that thin little "skin of the apple" of the sun (and other stars) from which comes the light by which we analyze the spectrum and learn all the wonderful things we described in Act III Scene 3, including the chemical composition, temperature, density, and so on.

---

2. Through what fraction of the radius of the sun does the photosphere extend? Measure the thickness of the skin on an apple, and see whether it is a good analogy to compare the photosphere to the skin of the apple.

Most of the chemical elements known to occur naturally on earth have been identified in the solar spectrum, and those which haven't are so rare that they would not be expected to be observable spectrographically. About three-quarters of the solar mass is in the form of hydrogen, and from 96 to 99 percent is hydrogen and helium. Among the more abundant of the remaining elements are oxygen, carbon, nitrogen, silicon, magnesium, neon, iron, and sulfur. The solar composition is similar to that of very many other stars. Appendix 16 gives the observed relative abundances of the elements in the universe.

When we look at the sun it does not appear uniformly bright all over its disk. The reason is that light we see from the center of the disk can escape from deeper, hotter, and thus brighter layers in the sun (point *A* in Figure IV.3). But we would not see light from that depth coming from a typical point partway to the limb (or apparent edge), *B*, because to reach us from there the light would have to traverse a slanting path through the photosphere, and any emitted from so deep a layer would almost certainly be absorbed before getting out in our direction. The only light that can reach us from the very limb of the sun *(C)* is that emitted from the uppermost level of the photosphere, which is cooler and thus darker. The phenomenon is called *limb darkening*.

Because of limb darkening, when we analyze light from different parts of the sun's disk, we are, in effect, probing to different *depths* in the photosphere. This enables us to learn something about what it's like in there. We find that the temperature of the gases at the top of the photosphere is about 4500°, and that of gases at the bottom is about 6800°. About half the sun's radiation emerges from layers hotter than 5700°. The density of the gases in the photosphere also increases in depth, but throughout the whole region it is

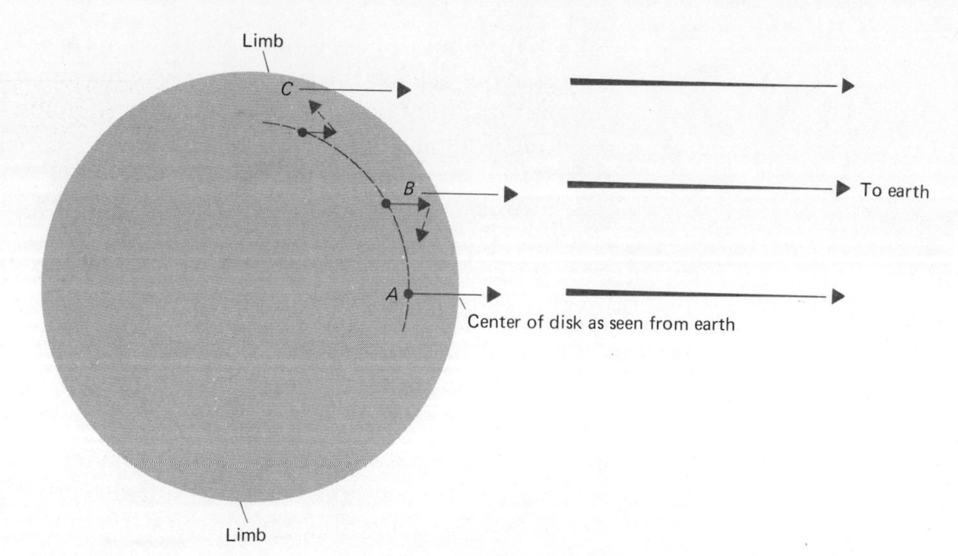

FIGURE IV.3
Limb darkening.

nevertheless a pretty good vacuum; the typical density is only about one ten-thousandth that of the earth's atmosphere at sea level.

On the other hand the *mean* density of the sun as a whole is nearly one and one-half times that of water, which tells us something quite interesting. The parts of the sun we observe are only the very tenuous outermost levels; practically all the actual mass of the sun lies far below the photosphere.

The Chromosphere

Just above the photosphere the gases are transparent to most visible light, but the atoms there still absorb light at some wavelengths, and some of the light they reemit is in the form of bright visible spectral lines. (In Act II Scene 3 we described how atoms do this sort of thing.) For example, the hydrogen present just above the photosphere is emitting light in the Balmer series of spectral lines. The first and strongest emission line in that series is in red light, which means that if we could see those gases they would look red.

Well, we *can* see them, but it's not easy. Until the twentieth century this was possible only during total solar eclipses; then, sure enough, just at the beginning of totality you could catch a glimpse of a red fringe in front of the approaching limb of the moon. Because of its pretty color, that part of the sun is called the *chromosphere.* After the eclipse of 1842, astronomers became interested in the chromosphere, and by 1868 they had observed its spectrum. Over the next quarter century a set of lines was observed in the chromospheric spectrum that could not be associated with a known chemical element. These lines were presumed to be due to a new element and it was named *helium* (from *helios,* Greek for "sun"). Subsequently, in 1895, helium was found also on earth.

The bottom of the chromosphere has the same temperature as the top of the photosphere, about 4500°, but at high elevations, the temperature *increases,* until it reaches 100,000° at the top of the chromosphere, 2000 to 3000 km above the photosphere. There the chromosphere breaks up into a vast forest of jetlike spikes sticking up into the corona. These spikes are called, imaginatively, *spicules.*

In the old days the chromosphere and corona could be observed *only* at times of solar eclipses. Now, however, they can be observed at other times

FIGURE IV.4
Solar spicules, photographed in the light of Hα at Big Bear Solar Observatory. *(Hale Observatories)*

with a *coronagraph,* an instrument that produces an artificial eclipse with an occulting disk at the focal plane of the telescope. The first successful photograph with this instrument was made in 1930 by Bernard Lyot, its inventor, at Pic du Midi, France. Today we can also observe the corona from space. Solar eclipses, however, still provide the best opportunity to observe the outer, very faint part of the corona, which can sometimes be detected millions of kilometers from the sun.

The Corona

Temperatures in the corona are even higher than in the upper chromosphere; they get up to millions of degrees. We knew about this decades ago, because spectra taken during eclipses showed atoms in a very high state of ionization. In 1973, observations from Skylab more than confirmed the high coronal temperatures by detecting the spectrum of iron 16 times ionized. Because of its high temperature the corona is much brighter in ultraviolet radiation than in light, and it is now studied extensively with rockets and space probes. We also observe radio waves emanating from disturbances in the corona, and we can tell the height above the photosphere of their origin from the wavelengths of the radiation that get out of the corona to reach us.

We think that a major source of the heat of the corona is from shock waves due to matter surging up from the photosphere or its environs. Despite its high temperature, the heat content of the corona is very low because of its extremely low density. If you were to stand in the midst of the corona, you would be very hot, not because of its hot gases around you but because of the large amount of energy your body would be absorbing from the photosphere

FIGURE IV.5 The solar corona, photographed during the total eclipse of March 7, 1970, near a time of maximum sunspot activity. Photograph by Gordon Newkirk Jr. *(See Figure IV.6 for complete credit.)*

FIGURE IV.6 The solar corona, photographed during the total eclipse of 30 June 1973, near a time of minimum sunspot activity. Both this photograph and the preceding one (Figure IV.5) were obtained by a special technique developed by Gordon Newkirk, Jr. *(High Altitude Observatory, a Division of the National Center for Atmospheric Research. NCAR is operated by the University Corporation for Atmospheric Research under Sponsorship of the National Science Foundation.)*

of the the sun. Were it not for the nearby sun, the low-density hot corona would take an exceedingly long time to warm you up.

Solar Activity

The very fact that the chromosphere is hot and the corona is hotter has probably given you an indication that this story cannot be finished. Indeed it is not! We divide the humdrum existence of the sun into its "quiet" times and its "active" times. Both are about the same so far as most of us are concerned in our daily lives. They are also about the same so far as the sun is concerned, because its active times might be compared to a man having a few annoying pimples on his face. But even the "quiet" sun is not all that quiet.

Take the photospere, for example. It is in a continual state of seething and heaving. Convection currents constantly carry bubbles of hot gas up to the top of the photosphere, where they cool, break up, and their matter falls inward again. These "bubbles" are anywhere from a few hundred to a few thousand kilometers across and are called *granules,* because telescopic photographs show them to look like a mass of rice grains in a pot. The granules, in turn, seem to be parts of larger units about 30,000 km across (called *supergranules*) within which matter seems to flow slowly from the center to the edge.

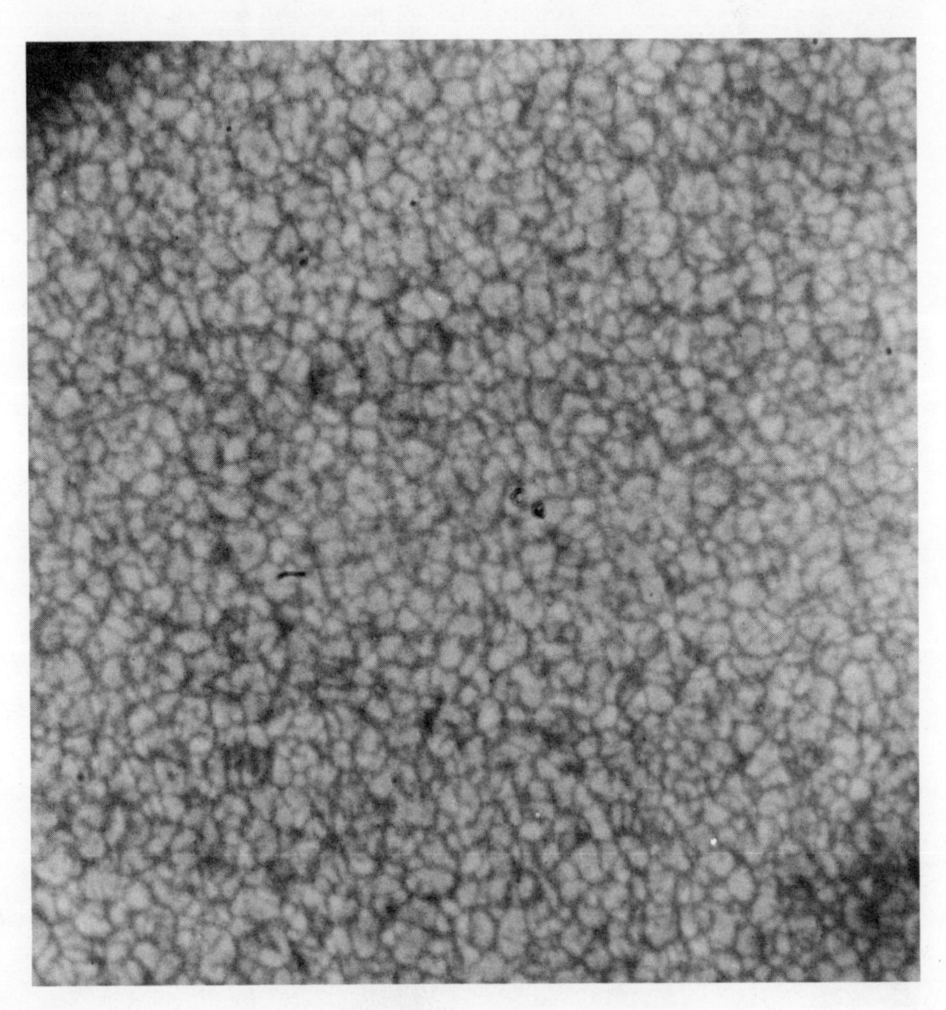

FIGURE IV.7
Solar granulation, revealed clearly on this photograph from a balloon at an altitude near 25,000 m. *(Project Stratoscope, Princeton University)*

The "active" sun is even more interesting. Most conspicuous in the active regions of the sun are sunspots. Since antiquity rare, large naked-eye spots have been seen. Galileo observed many much smaller sunspots with his telescopes, and showed that they are on the surface of the sun. They look like large pores through the "surface" of the sun, and even Herschel suggested that they were "holes," through which we could see into the cool, dark, probably inhabited interior of the sun.

So convincing did Herschel's speculation seem that a few decades ago a rich German citizen offered a substantial prize to anyone who could disprove the idea. Well, it is easy today to see that the sun cannot be hollow, so a group of eminent German astronomers rose to the challenge and proved that the hypothesis was untenable in light of modern knowledge of the most elementary physics. The good citizen, however, being unconvinced by the methods of modern science, refused to pay up, and was taken to court. He lost, and the money was used to support graduate students studying astronomy. I suppose it could happen only in Germany.

---

3. Since antiquity, sunspots had been observed with the unaided eye, especially when the sun was shining through smoke or haze so that it could be looked at directly. Generally the spots were thought to be caused by matter in the earth's atmosphere projected against the sun's disk. Suggest a way you might test this idea.

Sunspots are not holes in the sun; they are regions where the gases have expanded somewhat, and thus have cooled by about 1500°, so that they appear dark by contrast against the brighter photosphere. Held against a cooler star, sunspots would shine as bright spots of light. Typically, individual sunspots persist on the surface of the sun for a few days or a few weeks; on rare occasions one will last for a few months. However, sometimes there are many spots on the sun, and other times very few. In fact, in 1851 the German apothecary and amateur astronomer Heinrich Schwabe published the results of a study he made showing that whereas individual sunspots come and go

FIGURE IV.8
Bipolar spot group photographed 21 May 1972 at Big Bear Solar Observatory; *(left)* in white light; *(right)* in red light of Hα. *(Hale Observatories)*

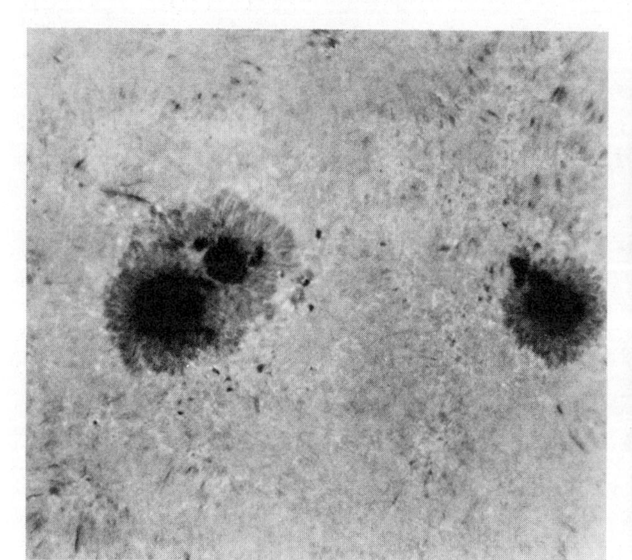

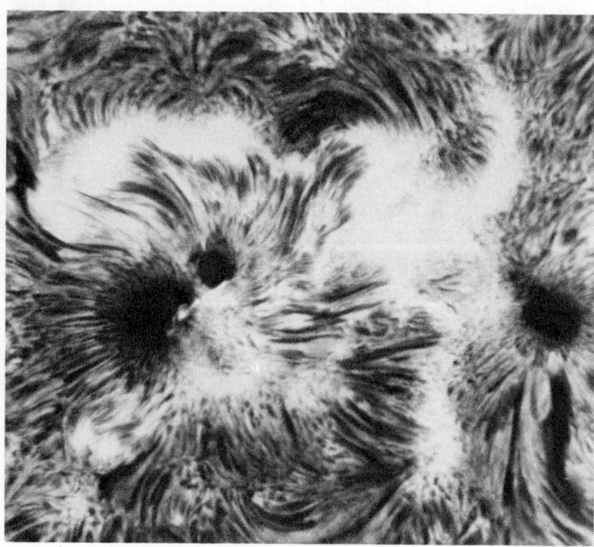

in days, weeks, or months, the average numbers of them on the sun increase and decrease in regular cycles.

Since Schwabe's time it has been found that the actual interval between *sunspot maximum*—the time of the highest average number of spots—and *sunspot minimum*—the time of the smallest average number—and back again has varied from 8 to 16 years, but has had an average of 11.1 years. This time has been called the *sunspot cycle*, or sometimes *solar cycle*. Coming and going with the sunspot cycle are lots of other phenomena characteristic of the active sun. These include *plages*, regions in the chromosphere above sunspots where ionized gas is changing state and becoming neutral and is emitting light in certain emission lines in the process. *Prominences* also occur in the neighborhood of active regions around sunspots. Prominences appear as great jets of gas rising above the solar surface and up into the corona. Actually they too are regions where gas in the coronal regions is changing state and emitting light. Prominences sometimes appear as great arches over sunspot areas, and extend to heights of hundreds of thousands of kilometers into the corona.

Plages and prominences are regions in the sun's upper atmosphere where the gases are emitting light at certain wavelengths. They are most easily observed, therefore, when the sun is photographed through special filters that transmit light only of those and the immediately neighboring wavelengths. Such photographs are called *filtergrams*. Many of the spectacular photographs of active regions of the sun reproduced in this book are such filtergrams.

FIGURE IV.9
Solar prominence of 31 March 1971, in the light of Hα, photographed at the Big Bear Solar Observatory. *(Hale Observatories)*

The most spectacular of the phenomena of the active sun are the *solar flares*. A flare is a sudden emission of energy from a localized region of the sun just above the photosphere, typically a few thousand or at most a few tens of thousands of kilometers across. Usually most of the flare energy is in the

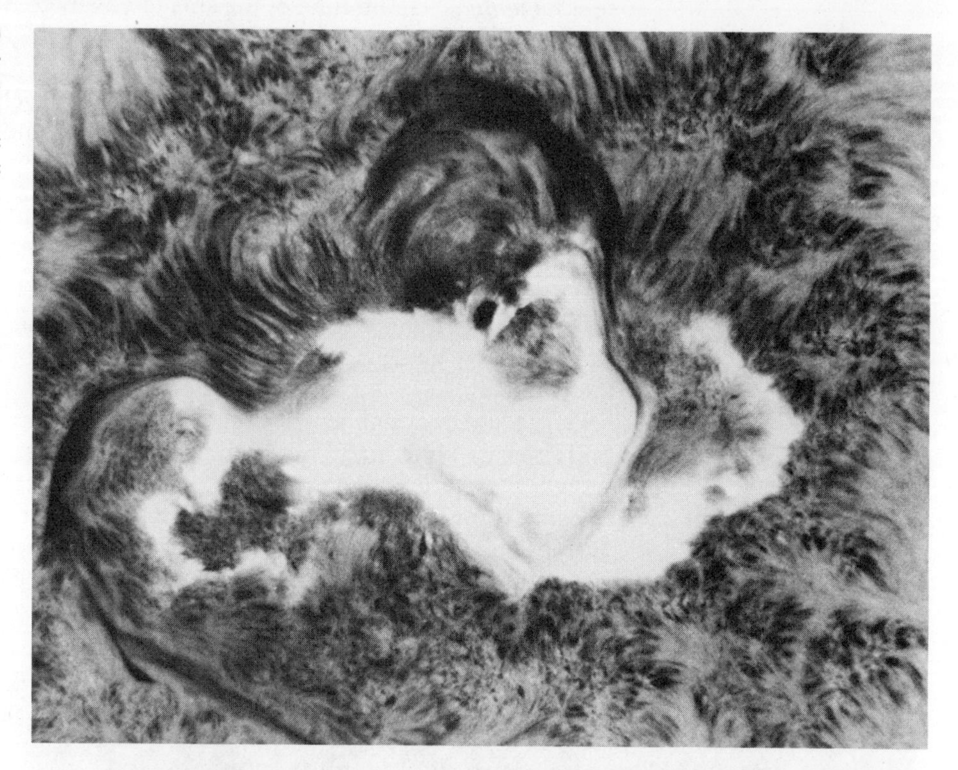

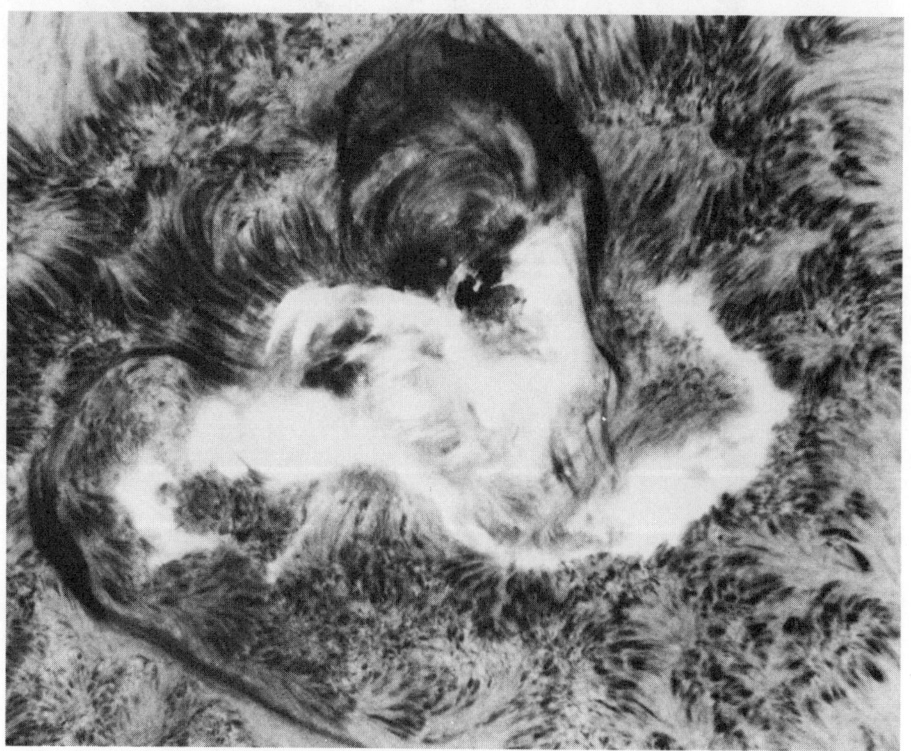

FIGURE IV.10
Two photographs of a solar flare on 7 August 1972, made in the light of Hα at the Big Bear Solar Observatory; *(upper)* at maximum intensity; *(lower)* 25 minutes later. *(Hale Observatories)*

wavelengths of certain emission lines, but sometimes a flare is observed in the continuous spectrum of white light as well. The average flare rises to its maximum brightness in a few seconds or minutes, and then fades out over the next several minutes or hours.

During a flare an enormous amount of energy is released. It is seldom noticeable in visible light against the bright light of the sun, but rocket and satellite observations show that in ultraviolet and X-ray radiation a flare may outshine the sun for a few moments. Also, coronal radio bursts are common near flares. During major solar flares large numbers of electrons and ions are ejected from the sun at high speed; they reach the earth and can be observed here, on the average, about 50 hours later.

Solar Magnetism

These phenomena of the active sun are now known to be associated with solar magnetism. Magnetic fields in the sun were first observed in 1908 by the American astronomer George E. Hale. It is possible to observe magnetic fields spectroscopically because of an effect discovered by Pieter Zeeman, and now known as the *Zeeman effect,* wherein the energy levels of atoms in a magnetic field are split up into several distinct, but closely spaced levels. Thus the spectral lines produced by those atoms are also split up into groups of closely spaced lines. Hale's pioneering study of solar magnetism was later taken up in detail by H. W. Babcock and his son H. D. Babcock (the latter later became director of the Hale Observatories).

It turns out that active regions of the sun, those with sunspots, plages, prominences, and flares, are always those where there are strong magnetic fields. The fields in sunspots range from 100 to 4000 gauss—as great as the field around a very strong laboratory magnet. A *gauss,* named for the same physicist and astronomer who calculated the orbit of Ceres, is a unit used in the measurement of the strength of magnetic fields; the field of the earth has an average strength of about 1 gauss.

We have seen (Act II) that magnetic fields are produced by moving charges. In the sun they are thought to be produced by ions moving in the sun itself. Now the lines of force of a magnetic field interact with the charged particles in an ionized gas. If the gas density is great enough, it tends to dominate and causes the magnetic lines of force to tag along with its ions and electrons. On the other hand, if the gas density is very low and the field strength fairly high, the magnetic field controls the motions of the charged ions and electrons in the gas.

In the photosphere, the gas wins out, and drags the magnetic field lines with it as it follows the sun in its rotation. Above the photosphere, the magnetic fields exert an important influence, and many of the chromospheric and coronal phenomena are dominated by solar magnetism. The plages, for example, are regions in the chromosphere where the magnetic field tries to expand, lowering the gas density, forcing the gas to cool so that some of its ions recombine with electrons, giving rise to the emission of light. Some prominences are similar phenomena, but occurring in great looped fields above the sunspot areas. In other prominences, the ions and electrons of the coronal gas flow along the lines of force of the magnetic fields. The boundaries between supergranules seem to be regions of relatively high magnetic field intensity, along which spicules jet upward into the corona. And the corona itself is organized by and outlines the outer magnetic field of the sun.

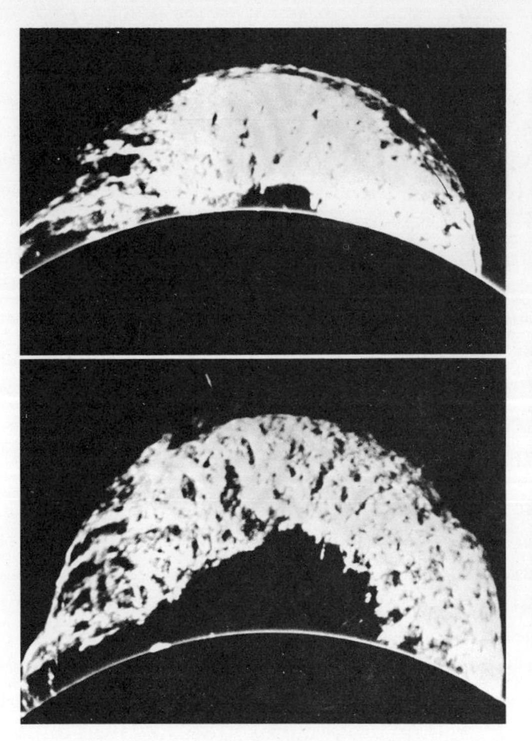

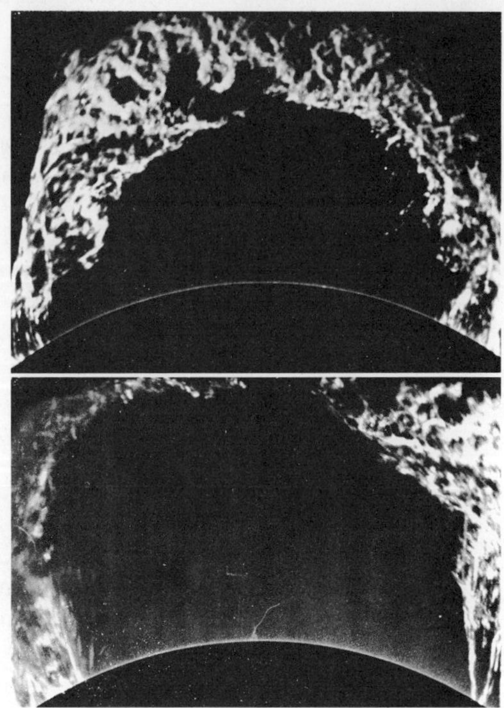

FIGURE IV.11
The great explosive prominence of 4 June 1946. The ellapsed time between the first *(upper left)* and the last *(lower right)* photographs was one hour. *(Harvard Observatory)*

Theory of the Solar Cycle

We do not by any means understand all about solar activity, but the more we learn, the more it is clear that solar activity is inextricably associated with solar magnetism. In fact, the favored theory of the sunspot cycle (due to H. D. Babcock) involves the solar magnetic field. Babcock's idea is that there is a general, although rather weak, solar magnetic field of roughly the intensity of (or perhaps a little greater than) that of the earth. Since the 1950s Babcock and his father established the existence of just such a solar magnetic field. Most of the field lines are in the photosphere and are forced to follow the motions of the photospheric gases. But you recall that the sun rotates differentially, the equatorial parts moving faster than those regions at higher latitudes.

A look at Figure IV.12 shows what must occur. The low-latitude regions of the sun draw the field lines forward, compared to their positions in the polar regions, and after a few rotations, stretch them around the sun so that they wind up close to themselves. As the differential rotation continues, the field lines wind up tighter and tighter, and as they do so, adjacent field lines, wound ever closer together, build up strong local magnetic fields. Local turbulence can cause the lines to become twisted together. The matter in such a region tries to expand and lower the field strength, and as it does so, it rises. The rising currents can carry "cables" of field lines out through the photosphere into loops above it. Sunspots, in this theory, occur where the

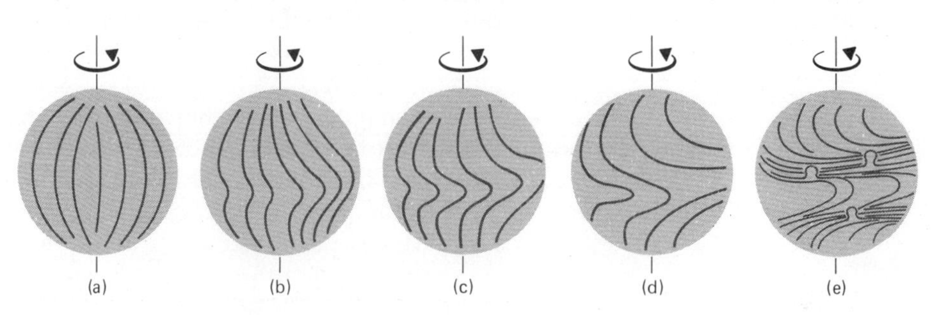

FIGURE IV.12
The Babcock model of the solar cycle.

field cables intersect the photosphere, and expanding gases in those regions cool and darken. The phenomena in the chromosphere and corona are associated with those same loops above the active sunspot regions. Eventually, as the differential rotation continues, the field lines coalesce and disperse, dissipating the strong field, and the process starts over again.

I have oversimplified things here; for example, I have completely ignored the reversal of the polarity of the sun's field with each sunspot cycle. But all I really want to do is paint a general picture of the many complex, chaotic, and often violent happenings that are constantly going on in that outer, absolutely trivial part of the sun—its outer wisps.

Solar-Terrestrial Relations

But trivial as they are insofar as the sun is concerned, they are of considerable importance to us. Were it not for absorption by the earth's atmosphere, those sudden bursts of ultraviolet radiation and X rays that occur during solar flares would most certainly be devastating to terrestrial organisms. Even as it is, we get some message of the solar phenomena, because during times of unusual solar activity the continual rain of ions coming from the sun (the *solar wind*) is enhanced.

It has long been known that sunspot activity is correlated with certain things on earth. The most apparent of these are the displays of Aurora, or northern and southern lights, caused when the solar ions strike the upper atmosphere and enhance its ionization. The earth's magnetic field is disrupted slightly at these times too, causing what are called magnetic storms. Moreover, the rain of ions from the sun, as well as high-energy electromagnetic radiation, disrupt the ionized layers high in the atmosphere—the ionosphere—and change their ability to reflect radio broadcast waves back to the ground. These changes result in interference with long-range radio communication—especially important before the time of the communications satellites.

Other effects have been suggested as well. Some may be real; for example, there is now serious study of the possibility that certain short-period weather changes may be correlated with solar activity. Correlations of solar activity and planetary configurations have been claimed but never confirmed. Nor is there any evidence that solar activity has anything to do with the length that wool grows on sheep in the Northern Hemisphere, nor with stock market ups and downs. Some people derive great pleasure in searching for correlations between natural and almost certainly unrelated human events. Nevertheless I know of no expert who will deny that in various real ways solar activity can be profoundly important to us all.

---

4. Suppose you look for a correlation beween the numbers of spots visible on the sun and the directions of the planets in the sky. You find nothing except what appears to be a weak correlation between the sunspot number and the position of Uranus in its orbit. You do a statistical test and find that for any given planet there is only one chance in seven that such a correlation would occur by chance. Would you be able to conclude that there is a real relation between Uranus and and solar activity? Why or why not? [Hint: How many planets are there?]

It is particularly interesting, therefore, that there is considerable evidence that solar activity actually ceased entirely, or nearly entirely, for about 70 years, between 1645 and 1715. First, there are few if any records of sunspots in those years. To be sure, sunspot records before 1645 are none too systematic or complete, but such data as exist indicated at least two spot cycles in the early seventeenth century, both of which happened to have periods somewhat longer than 11 years. But also, there is no record that the solar corona was visible during eclipses that occurred in those 70 years. At the time of this writing the matter is still unresolved, but the possibility that solar magnetism was absent or was at a very low level at that time is most intriguing.

The Solar Interior

But, as I have said, solar activity is of no consequence whatsoever to the great, dependable, stable solar interior. At least it is stable for the time being. It will not always be so; in a few thousand million years there will be developments that will spell curtains for the earth and any left on it. But for our lifetimes, and for any future that we can possibly imagine for mankind, we can pretty confidently predict that the sun as a whole will behave itself. Much of this act of our drama is concerned with the future of the sun. Let us see how in the world we can ever hope to know what is going on inside it.

Our goal in the study of the interior of the sun (or of any star) is to learn the temperature, pressure, density, and composition of the material, as well as the amount of energy generated at every point inside that body. When all this information is put together, we have what is called a *model* for the interior of the sun (or star).

Consider the cross section of the sun shown in Figure IV.13. Inside it I have shown a spherical shell of the solar material. That portion of the sun has mass, as does the part of the sun interior to it. The two must pull on each other because of their mutual gravitation. Now, it is easy to prove (Newton did it) that the pulls of all the parts of the sun outside the shell cancel out each other, so as far as gravitation is concerned the only part of the sun affecting the shell is that part interior to it. Well, then, why doesn't the shell fall inward?

FIGURE IV.13
A spherical shell in the solar interior.

Chicken Little was concerned about the same thing. "The sky is falling!" she is reported to have said, in great agitation. But she was wrong; the sky doesn't fall for exactly the same reason that the shell in Figure IV.13 doesn't collapse inward. The gas all around the shell exerts pressure on it, but the gas inside, being a bit hotter and denser, exerts a bit more pressure than that outside. The difference between the pressure of the gas inside the shell and that outside is just enough to balance its weight (the pull of gravitation on it).

What if the pressure difference were greater than the gravitational force? The shell would move out some and expand; so would the gases on either side of it. The expansion would make all the pressures drop and the expansion would continue until the pressure difference across the shell had lowered to the point where it could just balance the shell's weight. What if the pressure were too little to support the shell's weight? Then the shell would fall in, thus compressing the matter and causing the pressure to build up until it could balance the gravitational attraction. Do you see how the entire sun must adjust itself until every part of it has reached this state of equilibrium wherein the weight of that part is just balanced by the net outward pressure on it? This condition is called *hydrostatic equilibrium.*

---

5. Discuss the analogy between hydrostatic equilibrium and an inflated balloon. In the balloon, what force corresponds to gravitation in the sun? What happens if the balloon is placed in a warmer environment? What if it is placed in a refrigerator?

6. If the earth's atmosphere were suddenly cooled to absolute zero, what would happen to it?

7. Each square centimeter of the earth's surface experiences an atmospheric pressure of about 10^6 dynes (or equivalently, there are 14.7 lb of pressure per square inch). Suggest how the total weight of the atmosphere can be calculated.

Now if the gases within the sun were completely transparent, they would absorb no energy and the radiation generated deep in the solar interior would just flow right out through the sun. But then there would be nothing to heat up those gases; hence they would have no pressure, and the whole sun would collapse on itself. In fact, however, the solar material is highly opaque and absorbs radiation with high efficiency. It is this opacity that traps energy inside the sun and keeps its gases hot. On the other hand, each region of the sun must constantly reradiate energy at precisely the rate it absorbs it; if it did not, it would heat up or cool off until it did reach that state of *thermal equilibrium.* This condition of thermal equilibrium gives us additional information about the state of the gases in the solar interior. But to use that information, we must know in detail how the gases absorb energy, that is, their source of opacity.

---

8. A poker is placed in the burning coals of a furnace and left there for a long time. It is seen to produce a steady red glow. Is the poker in thermal equilibrium? When it is removed from the fire and held in the cool room air, is it then in thermal equilibrium? Explain.

The theory of the opacity of hot gases is extremely complex; it involves all those various atomic processes described in Act II Scene 3. We understand the basic physics involved, but the calculations are very nasty, and require large computers. Nevertheless, such calculations have been carried out for conditions appropriate to those in stellar interiors, albeit with a number of simplifying approximations. It is interesting that, as happens, one of the best and most extensive sets of calculations of opacities of hot gases has been provided by a rather unexpected source.

That source is the Los Alamos Scientific Laboratory. The only places on earth where we encounter conditions anything like those in stellar interiors are in the fireballs of nuclear bomb explosions. Because the nuclear bomb people had to understand these fireballs, years ago they hired astronomers who understand gases under such unusual conditions. A cadre of astronomers still at Los Alamos spend part of their time studying the nature of gases in stellar interiors; they have made very important contributions to our knowledge of stellar structure and evolution. It is a fascinating peaceful fallout from a laboratory mainly concerned with hellish devices.

Thus the conditions of hydrostatic and thermal equilibrium provide us with important information needed to construct a solar or stellar model. We do not yet, however, have all the pieces of the puzzle, because some of the regions of the sun must contain sources of energy. Our next problem, therefore, is to find where the solar energy comes from. It cannot be simple combustion, as some early philosophers thought, because even if the sun were pure carbon (like coal) and there were enough oxygen around to burn it all, to provide its present luminosity the sun would be burned up in 10 thousand years or so. *Never* describe the sun as a great ball of burning gas or a ball of fire; that's nothing compared to reality.

The Problem of the Sun's Energy

About the middle of the nineteenth century astronomers thought they had discovered the source of the sun's energy. The physicists H. von Helmholtz and Lord (W. Thompson) Kelvin hypothesized that the sun's energy comes from its gravitational contraction. Here's the idea: The sun is hot inside, so it radiates some of its internal energy into space. But that tends to cool off the interior, lowering the pressure of the gases there. Thus the sun's gravitation takes over and contracts the sun a bit. But the contraction builds up the pressures again, keeping the sun always in hydrostatic equilibrium. But the higher gas pressures result in even higher temperatures than before, so the sun's internal temperature actually increases, even though it radiated some heat away. Thus it radiates away more energy, and the process continues. The sun keeps shrinking, heating up, and radiating.

But where does the *energy* come from? It's *gravitational energy*. From everyday experience we all know what gravitational energy is; in physics courses it is usually called *potential energy*, or *gravitational potential energy*. Since the earth is pulling on things, if we drop them, they fall. Before we let them go, they are stationary, but by the time they hit the floor, they may be moving pretty fast. Where did they pick up that energy? It was the gravitational energy we gave them by holding them above the earth. Gravitation can make things move; therefore it has potential energy. The closer together things fall, the more they speed up and the more of their gravitational energy they convert to kinetic energy. When things are infinitely far away, they have the maximum amount of gravitational potential energy, because then they can give up the most possible energy before coming together.

---

9. You ascend a high staircase and drop a blackboard eraser down the stairwell. As it falls, it releases gravitational potential energy and picks up speed until it strikes the floor at the bottom. What then happens to that gravitational energy?

When the gases of a star (or the sun) fall together, their gravitational energy must go somewhere. The detailed calculations show that for gases consisting of simple atoms, as in the sun, exactly half the released gravitational energy goes into heating up the interior and half is radiated away as luminosity. Further calculations by Helmholtz and Kelvin showed that the sun need shrink only exceedingly gradually to account for its present output of energy. In fact, during recorded history, man would not have been able to measure the change in the size of the sun.

---

10. If you could magically force the central 25 percent of the sun to cool off suddenly, what would happen to the temperature of the earth in the immediate future? Explain.

In the initial stages of its life, a star does shrink under its own gravitation, and the gravitational energy released does account for its luminosity. In more advanced stages of evolution, stars also undergo such contraction in certain parts of their interiors, and release energy in the process. Many astronomers and physicists think that some of the most luminous objects in the universe—quasars, for example—receive their energy from great masses of matter in gravitational collapse upon itself. If stars become black holes, as we think many probably do, they must release great amounts of gravitational energy. We shall soon be discussing some of these things. Rest assured that this idea of Helmholtz and Kelvin is a good one and is very much in the forefront of modern astrophysical thinking.

But it does not account for the energy of the sun today. Helmholtz and Kelvin found that from the time the sun began shining, it could have delivered its present luminosity for a few *tens of millions of years* before shrinking to its present size. In Helmholtz and Kelvin's time, that seemed very long indeed, and thus they thought gravitational contraction of the sun must surely be its source of energy.

Then, less than a half century ago, the geologists discovered that the earth, and presumably the solar system, has been around for several *thousand* million years. So much for the Helmholtz-Kelvin hypothesis! Before we finish our story of how we make a solar model, therefore, we must explore the real source of the sun's energy.

SCENE 2

The Atomic Nucleus

The Greeks had speculated that matter was composed of particles, which they called *atoms*—or at least the Greek word for them. But those ancestors of ours were *guessing;* they had no special knowledge about the most basic constituents of the stuff that we are made of. We, in contrast, have *learned* a great deal about atoms. I have already described something of our knowledge of the atom—its nucleus and electrons, bound together with the electromagnetic force, but obeying the strange laws of quantum mechanics. But now we probe deeper than the atom, into the atomic nucleus itself, and our probe brings us to one of the most exciting frontiers of modern physics.

But, you may well be wondering, if the atom is too small to see even with our most powerful microscopes, how can we ever hope to examine those even smaller particles that make it up? We don't see them, of course, but we examine them by learning about their properties—by learning how they behave. Take the electron, for instance. No one has ever seen an electron, yet we have studied it and played with it in a host of ways. For example, we "boil" electrons off hot wires, and attract them away from those wires with electric fields, giving them high speeds. Then we deflect them with magnetic fields (remember, charged particles, like electrons, are deflected by magnetic fields) and make them follow rather precisely determined paths, so that they strike targets: phosphor substances on the insides of our television screens. These phosphors glow in various colors, producing the pictures we enjoy every evening.

So you see, we know a lot about electrons without ever seeing them; we understand them so well, in fact, that we manipulate them to serve our business and pleasure in very many ways. But how can we study the structure of the atomic nucleus?

Long ago physicists had figured out that the atom of hydrogen has a nucleus (a proton) with a positive charge equal to the negative one on the electron. (By the way, the labels positive and negative are completely arbitrary; we could have easily given each the other's name.) But then it was found that the nucleus of helium, with twice the charge of the hydrogen nucleus, has four times the mass. The helium nucleus must, therefore, contain something else besides protons—something that gives it extra mass but not extra charge, a neutral mass. Thus the neutron was hypothesized. The neutron was first isolated in the laboratory in 1932 by James Chadwick. So we knew, back then, that there were protons and neutrons, in addition to electrons. When I was in high school, we were told that those were the only basic particles of which all matter is formed. Things have changed.

Cosmic Rays

One hint that other subatomic particles exist was provided by *cosmic rays.* Cosmic rays first revealed themselves by removing electrical charges from insulated objects, such as charged electroscopes. A simple electroscope (Figure IV.14) consists of a conductor extending into an insulated flask with thin metallic leaves at its end. If the upper end of the conductor is charged with static electricity, say by rubbing it with cat's fur, the charge is transmitted to the leaves; since both are then given the same charge, they repel and separate. Any charges of the opposite sign in the air, on the other hand, will be attracted to the upper end of the conductor of the electroscope, thus neutralizing it and causing the leaves to gradually come together again. Early in the century it was learned that when electroscopes were charged up in this manner, inevitably they discharged again after a time. This shows that there is charge in the air. Evidently something was continually ionizing the atoms of the air, and the ions formed, attracted to the electroscopes, caused them to discharge.

When electroscopes were carried aloft in balloons, it was found that they discharged even faster, showing that the charge of the air was greater at high altitudes. Thus it was concluded that some high-energy radiation was coming from space and ionizing the air; at higher altitudes, where more of this radiation would be absorbed, the ionization would naturally be stronger. The mysterious radiation was called *cosmic rays;* at first it was thought to be

FIGURE IV.14
A simple electroscope.

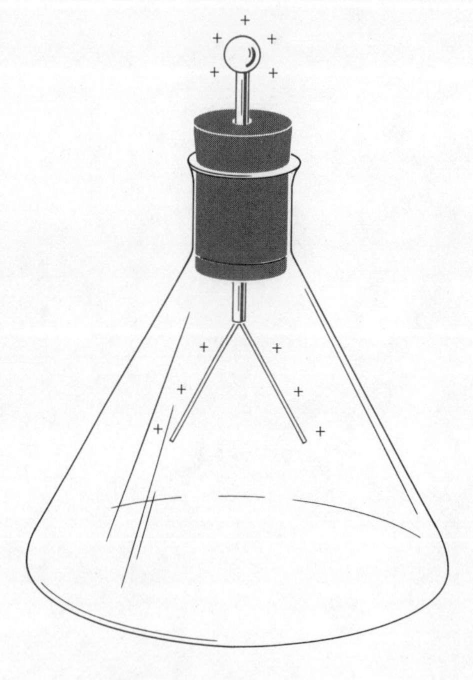

very high energy electromagnetic radiation, of even shorter wavelength than gamma rays.

Several decades ago, however, it was found that cosmic rays cannot be electromagnetic radiation, because they are deflected by the earth's magnetic field. Electromagnetic radiation is not affected by magnetic fields; the motions of charged particles are. Some cosmic rays, those of high energy, strike the atmosphere at all latitudes, but those of more moderate energies are deflected by the earth's magnetic field, back into space again, unless they encounter the earth at places near enough to the magnetic poles to follow the field lines down into the earth's atmosphere.

Investigations in the 1930s and 1940s showed that cosmic rays are actually the nuclei of atoms striking the earth from all directions in space. Most of them are the nuclei of hydrogen atoms; most of the rest are the nuclei of helium atoms. A few percent are the nuclei of other atoms, in roughly the relative abundances that are found elsewhere in the universe.

The interesting thing is that cosmic-ray particles are approaching the earth at extremely high speeds—nearly the speed of light. Thus they carry a good deal of energy. The total amount of energy striking the earth each day in the form of cosmic rays is, in fact, roughly the same as that striking it in the form of starlight. It is obviously important to astronomers to understand the origin of these cosmic rays from space. A few, we find, come from the sun, in particular when there are intense solar flares. Most, however, especially those of the highest energy, come from beyond the solar system. We shall return later to this important radiation.

---

1. It is likely that cosmic rays are responsible for some cases of cancer. If so, what places on earth would you expect to be high in cancer risk from this cause? Why? (In any case, the risk is exceedingly small compared to, say, inhaling your own or other people's cigarette smoke.)

FIGURE IV.15 Air view showing part of CERN facilities. *(CERN)*

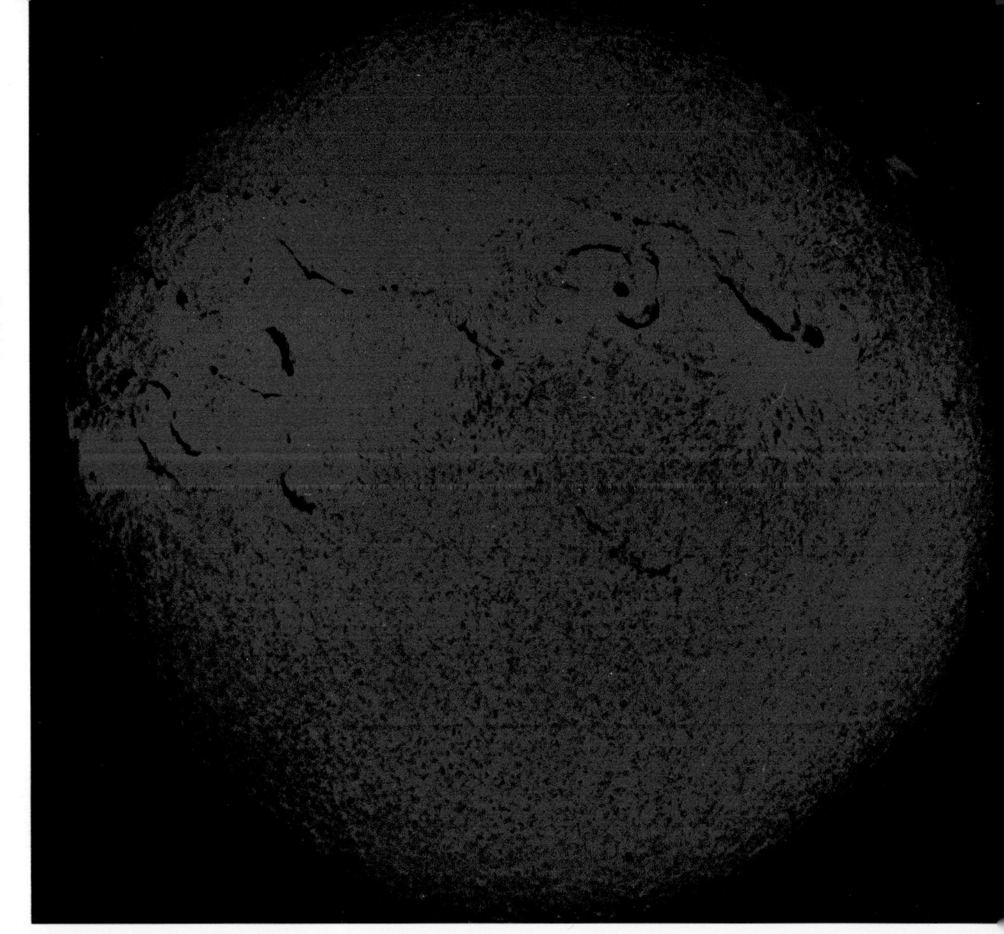

The sun in the red light of the first Balmer line of hydrogen. (*Carl Zeiss*)

Ultraviolet photograph of a solar flare, taken by Skylab astronauts. The different colors indicate relative ultraviolet intensity. (*NASA*)

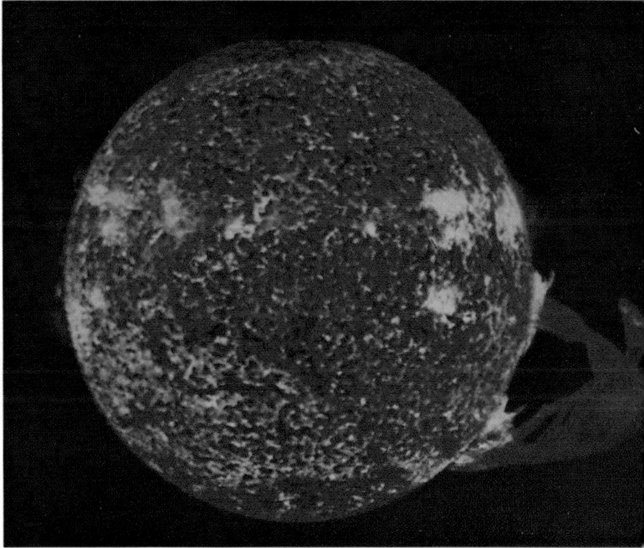

Ultraviolet photograph of the entire sun, showing several flares and a large prominence, taken by Skylab. (*NASA*)

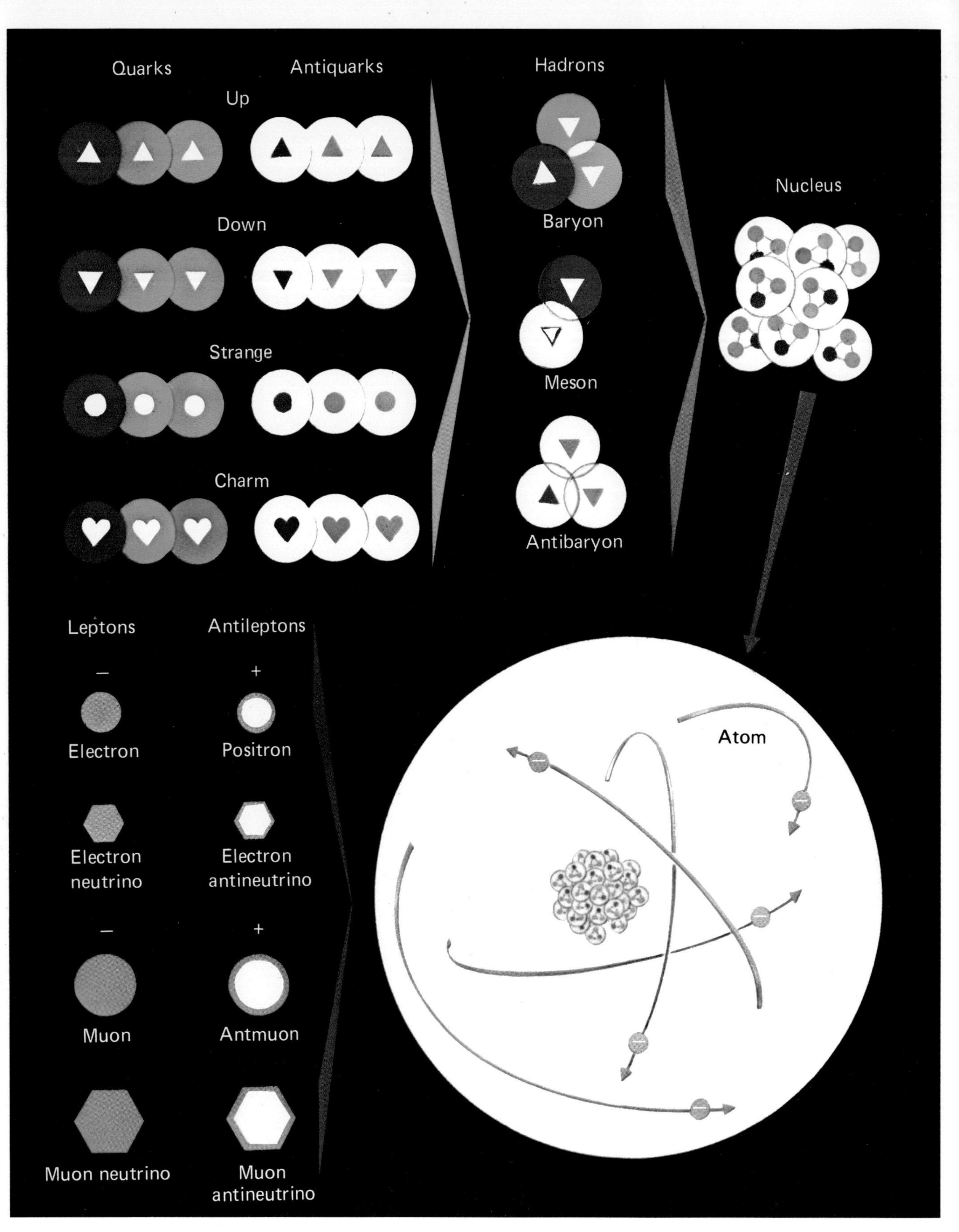

A current model of the subatomic and subnuclear particles.

During the 1930s and 1940s, cosmic rays were also of great interest to physicists. When those atomic nuclei strike the earth's atmosphere far above the ground, atmospheric atoms are broken up into multitudes of other subatomic particles that rain to the earth as *showers* of secondary cosmic rays. Prominent among them are the muons, described briefly in Act II Scene 4. It was found that other new kinds of particles were produced as well by the high-speed cosmic ray particles from space. To physicists, therefore, the cosmic rays provided a means of seeing what happens to atomic nuclei when they are smashed into by other atomic nuclei with extremely high speeds.

Since World War II, we have been able to build laboratory accelerators, devices in which we can accelerate subatomic particles to speeds and energies comparable to some of those commonly encountered in the relatively less energetic cosmic rays. Thus we can observe in the laboratory what occurs as the result of very high speed collisions between atomic nuclei. What we find, in a nutshell, is that the nuclei break up into all manner of different kinds of particles.

Subnuclear Particles

One kind of particle is the photon—the basic unit of light energy, which we have already come to know and love so well. The photon, of course, is all energy, and has no rest mass. But it can never be at rest anyway (always moving with the speed of light), and *all* energy has a mass equivalent. As we have seen, the photon behaves like a particle in many experiments. The other kinds of particles that have been discovered consist of the *leptons,* or "lightweights," and *hadrons,* or "heavyweights."

There are four kinds of leptons. They are the electron, the muon, and two kinds of neutrinos, one associated with the electron and one with the muon. The muon is like the electron, but has about 200 times as much mass. We have mentioned neutrinos before; they are particles that have energy but no rest mass, and thus travel with the speed of light. They carry no charge, so, like the neutron, are electrically neutral. Since they are massless, they are rather like "little" neutrons; thus they were dubbed "neutrinos" by the Italian physicist Enrico Fermi. Neutrinos were first hypothesized in 1933 by Wolfgang Pauli to account for small amounts of energy that appeared to be missing in certain nuclear physics experiments. Neutrinos are elusive little fellows, and are extremely hard to capture and detect in an experiment. Yet, the neutrino was detected in 1953; it has subsequently been well studied in the laboratory, and nuclear physicists today understand the properties of both kinds of neutrinos quite well. Of the four leptons, the electron and the neutrinos are stable particles; as we have said earlier, however, the muon decays, on the average, in about 2 microseconds, turning into a stable electron (or positron), a neutrino, and an antineutrino. Otherwise, the muon behaves just like an electron.

---

2. In what important respect (or respects) is a neutrino very different from a neutron?

For each lepton, there is also the corresponding antiparticle—the positron (antielectron), the positive muon, and antineutrinos of both kinds. Recall from Act II Scene 3 that if a particle and its antiparticle come into contact, they annihilate each other, thereby turning into energy or other particles.

Whereas there are only four kinds of leptons, very many kinds of hadrons have been discovered—about 300. They can be classed, however, into three groups: the *baryons,* which include (among others) the proton and neutron, the *antibaryons,* or antiparticles corresponding to the baryons, which include the antiproton and antineutron, and the *mesons,* which include such particles as the *pion.*

Of the 300 or so known hadrons, only the proton and antiproton are stable. The others spontaneously decay (change) into other particles and/or energy, and those other particles also decay, and their products decay, until they have all turned into stable particles and energy. The time a typical hadron lasts before decaying into something else is typically about 10^{-23} s. Some, however, hang around in this world a bit longer, especially the neutron, which, in the free state (outside an atomic nucleus) lives an average of 15 minutes. Then the neutron turns into a proton, an electron, and an antineutrino.

---

3. What do you suppose are the decay products of an antineutron?

The "zoo" of hadrons has been investigated at great length. The properties of the various kinds of particles have been measured and cataloged as regards their masses, their charges, their spins, and other ways they behave. The analysis has been aided by the discovery of certain rules that seem to govern the kinds of reactions hadrons can undergo with other particles.

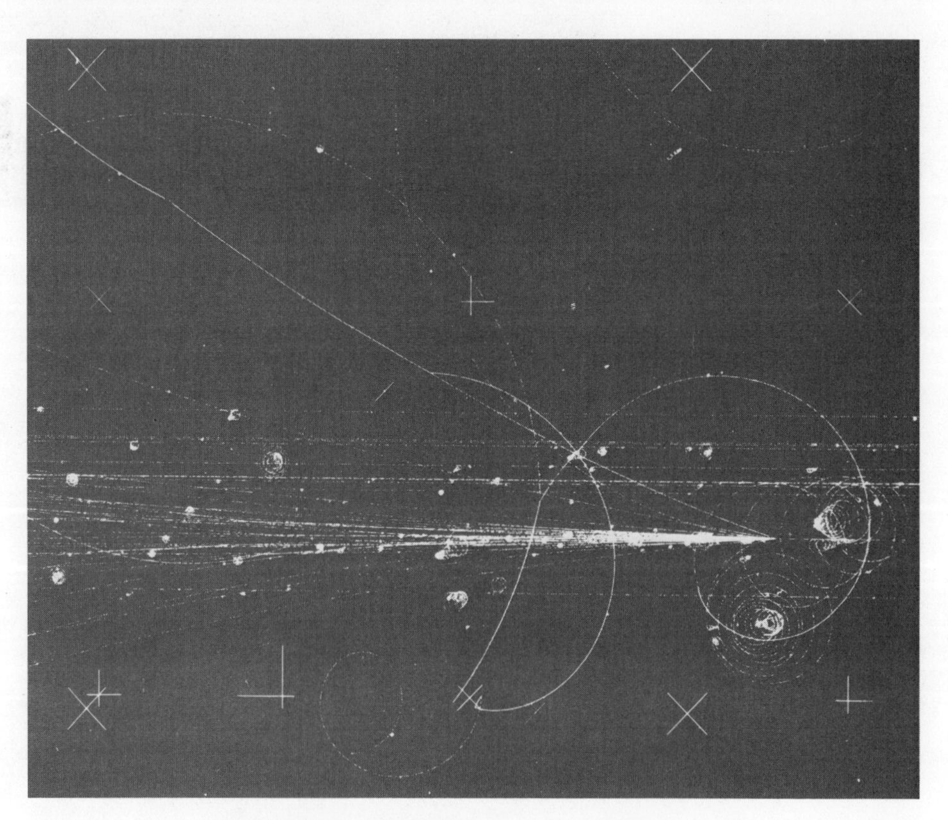

FIGURE IV.16 Bubble chamber tracks. *(National Accelerator Laboratory Photograph)*

Although some rules and some order have been discovered in the past 25 years, during which nearly all these new particles (other than the proton, neutron, and very few others) were discovered, the subject is very complicated. It's not like Mother Nature to be so complex. So physicists have wondered if there were not some underlying simplicity waiting to be discovered, which would account for all the observed complexity. It is very premature to say that the basic simplicity has finally been unearthed, but we are getting a glimpse of it—or at least we think so. The new simplicity was suggested independently in 1963 by Murray Gell-Mann and George Zweig, both at the California Institute of Technology.

The Quark Theory

Gell-Mann and Zweig figured out that the various properties of the many hadrons could be understood if all hadrons were composed of three different kinds of more basic particles. Those particles, never actually observed or detected, are called *quarks.* (The name comes from James Joyce's *Finnegan's Wake:* "Three quarks for Muster Mark . . .") Originally three kinds of quarks were postulated; they were called "up," "down," and "strange," each with its corresponding antiquark. Baryons are made of three quarks each, and antibaryons of three antiquarks; mesons are made of one quark and its own anti in combination. That was the original idea. Then the 300 different hadrons could be understood as just 300-odd arrangements of the three kinds of quarks (and their antiquarks).

The three kinds of quarks have also been referred to as three "flavors" of quark. It turns out that to satisfy certain quantum-mechanical laws, we really have to assume that each flavor of quark also comes in three different "colors"—three different states that are indistinguishable from each other in other respects. According to this idea, one quark of each of the three "colors" must be present in each baryon. It all sounds sort of mystical, but it turns out that the nuclear physicists have predicted results of nuclear experiments, based on these theories, that have actually been observed.

In the mid-1970s, the quark theory had a great psychological boost. If the whole idea made sense, it seemed that there were three basic kinds (flavors) of quarks, along with their various colors and antiquarks, that accounted for all hadrons and their many varied properties of charge, mass, spin, and so on. But there were *four* known kinds of leptons. It seemed only fair, for aesthetic reasons, that there should be *four* kinds of quarks. (Indeed, it would be *charming.*) In 1964 Sheldon Glashow and J. D. Bjorken predicted such a fourth quark. Glashow called it the *charmed quark,* and predicted certain "charmed" properties it should give a hadron that contained it. In 1974 two different laboratories, Stanford Linear Accelerator Center and Brookhaven National Laboratory, independently discovered a massive hadron that had the properties predicted for one containing charmed quarks. In fact the physicists thought this new particle, called the *J* by one lab and the *psi* by the other, should be made of one charmed quark and one charmed antiquark; they dubbed the thing *charmonium,* and predicted further properties for it and ways it should decay. More recently, additional particles have been discovered that seem to fit those predictions.

In short, the quark theory seems to be standing up, even to the point of rounding it out with a fourth kind of quark. It's only a model, of course, like everything else in science, and is on extremely preliminary footing compared to, say, electromagnetic theory. Even now, some physicists are suggesting that maybe there should be *six* kinds of quarks! Yet, we are making progress in understanding the atomic nucleus; at least it seems that the many kinds of short-lived particles that have been discovered in the past quarter century can be described as manifestations of four (or maybe six) even more fundamental units.

Nuclear Forces

Now what are the forces that hold everything together and govern the laws of the universe? As we have seen, the electrons of the atom are bound to the vicinity of the nucleus by the electromagnetic force. The electromagnetic force has no effect on neutral particles, and has nothing to do with the mass of its subjects. These reasons alone would rule out this force as the force that binds the nucleus together. Moreover, the nuclei of most atoms contain several positively charged particles (protons) that repel each other; the force that holds together the nucleus of the atom must be strong compared to the electromagnetic force.

This force is, in fact, about 100 times as strong. The *strong nuclear force* that binds the hadrons in an atomic nucleus has no effect on charges and no effect on leptons. Moreover, it is a very short range force; it is extremely strong if the hadrons are in virtual contact, but it drops off to almost nothing if they are even the least bit separated. The strong force, therefore, is sort of a nuclear "glue."

---

4. If the nuclear force is the strongest of the known natural forces, why do you suppose it escaped discovery until the twentieth century?

There is also another nuclear force. The neutrino is a lepton, and hence is not affected by the strong force; it is also neutral, and so is not affected by the electromagnetic force. It feels only that other force, which is a few hundred or thousands of times weaker than the electromagnetic force, and is thus called the *weak nuclear force*. The weak force is involved in every reaction when there is an absorption or emission of a neutrino. For example, it holds the neutron together.

We have now surveyed all four of the known forces of nature. The strong and weak nuclear forces and the intermediate electromagnetic force, together with the far, far weaker gravitational force, are believed to be sufficient to account for all events known in nature. In fact, physicists hope to find an underlying unity connecting these four forces. Some theoretical progress has already been made in understanding connections between the strong and weak nuclear forces and the electromagnetic force.

To summarize: An atom consists of a positive nucleus and negative electrons, united by the electromagnetic force. The nucleus itself, bound together by a very much stronger nuclear force, is made up of hadrons. Each hadron, in turn, is thought to be composed of either two or three quarks selected from a family of four (or possibly six) different kinds of quarks; various combinations of different quarks account for the 300-odd different kinds of hadrons known. All, however, except the proton and antiproton, rapidly decay. Some of these decays, and other nuclear reactions, involve the absorption or emission of neutrinos, which invokes the far less intense weak force.

It seems to me to be a fantastic intellectual achievement to have been able to discover so much about the sub-submicroscopic nature of matter—all matter. Let's say that after just finishing dinner, you first look out of your window to the moon, above the brick house across the street, and then come back to this wonderful book in your hands. All that stuff—your roast beef and gravy, the bricks and mortar in that house, the moon, the pages of this book, the ink on that page, to say nothing of the hands that typed the first rough draft in a cold flat in Edinburgh in January, and the snow outside that flat—all of it is made up, ultimately, of quarks and leptons, all jiggling about at tremendous speeds, and exerting forces on each other. Some other quarks and leptons are jiggling about, causing electrical currents imbedded in still larger aggregates of quarks and leptons, stimulating some of those ensembles of quarks and leptons to generate still more currents—with the ultimate result that an unbelievably huge bulk of quarks and leptons called "you" is consciously thinking at this moment about all those quarks and leptons.

It's funny, though, that so far as we now know, the only particles we really need to make the universe work are the proton, neutron, electron, electron neutrino, and their antis. Why did Nature bother with all those other quarks and leptons? Are we missing some point somewhere? At present, we don't know. It seems, though, that we can account for almost everything observed naturally and everything necessary, including the generation of the energy of the sun and stars, with that economical collection of just four very basic, fundamental particles and their antis. Which brings us back to where we left off in our drama: What is the source of energy of the sun? Because it can't be the chemical energy of burning (involving the electromagnetic force) or the gradual contraction of the sun (involving the gravitational force), we must look to the atomic nucleus.

---

5. Consider a nucleus of 11 protons. The total electrostatic repulsion between the 11 protons is expected to be about 10 times as strong as that between two protons of comparable separation. If the strong nuclear force is about 100 times as strong as the electrical force, but acts only between adjacent particles, what is the largest number of protons you would expect a nucleus to be able to have and still remain stable? What do you know about nuclei that actually contain larger numbers of protons?

Binding Energy of the Nucleus

The hadrons of most nuclei between hydrogen and uranium cling together with the *strong force*. It takes energy to separate them, and in doing so, we would have to give them potential energy, just as we do to a rock when we lift it. By the same token, if those hadrons come together to make the nucleus, they give up that energy, just as the rock does when it falls back to earth. Thus the energy of a nucleus is less than that of its parts when they are separate. Since energy and mass are equivalent, the mass of a nucleus is less than the sum of the masses of its parts. This strange-sounding statement does not express a breakdown in the conservation of mass energy. The "missing mass" is just the energy given up by the hadrons in coming together. It is equal to the energy it would take to separate them again, and is called the *binding energy* of the nucleus. The mass equivalent of that binding energy is called the *mass defect.*

The mass defect is zero, of course, for the nucleus of the ordinary hydrogen atom, because it consists of only one hadron—the proton. The mass defect is very small for the most massive nuclei that are just barely stable. It is largest for the nuclei of atoms near iron, cobalt, and nickel in the periodic table, and is less for both lighter and heavier nuclei. Iron has 26 protons in its nucleus and usually 30 neutrons, giving it 56 hadrons. If light nuclei can come together to make a new nucleus of iron or something lighter, energy will be released in the reaction; this is called nuclear *fusion*—combining small nuclei into bigger ones. On the other hand, it *takes* energy to combine a nucleus of iron or something heavier with another nucleus to make a still heavier one. Energy *is* released, however, if a very heavy nucleus can split up into lighter ones (leaving something heavier than the nucleus of iron); that is called *nuclear fission.*

Natural radioactivity is one way a heavy nucleus can split up and release energy. Uranium nuclei spontaneously change into lighter ones by emitting helium nuclei and energy. Eventually, a sample of uranium ends up as lead, although many thousands of millions of years are required for the conversion. At one time people thought that natural radioactivity might be responsible for keeping the sun shining. However, even if the sun's uranium could be turning into lead fast enough to produce the sun's present luminosity, that uranium would be used up in a few months; uranium is very rare on the sun (as it is on the earth).

Nuclear Fusion in the Sun

If nuclear energy is the source of the sun's radiation, we must look to the fusion of the light elements that are plentiful in the sun—hydrogen and helium. The helium nucleus consists of four hadrons—two protons and two neutrons. It would take four hydrogen atoms to produce this much matter. Now let's see how it works out. For convenience, we express the masses of these tiny particles, not in grams but in terms of the atomic mass unit *(amu).* One amu is defined as one-twelfth the mass of the nucleus of the most common isotope of carbon; it turns out that 1 amu is about 1.67×10^{-24} g. One hydrogen nucleus (proton) has a mass of 1.00813 amu; so four of them have a combined mass of 4.03252 amu. A helium nucleus has a mass of only 4.00389. The difference in mass of 0.02863 amu is the mass defect. This is precisely the amount of mass lost in the transaction if four hydrogen atoms turn into a helium atom.The mass defect is about 0.7 percent of the original mass of the hydrogen.

Suppose just one gram of hydrogen could be turned into helium. This means that 0.007 g of matter would disappear and would be released as energy. According to Einstein's equation, $E = mc^2$, that 0.007 g creates 6.4×10^{18} erg, enough energy to raise a 1000-ton ship 65 km into the sky. From one gram! But a 100-megaton H-bomb (which works on the same principle) releases nearly a million times as much energy as this. The binding energy of the nucleus is nothing to sneeze at! I should think it would make one shudder a bit.

But will nuclear fusion work to keep the sun shining? All the sun has to do is convert some 600 million tons of hydrogen into helium each second, with the consequent conversion of 4 million tons of mass into energy, to account for its radiation of 4×10^{33} erg/s. That's a lot of hydrogen going down the tube, but a good supply is there in the sun. In fact, even if only a small fraction of the sun's hydrogen could turn to helium, it would still be enough to keep the sun in business for thousands of millions of years. Today it is generally accepted that the conversion of hydrogen into helium is what keeps the sun and most stars shining for most of their lives. There are limits, of course, to even this source of energy, and we shall see what they imply later in this act.

---

6. The sun's luminosity is 4×10^{33} erg/s, and its mass is 2×10^{33} g. What if the sun began as pure hydrogen and could convert all that hydrogen into helium? How long could the sun continue to shine at its present luminosity before all the hydrogen was converted into helium? (Give answer in years; one year contains about 3×10^{7} s.)

We have not yet explained exactly *how* four atoms of hydrogen go about converting themselves into helium. Indeed, if they were prone to do that, one might think it would be risky to store hydrogen around the house. Actually, there are several series of nuclear transformations that can lead to the conversion of hydrogen into helium under conditions that exist in stellar interiors. These reactions occur, however, only at very high temperatures—well over 10 thousand million degrees—so the atomic nuclei are moving about at high speeds. The reason they have to move fast and have high energy is that the nuclei all repel one another (because of their positive charges). The strong nuclear force will only bind two nuclei that are in close contact, but for them to get together where the nuclear force can take over, they must overcome the strong electromagnetic force pushing them apart from each other.

In fact, even at the high temperatures in the interior of the sun, the protons aren't moving fast enough to crash close enough together to overcome the electromagnetic repulsion. Consider two protons on a collision course (Figure IV.17a). Because of the electromagnetic force, rather than crash they repel each other and veer away as in Figure IV.17b).

But now let's look in more detaol at their point of nearest approach, with protons 1 and 2 at *A* and *B*. The trajectories shown in Figure IV.17b are actually only the most probable paths for the protons. Because of that uncertainty principle we described back in Scene 3 of Act II, however, there is always an innate uncertainty of the precise position of each proton. In Figure IV.17c, we show, by the dashed lines, the probability of each being in various places. We see that if *A* and *B* are not too far apart, there is a finite chance of the protons actually being at *C*—that is, both in the same place. In that case, the nuclear glue can stick them together and a nuclear transformation can occur. Throughout the great volume of the sun's hot central region, enough protons are getting together in this way all the time to account for the sun's output of energy.

In the sun, the most important of the chains of events by which hydrogen combines to make helium is the following. (1) Two protons combine, form a nucleus of deuterium (a hydrogen nucleus with one proton and one neutron), and a positron and a neutrino are emitted. (2) The deuterium nucleus combines with another proton to make a nucleus of a light isotope of helium (two protons and one neutron), with some energy coming off. (3) Two of the light

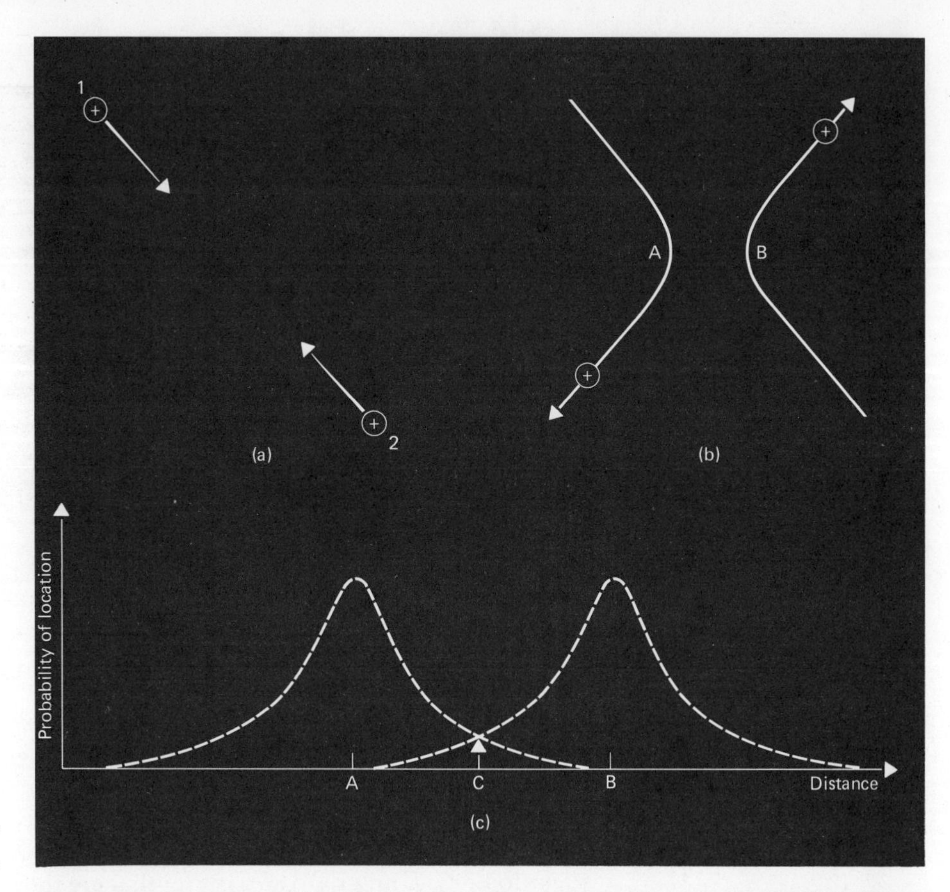

FIGURE IV.17
An encounter of two protons.

helium nuclei combine to make ordinary helium, releasing two protons and more energy. There are also some alternative possibilities, but they all end up with the same result. These various nuclear reactions proceed at rates that depend on the temperature and densities of the gases in the solar (or stellar) interior, and generally go very much faster where the temperatures are highest. Thus most of the sun's nuclear energy is produced near its center, where the gases are hottest. In the next scene we shall see what happens when the hydrogen gets used up in the center of the sun or a star.

The Strange Case of the Missing Neutrinos

Just so you won't go out to intermission after this scene, thinking that everything is completely cleared up, I should mention the solar neutrino problem. At least it is (or seems to be) a problem at the time of this writing. Anyway, you ought to know about the kinds of snags astrophysicists run into. When hydrogen goes to helium in the sun, about 3 percent of the energy is carried away by neutrinos, and does not appear as electromagnetic radiation. Neutrinos are hard to detect, so you might expect that we may as well write off that 3 percent as a sort of solar energy tax. It is just possible, though, to detect some of them, or at least it should be.

Thus, physicist Raymond Davis and his colleagues have assembled a giant neutrino detector in South Dakota. It consists of a 378,000-liter tank of cleaning fluid. The nuclei of a few of the chlorine atoms in the cleaning fluid should absorb some of the solar neutrinos and produce radioactive argon at the rate of about one argon nucleus per day. Davis' apparatus is so sophisticated that

FIGURE IV.18
Raymond Davis, Jr.'s, 378,000-liter neutrino detector in a mine at Lead, South Dakota. *(Brookhaven National Laboratory)*

it can detect those rare argon atoms, and thus indirectly record the absorption of solar neutrinos by the cleaning fluid. Careful laboratory tests show that the system works according to plan, but so far, after several years of observation, it does not seem to be picking up solar neutrinos, or at least not enough to be sure that those produced in the sun are passing by the earth as expected. Possibly there is something wrong with the experimental setup, or perhaps there is a minor slip somewhere in the theory. I think most likely the matter will clear up soon (maybe it will have by the time you read this).

But what if the problem *doesn't* go away? Then new experiments will have to be tried, and if the neutrinos are still missing, the physicists, and maybe the stellar interior astrophysicists, or both, must go back to the "drawing board." This is science at work. It's a marvelous, exciting game; that's why it's such a great satisfaction when we find out something new or get an idea that works out—thereby scoring.

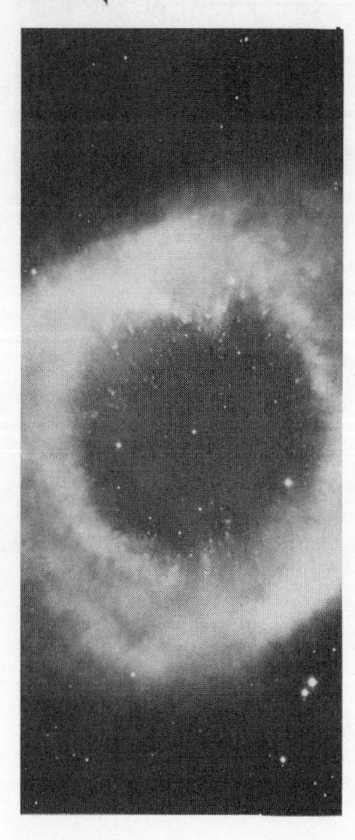

Stellar Evolution and Nucleogenesis

Now that we understand the source of energy of the sun and stars, we have enough clues to calculate a stellar (or solar) model. Remember that hypothetical shell inside the sun we talked about in Scene 1? We want to know the temperature, pressure, and density of the gas within that shell, as well as the amount of energy generated inside it. Our physical clues are (1) the condition of hydrostatic equilibrium—that the shell isn't sinking into the sun or pulsing out; (2) the condition of thermal equilibrium—the shell reradiates the energy it absorbs; (3) the gas law—the pressure of the gas is proportional to the product of its temperature and density; (4) the physical theory of opacity—the mechanisms by which gas absorbs radiation; and (5) the source of energy—how much nuclear energy is generated by the gas, given its composition and physical conditions. We use the same physical principles to calculate the conditions for every shell in the sun (or star), and if we add all the shells together, of course we get the whole sun. That's what's involved in calculating a model for the solar interior.

---

1. You have possibly encountered Boyles' law and Charles' law in high school. How do these laws relate to the gas law discussed above?

Calculation of a Stellar Model

Here's one way the calculations can be done: We know the conditions at the surface of the sun. We thus use all the above physical principles—expressed as mathematical equations—to figure out what the conditions must be just beneath the outermost layer to account for its conditions. For example, the first shell beneath the surface has to support the weight of the surface shell, so its gas pressure is higher. After carrying out the calculations, we know what it must be like in that first shell down; then we figure out what the shell beneath it has to be like. And so we go, step by step, until we reach the center. If we have put in the right numbers and haven't made a mistake in arithmetic, the combined mass for all those shells should add up to the entire mass of the sun. If not, we start over and try again until everything comes out right. With big computers, we calculate the conditions of all the shells simultaneously, but the ideas are the same.

As an example, Table 1 gives the result of a solar model calculated by R. Ulrich, at UCLA. In the future, as we improve our knowledge of opacity and of the details of nuclear energy generation, we will doubtless calculate slightly improved and, hopefully, more accurate models; however, the data in the table are probably pretty close to the real conditions inside the sun. The first column shows the fraction of the distance from the center to the surface of the sun. The next two columns give the fraction of the sun's mass contained within that radial distance from the center and the fraction of its total luminosity generated within that distance. The next two columns give the temperature and density of the gas at that distance from the sun's center, and the last column shows what fraction of the solar material is hydrogen.

TABLE 1 Model for the Structure of the Sun*

Fraction of Radius	*Fraction of Mass*	*Fraction of Luminosity*	*Temperature (millions of °K)*	*Density (g/cm³)*	*Fraction Hydrogen (by mass)*
0.00	0.000	0.00	15.0	148	0.38
0.05	0.011	0.10	14.2	125	0.47
0.10	0.076	0.45	12.5	86	0.59
0.15	0.19	0.78	10.7	56	0.67
0.20	0.33	0.94	9.0	36	0.71
0.30	0.61	1.00	6.5	12	0.73
0.40	0.79	1.00	4.9	4	0.73
0.60	0.95	1.00	3.1	0.5	0.73
0.80	0.99	1.00	1.3	0.1	0.73
1.00	1.00	1.00	0.0	0.0	0.73

* Adapted from R. Ulrich.

2. What radius of the sun, according to the Ulrich model, contains half the solar mass? What fraction is this of the sun's total volume?

Note how strongly the mass is concentrated to the sun's center, and also that almost all the luminosity is generated from the very inner hottest parts. These points are shown graphically in Figure IV.19. The surface of the sun in this model corresponds roughly to the photosphere. Of course the temperature and density are not precisely zero at the photosphere, but they are essentially zero compared to their much higher values inside. Note that the density at the center of the sun is well over 100 times that of water—greater than is possible for any solid. In a solid, the nuclei and electrons of the atoms are bound in a rigid lattice structure with enormous amounts of empty space between the particles. The atoms at the sun's center are almost completely ionized (most atoms have lost all their electrons) and are in the gaseous state, so they can be crammed very much closer together. Even so, the center of the sun is mostly empty space; we shall see that gases can have incredibly higher

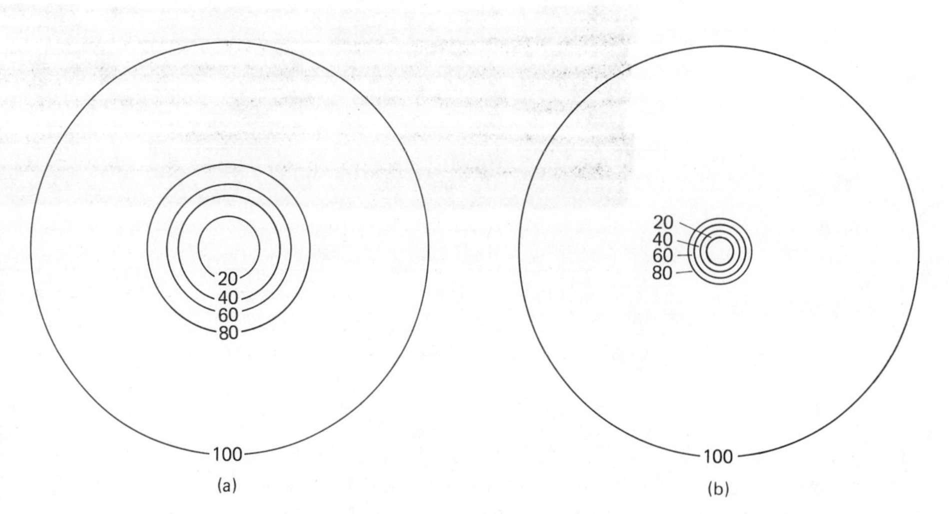

FIGURE IV.19
Distribution of *(a)* mass and *(b)* energy generation within the sun, according to the model in Table 1. The numbers indicate what fractions of the mass or energy generation are included within the radial zones shown.

densities than they do there! Finally, note that at the center of the sun nearly half the original hydrogen (which was 73 percent of the sun in this model) has been used up in conversion to helium.

Explanation of the Main Sequence

Every star, once it has reached the point in its life where it derives its energy from nuclear reactions, must obey the same laws of physics that apply to the sun, and must adjust itself accordingly. Now every star is born with a certain amount of material, which contains the particular distribution of elements—or *chemical composition*—of the cloud of gas from which it condensed. Except (we think) for the first stars to form in the Galaxy, most stars have chemical compositions not very different from the sun's. Suppose we calculate models by the procedures described above for a whole bunch of stars, all of composition like the sun's, but all of different mass—spanning the entire range of known stellar masses.

What we find is that the most massive stars turn out to be very luminous and hot on the outside, and that the least massive ones are relatively dim and cool. In fact, we can calculate the absolute magnitudes and temperatures for each of those hypothetical stars and plot them on a Hertzsprung-Russell diagram. And do you know what? They line up right on the main sequence! The *observed* main sequence is just the locus of points representing absolute magnitudes and temperatures of stars with about the same distribution of elements as in the sun, but having different masses. It's nice that the physical theory, coming from such varied sources as the nuclear physics laboratory and Newtonian gravitation, accounts so beautifully for the known properties of stars.

3. Interpret the mass-luminosity relation (Act III Scene 3) in terms of stellar structure theory.

Early Stages of Stellar Evolution

Of course stars don't end up full blown on the main sequence. First a star has to contract from the prestellar cloud of gas. The time required to do so is the time required for it to fall together of its own gravitation. It took the sun a few million years. Once the material is dense enough to become opaque, the energy released by the in-falling matter is trapped inside, building up the gas pressure until it stops the free fall inward; then the star is in hydrostatic equilibrium. Next it contracts only gradually, converting gravitational energy into luminosity and heat in its interior. In this period of its youth, the star is shining by the mechanism suggested by Helmholtz and Kelvin to account for the sun's luminosity. This period of the sun's life lasted some tens of millions of years. Finally, when the temperature near the star's center is high enough (10 to 15 million degrees), nuclear reactions begin and supply new energy as fast as the star radiates energy away. The star now begins its adult life, which it spends as a main-sequence star. For the sun, that life should last about 10 thousand million years.

How fast a star goes through these stages, though, depends on its mass; everything is faster, the higher the mass. The most massive stars fall together from their protostellar clouds in less than a million years, and after becoming opaque reach the main sequence in another few tens of thousands of years. We shall see later why their main-sequence lifetimes are under a million years. The least massive stars, on the other hand, take several times as long as the sun to fall together, and hundreds of millions of years to reach the main sequence. In fact, a star less than about one-twelfth the mass of the sun never heats up inside enough for nuclear reactions to take over, and so just keeps contracting slowly, releasing gravitational energy, for hundreds of thousands of millions of years, until it eventually dies. We'll see how it ends up in Scene 7.

4. If stars of less than 0.085 solar mass never get hot enough in their centers to fire up nuclear reactions, can such objects rightly be called "stars"? If not, what are they then?

Life on the Main Sequence

Hydrogen is the fuel of main-sequence stars, but their hydrogen is not inexhaustible. Already, after some 4.6 thousand million years, the sun has used up nearly half the hydrogen at its center. After another five or so thousand million years, the hydrogen will be used up in an inner core containing about 10 percent of the sun's mass. Then there will be a core of nearly pure helium, contaminated with that few percent of the heavier elements also present in the sun.

Now a star of 100,000 times the sun's luminosity is converting hydrogen to helium 100,000 times as fast as the sun is. Such a star starts off with more hydrogen than the sun did to begin with, but less than 50 times as much; consequently it achieves that helium core in only a couple of million years. However, the least massive stars take thousands of times as long as the sun takes to reach that helium core stage.

Evolution to Red Gianthood

You can probably guess what happens to that helium core with no more nuclear fuel to supply energy to it; it starts to contract, just as stars do before they reach the main sequence. Now two things happen. First, that contracting core releases energy into the surrounding gases, causing the star's outer parts to distend enormously—*the star becomes a red giant.* But only half the gravitational energy given up by the core is radiated into the outer envelope of the star; the other half goes into heating the core itself. Thus, while the outer part of the star expands and lowers in density, its core contracts, and becomes extremely dense and hot.

---

5. Why does the helium core of a star contract?

Meanwhile, within the shell of material just outside the core, not yet devoid of hydrogen, nuclear energy is still being generated. In fact, because the core keeps heating it up, the hydrogen burning accelerates. (Astrophysicists talk about the conversion of hydrogen to helium as hydrogen "burning," but of course it is not burning in the conventional chemical sense.) The shell of hydrogen burning, therefore, keeps adding more and more helium to that shrinking core, so while it gets physically smaller, it increases in mass. The process doesn't stop until a new source of nuclear energy appears in the core.

---

6. Why does a new source of energy stop the contraction of a star's core?

Helium Burning

That new nuclear reaction is three helium nuclei (called *alpha particles*) combining to make a nucleus of carbon. But until the 1950s no one thought this could happen. The trouble is that a nuclear transformation doesn't necessarily occur just because two or more nuclei come together and stick. Three alpha particles coming together, for example, have the right ingredients—two protons and two neutrons each—to make a carbon nucleus, but before that compound nucleus of three alphas becomes carbon, it must emit some energy. It is analogous to an excited atom emitting energy to jump to its ground state. But it takes a bit of time for the compound nucleus to do this, and it was expected that before having that chance it would scatter its three alpha particles away again.

However, the British astrophysicist Fred Hoyle (now Sir Fred Hoyle) pointed out that the three helium nuclei could make carbon if the nucleus of carbon happened to have an excited state 12.5-millionths of an erg above its ground state. Nuclear physicist William Fowler and his associates at Caltech tested the idea in the laboratory, and sure enough, that energy level of the ex-

FIGURE IV.20
Sir Fred Hoyle. *(Caltech)*

cited carbon nucleus is there! This means that the three alpha particles can stay together long enough to transform into carbon. However, it doesn't happen unitl the temperature of the helium core reaches about a hundred million degrees.

Hoyle was led to expect that that energy state of carbon exists because, first, it had been discovered that the cores of red giant stars can reach the necessary temperature. But mainly, if that energy state were *not* present, helium could not transform to carbon and no further nuclear reactions would occur after hydrogen burning. In *that* case, though, it would be very hard to imagine where carbon and the heavier elements came from, and things like planets, and us, and astronomy books wouldn't exist. Hoyle had the idea that the heavy elements in the universe could be built up in the hot centers of red giant stars.

Nucleogenesis

The details of how heavy elements can be built up from helium by nuclear reactions in stellar interiors were worked out in a now-famous study by Hoyle, Fowler, and astrophysicists Geoffrey and Margaret Burbidge, and were published in a paper now known among astronomers not by its title but as "B^2FH" (for Burbidge, Burbidge, Fowler, and Hoyle). They found that once carbon is formed, successive nuclear reactions can occur to build up oxygen, neon, magnesium, silicon, sulfur—in fact all the chemical elements up to iron—in roughly the relative abundances we find for them in the universe. The temperatures required for these reactions range from 200 million to 3500 million degrees; such temperatures are now believed to occur near the centers of stars that are advanced in their evolution.

With each new kind of nuclear fusion up to iron, energy is released, so each of the processes described by B²FH provides a new, if often very temporary, source of stellar energy. Elements farther along the periodic table than iron can also be built up by nuclear reactions, but these reactions *require* energy rather than release it. However, we think this can happen, too, during certain explosive stages of the evolution of massive stars, particularly in supernovae, which we shall discuss shortly.

For centuries the alchemists tried to transform the elements from one to another. They failed, but the stars have been doing it all along. The process of building up the chemical elements from hydrogen is called *nucleogenesis*. It is a tremendously exciting part of this drama of the evolution of the universe. And as you can see, were it not for that lucky energy level in the carbon nucleus, we probably would not be around to enjoy it!

Stellar Old Age

Eventually, then, the core of the expanding red giant gets hot enough for helium to begin burning into carbon. This new energy source causes the core to expand again, and the star, as a whole, to contract, ending its stint as a red giant—at least for a while. Soon, however—in fact in a time very short compared to the main-sequence lifetime of the star—a core of pure carbon develops at the center of the helium core. Now we are back to where we were before; the carbon core shrinks and the star expands again to a red giant.

Subsequent stages of evolution have been theoretically calculated in some detail for stars of various masses. These relatively rapid states of evolution produce changes that shove the star back and forth across the top of the Hertzsprung-Russell diagram several times, each change in evolution occurring as a new nuclear energy source fires up.

Pulsating Stars

During certain periods in these advanced stages of the evolution of massive stars, they become unstable and pulsate. There are several types of pulsating stars; the principal ones are listed in Appendix 12. Most known pulsating stars are giants and supergiants. Especially numerous are the long-period red giant variables, with periods ranging up to more than one year. A famous example is the star Mira, in the constellation of Cetus. Mira is usually at a magnitude of 8 to 10, too faint to see with the unaided eye. About every 11 months, though, it brightens to visibility for a few weeks. Another very common kind of variable star is the RR Lyrae star, which we met earlier when we described how Shapley found distances to globular clusters by measuring the magnitudes of RR Lyrae stars in those clusters. The famous Cepheid variables, so important to the determination of galaxian distances, are further examples of pulsating stars.

Most stars, like the sun, are stable, and even if you could start them pulsating (say, by squeezing them or by hitting them with a big stick), those pulsations would quickly damp out. Nevertheless, there is a natural period with

Raymond Davis, Jr.'s, "neutrino telescope." (*Brookhaven National Laboratory*)

The open cluster M16 and associated nebulosity in Serpens. (*Hale Observatories*)

The Pleiades and associated nebulosity in Taurus, photographed with the Palomar Schmidt telescope. (*Hale Observatories*)

The Dumbbell nebula, a planetary nebula in Vulpecula, M27. (*Hale Observatories*)

The Ring nebula in Lyra, M57; 5-m telescope photograph. (*Hale Observatories*)

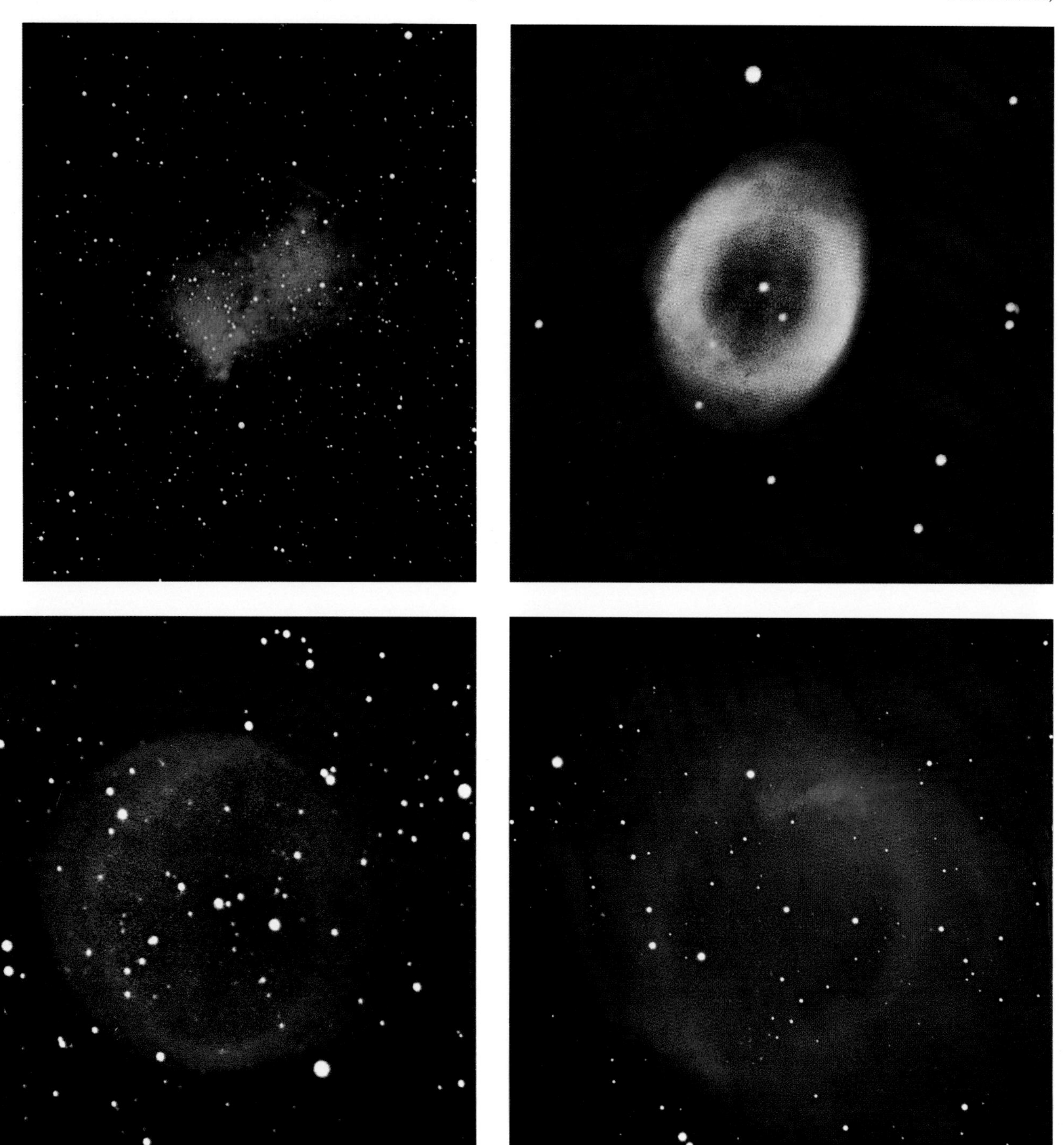

NGC 6781 in Aquila; Palomar Schmidt photograph. (*Hale Observatories*)

NGC 7293 in Aquarius; 5-m telescope photograph. (*Hale Observatories*)

The Orion nebula (*Hale Observatories*)

The Omega nebula. (*Kitt Peak National Observatory*)

The Rosette nebula in Monoceros, NGC 2237, photographed with the Palomar Schmidt telescope. (*Hale Observatories*)

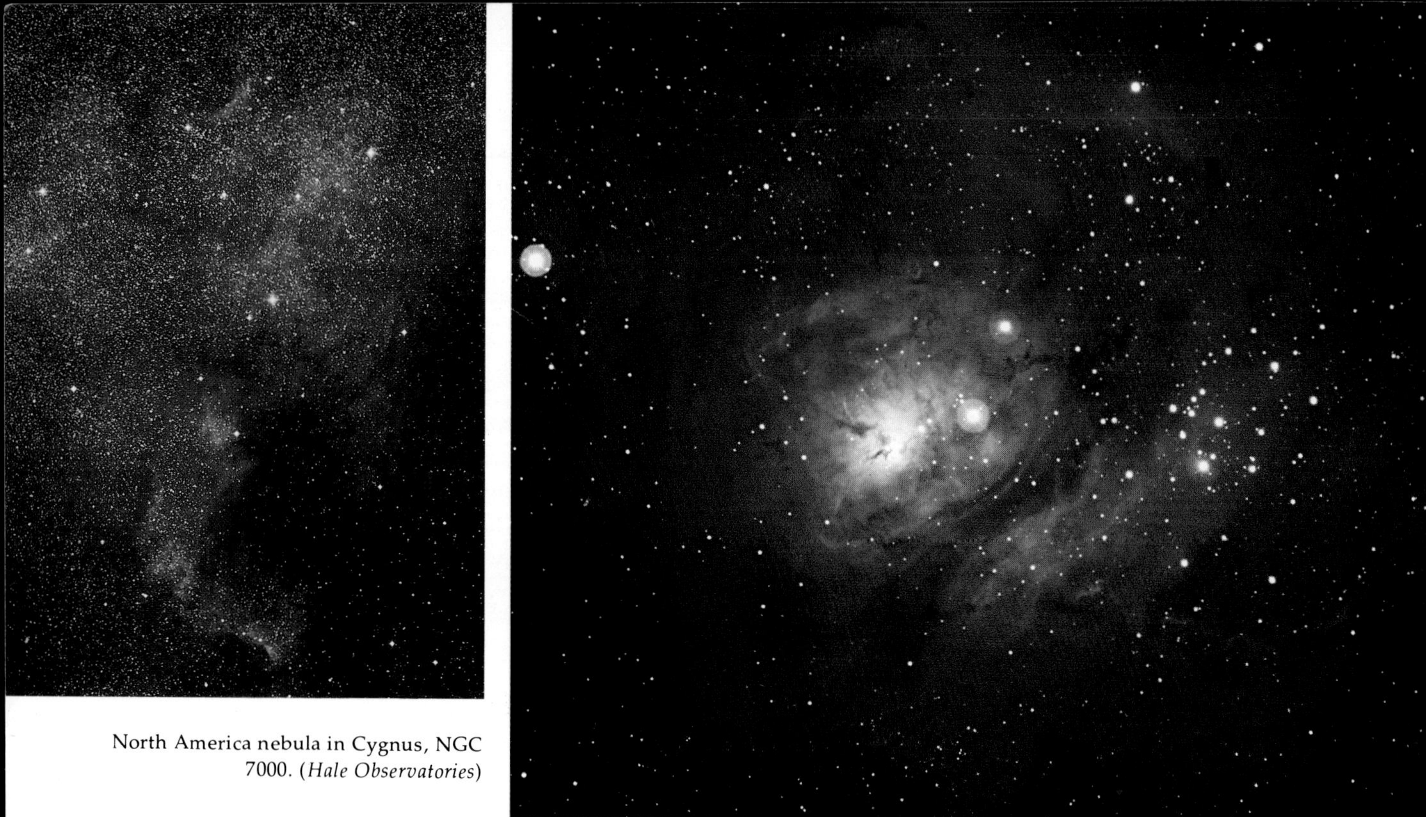

North America nebula in Cygnus, NGC 7000. (*Hale Observatories*)

The Lagoon nebula in Sagittarius, M8. (*Kitt Peak National Observatory*)

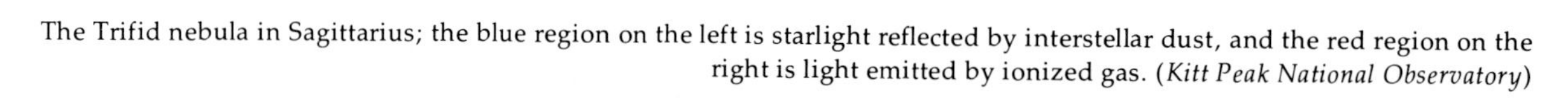

The Trifid nebula in Sagittarius; the blue region on the left is starlight reflected by interstellar dust, and the red region on the right is light emitted by ionized gas. (*Kitt Peak National Observatory*)

which any star would pulsate if it were not stable. It is easy to understand how that period of pulsation can be found. The time it takes the star to go from its largest to its smallest size is essentially the time it takes its surface layers to fall through that distance. That, in turn, depends on the gravitational acceleration at the surface of the star and how it changes as the star shrinks.

As you might expect, if a star were very large and of low density, the force of gravitation at its surface would be relatively low, and the surface, if unbalanced, would fall relatively slowly. Thus giant stars of low density would pulsate slowly—with long periods. Conversely, small dense stars have very strong local gravitation at their surfaces, and would pulsate rapidly if they were unstable.

If the sun could pulsate, its period would be somewhat under a half hour. It turns out that the shortest period with which a star (or anything) can rotate and still hang together of its own gravitation is just a few times as great as that with which it would pulsate. The sun, in fact, could not rotate with a period of less than about 2.7 hours.

Now consider stars of various sizes, but all of the same mass as the sun (most stars, as we have seen, have masses not very many times greater or less than the sun's anyway). Table 2 gives the natural pulsation periods of various kinds of stars. By giving "examples" in the table, I am tipping my hand, and giving away some of the thunder of Scene 7. Of course you are not surprised that a supergiant star pulsates in a period of more than a year. But what kind of object, with the mass of the sun or more, could pulsate or rotate in a tiny fraction of a second?

TABLE 2 **Pulsation Periods of Various Stars of One Solar Mass**

Radius (solar radii)	*Period*	*Examples*
1000	2 yr	Red supergiants
100	1 mo	Cepheid variable
10	0.7 d	RR Lyra star
1	0.5 hr	Sun
0.1	1 min	
0.01	2 s	White dwarf
0.00001	0.00006 s	Neutron star

7. Why are most known pulsating variable stars giant stars?

Mass Exchange in Binary Stars

Stars that belong to close binary systems (binary stars in which the two members are close together) may undergo rather unusual evolution. On a line between the centers of the members of any double-star system, there exists a point at which a particle of matter feels itself attracted equally to both stars, and doesn't know to which star it belongs. There are many binary stars with their members so close together that when one of them begins to expand on the way to becoming a red giant, its surface passes through that point of confusion. When this happens (and it must happen rather often) some of the material from the growing star flows onto its companion star, and some other part of the matter is lost into space. Thus the mass of each star in such a system changes in time, with the result that the evolution of the stars in a close binary can be very complicated. Today there is a lot of interest in what happens to the stars in such a close binary as they exchange mass, or, alternately, as they lose matter from the system entirely.

All stars, however, must eventually exhaust their store of nuclear fuel and other energy, and die. We return to that stellar graveyard in Scene 7. We anticipate the gems of that scene here only enough to mention that certain stars seem to have too much mass to die. Those stars must get rid of some of their matter by ejecting it into space.

Mass Ejection

The sun ejects some matter continually in the form of the *solar wind*, which is a more or less steady flow of ionized gas into space. Similarly, many or most stars are expected to emit *stellar winds* into the interstellar medium. At present, however, the solar wind is too small a breeze for the sun to get rid of much of its matter during its lifetime. For some stars, however, stellar winds are expected to be significant mechanisms of mass ejection.

Variable stars sometimes eject a bit of matter into space with each pulsation. Some stars eject matter explosively in *nova* outbursts; the amount of matter a nova ejects is typically about a hundred-thousandth or so of a solar mass. All novae are now thought to be members of very close binary star systems. We shall return to them in Scene 7.

Planetary Nebulae

One major mechanism by which stars eject matter is the planetary nebula phenomenon. A planetary nebula is a shell of gas ejected from a star—probably a very large red giant—late in its evolution. Typically a planetary nebula is one-half light year or more across and has an amount of matter equivalent to about one-fifth of the mass of the sun. The shell of gas goes off in such an orderly fashion that I think its parent red giant just sort of sloughs it off, rather as one blows smoke rings. Indeed, planetary nebulae often look like smoke rings, but most are really shells of gas, absorbing radiation from their parent stars and reemitting it in all directions. Because a planetary shell is usually hollow, our line of sight encounters more gas emitting light along the periphery of the nebular images in the sky (at points *A* in Figure IV.22) than it does when we look at the middle of the planetary (*B* in the figure) and see through the thin dimensions of the front and rear parts of the shell.

Planetary nebulae are relatively rare in the Galaxy; only about a thousand are known. However, when we take account of the interstellar dust that probably hides 90 percent of them from our view, we estimate that there may be more than 10,000. On the other hand, after a few tens of thousands of years a planetary nebula expanding at a few tens of kilometers per second disperses into space and loses its identity as such. Thus these nebulae are very temporary things, cosmically speaking. When we take into account their short lifetimes we realize that planetary nebulae are common enough that they may well be ejected from nearly all stars on their way to that happy hunting

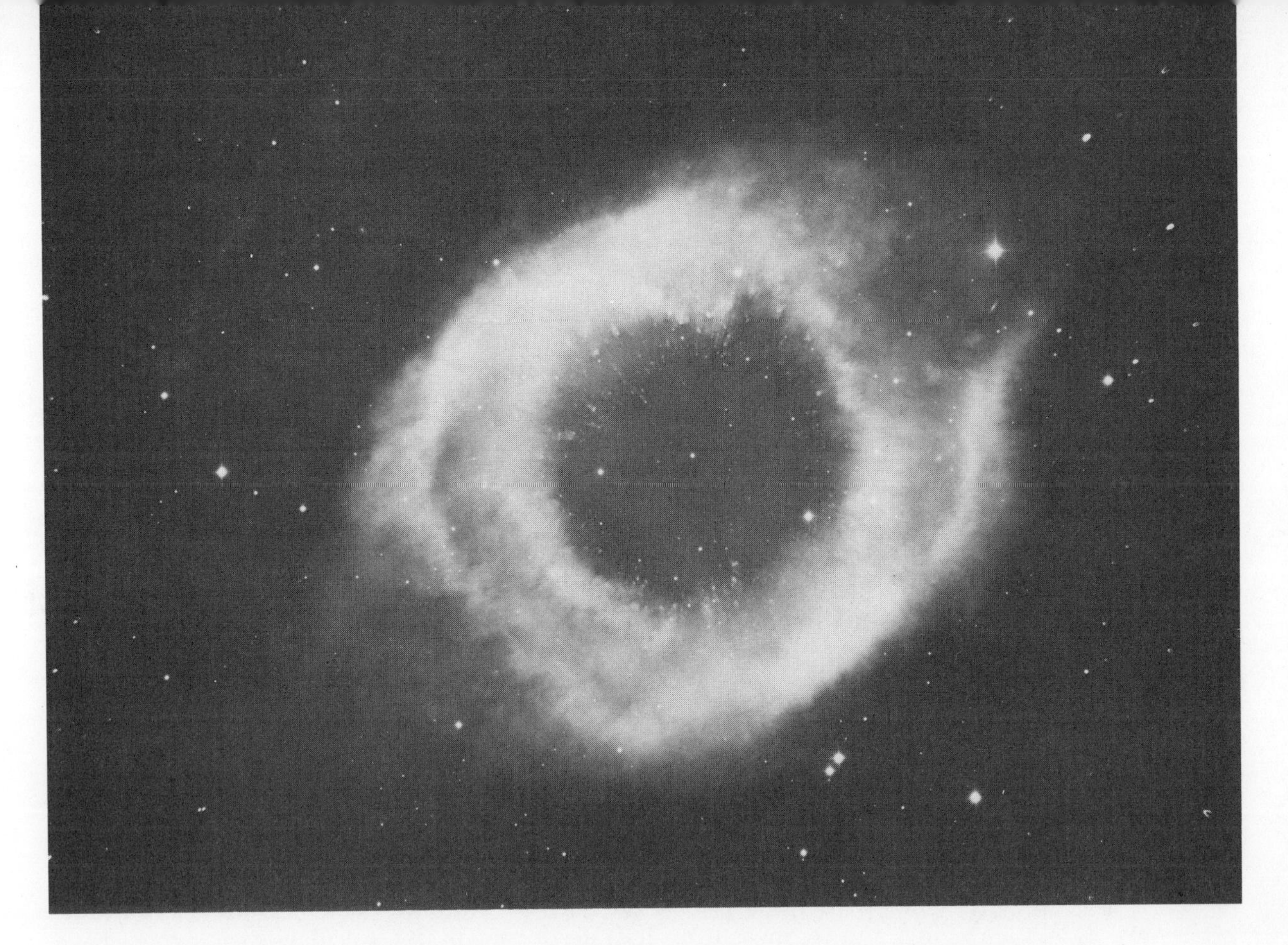

FIGURE IV.21 Planetary nebula in *Aquarius,* NGC 7293. (*Hale Observatories*)

FIGURE IV.22
Along the periphery of a spherical shell our line of sight encounters more material than when we look through the front and rear surfaces.

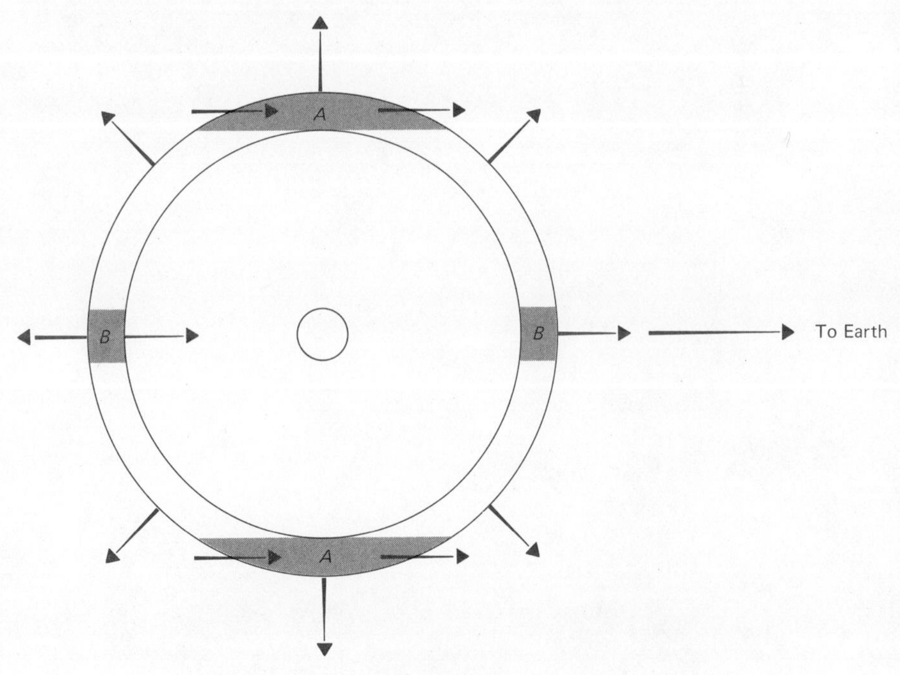

ground up in the sky where stars end up. It may be that the sun will someday eject a shell of its outer gases as a planetary nebula.

Supernovae

The most spectacular thing that has been observed to happen to a star is to become a *supernova*. We have already noted that supernovae have been observed in other galaxies. Several have been recorded in our own Galaxy, and there are quite a few gaseous remnants that are believed to be ejecta from prehistoric galactic supernovae. We estimate that on the average there are two or more supernovae in our Galaxy each century. In such a cosmic outburst, a star blows a large part of its mass off into space and at initial speeds of up to 10,000 km/s. There appear to be several kinds of supernovae, some of which may occur only in close binary-star systems. In all cases they reach temporary luminosities of many millions of times that of the sun at maximum light; the most luminous supernovae are more than a thousand million times the solar luminosity. Supernovae rise to maximum very rapidly—in just a few days, and then fade out again over succeeding months.

The Crab Nebula

The most famous supernova in our Galaxy is the one recorded by the Chinese in 1054 A.D. It appeared on July 4, and for a few weeks it was visible in daylight (as "bright as Venus") and it could be seen until 17 April 1056 in the nighttime sky. Today, in that same part of the sky is an expanding chaotic mass of gas known as the Crab nebula. The Crab nebula is seen to be expanding, and by measuring its rate of growth, we calculate that it must have

FIGURE IV.23 The Crab nebula. *(Hale Observatories)*

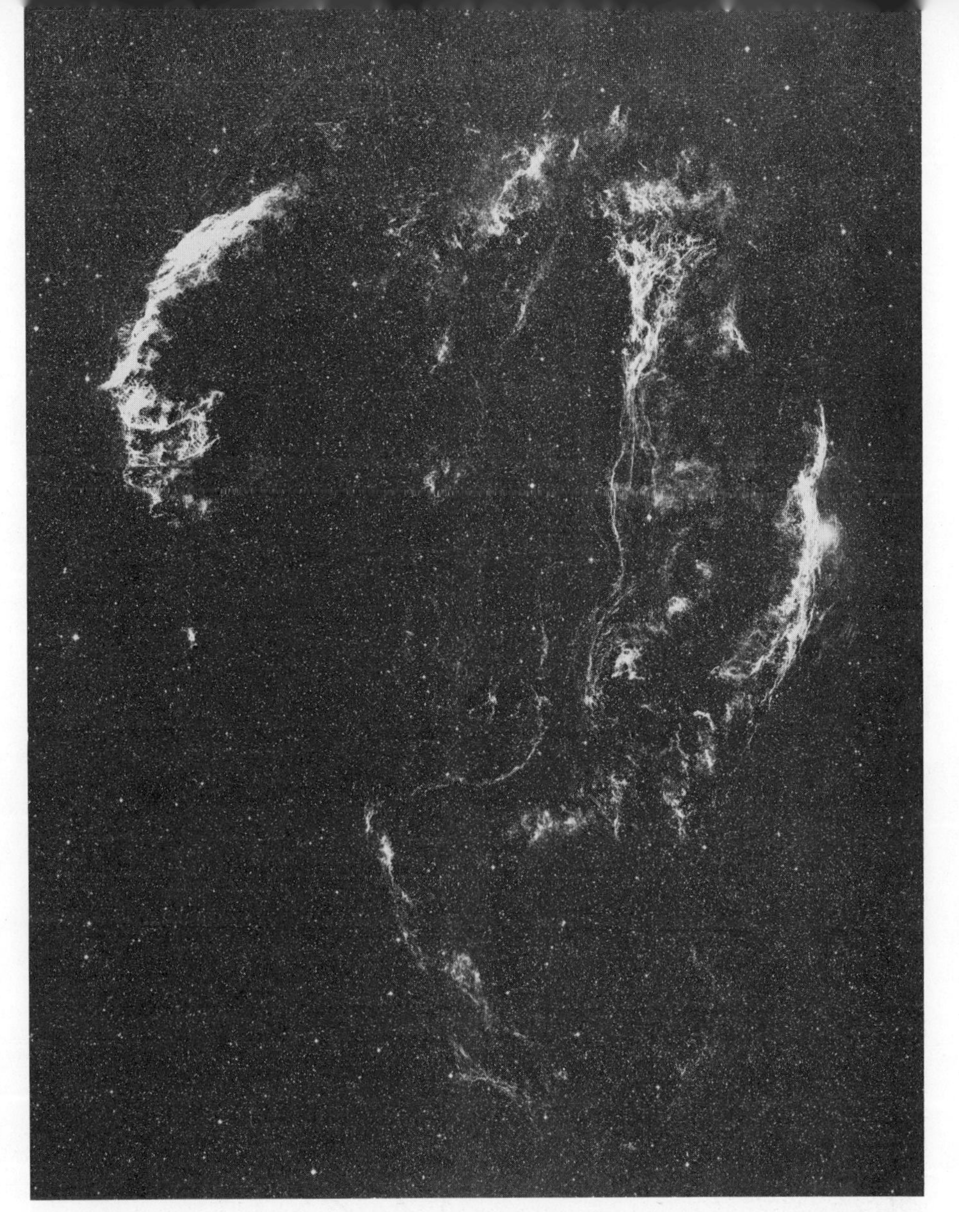

FIGURE IV.24
The Veil nebula in *Cygnus*—the remnant of a prehistoric supernova. *(Hale Observatories)*

begun from that supernova outburst, and hence must be the remnant of it. Its distance is about 2000 pc. It is a strong source of radio waves, ultraviolet radiation, and X rays, as well as of visible light.

Much of the radiation from the Crab nebula originates from electrons moving at nearly the speed of light and spiraling around magnetic lines of force, and is known as *synchrotron radiation* (we shall return to it in Act VI). Its origin in the Crab nebula shows that the supernova outburst not only involved magnetic fields but also gave rise to high-energy particles. The Crab nebula supernova, and probably other supernovae as well, are believed to be the origin of many of the cosmic rays continually bombarding the earth from space. The star at the middle of the Crab nebula is especially interesting, because it is a pulsar—the first one found at optical, as well as at radio, wavelengths. We discuss it and other pulsars in Scene 7.

The present thinking is that in supernovae enough energy is released to result in a burst of nucleogenesis of elements heavier than iron. It may even be that virtually all those heavy elements in the universe have been cooked up in supernova explosions.

In any case, it is clear that planetary nebulae, supernovae, ordinary novae, some variable stars, the stellar winds of normal stars, and some binary stars are spewing matter, once part of stars, into the interstellar space. Moreover, that stellar matter, as we have seen, is continually being changed in chemical composition by virtue of the very processes that keep stars shining. Those stars which evolve most rapidly, and are the seats of the most nucleogenesis, are the ones that burn themselves out soonest and eject matter first into space. Thus the interstellar medium, from which new stars are continually forming (as we shall see), is continually being enriched in its content of those elements heavier than hydrogen and helium by the stuff ejected from evolved stars.

We think that this enrichment has been going on since the first few million years after the first stars began shining in our Galaxy, probably more than 10 thousand million years ago. By the time the sun and solar system formed, less than 5000 million years ago, the interstellar medium contained the heavy elements needed to make the planets and us.

As you sit there reading about the evolution of stars thousands of light years away that existed thousands of millions of years in the past, bear in mind that you are made of the very atoms that were created in the centers of those stars.

SCENE 4

Between the Stars

Between the stars is a terrifically good vacuum; in fact, with minor exceptions it is a far, far higher vacuum than any laboratory vacuum on earth. That vacuum, however, is not perfect. If it were, we wouldn't be here to ask why, because as the older stars eject part of their matter, enriched in heavy elements, into that interstellar medium, new stars, like our sun, form from it.

The Content of the Interstellar Medium

The interstellar space contains a mixture of gas and what we call dust. That gas and dust make up a few percent of the mass of the Galaxy, the largest concentration of it being in the Galaxy's spiral arms. The gas has a typical density of about 1 atom/cm^3 in space, although in places there may be a few thousand atoms per cm^3. In contrast, the air you are breathing has about 2.7×10^{19} molecules of gas per cm^3. About three quarters of the mass of the interstellar gas is in the form of hydrogen, and from 96 to 99 percent of it is hydrogen and helium; the remainder is made up of the other elements in roughly their usual abundances (Appendix 16). The gas is quite transparent to radiation of most wavelengths, and consequently is usually invisible.

The interstellar dust consists of microscopic solid grains. The dust is usually intermixed with the gas, but in a typical place in space the mass of the dust is only about 1 percent that of the gas. On the average there may be only a few thousand grains per cubic kilometer—hardly a dust storm by Oklahoma standards (even though dust storms are rare in Oklahoma these days). Yet a few thousand parsecs of the interstellar grains are almost completely opaque to visible radiation. It should be no surprise to us that gas can be invisibly transparent, whereas a tiny concentration of solid particles or liquid drops can be opaque. Note what happens when some of the invisible water vapor in the air condenses into clouds, or when relatively few particles of smoke are suspended in the atmosphere.

In the nineteenth century astronomers knew that there was at least some gas in interstellar space, because, although it's transparent, it is not always invisible. Sometimes, as in the sword of Orion (the famous constellation of

FIGURE IV.25 The Orion nebula. *(Lick Observatory)*

the Hunter in the winter sky), the gas is emitting light, and you can see it through a telescope as a glowing cloud. In fact, seen through binoculars, the Orion nebula is quite beautiful, and it is faintly visible even with the unaided eye. We saw in Scene 5 of Act III that Herschel and others cataloged thousands of luminous patches or nebulae. Although most of these turned out to be galaxies, a few, especially those in the direction of the Milky Way, had spectra consisting of bright emission lines and were recognized as clouds of low-density luminous gas in space.

Interstellar Dust

But the dust was not so easily recognized, and we have already seen how its absorption of starlight led astronomers before Shapley to think the Galaxy thinned out within about a thousand parsecs of the sun. Since the interstellar gas and dust are concentrated in the spiral arms of the Galaxy, we encounter not only the glowing gas clouds but also the obscuring dust primarily in the direction of the Milky Way. Were it not for that interstellar obscuration, the Milky Way would be bright enough to read by at night.

The most conspicuous way in which the dust manifests itself is in the *dark nebulae*, which appear as opaque curtains hanging in the Milky Way. These are relatively nearby regions in which the concentration of dust particles is relatively high—several times average—so that most starlight is absorbed in passing through only a few tens of parsecs of the dust. Many examples can be seen on the photographs of the nebulae and star fields shown in this book. A particularly fine one is the "dark rift" in Cygnus (Figure III.28), a long dark cloud of dust that seems almost to split the summer Milky Way down the middle. The Coalsack in the Southern Cross is another famous dark nebula. You might think that the dark nebulae are so obvious that astronomers would have known at once about the dust in space. They are obvious, of course, but they were misinterpreted as holes through which we could see past the Milky Way into extragalactic space. They account, in fact, for most of the irregularities in Herschel's edge-on "grindstone" map of the Galaxy (see Act III Scene 4).

In most places in interstellar space the dust is more insidious. Rather than clump up enough to make obvious dark nebulae, it just sort of pervades

FIGURE IV.26
Dark nebulosity in *Norma*. *(Courtesy of the UK Schmidt Telescope Unit, Royal Observatory, Edinburgh)*

space like a subtle fog or like the smog in our large cities. One way to detect the interstellar dust in a particular direction is to look for distant galaxies there. Galaxies are seen in very large numbers near the poles of the Galaxy (in directions 90° from the Milky Way); in fact, large telescopes reveal more remote galaxies than foreground stars in those parts of the sky. Statistically, the numbers of visible galaxies per square degree in the sky become less and less as we look closer and closer to the directions of the Milky Way. Along the Milky Way, the dust hides all external galaxies; Hubble called this belt around the sky the *zone of avoidance.*

---

1. Explain why and how general absorption of light by dust in interstellar space makes stars seem to thin out in the distance.
2. Early in the twentieth century, distances to many Cepheid variables were estimated by statistically analyzing their proper motions. Then the absolute magnitudes of those Cepheids were calculated from their approximately known distances and their apparent magnitudes. These Cepheids lie in the directions of the sky toward the Milky Way where, although it was not known at the time, they are partially obscured by interstellar dust. Were the absolute magnitudes found for them too large or too small? Explain.

Interstellar Reddening

It is fortunate that the interstellar dust is not equally opaque to all wavelengths of light. It absorbs *selectively*, cutting out more of the blue and violet light of short wavelengths than of the red light of long wavelengths. Because red light penetrates the dust better than blue light, stars that are partly obscured by dust appear redder than normal, a phenomenon called *interstellar reddening*. Early in the twentieth century astronomers were confused by finding stars of spectral type B, which should be very blue, appearing as red as relatively cool stars.

The earth's atmosphere does a similar thing. Molecules of the air, although very transparent compared to dust grains, scatter the short-wavelength violet and blue sunlight far more efficiently than they do longer wavelengths.

---

3. How does the scattering of sunlight by air molecules affect the apparent color of the sun itself? Explain. When is the effect most apparent?

Now you recall that the way we find distances to most stars is by knowing their absolute magnitudes (say, from their spectra), measuring their apparent magnitudes, and calculating their distances with the inverse-square law of light. However, if a star is partially obscured by dust, its apparent magnitude is greater (fainter) than it would be in the absence of dust, and we will calculate too great a distance for it. Fortunately, interstellar reddening comes to our rescue. We have determined the law of interstellar reddening, and find it to be almost the same in all directions in the Milky Way. This means that the amount a star is dimmed is directly related to how much it is reddened. From its spectrum we know what color it should be. We observe its actual

FIGURE IV.27
The star density appears to vary across this field in *Norma* because of differing general absorption. *(Courtesy of the UK Schmidt Telescope Unit, Royal Observatory, Edinburgh)*

color (for example, by measuring its light through different color filters) and find how much it has been reddened by dust. This tells us by how much the star is dimmed, so we can correct its observed magnitude to what it would be if the space between it and us were completely clean.

---

4. Two stars appear equally bright. Both are main-sequence stars of spectral type B. One is blue and the other is red. Which is more distant and why?

The dust grains, it turns out, don't really absorb much of the light they interfere with; most of it they simply *scatter*—that is, reflect helter-skelter. But they do actually absorb a little starlight, so they must heat up until they reemit that light as fast as they absorb it. Because they don't absorb very much energy, they remain pretty cool, and most of the radiation they reemit is in the infrared. Sometimes we can observe this radiation. For example, the central parts of many galaxies (including our own) emit a lot of infrared radiation that we are pretty sure is glowing dust reemitting the energy it absorbs from stars. Shortly we shall see that some stars are expected to be surrounded by rather dense shells of dust; these, we think, are either very young or very old stars. In any case we should see enhanced infrared emission from those stars due to the absorption and reemission of energy by the dust. Such circumstellar dust is, in fact, believed to account for the strong infrared emission from a number of stars discovered in recent years.

Reflection Nebulae

But how, you might be wondering, can we possibly know that the interstellar dust mostly just scatters starlight rather than truly absorbs it? Because we can observe that scattered light, just as we observe the blue daytime sky. It is estimated that a third of the light of the Milky Way is diffuse starlight scattered by dust. But in particular we often see a glow of light around bright stars embedded in dust. Since the dust scatters short wavelengths (allowing the longer, red wavelengths to pass on through more readily), the scattered light appears blue—for the same reason the sky is blue. A very fine example is the diffuse glow of light around the brighter stars in the Pleiades, a cluster of stars formed only a few hundred million years ago that are still buried in the

FIGURE IV.28 Pleiades. *(Lick Observatory)*

leftover interstellar material from which they were formed. These luminous dust clouds are called *reflection nebulae*.

Nature of the Grains

Edwin Hubble (whom we met in Act III) studied the brightnesses of reflection nebulae compared to that of the stars that illuminate them. His investigation showed that the dust is highly reflecting, like snow; we say it has a high *albedo*.

---

5. How do you suppose we have been able to learn that the interstellar grains are highly reflecting—that is, of high albedo?

The clues from reflection nebulae, interstellar absorption, and interstellar reddening give us considerable information about the grains themselves. They must be small, not too different in size from the wavelength of light, although they doubtless have a considerable range in sizes. They must be mostly of dielectric (nonconducting) substances. Yet they must be composed of those elements we think are common in interstellar space.

One of the experts in the subject is George Field, director of the Center for Astrophysics at Harvard. Field suggests the model of the interstellar dust grain shown in Figure IV.29. The grain in his model is, typically, 1000 Å in diameter. It contains a core of iron and silicates. Surrounding the core are ices—water, carbon dioxide, and other compounds. On the outermost surface is

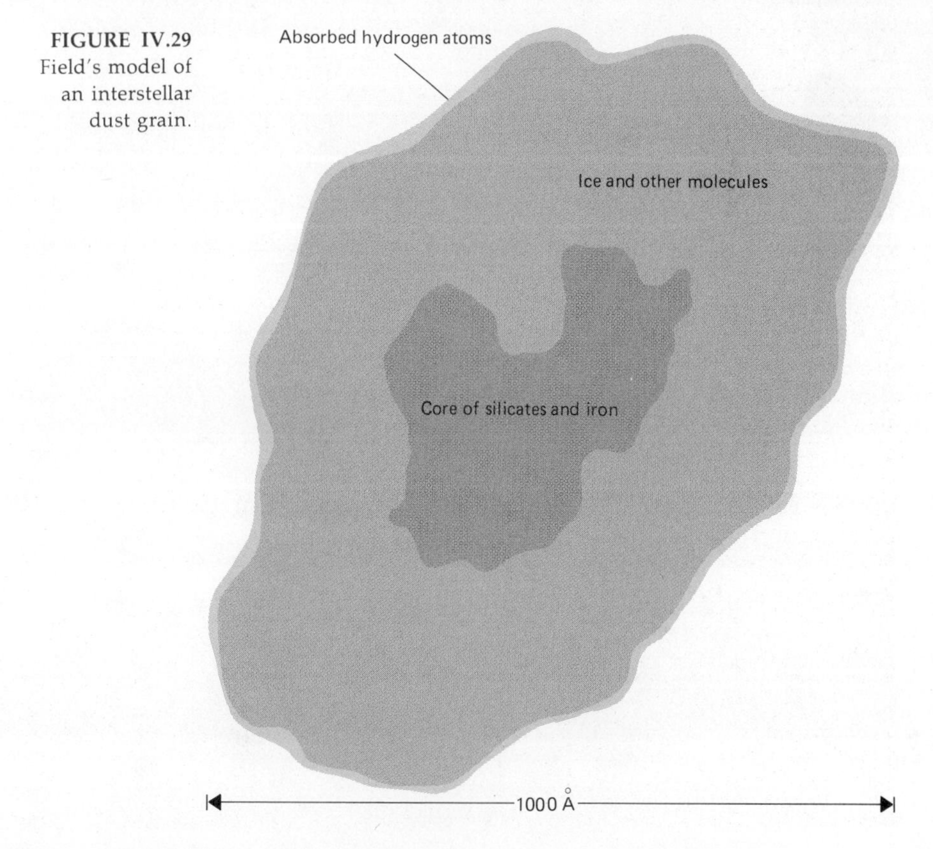

FIGURE IV.29 Field's model of an interstellar dust grain.

a layer of atoms of hydrogen, and a few other kinds of atoms, that strike the grain from interstellar space and stick. Other atoms are continually striking the grain and undergoing chemical reactions with the atoms on its surface, thus giving rise to molecules of hydrogen gas (H_2), OH, water, and more complex molecules. It is, in fact, an ideal gathering point of atoms for the formation of chemical compounds. Moreover, as Field points out, it sort of resembles a miniature planet, with its rocky and metallic core, its ice mantle, and its "atmosphere" of hydrogen and other gaseous elements.

The relatively difficult thing is to understand where the grains came from in the first place. Detailed calculations show that it is almost (if not entirely) impossible for grains to condense directly out of interstellar gas, because the gas is too diffuse—of too low a density. Two possibilities, however, present themselves. One is that the grains form from matter ejected from red giant stars. As the material moves out it cools, and some of it may be able to condense into solid grains before it disperses to too low a density. Another way is in the formation of new stars. As gas falls together to higher and higher densities, part of it should condense into solid particles, making a shroud of dust surrounding newly formed stars. In fact, one modern way of looking for such new stars is to search for the infrared radiation expected from dust surrounding those embryo stars.

Interstellar Lines

Whereas the interstellar gas, unlike the dust, is transparent to light, its atoms can still absorb starlight at certain discrete wavelengths. Thus we often see dark *interstellar absorption lines* superimposed on the spectra of stars whose light has had to shine through the interstellar gas to reach us. There are several ways we can tell which few of the many lines in the spectrum of a star are produced by the intervening gas. First, since the gas is usually quite cold, its atoms are in different physical states from those in the star and produce different sets of lines. The interstellar lines are also quite sharp, whereas the star's lines are broadened by the individual Doppler shifts of the rapidly

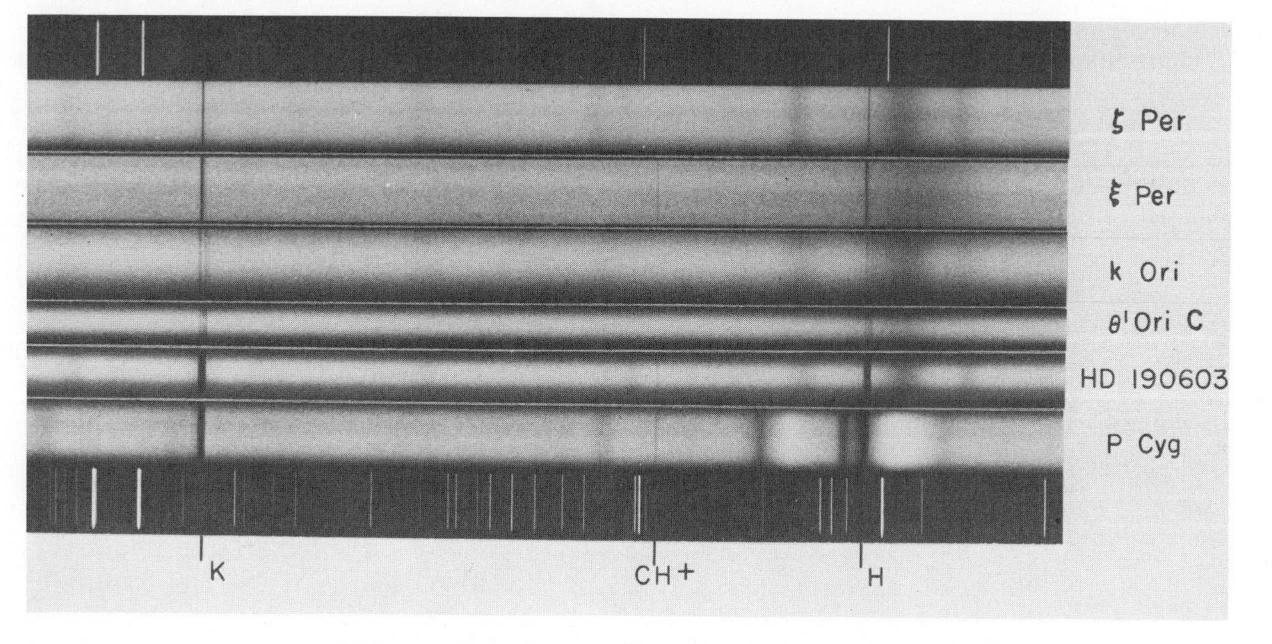

FIGURE IV.30 Interstellar H and K lines of ionized calcium and a line of the CH radical showing in the spectra of several stars. Note how especially the H and K lines are multiple in some spectra because of several discrete gas clouds with different radial velocities. *(Lick Observatory)*

moving atoms in its atmosphere. Moreover, a random gas cloud in interstellar space generally has a radial velocity different from that of the star, so the interstellar lines show a different Doppler shift; in fact sometimes the line of sight to the star traverses two or more discrete gas clouds, and we see several sets of interstellar lines slightly displaced from each other by the different Doppler shifts of the different clouds.

We find interstellar lines of most of these elements that we would expect to be able to produce lines in the visible spectrum under conditions like those in space. Especially conspicuous are lines of sodium and calcium. Also a few molecules and radicals are observed, such as cyanogen (CN) and the carbon-hydrogen radical (CH). The strengths of interstellar lines imply relative abundances of elements in space that are typical of abundances elsewhere.

H I and H II Regions

In most places the gas is cold and its most abundant element, hydrogen, is neutral. We call these regions of the interstellar medium *H I regions,* because H I is the symbol for neutral hydrogen. Here and there, however, are those relatively rare very hot stars of spectral types O and B. These stars emit most of their radiation in the ultraviolet, and that high-energy ultraviolet ionizes the hydrogen in the surrounding gas. Thus around each of these hot stars is a more or less spherical region of ionized hydrogen called an *H II region* (H II is the symbol for ionized hydrogen). H II regions are not precisely round, because the gas is not spread perfectly smoothly in space, but rather has an irregular, lumpy distribution.

The greater the amount of ultraviolet radiation emitted by the star, the greater the size of an H II region; sometimes a small group or cluster of O and B stars makes a very big H II region. But the denser the gas, the smaller the H II region, because the denser gas has more hydrogen, which, of course, requires more ultraviolet energy to keep a given volume ionized. On the other hand the hydrogen ions (protons) are always capturing electrons and becoming temporarily neutral—until they are reionized by absorbing additional ultraviolet photons. As the protons capture electrons, and the electrons cascade downward through the allowed energy levels of the hydrogen atoms, the atoms emit light in the form of emission lines. So the H II regions glow. The denser the gas in them, the more hydrogen atoms there are emitting per unit volume and the brighter the glow.

The brighter glowing H II regions are the *bright nebulae,* or *emission nebulae,* such as the Orion nebula. In a bright nebula like that in Orion there are a few thousand atoms per cubic centimeter—relatively dense for the interstellar gas, but still an extraordinarily high vacuum. The emission nebulae, fluorescing stellar ultraviolet radiation into visible light we can photograph, often make very pretty pictures—especially in color. You will note that in the color photographs in this book, many nebulae appear red. This is because one of the brightest visible spectral emission lines is the Balmer alpha line produced by hydrogen. The greenish color you sometimes see is produced by a line of doubly ionized oxygen. The white in the color photographs of some of the nebulae is not real, but is the result of overexposure; the photographs were exposed at the telescope a long time to bring out the fainter details.

The gas and dust in space are, of course, intermixed. Thus wherever there is a bright emission nebula produced by a star, the dust that is also around the star should be reflecting some starlight as well. However, around a hot

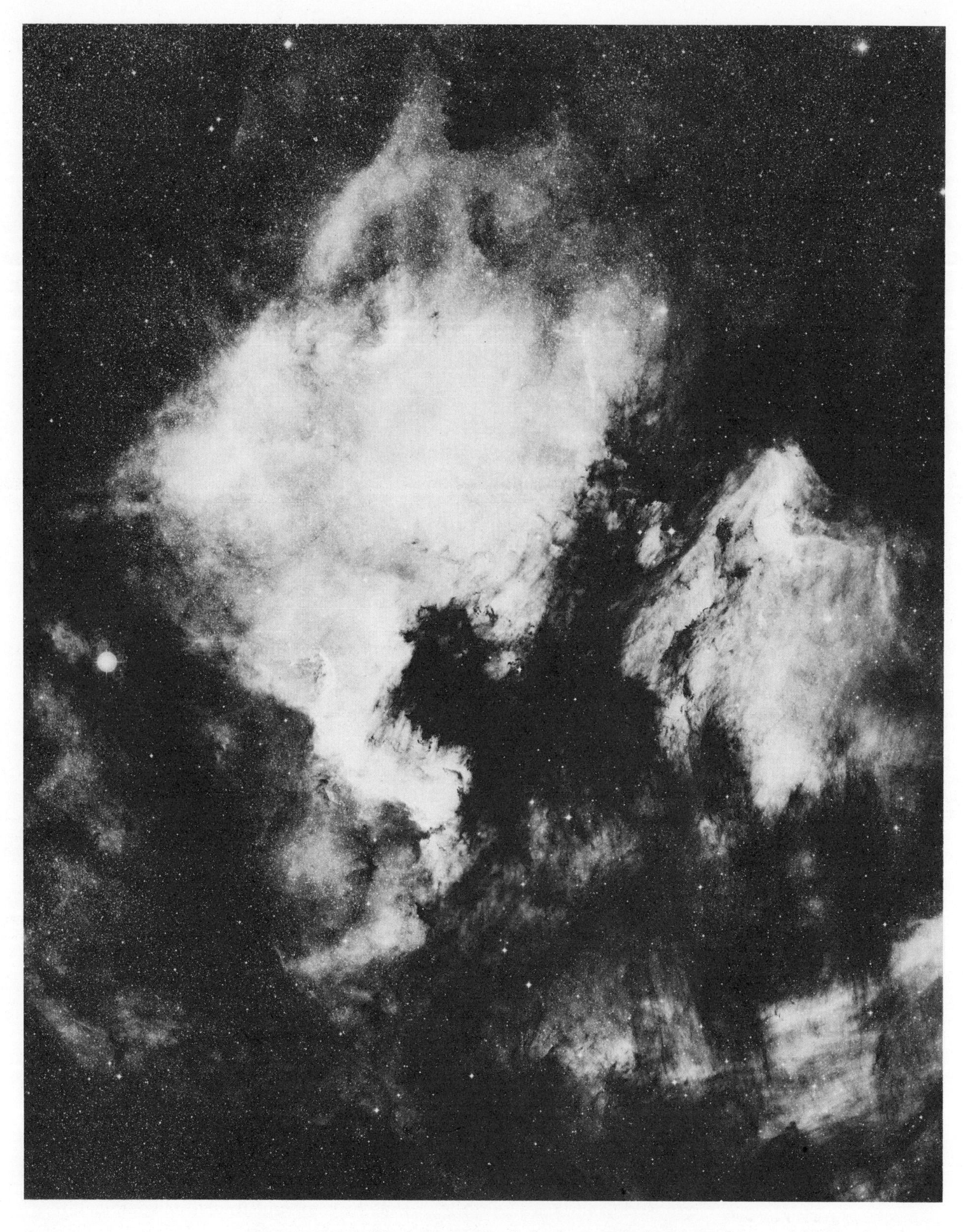

FIGURE IV.31 "North America" nebula photographed with the Palomar Schmidt telescope. Note how foreground dust silhouettes itself in front of the emission nebula to seem to form the Atlantic coastline. *(Hale Observatories)*

star the glowing gas is generally so brilliant that we don't notice the dust in the photographs. The light from the reflection nebulae can, however, be measured.

Now look at the Kitt Peak National Observatory color photograph of the Trifid nebula in Sagittarius. The Trifid is the big red thing; the red is glowing hydrogen gas, ionized by the hot star in the middle of the nebula. The black irregular streaks on top of it are foreground dust clouds obscuring some of the nebular light. But now look *beside* the red emission nebula; that blue glow around the bright star is a reflection nebula. The star is pretty bright but not hot enough to ionize much gas, so we don't see an emission nebula there. The entire region covered by the photograph is filled with intermixed gas and dust.

---

6. Are the Pleiades surrounded by a conspicuous H II region? How do you know? What can you say about the most luminous stars in the Pleiades?

Dynamics of the Interstellar Medium

The ionization of the hydrogen heats the gas, so H II regions are hot—around 10,000°. The H I regions are cold, usually around 100°, but in some places even below 20°. You can easily imagine what must happen where a region of hot gas is surrounded by cold gas of the same density. The hot gas is at a pressure about 100 times as great as the cold gas, and it thus pushes outward—that is, the H II region *grows.* The densities of the gas and dust naturally build up at the outer boundary of the expanding H II region. That's what produces the bright edges you can see in some of the emission nebulae whose photographs adorn these pages.

If the interstellar material started off perfectly evenly distributed, all expanding H II regions would appear as nice round bubbles with bright edges. But the gas and dust are lumped up irregularly, so the hot gas expands faster in some directions than others, pushing faster against the least dense material. Here and there denser patches of gas and dust resist the expanding front of the H II region, and the hot gas just flows on by and around them. Thus long, dark tongues of cool matter can often be seen, looking like intrusions into a bright nebula and pointing to its center. Donald Osterbrock, now director of the Lick Observatory, and a specialist in the study of gaseous nebulae, has colorfully dubbed these dark intrusions *elephant-trunk structures.* Look at the color photograph of the Horsehead nebula in Orion. The bright half of the picture is an expanding H II region. The dark thing that looks like a horse's head silhouetted against the bright nebula is an elephant-trunk structure.

Now I want to call your attention to the nebulosity (the H II region) surrounding the star cluster M16. There are two photographs; the color one is prettier, but the black and white one shows better what I want to talk about next. Note that there are many elephant-trunk structures. But also note the several small, dark blobs completely surrounded by light. These enclaves, called *globules,* are places where the expanding hot gas has filled in parts of elephant-trunk structures and isolated their tips.

FIGURE IV.32
M16. *(Hale Observatories)*

The hot gas is pushing in on all sides of a globule, making its gas and dust ever denser. The interfaces of elephant-trunk structures and H II regions, and especially globules, are regions of particularly dense interstellar matter, and some astronomers believe them to be sites of future star formation. The famous astronomer Bart Bok, an expert in the study of the Galaxy, estimates that there are 25,000 globules in our Galaxy, each of them dense enough for its matter to collapse under its own gravitation into stars or star clusters in periods of a half million years or so.

Radio Emission from the Interstellar Medium

We have described the absorption and emission of light by the interstellar medium, but have yet to mention radio radiation. Radio waves are very long compared to the size of the dust grains, so are not absorbed by it. Thus the interstellar material is transparent to radio radiation, and we observe it from all parts of the Galaxy.

The gas in space can emit and absorb radio waves at certain discrete wavelengths and can produce radio spectral lines. We have already discussed the 21-cm line of neutral hydrogen, observed both in emission and in absorption, and how observations of the Doppler shift of that line originating in gas in different parts of the Galaxy have helped us map out the locations of the spiral arms (Act III Scene 4). Many other radio spectral lines of hydrogen and other gases have also been observed, although none is as strong as the 21-cm line.

There is also a lot of continuous radio emission coming from the gas in space. This is of two types. Most of it is that which is normally emitted by a hot gas when electrons pass ions and undergo free-free transitions (see Act II

FIGURE IV.33
(Upper) A painting of the sky with the Milky Way along the middle as an "equator." *(Lund Observatory) (lower)* A painting, with the same scale and orientation, showing the relative intensity of radio radiation from the sky, with the same scale and orientation. *(Griffith Observatory, Lois Cohen)*

Scene 3); that is called *thermal emission* or *radiation*. Thermal radio radiation from the gas in the Galaxy was the first radio energy to be detected from space. This was what Jansky observed in 1931 (Act II Scene 2).

The other kind of continuous radiation we observe at radio wavelengths (and in other parts of the electromagnetic spectrum as well) is that produced when charged particles (electrons) moving at nearly the speed of light are accelerated in magnetic fields. We observe this radiation in such laboratory accelerators as the synchrotron, so it is called *synchrotron radiation*. Unlike thermal radiation, it is nothing like that from a black body, but has a very characteristic distribution of energy over different wavelengths; it is thus also called *nonthermal radiation*.

The fact that we observe nonthermal radiation from many parts of the Galaxy indicates two things. First, there is obviously a source of electrons with extremely high energy (supernovae may supply some of these). Second, there are galactic magnetic fields capable of accelerating those high-speed electrons.

In fact, magnetism plays an important role in the Galaxy. The average field strength is very weak compared to terrestrial magnets—in fact it is only a millionth to a hundred thousandth as strong as the earth's magnetic field. But the Galaxy is very big, so there is a lot of magnetic energy out there. The field lines of force appear to run along the spiral arms. We think that the magnetic fields are also responsible for speeding up some of the atomic nuclei to the incredible energies they have as they strike the earth in the form of cosmic rays.

Interstellar Molecules

Until recently we thought that the neutral hydrogen in interstellar space was all in the atomic form, but in 1970 George Carruthers observed the band of absorption lines produced by molecular hydrogen (H_2) in an ultraviolet stellar spectrum obtained from a rocket. In 1972 the United States launched the *Copernicus* satellite, an orbiting astronomical observatory named for the Polish astronomer on the occasion of the 500th anniversary of his birth. The ultraviolet observations obtained from space by the Copernicus satellite confirmed that there is a good deal of molecular hydrogen in space.

Since 1972, many other molecules have been discovered in interstellar space, most of them by radio emission lines they emit. In fact, several dozen kinds of interstellar molecules are now known to exist. They include molecules of water, ammonia, silicon monoxide, and hydrogen sulfide. But more surprisingly, there are also many organic molecules, some of which are quite complex, including cyanoacetylene and acetaldehyde—regarded as the starting points for the formation of amino acids, necessary for living organisms.

We are quite sure that dust is very much involved with the formation of these relatively complicated molecules. First they are found in dusty regions of space. In fact, many (including molecular hydrogen) would soon be decomposed by the ultraviolet light from hot stars if they were not shielded from that radiation by the surrounding dust. Moreover, we cannot imagine how such complex molecules could form in the very low density of interstellar space except on the surfaces of dust grains.

Do not think that because we have discovered prebiological organic molecules in interstellar space, it means there is life there; far from it. What it does mean is that the smallest building blocks—the first steps—in the formation of living organisms can form rather easily, because they are found even in the extreme vacuum of space. It makes it all the easier to understand how those first steps toward life can occur under the far more favorable conditions expected on the surfaces of planets. We shall see in the next act, in fact, that given the correct circumstances, the development of basic organic molecules such as amino acids is almost inevitable. The dark and cold of interstellar space may seem remote from earth, but we have seen how inextricably involved it is with our own origins.

SCENE 5

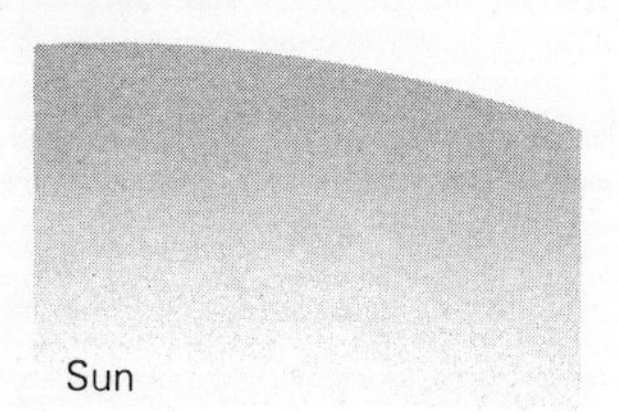

Origin of the Solar System

We know pretty well when it happened. It wasn't a dark and stormy night; nights hadn't been invented yet. But it wasn't the *beginning,* either. The beginning was at least 5000 million years earlier, and probably farther back than that. There was plenty of time for stars to form, evolve, and die, and for many to spew their evolved matter into the interstellar medium. And there was plenty of time for the interstellar matter to change from pure hydrogen and helium (if, in fact, the matter began that way) to a composition with 2 to 4 percent of its mass contained in the heavier elements.

Radioactive Dating

It happened when the time was ripe for the formation of the solar system. We can tell when it was by *radioactive dating.* We have already seen (Scene 2) how the nuclei of certain heavy elements are radioactive, emitting helium nuclei, photons, and other particles, and decaying to lighter elements. Uranium, for example, decays through a sequence of other elements, including radium, and ends up as lead. On the average, though, it takes a long time for a nucleus of uranium to decay. It has just one chance in two of doing so in a period of 4500 million (4.5×10^9) years. This means that in a sample of the most common isotope of uranium, half of the original nuclei will have decayed after 4.5×10^9 yr. In another 4.5×10^9 yr half of what remains will have decayed, leaving only one-fourth of the original sample, and so on. We call that interval 4.5×10^9 yr the *half-life* of uranium.

Suppose there is some uranium in a piece of rock. As it decays, the end products are certain forms of lead and helium. Thus by examining the relative concentrations of uranium, helium, and lead in a rock containing radioactive uranium, we can tell how long the decay process has been going on, and thus the age of the rock. We find that the oldest things on earth are 4.6×10^9 yr old. We find the same age for the oldest moon samples and for the oldest meteorites that have struck the earth and been analyzed. Theoretical studies of the solar evolution suggest a similar age for the sun. We can pretty confidently say, therefore, that *when* it happened was just a little farther back than 4.6×10^9 yr.

1. In a hypothetical sample of uranium ore only 1/16 of the original supply of uranium remains. How old is the ore?
2. Very roughly, what fraction of the earth's original uranium supply has radioactively decayed?

The Early Solar Nebula

Exactly *how* it happened is a little harder to say with such confidence, but from what we know of the nature of matter, of physical laws, and of the present state of the solar system, we think we can describe the general picture. It must have started somewhere in the interstellar medium, perhaps at the boundary of an elephant-trunk structure, perhaps in a globule, or perhaps somewhere else. Anyway, over a region perhaps a light year across, interstellar gas, probably with a certain amount of intermixed dust, began to fall together of its own gravitation. The *solar nebula* was born.

Now the Galaxy is rotating, and there was some small amount of rotation of the gas cloud that became the solar nebula. As the nebula contracted, it had to conserve its angular momentum. Thus it had to spin faster and faster as the gas came together, getting closer and closer to the axis of rotation. What was very leisurely rotation of the original gas cloud became relatively rapid spinning of the much smaller ball of gas. Presently its equatorial regions were moving fast enough to stay in circular orbit, and they could fall in no farther. As the nebula kept shrinking, it left behind more and more of its equatorial matter in a rotating disk, each part moving in an approximate circular orbit about the center. Eventually, then, the solar nebula became a rotating disk, from which the planets formed, with a large blob of gas at the center, to form the sun.

We know all stars don't form in just this way. Often the original protostar fissions into two condensations, to become a double star, and the original angular momentum is conserved in the orbital motions of the two stars rather than in one star with its system of planets. Other times the clouds break up into clusters of stars. We don't know how often planetary systems can form. At most it's only half the time, because about half the stars around us are members of binary-star systems. Some recent studies of the incidence of duplicity among stars have suggested that the formation of planetary systems might be relatively rare. Alas, we have no direct evidence of a single other planetary system, but you must realize that as yet we know of no way to detect planets like the earth in revolution about other stars. At least it happened here, and most of us would be very surprised if it is so rare that it has not happened many millions of times in our Galaxy.

As the solar nebula contracted and its density increased, many dust grains must have condensed from its gases. Some of those grains must have stuck together to make larger ones, so by the time the rotating disk was formed, it was pretty dirty. Meanwhile, the gases at the center of the nebula became opaque, trapping energy inside, and the sun began its youthful experience as a star shining by its slow contraction and release of gravitational energy. Theoretical studies show that in the early stages of a star's slow gravitational contraction, the star is quite luminous, temporarily hundreds of times brighter than when it reaches the main sequence. It took the sun, though, only a few million years to drop in luminosity to something like its present value. Subsequently, its luminosity remained relatively constant, while it gradually contracted, heating its interior, until sources of nuclear energy fired up, some tens of millions of years later.

Formation of the Planets

In the inner part of the solar nebula the high luminosity of the young sun would have evaporated the grains that were composed of volatile substances. Thus particles of water ice and frozen carbon dioxide could exist only far out in the disk. Rocky and metallic grains, on the other hand, could survive throughout the disk. In all parts of that rotating disk, though, the orbiting particles were constantly colliding and often sticking together, and many began to grow by accretion. A few began to get big enough to gravitationally affect those which came near. Sometimes smaller particles would pass close enough to bigger ones to bump into them and stick. But if they didn't pass close enough to hit, they could be gravitationally deflected to another part of the disk, or even out of the solar system altogether.

In this way, a few large chunks gradually won out over their neighbors, either capturing them or getting rid of them, thereby sweeping out ring-shaped swaths in the solar nebula, all centered on the sun. They became the planets. In the final stages of this accretion the young planets swept up the last of the solid chunks remaining in the disk. There must have been many crater-producing explosions as these chunks smashed home. On the planets without dense atmospheres, we can still observe the heavy cratering produced, we think, in this period.

---

3. Why do we not find extensive cratering on the earth?

FIGURE IV.34
Cratered terrain of Mercury photographed by Mariner 10. *(NASA/JPL)*

FIGURE IV.35
Jupiter, photographed with the 5-m telescope on Palomar Mountain. *(Hale Observatories)*

Those planets in the inner part of the solar system—Mercury, Venus, Earth, and Mars—were built up of rocky and metallic particles. They have a lot in common, and are called *terrestrial* (earthlike) planets. They could attract and hold on to none of the gases in the solar nebula. Their present atmospheres have outgassed from the rocks beneath their surfaces (next act); Mercury, however, is too small to retain even this kind of atmosphere.

Far out in the nebula, on the other hand, it was cool enough for icy grains to exist, as well as rocky and metallic ones. The planets that accreted out there —Jupiter, Saturn, Uranus, and Neptune—thus formed out of lots of ices as well as of rocks and metals. They too have a lot in common, and are called *Jovian* (Jupiter-like) planets. Jupiter and Saturn were large enough to even attract and hold a large amount of the gas in the solar nebula. Jupiter, in particular, has a present composition almost like the sun's; it is mostly hydrogen and helium, although in Jupiter those gases are compressed to the solid or liquid state.

The Minor Planets

In between the terrestrial and Jovian planets is an open space where there is no planet. We have already seen (Act I Scene 5) that the minor planets are found there instead. Ceres, the first discovered, is the largest. It and a few other minor planets are big enough to present measurable disks when viewed telescopically. Thus we can calculate their diameters from their observed angular diameters and known distances, just as Shapley found the sizes of globular clusters (Act III Scene 4). Most are too small for us to see their sizes in the telescope, so we must calculate their sizes from the amount of light they reflect.

---

4. To find the size of a minor planet from the amount of light it reflects, we must know its albedo. We can determine its albedo from the amount of infrared radiation the minor planet emits. How do you suppose this is possible? [Hint: Where did that energy in the infrared originally come from?]

Ceres is just over 1000 km in diameter. Pallas is second in size, with a diameter of about 600 km. Then comes Vesta, 550 km across. Vesta comes close enough to earth that it can sometimes barely be seen with the unaided eye. About a dozen (or fewer) minor planets are over 160 km in diameter. Most of those which can be observed are only a kilometer or so in diameter.

An expert on minor planets is astronomer Tom Gehrels. He was an office mate of mine when I was a graduate student at Caltech; he had just come from his native Holland. Gehrels has made systematic searches for minor planets and estimates that there may be 100,000 or so big enough to observe. There must be many more smaller ones, but they can't contribute much mass. So the total amount of matter in the minor planets doesn't amount to very much. A good estimate of the combined mass of all minor planets is about 1/20 that of the moon, or about 1/1600 that of the earth. Many people have speculated that the minor planets were formed by a planet that broke up. But there is no reason why such a planet *should* break up, and anyway, the total mass of the minor planets is too small to make even an earthly moon, let alone a self-respecting planet.

On the other hand, results of the analyses of the orbits of minor planets strongly suggest that early on in the history of the solar system, there were only a few of them, possibly of mass comparable to that of Ceres. Then, as time went on, a collision between two of them occurred, making many more much smaller particles. With all those fragments out there, subsequent collisions would have become far more common, so that the fragmentation of minor planets would accelerate in time.

Meteoritic Matter

If the minor planets are fragmenting, many should be colliding with the earth from time to time. We think that they do. Most of the fragments, though, are very tiny and burn up in the atmosphere, producing *meteors*. A typical observer, on a dark clear night, can count an average of six or so meteors, or "shooting stars," per hour. Most of these moving starlike spots of light are

FIGURE IV.36
Photograph of the trail of a bright meteor that happened to cross the field of view of the telescope during an exposure on the Andromeda galaxy. *(F. Klepesta)*

produced by the collision of the earth with the debris from comets (as we shall see in a moment). Some of these meteors, however, are caused, we think, by fragments of minor planets. A particle in space, revolving about the sun, before it collides with the earth is called a *meteoroid.*

Although minor planets are believed to account for only a small fraction of the meteoroids, they nevertheless may account for nearly all those rare ones that survive their passages through the earth's atmosphere and land. We commonly find these "interplanetary visitors" on the ground, and call them *meteorites.* The largest ever recovered is the *Hoba West,* near Grootfontein, South-West Africa; it has an estimated mass of 45 tons. Many smaller meteorites are displayed in museums. Meteorites range in size down to the very numerous *micrometeorites* (only a few thousandths of a millimeter in diameter) that filter through the atmosphere and can be collected from ocean floors and even from rooftops. Most, if not all, of those meteorites in museums, and which have been studied extensively in the laboratory, are probably tiny fragments of minor planets broken up in prehistoric collisions in space.

Meteorites are made, of course, of the familiar elements we know on earth. Most of those which fall are stony in composition. A few, however, and at least half of those found accidentally in the ground, are metallic—consisting mostly of iron and nickel. Stony meteorites weather and look like terrestrial rocks, and thus are easily overlooked; the metallic ones attract attention. Now that the public is meteorite-conscious, people often bring unusual-looking pieces of metal to experts for identification. The late Frederick Leonard, who specialized in classifying meteorites, had a special class for those frequent rusty bolts and old cannon balls brought to his attention: *meteorwrongs.*

It is estimated that each century several meteorites strike the ground with enough force to produce a crater 10 m or more across. To produce such a crater requires a fall of a meteorite with a mass of at least 2 tons. A number of larger craters on the earth were produced by meteorites that must have struck it in prehistoric times. The most famous is Meteor Crater, near Winslow, Arizona. That crater is 1300 m across and 180 m deep. Over 25 tons of fragments of iron meteorite have been found in an area up to 7 km around the

FIGURE IV.37
Meteor Crater in Arizona.
(American Meteorite Museum)

crater. Evidently the large mass, estimated to have hit the earth about 22,000 years ago, disintegrated on impact. Other larger craters, or highly weathered fossil craters, are known that are believed to have been meteoritic in origin.

So we see that the earth is still involved in physical activities in the solar system. It is interesting to contemplate what might happen if a typical minor planet, about 1 km in diameter and weighing some 10^{15} g, should strike the earth today. It would hit with the energy of at least 20,000 megaton hydrogen bombs. It could probably wipe out an entire country the size of Switzerland. Astronomically, we lead pretty fragile lives.

Satellites

Most planets have satellites. The earth has only one natural moon, but Jupiter has more than a dozen. The satellite systems of the Jovian planets resemble miniature solar systems. In each of these systems, most of the satellites revolve about the planet in its equatorial plane, in nearly circular orbits, and in the same direction as the planet rotates. They could well have been formed from a protoplanet nebula, analogous to the solar nebula, that condensed from it just as the solar nebula condensed from the gas in the rotating Galaxy. Then the satellites could accrete from a rotating disk within the solar nebula,

FIGURE IV.38
Jupiter and its four brightest satellites emerging from behind the moon following an occultation. *(Photograph by P. Roques, Griffith Observatory)*

and their planet, from the material at the center of that disk. Whereas the large inner satellites of Jupiter (including those discovered by Galileo) fit into this picture, the outermost satellites of Jupiter do not; they are believed to have been former minor planets that fell under Jupiter's gravitational domination.

The origin of the earth's moon, on the other hand, was probably different. The moon is the largest satellite in the solar system in comparison to its planet. The earth and moon may have been formed together in the solar nebula, or they may have been formed in different places, but at roughly the same distance from the sun, and then later fallen into each other's gravitational influence and captured each other in mutual orbital revolution.

---

5. If the moon formed in the solar system separately from the earth, why do we think it nevertheless formed at roughly the same distance from the sun as the earth?

Comets

The members of the solar system most difficult to explain are the comets. As we described in Act I Scene 5, there is evidently a tremendous cloud of them —perhaps hundreds of thousands of millions, revolving about the sun in orbits tens of thousands of astronomical units in radius, waiting for passing stars to perturb their motions and to send some into the inner solar system, where we may observe them. Most astronomers think the comets were probably formed out there, condensing from the solar nebula early in its history—perhaps before any of the planets were formed. Some, however, think they may have been formed in the rotating disk, and were later ejected from it by planetary perturbations. I think the latter idea is much harder to understand in terms of Newtonian physics, but perhaps it is not impossible. On the other hand, radio observations of Comet Kohoutek in 1973 revealed the molecules hydrogen cyanide (HCN) and methyl cyanide (CH_3CN). Both are observed in the interstellar medium but not naturally in the inner solar system

FIGURE IV.39 Comet Kohoutek, photographed 1 December 1973 with the Palomar Schmidt telescope. *(Hale Observatories)*

(where they are easily broken up by sunlight), which lends support to the idea that comets were formed early in the solar nebula.

Comet Kohoutek, despite the fact that it did not brighten as much as expected and was a visual disappointment to many, was a tremendously important comet to astronomers. It was discovered nearly a year before it passed nearest the sun (perihelion), providing ample time to prepare for many detailed observations of it. It was observed from rockets, from Mariner 10, by radio, extensively from ground observations, from an airplane, and even by the Skylab astronauts, and a great deal was learned about it and about comets in general.

The discovery of Comet Kohoutek is quite interesting in itself. Lubos Kohoutek, for whom it is named, is a very modest and friendly person, who is quite embarrassed about the fame (or infamy) his comet brought him. He is primarily interested in planetary nebulae (those shells of gas ejected from evolved stars); in fact, I first met him in Czechoslovakia at a conference on planetary nebulae. However, he also has an interest in comets, and in 1971 he was working with the Schmidt telescope at the Hamburg Observatory, trying to find the remains of Comet Biela, which broke up into two comets in its approach to the sun in 1846. Both comets returned on about the same orbit in 1852, but neither has been seen since. On 27 November 1872, however, and again on 27 November 1885, on both of which occasions the earth passed through the orbit of the late Comet Biela, spectacular showers of meteors were observed—evidently caused when the earth collided with debris of the comet (or comets) still revolving about the sun in its old orbit.

Well, Kohoutek had hoped to discover something of the remains of the original comet, Biela, but he found nothing. Instead, though, quite by accident he discovered about 50 new minor planets, some of which had rather unusual orbits, in that they passed very close to the earth. It happened that early in 1973 those most interesting minor planets were in a favorable position for observation. So Kohoutek was trying to get more observations of them, and he took a number of photographs of the appropriate region of the sky. On two of these photographs, one taken on February 27 and the other on March 7, he discovered new comets, the latter of which became the famous Comet Kohoutek.

Relation of Comets to Meteors

From the above tale, you can appreciate that comets are extremely flimsy and often disintegrate. When they do, they leave a lot of mess behind them in their orbits. The earth constantly collides with this stuff, and that debris is responsible for the overwhelming majority of the meteors we see in the sky. In fact, at certain times of the year, when the earth passes through the orbits of recently disintegrated comets, we have unusually good displays of meteors. The best regular display is that of August 11 of each year, when the earth passes through the trail of trash left from Comet 1862 III. On that date we typically see as many as 50 meteors an hour seeming to radiate from the constellation of Perseus in the sky. The *shower* of meteors, therefore, is called the *Perseid shower.* Much more spectacular showers have occurred, and certainly will occur, but the Perseids are the most predictable.

FIGURE IV.40
Ganymede, Jupiter's largest satellite, photographed by Pioneer 10 at a distance of 751,000 km. *(NASA)*

6. Some meteor showers, such as the Perseid shower, are about equally spectacular every year, while others are poor most years but may occasionally produce good shows. For example, the Leonids, which occur on November 16, are spectacular every 33 years. Can you suggest an explanation?

A typical meteor looks like a star moving across the sky in just a few seconds, and then fading out as the particle giving rise to it disintegrates and vaporizes in the atmosphere. Most meteors occur at heights of 95 km; a very few start as high as 130 km. Nearly all burn out by the time they descend to 80 km. (The particles that do *not* completely burn up and which consequently fall on the ground are, as I have already explained, mostly if not entirely of minor planet origin.) By measuring the total amount of light it emits, we can calculate how much energy a meteoroid brings into the atmosphere, and hence what its mass must be. It turns out that the mass of a meteoroid giving rise to a typical bright meteor is about one-quarter gram, roughly the mass of a small pebble.

7. A cubic centimeter of water has a mass of 1 g. If a typical meteoroid has a density twice that of water, approximately how large would a meteoroid of mass 1/4 g be?

Sweeping the Solar System Clean

I have not yet explained what happened to all that gas and dust in the disk of the solar nebula which did *not* end up in planets, their satellites, minor planets, and comets. There is today in interplanetary space a fine dust lying in the plane of the planetary orbits, but this is thought to be mostly debris of minor planet collisions, and not primordial. Indeed, if it were not continually replenished by new fragmentation of minor planets, it would soon disappear.

It would disappear because there are forces that continually clean out the solar system of its dirt. One force is the pressure of light. Photons, as we have seen, have a mass equivalence; thus when they are absorbed, they transfer some momentum to the object they hit. This transfer of momentum caused by the absorption (or reflection) of radiation is called *radiation pressure.* The pressure of solar radiation is enough to shove tiny particles out of the solar system. It works best on particles with sizes similar to the wavelength of the radiation, most of which (from the sun) is visible light. It blows dust out of comets, for example, and produces some of the comet tails that point generally away from the sun.

The solar wind also plays an important role in cleaning out interplanetary space. Between radiation pressure pushing out small particles and the wind of ions and electrons from the sun electromagnetically repelling ionized gas from the solar system, you can see how the original gas and dust that did not form into planets, minor planets, satellites, and the like, was removed from the planetary realm.

Earthrise, lunar farside in foreground; photographed by Apollo 17. (*NASA*)

Comet Ikeya-Seki, photographed by J. B. Irwin in Chile, October 1965.

Comet Humason (1961a), photographed with the Palomar Schmidt telescope. (*Hale Observatories*)

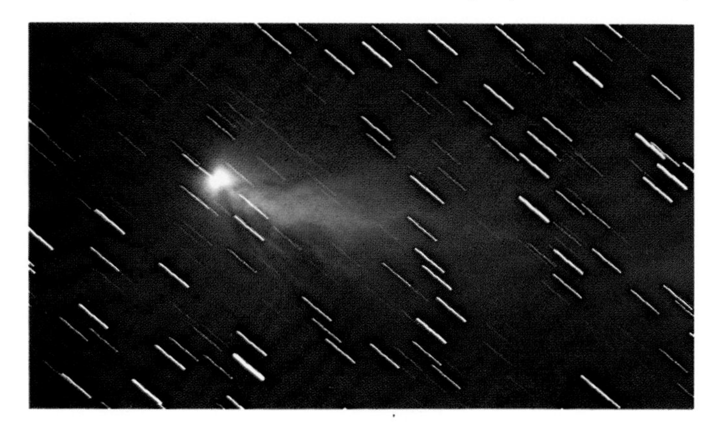

The trail of a meteor in the Perseid shower, photographed by Ronald Oriti, August 1963. (*Courtesy, Ronald Oriti*)

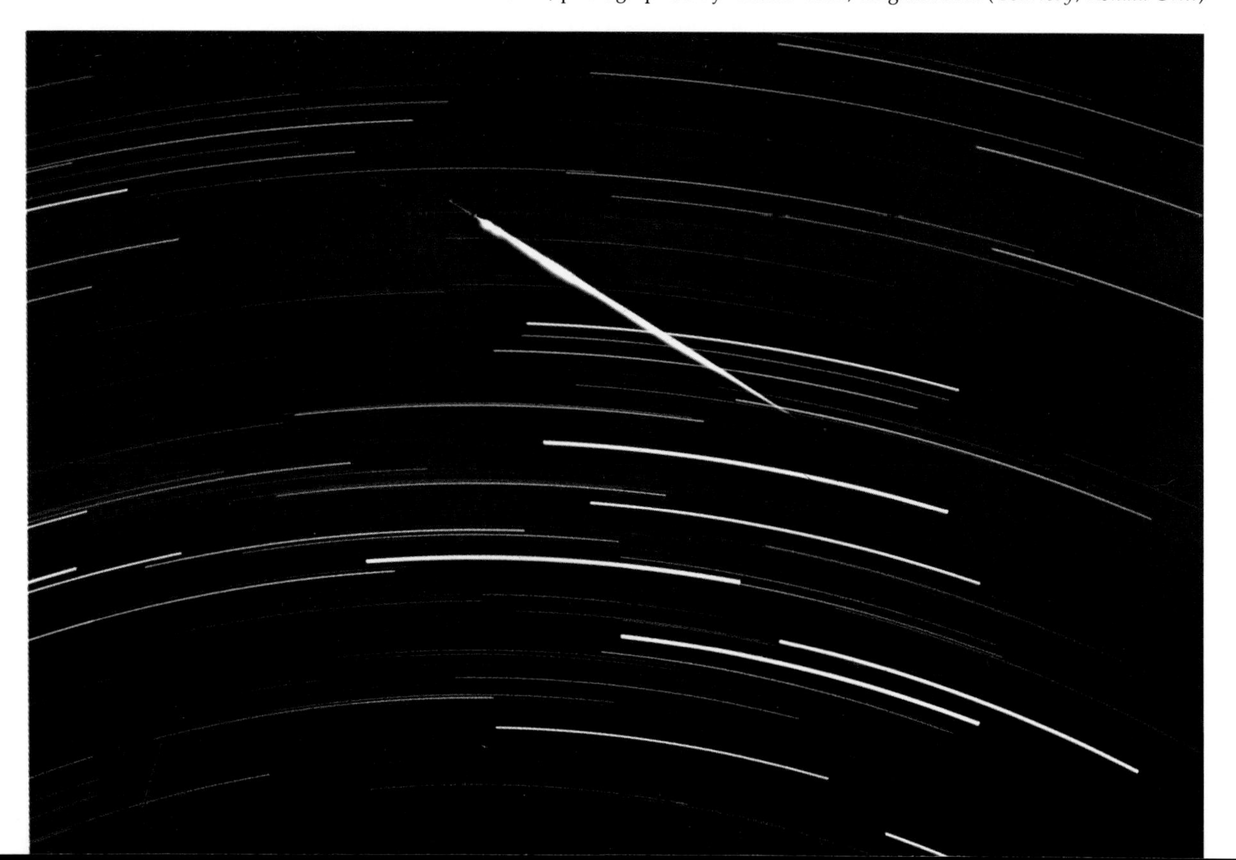

Slice of the Kamkas iron meteorite, which has been polished and then etched with a dilute nitric acid solution to show the criss-cross Widmanstätten figures. (*Photograph by Ivan Dryer*)

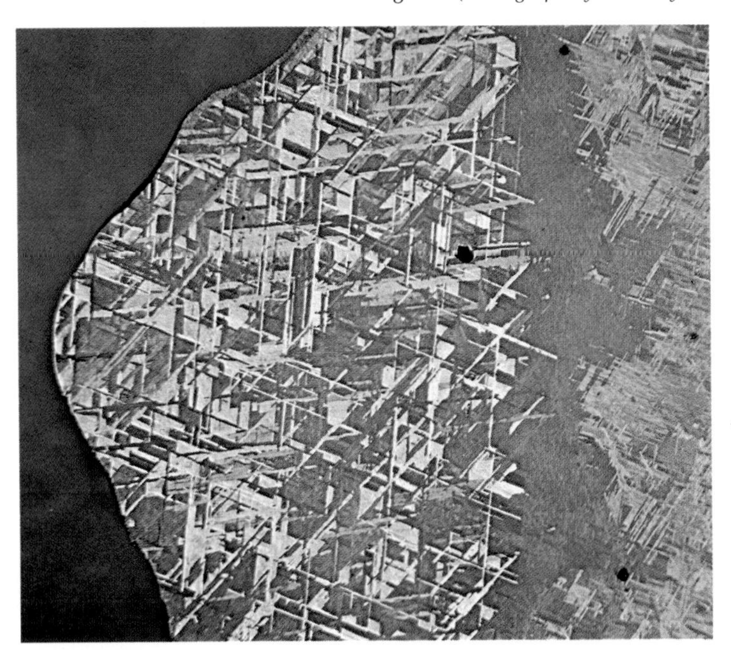

Stony-iron meteorite, Glorieta Mountains, N.M. The specimen has been polished and etched to show the metallic structure. (*Photograph by Ivan Dryer*)

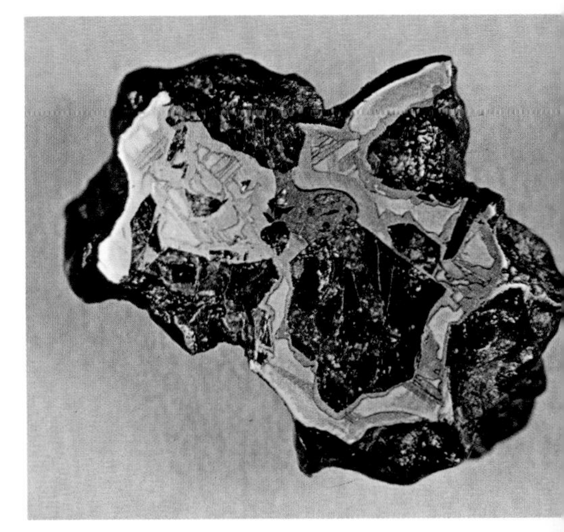

Polished slice of the Albin, Wyoming, stony meteorite. This type of meteorite consists of nickel-iron metal with inclusions of the green mineral olivine. (*Photograph by Ivan Dryer. All meteorite specimens on this page are from the collection of Ronald Oriti, and are reproduced by the kind permission of Mr. Oriti.*)

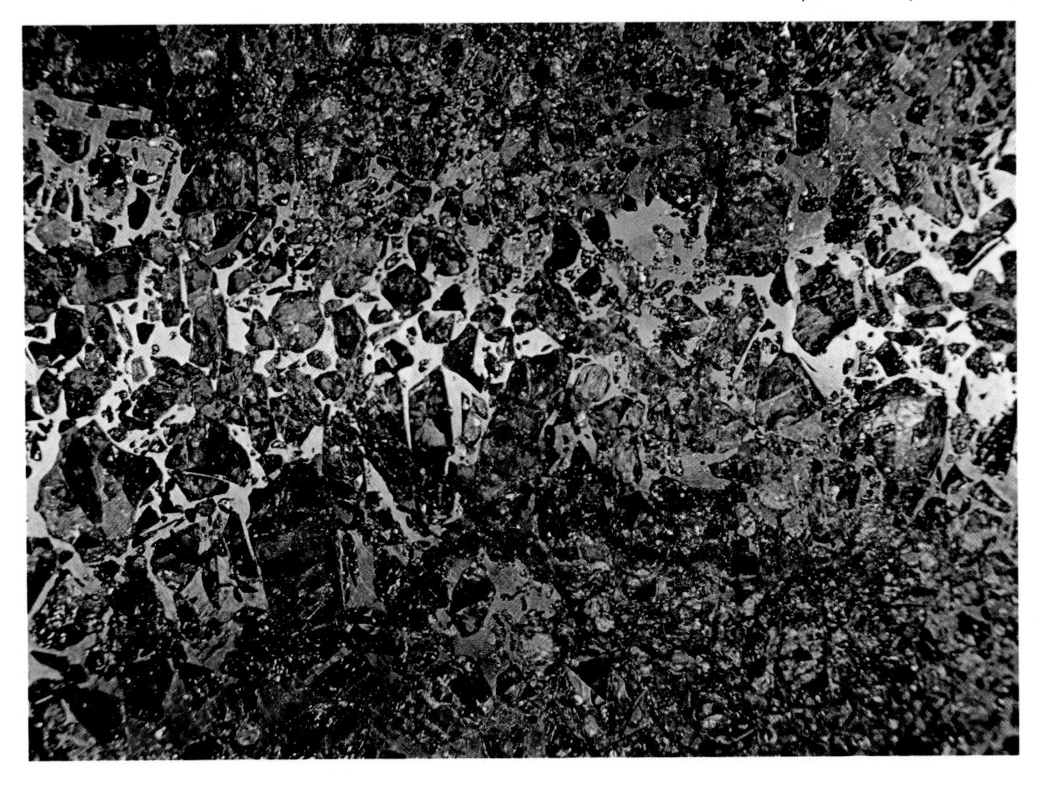

The Veil nebula, or Cygnus "loop," NGC 6992; photographed with the Palomar Schmidt telescope. (*Hale Observatories*)

The Crab nebula, photographed with the 5-m telescope. (*Hale Observatories*)

Summary

In the foregoing scenario we have followed the origin of at least one typical star, along with the stuff associated with it. Many details may be wrong in this picture. I think the general idea is probably close to being right, but as new research uncovers new information we can expect modifications. Indeed, all specialists would not describe the start of the solar system in precisely the same way. I have tried to choose a happy compromise between speculation and pretty sound theory, but I guarantee nothing.

On the other hand, I can give a pretty good idea of the present distribution of matter in the present-day solar system. It is given in the following table, which shows the fraction of the mass of each part of the solar system; even then, there are some pretty big uncertainties, as you can see by the question marks. The interplanetary medium is that temporary dust and the solar wind particles—all on their way out, but being constantly replenished.

Distribution of Mass in the Solar System

Object	*Percentage by Mass*
Sun	99.86
Planets	0.135
Satellites	0.00004
Comets	0.0003 (?)
Minor planets	0.0000003 (?)
Meteoroids	0.0000003 (?)
Interplanetary medium	less than 0.0000001 (?)

For now, at least, the solar system is relatively stable. Compared to human events it is extremely stable. On the cosmic scale, however, we are living in temporary comfort. Inevitably, unless we are very much mistaken, the sun will expand into a red giant. We think this will occur in about 5000 million years. How large and luminous it will become is less certain, and our current estimates depend on theoretical calculations, which may be wrong in detail. Our best guess, though, is that the sun will grow to a size that extends out far beyond the earth's orbit, nearly to that of Mars. The density of the solar matter will be extraordinarily low where the earth is, but the radiation within the sun will still heat the outer part of the earth enough to vaporize some of it. Then the friction between the partly vaporized earth and solar material will cause the earth to lose energy and slowly spiral into the hotter denser solar interior; calculations by R. Ulrich indicate that this end will come about 10,000 years after the sun's surface reaches out to the earth.

What will the end be like? The earth will simply turn into gas, and it, along with its mountains, grand canyons, cities, people, and astronomy books will merge with the material of the sun. But what of the sun itself? What if we, in that remote future, could leave the earth during that critical phase, and watch what happens? Well, the sun's remaining life after then would be short lived. The sun's energy, like everything else, is limited. But wait—that comes later, and I won't tell you about it yet. Anyway, it won't matter to man, because he is far too stupid to survive to that time. Man, in fact, seems to have programmed his downfall for the next few decades, unless . . . unless . . . — but that, too, must wait until we have explored more aspects of this drama of the universe.

SCENE 6

Checking Out the Theory-Evolution of M2001

Before we go on to describe the end of it all—that is, how stars die—I think I should give you some indication of the evidence that lets astronomers check out the theory. Well, we check out the theory as we do all theories, by using it to make predictions and by checking those predictions against experiment or observation.

First, how do we know that stars are even still forming? We think the answer is that we find about us in the Galaxy stars of spectral types O and B that have 100,000 to 1,000,000 times the luminosity of the sun. Such stars have all the characteristics that we expect of highly luminous stars if they had recently formed from interstellar matter. In fact, those particular stars are even found in the regions of space where we find interstellar gas and dust. We think they are newly formed, because at the rate they are pouring energy into space they would burn themselves out in only a million years or so; had they been formed with the sun, more than 4.6×10^9 yr ago, they would long since have exhausted their nuclear fuel and extinguished themselves.

How to Test the Theory

All right; but given that stars are currently forming in the Galaxy, how can we be so sure that they form and evolve in the way astrophysicists say they do? I think the best test is to make those astrophysicists predict the observable characteristics of stars in various stages of their evolution, and then tell us how to compare the predictions with reality. As we have seen, the characteristics most easy to observe are the luminosities (or absolute magnitudes) of stars and their temperatures. So, I say, let the theorists give us a set of Hertzsprung-Russell diagrams of star clusters at various stages of their evolution (according to their theories) so that we can compare these theoretical diagrams with actual ones of real star clusters.

Of course those crafty theorists are ahead of us; they have already done just that! A particularly nice set of calculations was carried out in the mid-1960s by Rudolf Kippenhahn and his associates at Göttingen. Kippenhahn wanted to follow, theoretically, the evolution of a hypothetical cluster of stars, all of chemical composition like the sun's but with a range of masses. He assumed as his starting point that all those stars had already become opaque and were in hydrostatic equilibrium, but had not yet contracted enough or raised their central temperatures enough for nuclear reactions to take over; in other words, they had not yet reached the main sequence, and had just begun deriving their energy by slow gravitational contraction.

---

1. Where on the Hertzsprung-Russell diagram would a star be when it was still in the solar nebula stage and had not yet begun to radiate away energy as electromagnetic radiation?

Well, Kippenhahn shoved into the computer all the data about the masses of the stars and their chemical compositions. He also told the computer all about the laws of physics we described in Scenes 1 and 2, and then asked it to plot an H-R diagram of the cluster, showing the temperatures and luminosities of its stars, at a number of different times.

---

2. The computer, of course, must be carefully programmed to perform these involved calculations. In practice the calculations provide the luminosity and radius of a star of assumed mass and chemical composition. How can this information be used to plot the star on the Hertzsprung-Russell diagram?

Now Kippenhahn wanted to emphasize the fact that the computer was generating the evolution of a hypothetical, or fictitious, cluster. Thus he gave his theoretical cluster a purely fictitious name: M007. At that time (in the 1960s) the James Bond spy character was known to almost everyone. But our fictional heroes have changed more than the theory of stellar evolution. Thus when I asked Rudolf if I could show you his H-R diagrams for M007 in this book, he said "sure," but suggested that no one would know about "007" anymore. I had already realized this, and so decided to rename his cluster M2001.

Early Evolution of M2001

It's very hard to predict how the cluster would look to an outside observer within the first few million years after its stars are in hydrostatic equilibrium, because for one thing we expect many or most of them to be surrounded by shrouds of dust that condensed from the in-falling gases of their prestellar nebulae. However, theory predicts that after about three million years the most massive stars in M2001 would have reached the main sequence, whereas those of two or three times the sun's mass would still be off to the right of it (in the H-R diagram), slowly contracting and heating up as they contract to that point where nuclear reactions will eventually take over. Figure IV.41 shows Kippenhahn's H-R diagram for M2001, at an age of 3 million years.

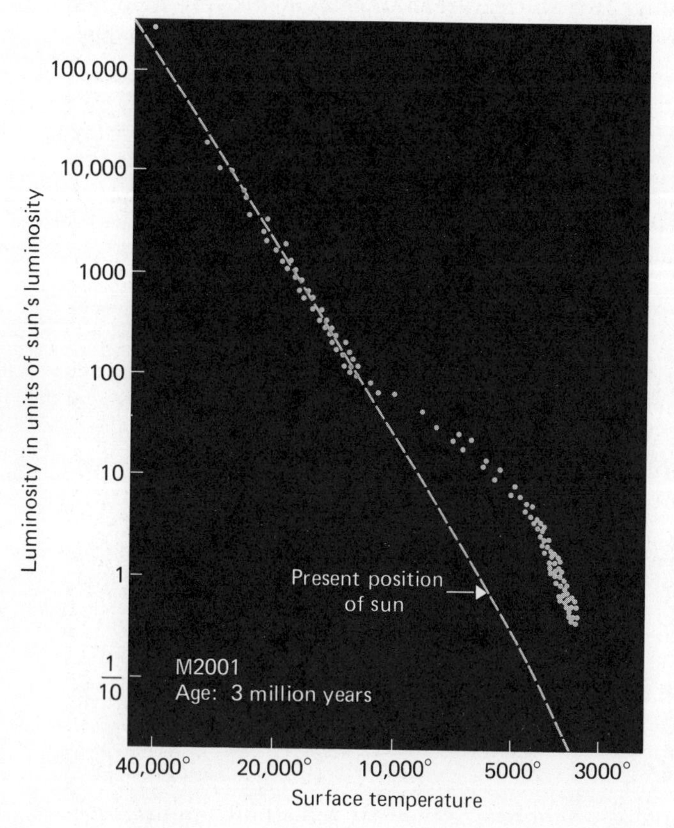

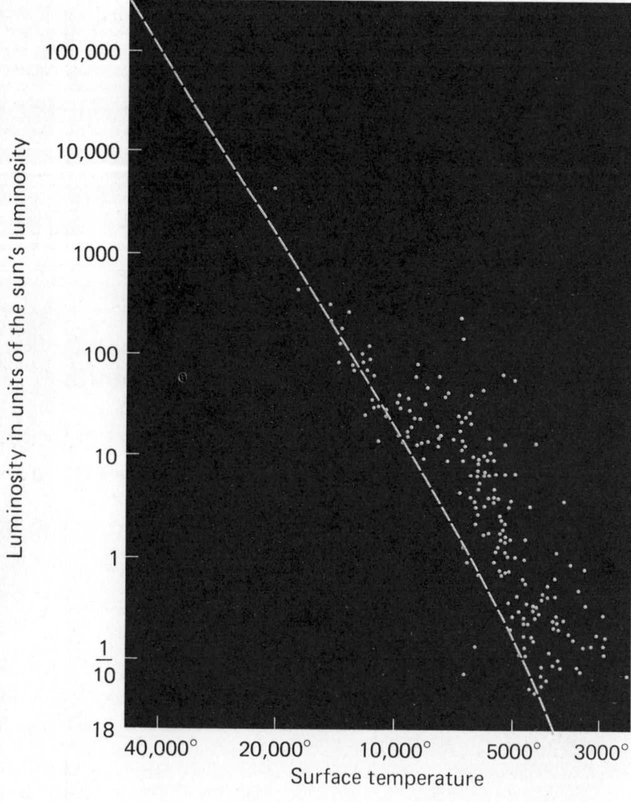

FIGURE IV.41
The H-R diagram of M2001 at an age of 3 million years.

FIGURE IV.42
The H-R diagram of cluster NCG 2264, from data by M. Walker.

Are there young star clusters that bear any resemblance to this theoretical prediction? Yes; since 1950 several clusters fitting that description have been found. One, and a fine example, is NGC 2264. It's H-R diagram, which I plotted from data gathered by M. Walker, is shown in Figure IV.42. The dashed lines in both Figures IV.41 and IV.42 show the location of the main sequence. Note how in both the theoretical and real clusters the brighter stars are on the main sequence, whereas the fainter stars are off to the right—the stars of lower mass that have not yet finished their gravitational contraction stage of evolution. There's a lot more scatter of the points in the real cluster diagram, because all stars would not really start contracting at precisely the same time, as in M2001. Yet the agreement is very good if NGC 2264 is a little older than 3 million years.

---

3. Why is it convenient to use the observed properties of stars in a cluster to check out stellar evolution theory, rather than use the properties of those stars surrounding the sun in space?

FIGURE IV.43
The very young open cluster NGC 2264, and associated nebulosity. *(Hale Observatories)*

Subsequent Evolution of M2001

As some more time goes on, all but the least massive and faintest stars in M2001 have reached the main sequence, but meanwhile the most massive ones have evolved away from it, having achieved that core of helium in their centers. The H-R diagram of M2001 at an age of 30 million years is shown in Figure IV.44. Eventually, after 66 million years (IV.45) the least massive stars are on the main sequence, and the most massive ones have evolved to their extinction. Stars that had main-sequence luminosities a few hundred times that of the sun have now evolved to the red giant phase of their evolution. Figure IV.46 shows the H-R diagram of the real cluster, M41, which I plotted up from data obtained at the telescope by my old high school chum Arthur Cox. M41 has an age a little under 100 million years, so its H-R diagram can be compared to that of Figure IV.45. (In this and in other diagrams of real clusters shown here, stars are not really missing from the faint end or bottom of the main sequence; it's just that at the distance of the cluster those stars are too faint to measure.) So far, everything seems to fit pretty well.

As more and more time goes on, the main sequence of a cluster terminates at fainter and fainter absolute magnitudes, like a candle burning down. The Hyades star cluster, whose H-R diagram is shown in Figure IV.47, is a few

FIGURE IV.44
The H-R diagram of M2001 at an age of 30 million years.

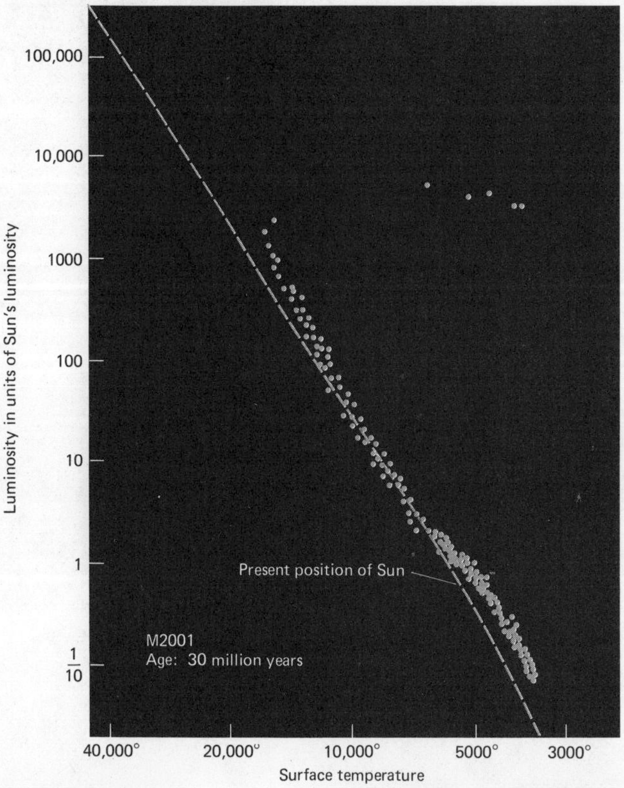

Luminosity in units of Sun's luminosity
100,000
10,000
1000
100
10
1
$\frac{1}{10}$
Present position of Sun
M2001
Age: 30 million years
40,000°
20,000°
10,000°
5000°
3000°
Surface temperature

FIGURE IV.45
(Bottom left) The H-R diagram of M2001 at an age of 66 million years.

FIGURE IV.46
(Bottom right) The H-R diagram of M41, from data by A. N. Cox.

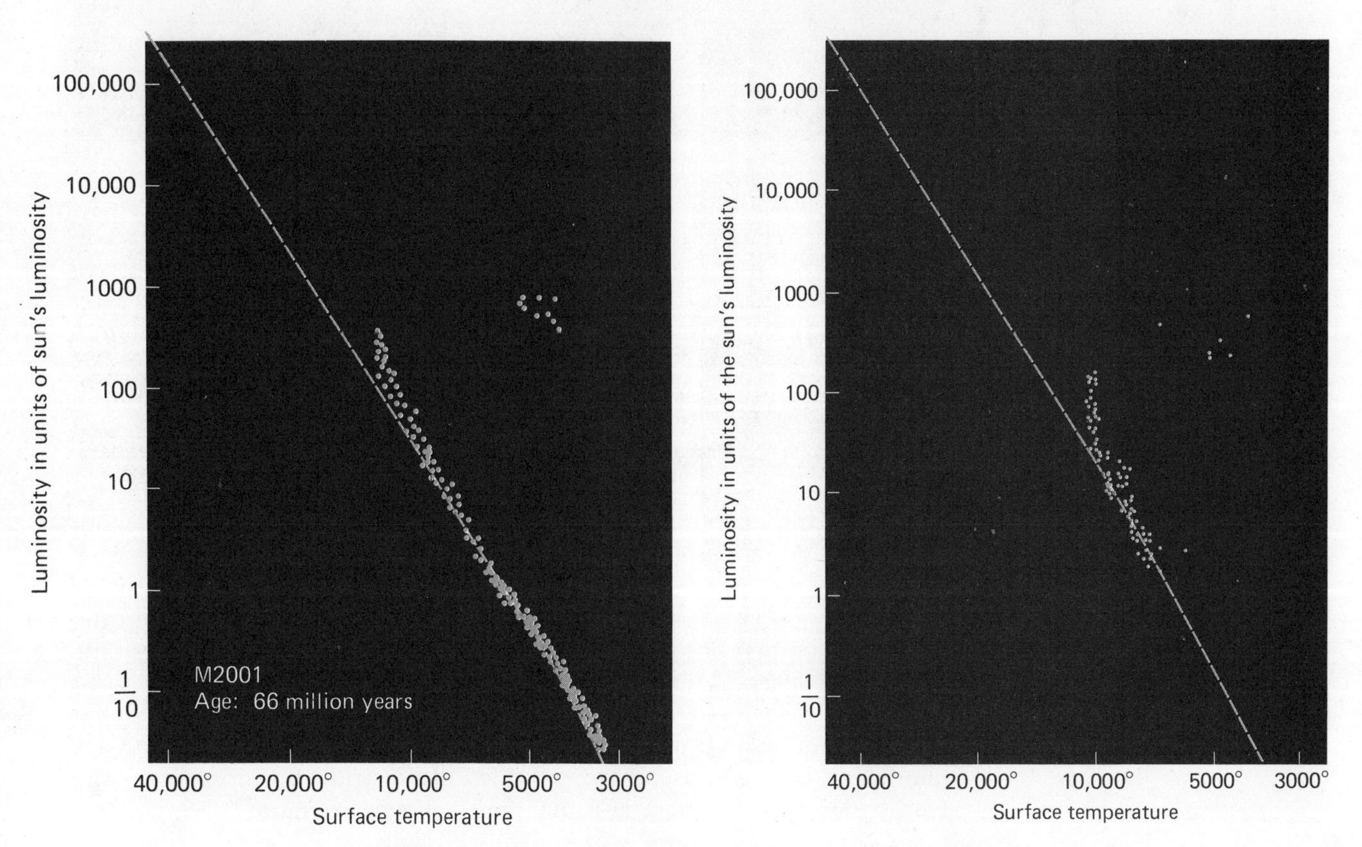

FIGURE IV.47
The H-R diagram of the Hyades star cluster.

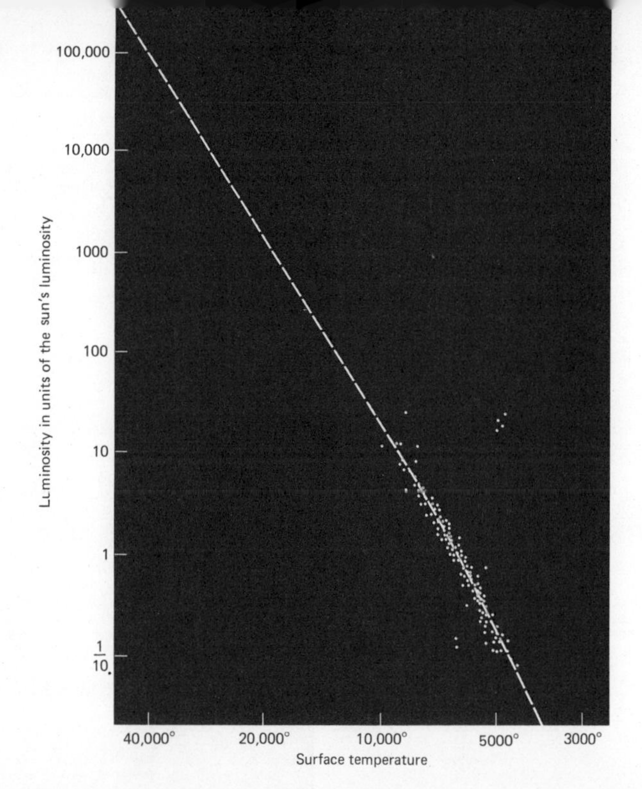

FIGURE IV.48
The H-R diagram of M2001 at an age of 4240 million years.

FIGURE IV.49
The H-R diagram of the globular cluster M3, from data by A. R. Sandage.

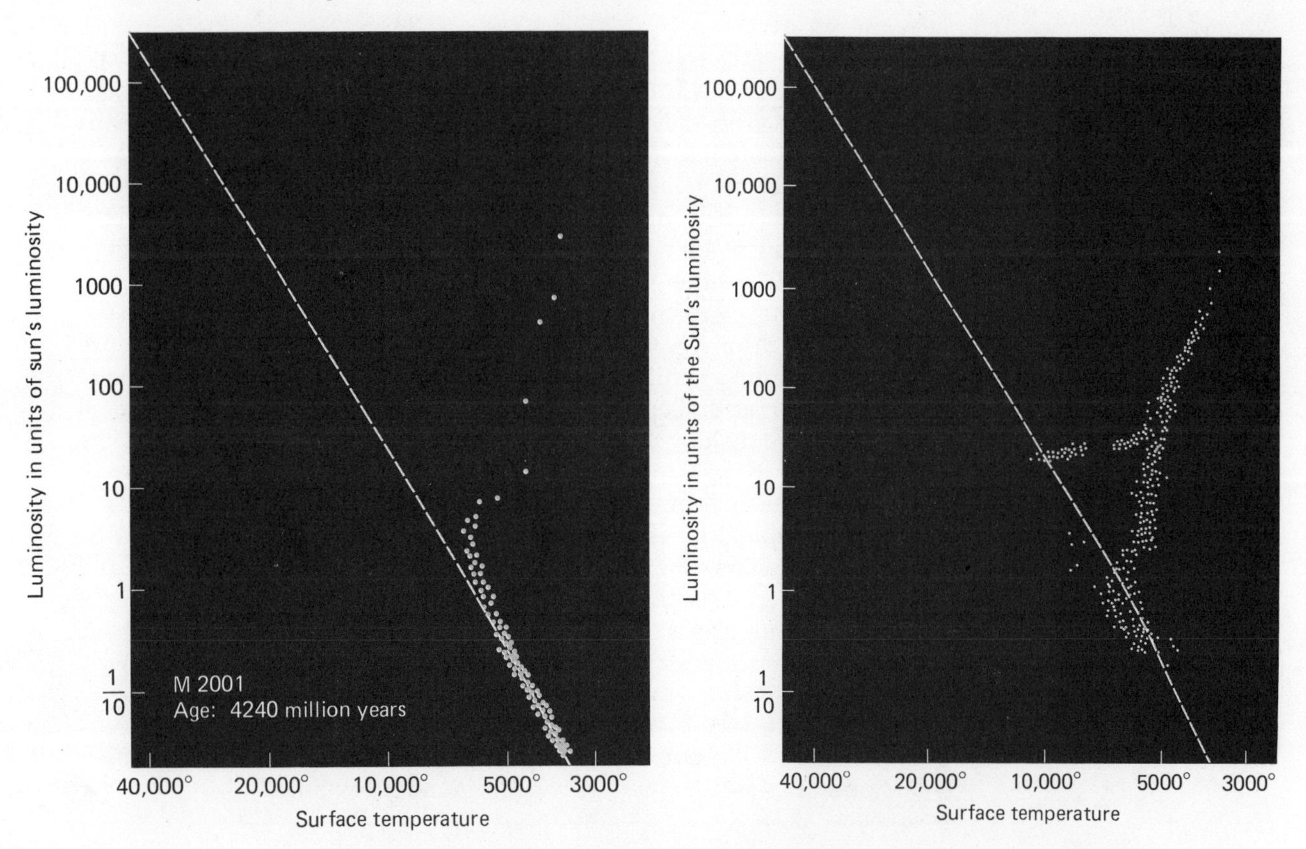

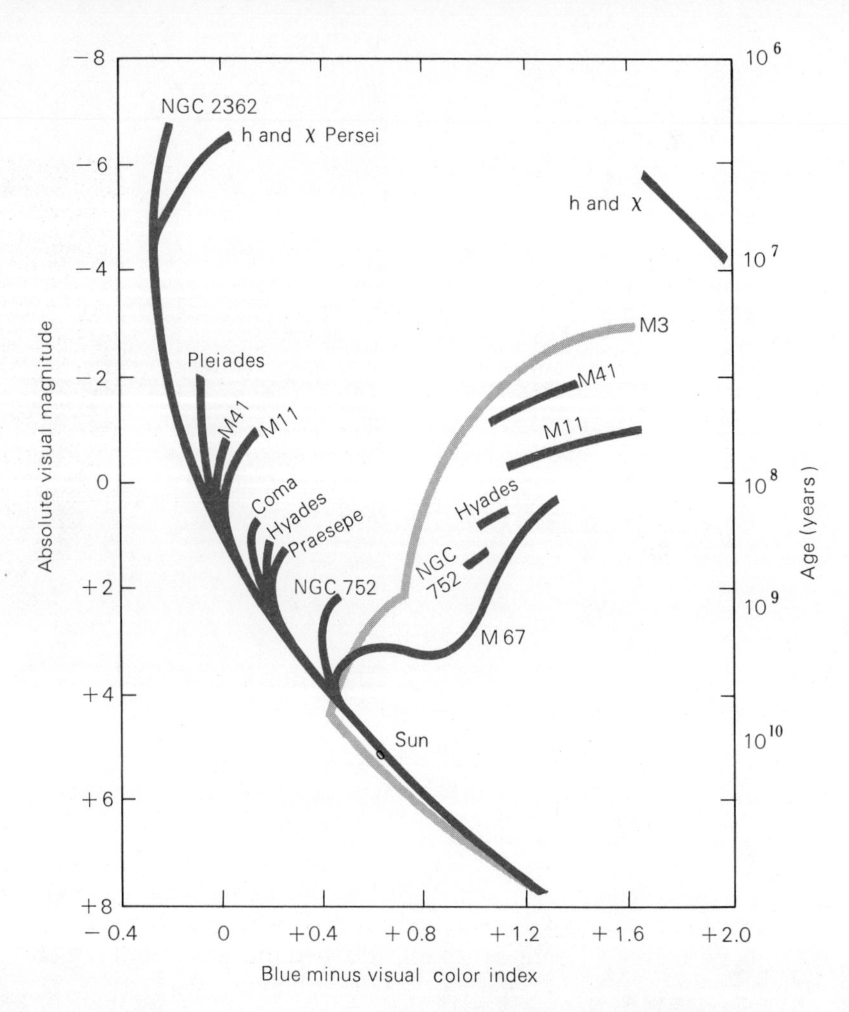

FIGURE IV.50
Composite H-R diagram for several star clusters of different ages. *(Adapted from a diagram by A. R. Sandage)*

hundred million years old. The trend continues until after 5000 million years, stars not much more massive than the sun have evolved away from the main sequence to red gianthood. The H-R diagram of the 4240 million-year-old cluster M2001 is shown in Figure IV.48. Compare it with my schematic H-R diagram of the even older globular cluster, M3, in Figure IV.49. M3 is probably at least 10,000 million years old.

Thus we see how we can compare the real and theoretical clusters, and see directly how well observations confirm the theoretical predictions. Figure IV.50 (adapted from a similar one by Sandage) shows a composite schematic H-R diagram for a number of real star clusters of various ages. You can tell how old they are by how much of their main sequences are left. In fact, the scale on the right side of the figure shows the theoretically computed ages of clusters the upper ends of whose main sequences terminate at those points.

I think you should be convinced by now that we understand pretty well the evolution of stars up to the point where they develop helium cores and leave the main sequence. As I said earlier in this act, the subsequent stages of evolution have not been very thoroughly studied theoretically. Still, we have a pretty good idea how stars of some masses evolve in their red giant stages.

---

4. According to the calculations for M2001, what, very roughly, was the age of the sun when it reached the main sequence?

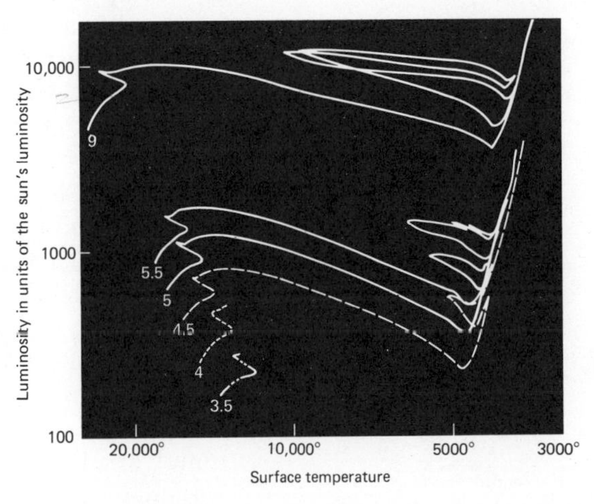

FIGURE IV.51
Theoretical tracks on the H-R diagram showing the evolution of stars of various masses from the main sequence to the red giant phase. *(Adapted from a diagram by E. Meyer-Hofmeister)*

Late Stages of Evolution

Look at Figure IV.51. This is a portion of the H-R diagram on which are plotted tracks of evolution of stars of mass from 3.5 to 9 times that of the sun. These tracks are theoretical ones, of course, calculated by Dr. (Mrs.) E. Meyer-Hofmeister, at the Max-Planck Institute in Munich. Notice how the higher mass stars are predicted to loop back and forth several times across the upper part of the H-R diagram, each change in direction occurring as a new nuclear reaction occurs in their interiors. All stars in a cluster are not *exactly* of the same age. Some reached the main sequence and started their nuclear burning slightly before others. So even those of the same mass cannot be expected to remain precisely in step with each other. Thus at any given time in the evolution of a cluster, stars of very slightly different mass will be sharing the same stage of evolution.

Thus, according to Dr. Meyer-Hofmeister, after about 57 million years a cluster like M2001 (which, I am afraid, she has never heard of) should have its most massive stars that are still living smeared about somewhat on their evolutionary tracks at the right side of Figure IV.51. The result, she predicts, should be an H-R diagram something like that shown in Figure IV.52. Now

FIGURE IV.52
(Below left) Predicted H-R diagram for the brighter stars in a cluster like M2001 at an age of about 57 million years. *(Adapted from E. Meyer-Hofmeister)*

FIGURE IV.53
(Below right) H-R diagram for the brighter stars in the star cluster NGC 1866. *(Adapted from E. Meyer-Hofmeister)*

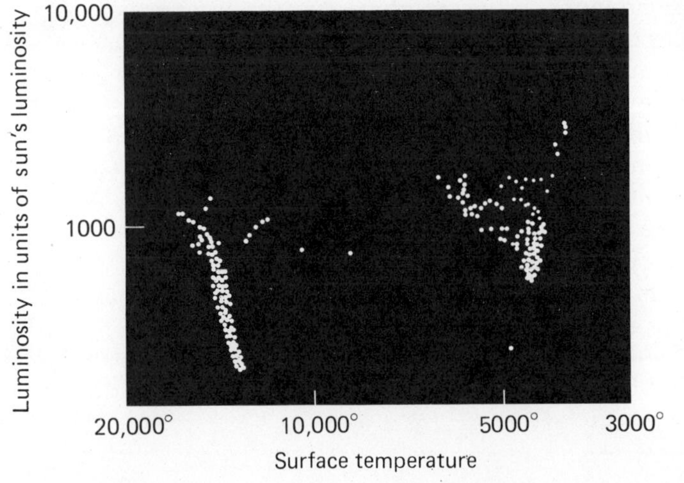

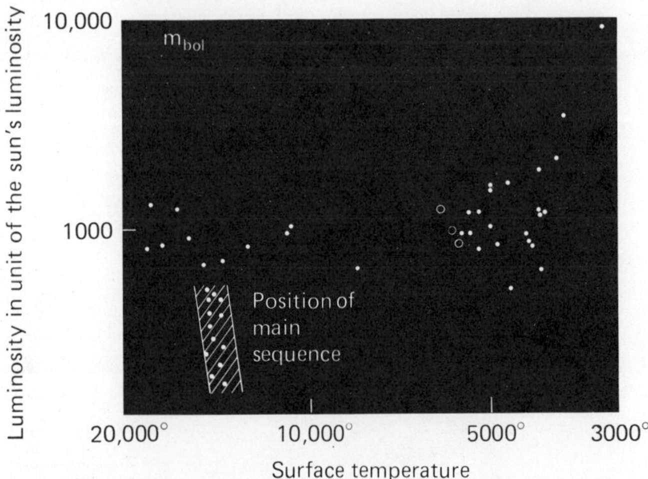

compare these predictions with the actual H-R diagrams of the brightest stars in cluster NGC 1866, shown in Figure IV.53. You can see that we have achieved some success in predicting even these more advanced stages of evolution of stars.

Summary

We now understand a good deal about stellar evolution. We can say with considerable confidence, for example, that clusters like NGC 2264 are very young, almost still in the process of formation from interstellar matter, whereas the very old globular clusters, like M3, are at least 10 thousand million years old. The main sequence of M3 has evolved away (or "burned down") to about the luminosity of the sun.

One day in 1966 when the British astronomer John Hazelhurst and I were having lunch with Rudolf Kippenhahn in Munich, we got onto the subject of limericks—a form of poetry (if you can call it such) peculiar to the English language. "But," said Rudolf, "there are no limericks on astronomical topics!" Well, this was a challenge to John Hazelhurst and me, and we both wasted some time composing limericks and sent our best (or worst) efforts to Rudolf, who was at that time a professor at Göttingen.

Well, some months later I was invited to give a colloquium at Göttingen. After my talk I was taken to the cellar of the observatory for a social session. On the walls were all kinds of mementos, including stolen signs and such irreverent items as toilet seats from major observatories about the world. Also on the walls were posted various verses, including a limerick by John Hazelhurst (which I cannot remember) and one by me. Mine reads:

> Said NGC 2264,
> "Let's go back to the club and dance some more."
> But said poor old M3,
> "You'd best go without me,
> "For I'm no longer as young as before."

(Well, I cleaned it up a little bit for this book.)

SCENE 7

The End

All good things must come to an end—even stars. The signal to the end of a star's life is when its available supply of nuclear energy is gone. That supply would be gone, for example, if the star were to end up as pure iron—or iron and other elements, near it in the periodic table, with about the maximum mass defect. The end of a star's source of nuclear power could, and probably would, come sooner, however, if it were unable to raise what fusionable material it had left to the requisite temperature to set off the necessary nuclear reactions. Anyway, one way or another, the end of nuclear energy must come. Then what?

The star still has thermal energy and gravitational energy. The thermal energy is just the stored heat inside it—the kinetic energy of its atoms and electrons. That energy will eventually be radiated away. But the star can still contract, and in doing so, release gravitational energy. At first you might think that when a star shrinks away to nothing, surely its gravitational energy is gone. But gravitational energy doesn't work that way. As the star gets smaller and smaller, its stuff falls closer and closer together, releasing more and more energy. The smaller the separation of the gravitating parts, the greater the gravitational force; as their separation approaches zero, the gravitational force approaches an infinite value, and the gravitational energy released becomes ever greater.

Does this mean, then, that there is no limit to how much energy a star can release as it shrinks under its gravitational contraction? In the classical gravitational theory of Newton, there is, indeed, no limit. On the other hand, no one assumed that matter could be packed to an infinite density. Surely, it was thought, eventually the condition is reached where the matter in a star is packed down to its basic parts and can be crushed no more; *then* its gravitational contraction and subsequent release of energy must stop!

But when it was learned that matter and energy are equivalent, and that particles of matter can be thought of as waves, it was not so clear why there should be a limit to packing of matter. If a block of matter has no real substance anyway, other than as waves of energy, what is to stop it from being crammed into a singular point of infinite density? Well, the same quantum theory that raises this question tells us that there is indeed a limit to the contraction of most stars. It comes about because of an interesting combination of the *Pauli exclusion principle* and the *Heisenberg uncertainty principle.*

---

1. The thermal energy of a star is roughly comparable (within a factor of two or three) to the gravitational energy it has released since its formation. Why should this be so? [Hint: According to the Helmholtz-Kelvin theory, what happened to that released gravitational energy?]

Electron Degeneracy

Fundamental particles fall into two classes, one in which an infinite number of particles can coexist in the same place with the same energy, and one in which no two particles in the same state can coexist in the same place at the same time with the same energy. The former kind of particles, which include nuclei of helium atoms and photons, are called *bosons;* the latter, which include electrons, protons, and neutrons, are called *fermions.* (They are named for the physicists S. N. Bose and E. Fermi, pioneers in modern physics.)

I realize that this may seem abstract and confusing. What it means is that the laws of quantum mechanics, which have been discovered to describe the behavior of particles very well indeed, preclude any two of certain kinds of particles (for example, electrons and neutrons) being in the same place at the same time doing the same thing. This prohibition is called the *Pauli exclusion principle,* after Wolfgang Pauli, another of those twentieth-century physicists who helped create modern science. (It was Pauli, in fact, who first hypothesized the neutrino. According to Sir Fred Hoyle, Pauli had bet astronomer Walter Baade a case of champagne that the neutrino would never be detected in the laboratory. In 1952, however, the neutrino *was* isolated. Sir Fred says he can attest that the bet was paid off; he personally helped Baade consume one of the bottles of champagne.)

More specifically, the Pauli principle states that no two electrons (say in a star) can occupy the same place and have the same momentum at once.[1] But what does "same" mean? Surely we can let one electron have a very, very slightly different speed or direction and occupy the same place as another electron. Ah, but if we say *that,* we have forgotten about the Heisenberg uncertainty principle. There is a limit to the precision with which the simultaneous position and momentum of any electron can be known, and hence with which it can exist. What the Pauli principle means is that no two electrons can exist in the same place with the same momentum within the range of uncertainty allowed by the Heisenberg principle.

The most common kind of particle in a stellar interior is a free electron, because almost all the atoms are completely ionized. If the gas in the star is compressed to a particular density, it means that there is a certain number of electrons in each cubic centimeter. Now the uncertainty principle tells us how large the range of momentum of those electrons must be in order to assure that no two will need to have the same position and momentum within

[1] Technically, there can be two electrons with the same momentum, for they can spin in opposite directions.

the allowed limits. The range of momentum present depends on the temperature of the gas; the higher the temperature, the faster some of the particles are moving and the greater is the range of momentum. But for a given temperature there is a given range of momentum, and hence a given limited number of electrons that can be crammed into a cubic centimeter. If in a given volume all the available states of momentum allowed by the Heisenberg uncertainty principle are occupied by electrons, that gas is said to be *degenerate.* This is the highest density to which a gas of a given temperature can be compressed without crowding its electrons past the point of packing which they resist with the most incredible pressure. It is the pressure of their *degenerate electrons* that prevents most stars from contracting forever.

---

2. If a star were pure hydrogen, all ionized, what fraction of its particles would be electrons? What if it were pure ionized helium? What if it were pure ionized iron?

Theory of White Dwarfs

In 1924 the British astrophysicist Sir Arthur Eddington suggested that stars could be compressed to very high densities if their atoms were completely ionized, because then the particles could be much closer together than in solids. Then, in 1926, the theory of degenerate electrons was worked out. In 1927 R. H. Fowler applied the theory to stars. He found that the more massive a star is, the more it heats up inside as it contracts, and the greater is the range of momentum available to the electrons in there. Thus, the more massive stars can get smaller before their electrons become degenerate and stop the contraction.

The full theory, including the effects of special relativity, was first worked out by the Indian-American astrophysicist S. Chandrasekhar. Chandra, as he is known to his friends, found that there is a relation between the masses and radii of stars which have contracted so far that their electrons are completely degenerate. That mass-radius relation is shown in Figure IV.54. Note that stars with masses like the sun's should end up with diameters about 1 percent that of the sun (and hence densities a million times as great!) That's the

FIGURE IV.54
The theoretical relation between the mass and radius of a completely degenerate white dwarf star. *(After S. Chandrasekhar)*

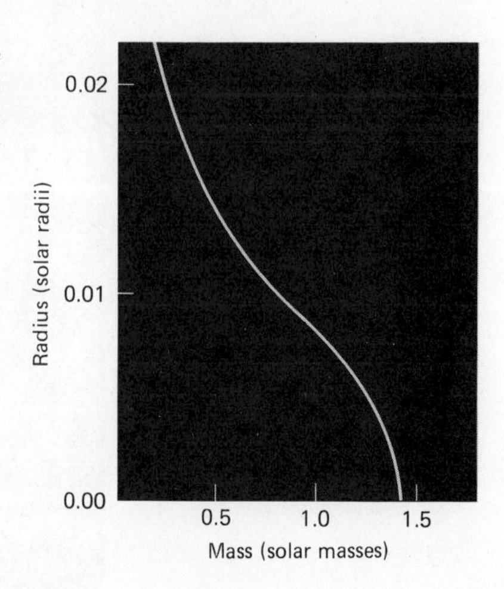

size of a planet like the earth. Recall from Act II Scene 3 that white dwarfs are stars fitting this description. Moreover, we know today that there are very many white dwarfs – probably enough to account for all or nearly all the stars which have had time so far, in the history of our Galaxy, to exhaust their nuclear fuel and evolve to that end state of evolution. Note in Figure IV.54, however, that there is a limiting mass for a white dwarf; the sizes of these compact stars suddenly approach zero at a mass of about 1.4 times that of the sun. Chandrasekhar showed that no star with a greater mass than this could exist as a stable white dwarf!

---

3. The sun has a mean density of 1.41 g/cm^3 (or 1.41 times that of water). What is the mean density, in terms of that of water, of a white-dwarf star of 1 solar mass?

Mass Ejection

On the other hand, the only stars that would yet have had time, in the history of the Galaxy, to exhaust their nuclear fuel and evolve to that state are those very stars whose masses are near or above the theoretical mass limit for white dwarfs. How could stars of mass *greater* than 1.4 solar masses evolve to white dwarfs of mass *less* than 1.4 solar masses? Obviously, only by ejecting some of their material. As we have seen, there are many known stars that are doing just this. For example, planetary nebulae, which are shells of gas ejected from

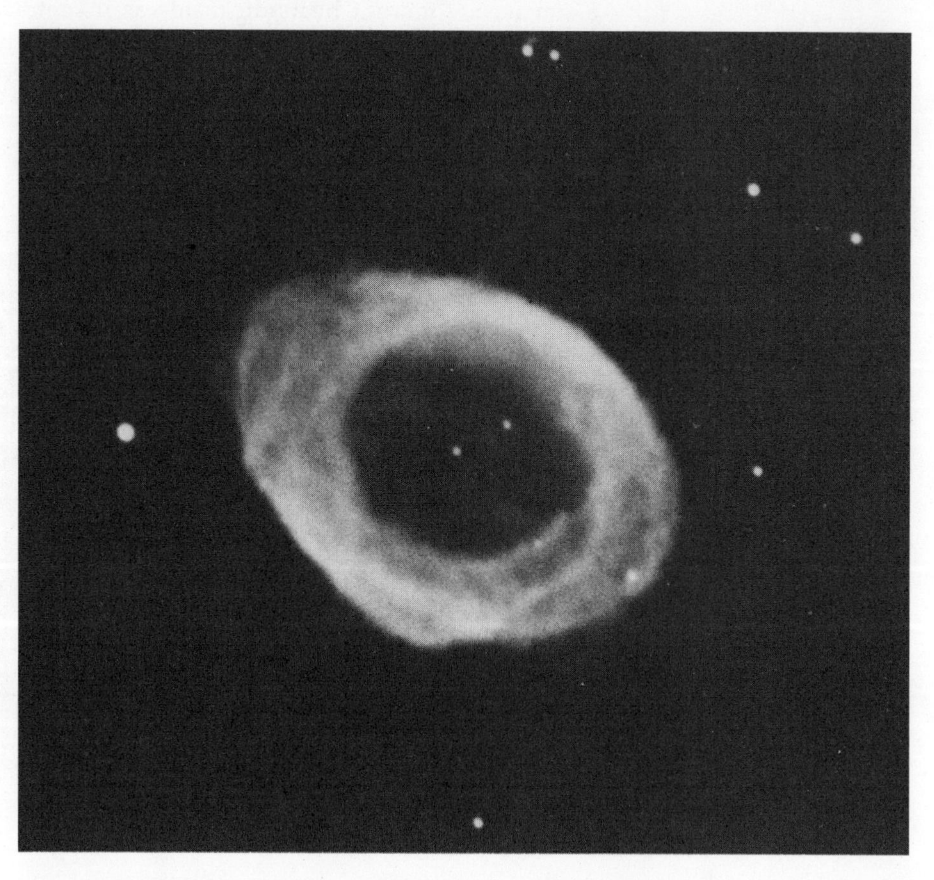

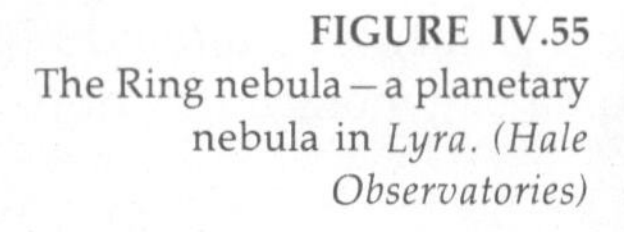

FIGURE IV.55
The Ring nebula – a planetary nebula in *Lyra*. *(Hale Observatories)*

(we think) red giant stars, are certainly mechanisms by which stars eject some of their matter into space. In fact, many planetary nebulae have central stars with the dimensions of white dwarfs. In other words, the evidence suggests that in one of the several points in its evolution back to the status of red giant, a star dislodges its outer part, which goes off as a planetary nebula, while the inner part collapses to a degenerate electron star—a white dwarf. Since there are enough planetaries to account for most or all old stars, it seems quite reasonable that most or all old stars eject a planetary nebula, thereby lowering their masses below the critical limit, so that they can settle down to white-dwarf retirement.

Novae

Novae, also stars that eject mass, are thought to be associated with white dwarfs as well. The California astronomer Rober Kraft found that many stars which are ex-novae are members of close binary-star systems. We suspect today that *all* novae are members of double-star systems, whose periods of revolution are very short—less than a day, which shows the stars in each such system to be very close together.

The popular theory of novae is that one star in such a system has already evolved to a white dwarf, while the other is beginning to evolve to a red giant. Eventually the surface of the growing star passes through that point between the stars where matter doesn't know to which star it belongs (see Scene 3). As a consequence some gas rich in "unburned" hydrogen transfers from the evolving star to the white dwarf. Compressed by the high surface gravity of the white dwarf, that new hydrogen soon becomes hot and dense enough to turn on its nuclear conversion to helium. The white dwarf thus heats up explosively at its surface, and ejects with a burst of energy that outer layer of material—typically a ten-thousandth to hundred-thousandth of a solar mass. With the unburned hydrogen gone, the white dwarf settles down again. After a while, however, new hydrogen piles up on it, and the process

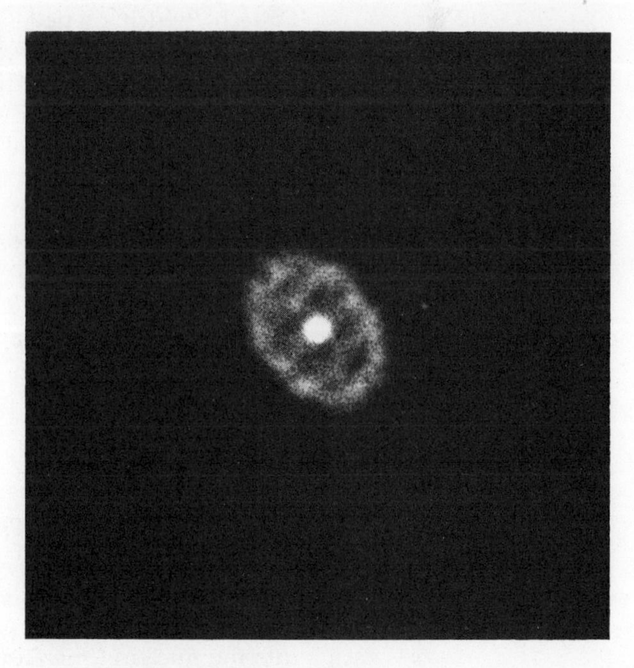

FIGURE IV.56
Nova Herculis erupted in 1934. In 1972 this photograph of its expanding shell was taken with the Lick Observatory's 3-m telescope. *(Courtesy H. Ford, UCLA)*

repeats. Indeed, many novae are known to have periodic outbursts. The brightest novae, however, those which brighten up to hundreds of thousands of times the sun's luminosity, are usually observed only once. They too, though, may occur periodically, but with longer periods (in the thousands of years) so that we have observed them only once in the history of astronomical observation and record keeping.

Neutron Stars

When I finished my graduate work, most people thought that we pretty well understood the end states of stars; those too massive to become white dwarfs simply ejected some of their outer gases until their masses were low enough, and then settled down to that state. On the other hand, a few people were suspicious that there might be alternatives. For example, what if a star *didn't* eject its excess mass? Could it find some way to end its active life other than becoming a white dwarf?

But what other way is possible? One possibility, predicted theoretically in 1932 by the Soviet physicist Lev Landau, is that a star could become a *neutron star*. The first neutron star candidates were suggested in 1934 by Walter Baade and Fritz Zwicky. Baade and Zwicky were both astronomers at the Mount Wilson Observatory (now part of the Hale Observatories).

I knew both Baade and Zwicky quite well, because I spent many nights with each of them at Palomar in the early 1950s. I was a young graduate student working my way by taking photographs for the National Geographic Society—Palomar Observatory Sky Survey with the 48-inch (1.2 m) Schmidt telescope, and during that time Baade and Zwicky were frequent observers at the 200-inch (5-m) Hale telescope. Baade was a German astronomer who had come to America in the 1930s. He worked a great deal in the study of galaxies, and I shall return to him again.

Zwicky was born in Varna, Bulgaria, but became a Swiss national and remained a Swiss until his death. In the 1930s he was a professor of physics at Caltech (which owned Palomar Observatory). I have been told the story (which I can believe, but which I cannot vouch to) that he got into astronomy as a consequence of a dinner conversation with Robert A. Millikan, then president of Caltech. Zwicky was reported to have said that any great scientist could change fields and make a name in five years. Millikan, rising to Zwicky's boast, said, "All right, Caltech is going to open a department of astronomy; I'll make you professor of astronomy and astrophysics, and give you five years to make your mark in that field!" As I say, I can't vouch for the details of the story, but I can vouch for the fact that Zwicky *did* make his mark in that field. He spoke at least seven languages, pioneered in the study of supernovae, studied galaxies and the distribution of galaxies, and discovered many clusters of galaxies. Further, he was a sincere humanitarian, and spent many dollars and hours in helping war orphans and in replenishing war-destroyed libraries.

Anyway, Baade and Zwicky suggested that the tremendous energy outburst in a supernova might be explained if a star collapsed to the very dense state of a neutron star. Their idea was close to prophetic, but it was 30 years before the astronomical community took it seriously.

The idea of a neutron star is this: The pressure becomes so great that the matter in the star is compressed to the point where the electrons are crushed into the nuclei of the atoms. Then only neutrons exist—all the fundamental

FIGURE IV.57
Fritz Zwicky.
(Caltech)

particles of the star are electrically neutral and the star has a density like the nucleus of an atom. What would such a star be like? A neutron star of the mass of the sun would be only 10 km in radius and have a density of some 5×10^{14} g/cm^3. A tiny raindrop, at the surface of such a star, would weigh over 500,000 tons. It happens that neutrons are Fermi particles, as are electrons, and are subject to the same quantum-mechanical limitations of packing as are electrons. However, having far greater masses, neutrons can be crowded to far higher densities.

But how can we imagine getting the matter of a star into such a state? We can think of at least two ways. One is if the star did *not* eject matter into space, but had more matter than the mass limit of a white dwarf. Now there is a mass limit for neutron stars, just as there is for white dwarfs, but we are not sure what that limit is. The best guess today is that it is between two and three times the mass of the sun. This means that a star of, say, 2 solar masses, exceeding the 1.4 solar mass white-dwarf limit, might be able to keep on contracting until the pressures pushed its electrons into the nuclei and made it a ball of neutrons; this might be one way to make a neutron star. The other way is to push the electrons into the atomic nuclei with a great implosion—perhaps, as Baade and Zwicky suggested, in a supernova, where the outer part of the star flies off while the inner part is crushed into a neutron star. We don't know for sure if that can actually happen; yet the main evidence we have for neutron stars today seems to be associated with supernovae.

Pulsars

The hint that neutron stars may actually exist has come through the discovery of the *pulsars*. The first pulsar was discovered in mid-1967 by Jocelyn Bell, a research student at Cambridge, who discovered a cosmic radio source in the constellation Vulpecula that seemed to come in pulses at the rate of one every 1.33728 seconds. Cambridge astronomers Anthony Hewish and Martin Ryle recognized the importance of the discovery and followed up on it (for which they shared the Nobel Prize). Subsequently, other such *pulsars* (for pulsating radio sources) have been found; they now number over 100. Their periods range from only 1/30 s to more than 3 s. The strengths of the pulses vary from one to the next, but they come at extremely regular intervals. The radio radiation coming in the pulses has the characteristic properties of synchrotron radiation, showing that the pulses of energy are emitted by electrons being accelerated by magnetic fields.

---

4. As observed from earth, the precise interval between the pulses of a typical pulsar vary periodically, and in step with the seasons. Can you suggest an explanation? [Hint: What is the direction of the pulsar with respect to the direction of the earth's orbital motion?]

One of the pulsars happens to be smack in the center of the Crab nebula—that gaseous remnant of the supernova of 1054 in Taurus. Remember that the Crab nebula is emitting radio waves and X rays as well as visible light. Moreover, the radio radiation from that nebula, as well as much of the visible light, is *also* synchrotron radiation. But what about the pulsar in the middle? That pulsing radio source is the one of shortest known period—1/30 s. When it was discovered, the X-ray people looked back over their data and discovered that a large part of the X-ray radiation is also coming in pulses and with the same period. Evidently, part of the X-radiation, as well as part of the radio radiation from the Crab nebula, is due to the pulsar at its center.

Long before the pulsar had been discovered in the Crab nebula, astronomer Rudolf Minkowski had noticed that the central star in the nebula had a strange spectrum—no apparent absorption lines. He thought it might be a white dwarf. However, that star at the center of the Crab nebula was subsequently identified as the pulsar. The clinching evidence is that the star itself is observed to turn on and off in visible light 30 times a second, the same as does the radio and X-ray pulsar. No one had known this before, because the star was observed only on photographs (or spectrograms) that were long time exposures. Now, though, we have modern electronic means of observing short-period light variations of such objects, and can produce a

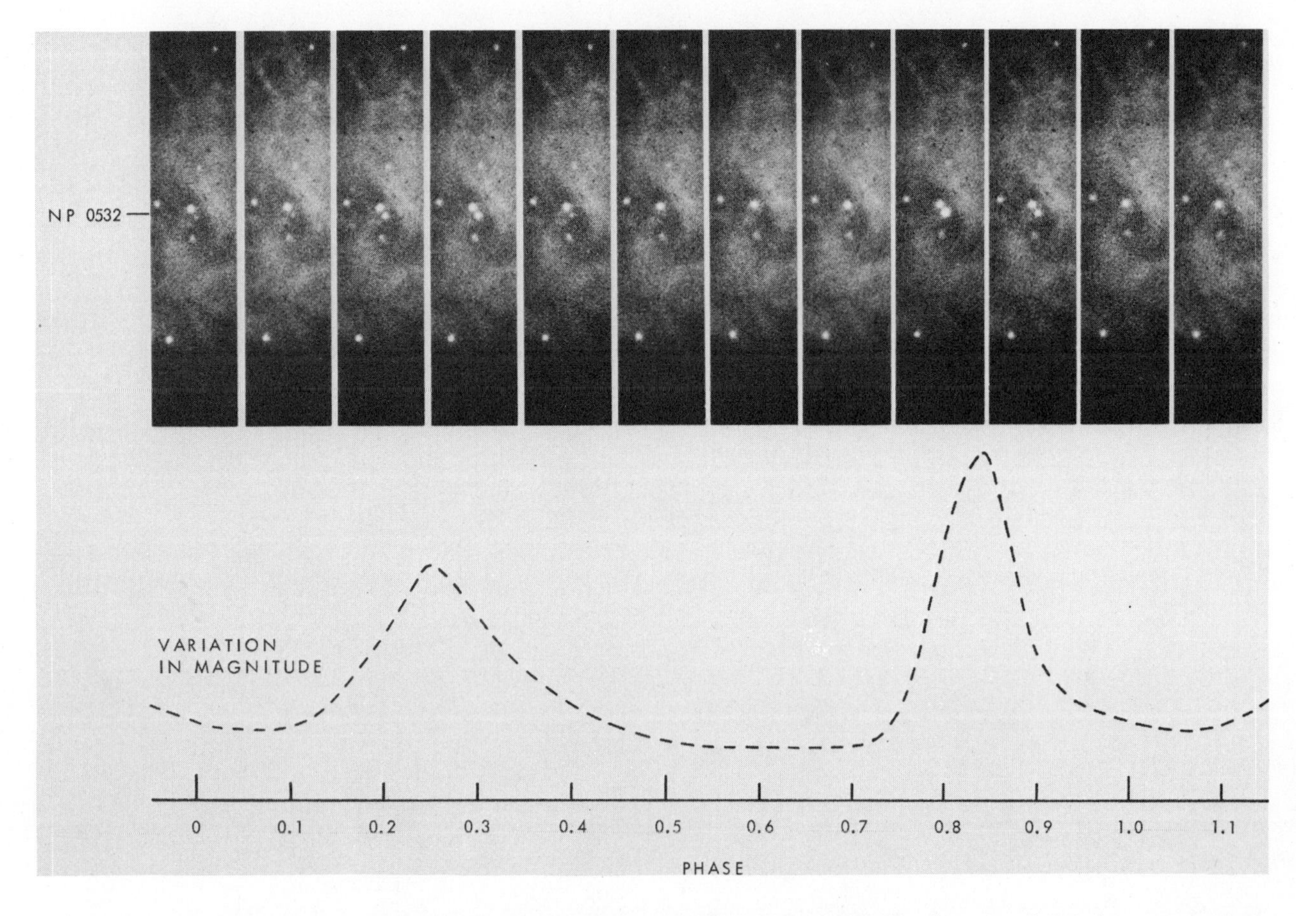

FIGURE IV.58
A series of photographs of the central part of the Crab nebula taken by S. P. Maran at Kitt Peak National Observatory. Note the star that seems to blink on and off; it is the pulsar, which has a period of 1/30 s. *(Kitt Peak National Observatory)*

series of photographs such as in Figure IV.58, showing how the star varies in light at the same rate that the radio waves and X rays pulse on and off.

So here is where we stand. The Crab nebula is known to be a supernova remnant. It is centered on a pulsar. A few other gaseous filaments, believed to be the ejecta of prehistoric supernovae, are also centered on pulsars. One, which pulses 11 times a second in the constellation of Vela, is also seen in visible light, although very faintly. None other, however, has so short a period as the Crab pulsar. Finally, there is a theory that supernovae might produce neutron stars. Can all these facts be related?

The relation becomes obvious when we try to imagine what kind of thing can make pulsars. The energy emitted is large, like that of a highly luminous star. Even though its light comes in short pulses, for example, the Crab pulsar puts out a lot more total energy per second than the sun does. It must be something like a star, or at least of stellar proportions, to produce enough energy to account for the pulsed radiation. So we must look for something with the energy of a star emitting very regularly timed pulses every second or so, and in one case 30 times per second.

A pulsar cannot be a pulsating star, however. If you look at the table in Scene 3, you will note that even white dwarfs cannot pulsate as fast as once a second or oftener. Nor can white dwarfs or anything larger rotate that fast. Neutron stars seem to be the only explanation. They would pulsate very much faster—in a few milliseconds; but they could rotate with any period greater than a few milliseconds. Most astronomers, therefore, are of the opinion that pulsars are rotating neutron stars.

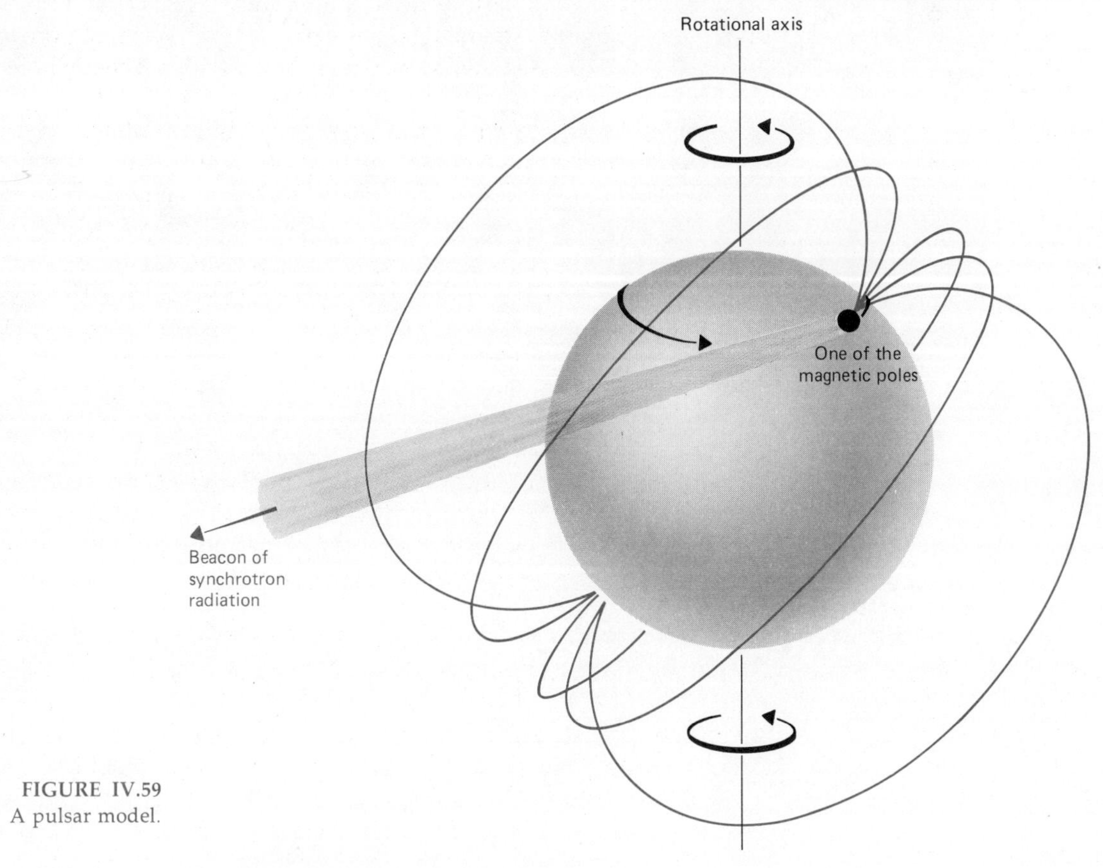

FIGURE IV.59
A pulsar model.

A Pulsar Model

We are not certain of the details of how it all works, but the general idea is thought to be something like the model illustrated in Figure IV.59. Shown is a neutron star, like the one postulated to exist at the center of the Crab nebula. Now most stars are presumed to have magnetic fields. The sun has a weak general magnetic field. If we were to compress the sun, however, by some 70,000 times in radius (to crush it to the size of a neutron star), its magnetic field would be stronger by more than 5000 million times and would comprise a very intense magnetic field. We thus expect that neutron stars should have such intense fields. Moreover, they should rotate rapidly, because as their parent stars contract, they speed up to conserve their angular momentum. Under relatively less pressure, at the stellar surface the neutrons are unstable and decay. Thus protons and electrons are available (as well as other particles), and these can interact with the magnetic fields; such interaction should be especially intense at the magnetic poles of the rotating neutron star.

We must now suppose that the magnetic axis of the neutron star does not coincide with its axis of rotation (no particular reason why it should; it doesn't in the case of the earth). This means that with each rotation one of the magnetic poles will be carried around more directly into our line of sight. Half a rotation later the other pole will be turned into our view, but, depending on the inclinations of the rotation and magnetic axes, more or less directly so than for the first pole.

Now, as I say, we don't yet know exactly how the synchrotron radiation originates from the rotating star; the best guess is that it is from a "hot spot" on or just above the surface of the star, or from some region in its atmosphere, probably associated with one or both of the magnetic poles. One possibility is electrons leaving the star (with the protons and other nuclear particles following along) and being accelerated in the intense magnetic fields, especially in the polar regions. In any case, synchrotron radiation is highly directed and depends critically on the direction of motion of the emitting electrons. Thus the radiation flowing away from the pulsar should be only in certain directions.

Current theory has it that with each rotation of the neutron star, when the motions of the electrons and the magnetic field are oriented just right, a beacon of radiation is flashed our way. Those rotating neutron stars whose axes are so oriented that we never receive such periodic pencils of radiation are never discovered as pulsars; they are reserved for other intelligent observers, if they exist, in other parts of the Galaxy.

If the axis of rotation of a neutron star is not too far off from being at right angles to our line of sight, we should expect to see *two* flashes of radiation coming our way, as alternate magnetic poles are turned into our view. We do sometimes see two pulses of alternately higher and lower amplitude from a pulsar. The Crab pulsar is such a case. The alternate pulses are usually unequal, however; the magnetic and rotational axes can seldom be expected to be so oriented in space that we see both magnetic poles with exactly the same aspect.

If all the above makes any sense, the rotating neutron star-pulsar, constantly ejecting matter into space along its magnetic field lines, should lose angular momentum. Thus the prediction is that pulsars should gradually spin down. In fact, the Crab pulsar is observed to be very gradually slowing in its spin, so its period is increasing over the years. We think that the other pulsars, of longer period, are older, and have already been slowed down greatly in their rotation. Young new pulsars, in this picture, spin rapidly and emit lots of visible, ultraviolet, and X-radiation as well as radio waves. Pulsars some thousands of years old, on the other hand, have lost most of their angular momentum (through radiation of mass into space), and also their energy; they spin only once a second or more slowly, and emit radiation from their magnetic "hotspots" only at radio wavelengths.

Whatever is the case with pulsars, they involve the acceleration of electrons moving at nearly the speed of light to account for their radiation. Many or most of these electrons must escape into interstellar space to produce more radiation in the interstellar magnetic fields. In particular, many electrons escaping the Crab pulsar must account for the synchrotron radiation from the entire nebula around it. Also the atomic nuclei that must necessarily escape the star with the electrons go into interstellar space. This emission of protons, alpha particles, and nuclei of heavier elements may well provide many of the cosmic rays observed at the earth.

---

5. Suppose five new pulsars are discovered with periods of
a) 1.32s, b) 3.04s, c) 0.05s, d) 0.97s, e) 1.92s.
From which would you most expect to be able to observe pulses of visible light? Explain.

Black Holes

We recall from Act II Scene 5 that spacetime is severely warped around dense objects. White dwarfs are pretty dense, and neutron stars far more so. In fact we have already seen how the redshift of light from white dwarfs was one of the classical tests of relativity. So was the deflection of light passing near the sun's surface—a very minor bending amounting to only about 1.7 seconds of arc. Near the surface of a white dwarf the bending of light, if it could be observed, would be about 1 minute of arc. Near a neutron star of one solar mass, the bending would be about 30°.

Figure IV.60 shows the effect of the warped spacetime at the surface of a neutron star on light leaving it. Light flowing out radially (that is, *normal* to the surface) is not deflected, although it is highly redshifted; photons emitted as violet light emerge as red light. Light rays directed at an angle slightly away from the normal to the surface, however, are deflected from that normal, and the greater the angle, the greater the deflection. Rays emerging at less than 30° to the surface are pulled back into the star. The light that leaves each point on the star and escapes into space is that aimed into a cone whose surface is about 60 ° from the normal.

Now imagine what would happen to the emerging light if that same neutron star could shrink to a still smaller size, as in Figure IV.61. As it becomes smaller and smaller, the deflection of light from its surface increases and the cone of emerging radiation that can escape shrinks in size. Eventually a point is reached where spacetime is so strongly warped that a light ray emerging at the tiniest angle from the normal is pulled back into the star, and the normal

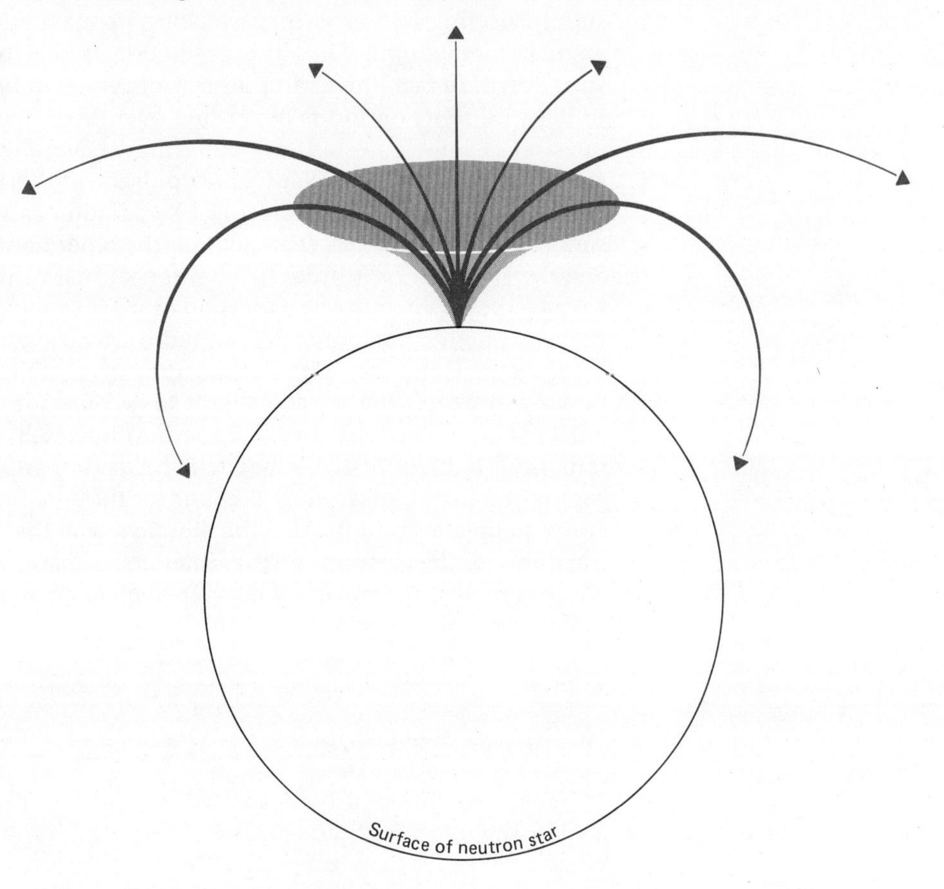

FIGURE IV.60
The paths of light rays emitted from a point on the surface of a neutron star.

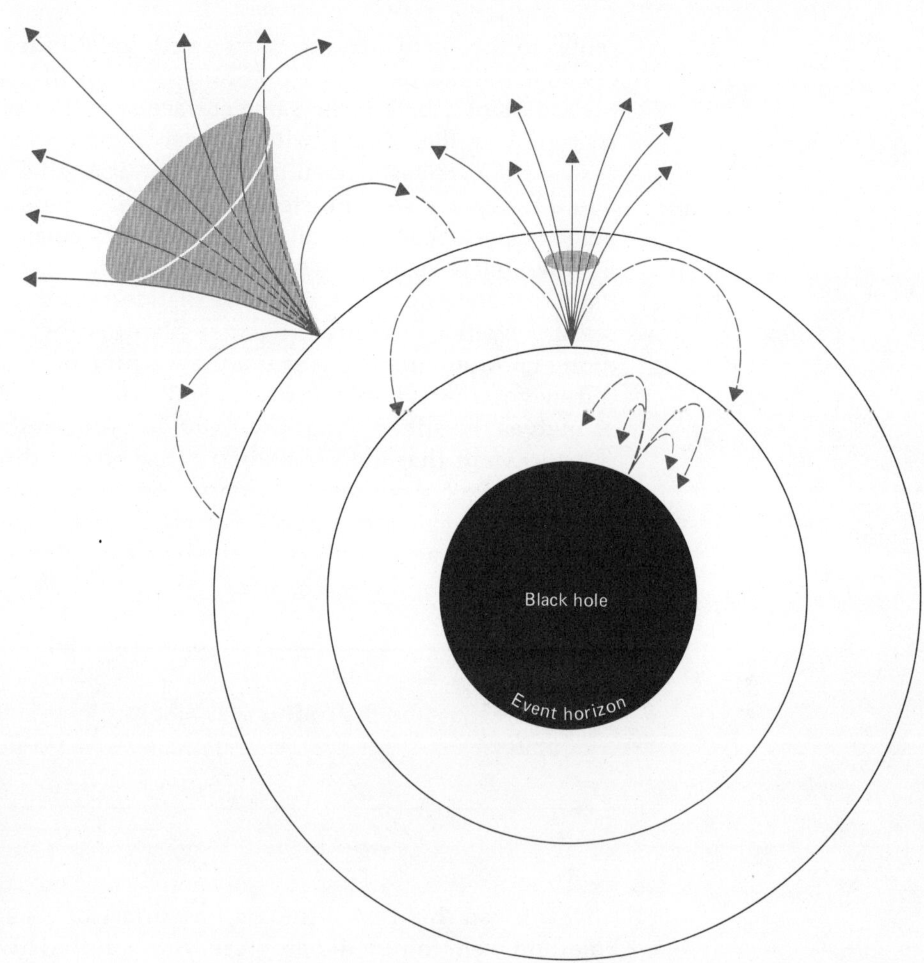

FIGURE IV.61
Radiation from the surface of a collapsing star can escape only if it flows out at not too great an angle to the normal to the surface, and that angle decreases as the star shrinks in size.

rays are gravitationally redshifted to infinite wavelength, so no radiant energy escapes the star at all. At that point, the surface of the star is said to define an *event horizon*. As the star shrinks further, its surface passing down through the event horizon, nothing escapes from it, not even light. The star simply disappears, and is said to have collapsed into a *black hole*.

We don't expect a neutron star to become a black hole, because it is a stable configuration, the tremendous weight of its matter being supported by the equally tremendous pressure of the degenerate neutrons. But what about a star of mass greater than the mass limit for a neutron star (probably about 2 or 3 solar masses)? If such a star does *not* eject matter into space and does *not* lower its mass below the white-dwarf or neutron-star limit, and is not rotating too fast, at present we know of nothing that will ever stop its contraction. Thus if massive stars can avoid mass loss, theory predicts that when they exhaust their nuclear fuel, they should contract, releasing gravitational energy, until they become black holes. At that point, they disappear from the universe.

Properties of Black Holes

The size of a black hole is usually meant to be the size of its event horizon. It depends, of course, on who's doing the measuring. Spacetime is so badly warped in and around the black hole that the length of a light path from the center to the event horizon can be very great indeed. It is easy, though, to calculate the radius of the event horizon according to Euclid's laws—that is, the radius of a ball of the same surface area, but of low density so that its spacetime is flat. That's what I mean here by "size," and it's called the *Schwarzschild radius,* named for Karl Schwarzschild, who described the situation a few years after Einstein introduced general relativity. Given a black hole of a particular mass, all we need do is calculate the radius it has to have in order that the speed of escape from its surface is equal to the speed of light. Rather interestingly, the same calculation was done late in the eighteenth century by the French astronomer Marquis Pierre Simon de Laplace. He didn't know about relativity and the warping of spacetime, of course, but he did suggest that if light is particles and if the speed of escape from a star were as high as the speed of light, we could never see the star.

It turns out that the radius of a black hole is directly proportional to its mass. A black hole of the sun's mass would have a radius of 3 km. Thus a star of, say, 5 solar masses—one we might expect to contract to a black hole—would have a radius of 15 km. The whole 5-solar-mass star would easily fit inside the limits of any big city.

---

6. What would be the radius of a black hole with the mass of the planet Jupiter?

Now if you could be sitting on the surface of a star collapsing into a black hole, and were to time its progress with your wristwatch, everything would seem more or less normal to you. You would find yourself passing on through the event horizon at the time you would expect to be. Once inside, you would not be able to communicate with any part of the exterior universe, although you could still receive messages from outside, because even though light cannot get out it can still get in.

On the other hand, an observer in a space ship well outside the black hole, watching you and your watch with his telescope, would see things quite differently. He would see your watch slowing down more and more as you approached the event horizon, because the warping of spacetime affects the rate of passing of time (an effect separate from that of special relativity). Just as you reach the event horizon your time, to him, would seem to stand still. The collapsing surface of the star would also appear to slow down as it approached the event horizon, and finally come to rest at that point.

In other words, to an outside observer a star never seems to actually pass on through the event horizon, but stops there. You might think, then, that no star can ever quite become a black hole, as seen from outside. Technically, you would be right; I suppose, then, we should call black holes "tattletale gray holes." On the other hand, the slowing of time corresponds to the gravitational redshift of the radiation coming away from the star. Very soon, radiation emitted as high-energy gamma rays comes out as extremely long wave radio waves, of energy far too feeble to measure. To all intents and purposes, therefore, the star *does* become a black hole, because as time goes on, it becomes as theoretically close to it as you like.

7. Why is the time dilatation in a gravitational field equivalent to a gravitational redshift? [Hint: What is the definition of frequency? How is the frequency of radiation affected by a redshift?]

There has been a lot of interesting theoretical study of black holes, looking into such things as what happens when they are charged or rotating. At one time, it seemed that matter falling into a rotating black hole might be able to reemerge in another place in space and time as a *white hole*. During that mysterious passage it was said to pass through an *Einstein-Rosen bridge*, also called a *worm hole*. The field is rapidly changing, though; last I heard the theoreticians were thinking that passage through an Einstein-Rosen bridge was *not* possible. I shall not try to be up to date on these speculative matters; they are bound to be out of date again before you are likely to read them. I refer you, instead, to *current* articles in authoritative publications. At any rate, at present we have no convincing astronomical evidence for such things.

Do Black Holes Exist?

On the other hand, we do have reason to think that black holes might exist. But is there any evidence that they *do*, in fact, exist? If so, how could we ever know, since they can't emit any light to us? The way we can detect them is by means of their gravitation. While the matter falls in, its gravitational field is left intact outside the event horizon. Perhaps if we saw a star describe a sudden change in direction on a hyperbolic orbit in space, we might conclude that it was being accelerated by the gravitation of an unseen object. But such a close encounter would be rare at best, and we would be unlikely to observe the event. On the other hand, lots of stars are binaries, and stars in those systems must use up their nuclear energy like all the rest. Perhaps we can detect a black hole by virtue of its being an unseen member of a binary star.

We saw earlier how the companion of Sirius was detected in just this way—by its acceleration of Sirius, about which it is in mutual revolution. But the companion of Sirius has subsequently been observed, and it is a white dwarf, not a black hole. To make it to black-holehood, a star must have a mass too great to become a white dwarf or neutron star—probably 3 solar masses or greater; to be safe we would like it to be 5 or more solar masses. We must confine our search to binary-star systems in which only *one* star is visible, but whose orbital motion shows it to have a rather massive companion. Moreover, we must be able to find the distance to the system—say, from the spectrum of the visible member—in order to calculate what the mass of that invisible companion must be.

But even *this* is not enough. If the double star is distant enough from us, the companion can give out a lot of light and still be too faint to see with our telescope, so its invisibility does not insure that it's a black hole. We must have a way of knowing that the invisible star is a *collapsed* object—one that has contracted to a very small size. Now here is one way that an unseen star might indicate itself to be very small: the collapsed star might well be surrounded by a disk of matter. Perhaps this could be material that reached circular velocity and stayed behind in orbit in the equatorial plane of the star

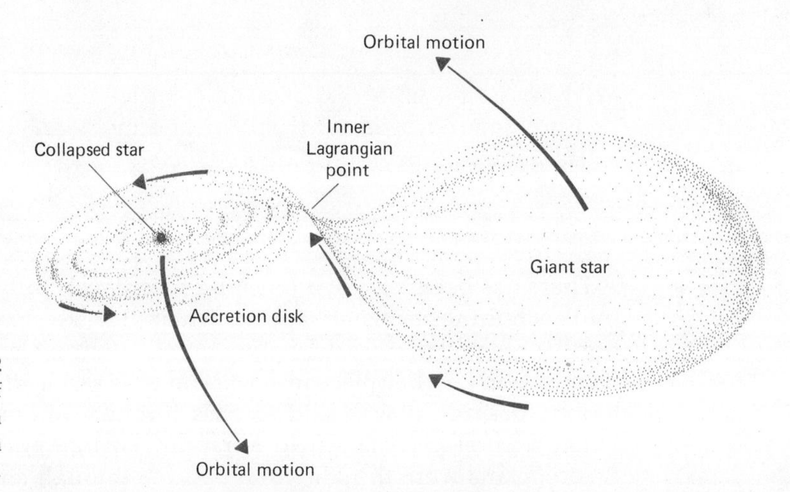

FIGURE IV.62
One model for the transfer of mass from a star to a collapsed companion in a binary-star system.

as it shrank. Or perhaps it is matter that transferred over to the vicinity of the collapsed star from the companion star.

One model for such mass transfer is shown in Figure IV.62. Because of the orbital motion of the system, matter does not fall directly onto the collapsed star, but spirals into an *accretion disk* around it. Some of the matter in the disk gains momentum from its surroundings and goes off into space, while other material comes in closer to the small dense star. Regardless of how the matter in that disk gets there, in the inner parts of the disk near the collapsed star it is moving extremely rapidly in its approximately circular orbits—up to 100,000 km/s. As the atoms of gas collide with other atoms at this speed, the gas becomes extremely hot, and should emit X rays. Ah! So we look for a massive unseen companion in a binary system from which we receive X rays.

---

8. Why would we not expect X rays from a disk of matter about an ordinary star, or even about a white dwarf?

Cygnus X-1

In December 1970, the first artificial satellite designed especially to detect X rays from space was launched from Kenya. It was named *UHURU*, Swahili for "freedon." Many sources of X rays were discovered. Some are evidently clusters of galaxies and other remote objects that we shall discuss later. Others are in the Milky Way. About a dozen (so far) appear to be in binary systems in our Galaxy. At least two of these X-ray sources in orbit about other stars emit short regular pulses, and are called *X-ray pulsars.* They are probably rotating neutron stars. Some of the others, though, may be candidates for black holes.

The most interesting one, and the best case to date, is Cygnus X-1, the first X-ray source discovered in the constellation of Cygnus. The companion star is a luminous spectral type B star, which probably has a mass at least 20 times the sun's. From the 5.6-day period of the B star, we estimate the mass of its invisible companion to be from four to eight times that of the sun. The X-ray radiation is in very short and irregular pulses, and they could well come from hot gas in a rotating disk of material around a black hole. The evidence is still only circumstantial, but many astronomers think that the dark companion in Cygnus X-1 probably is a black hole.

Some Ideas about Black Holes

There are a lot of popular misconceptions concerning black holes. One is that black holes go around sucking things up. However, the gravitational field far from a black hole is exactly like that around any other star of comparable mass. Light and particles can be captured into orbit about a black hole or sucked into it only if they approach very close to it, within about twice the radius of the event horizon. That's a pretty small target. Even collisions between stars (of radii hundreds of thousands of times that of a black hole) are expected to be extraordinarily rare.

If a 5-solar-mass black hole collided with the sun, it would certainly rip the sun apart, but even then, I suspect, most of the solar material would not be captured by the hole. A black hole can accrete interstellar matter it plows through, but far, far less than a full-sized star can. It would take an incredibly long time for a black hole to chance to bump into and absorb a million stars, and even then it would be only a few times bigger than a normal star. Of course at that mass it could have rather dramatic gravitational effects. On the other hand, there is a possibility that a stellar system like a galaxy could contain a very massive black hole that collapsed directly from a large amount of material that never condensed into stars.

---

9. If the sun could suddenly collapse to a black hole, how would the period of the earth's revolution about it differ from what it is now?

It is another misconception that black holes are necessarily the densest things in the universe. Recall that the size of the event horizon of a black hole is proportional to the hole's mass. A 10-solar-mass black hole would have a radius of about 30 km, and only about one-third the density of a 1-solar-mass neutron star. If the entire Galaxy could collapse to a black hole, it would be less than one-tenth light year in radius, but would have a density hundreds of thousands of times less than that of water. It is just possible, in fact, that the entire universe could be one huge black hole. We shall return to this idea in Act VI.

Small-mass black holes, on the other hand, would be enormously dense—if they could be built. The earth, if crushed to a black hole, would be about 1 cm in radius, and a typical minor planet would be about the size of an atomic nucleus. Perhaps such microscopic black holes *can* exist. The British relativistic astrophysicist Stephen Hawking has suggested a mechanism whereby mini-black holes just might have been produced in the primeval fireball that began the expansion of the universe (or the "big bang")—provided, of course, that the universe *did* start off that way.

Hawking himself is a remarkable person. While still a young man, he contracted a progressive nervous disease, and now he has lost the use of his limbs entirely; even his speech is comprehensible only to his family and closest acquaintances. Yet, rather than waste time on self-pity, he has taken advantage of the opportunity to do a lot of thinking, and has turned his extraordinary mind to some of the deepest problems in modern science.

Evaporation of Black Holes

Hawking suggests that black holes can *evaporate*. They do this, in his theory, in an esoteric but quite plausible way. Recall that if a particle and antiparticle come into contact, they annihilate, turning into energy. On the other hand, with a large enough amount of energy, both a particle and its anti can be *created;* it's done commonly in high-energy physics experiments, and is called *pair production*. In fact, the modern view is that all space is permeated with *virtual pairs* of particles. These pairs are constantly materializing from nothing. But of course the energy to create them cannot come from nothing, so the particles cannot last. Ordinarily they annihilate each other almost instantly, so that we never observe them. They can exist as separate particles only for a time close enough to zero to be permitted within the uncertainty of the Heisenberg principle. (However, if outside energy is available at just the right moment, these very temporary pairs of particles can interact with it and produce effects that are observed in certain experiments.)

Near a condensed object, such as a black hole, the gravitational tidal force can be extremely strong, strong enough, according to Hawking, to pull apart a particle and its anti during that brief period of their existence. When this happens, one of the pair may fall into the black hole. Sometimes its companion follows it, in which case there is no net change. Other times, however, that other particle without a companion to annihilate it escapes immediately into space. Later, those particles and antiparticles leaving the vicinity of a black hole do annihilate each other, producing energy that may be observable. But energy (or the combination of mass and energy) cannot be created out of nothing, even by a black hole. That energy must come from somewhere, and it comes from the black hole itself. Since mass and energy are equivalent, the energy lost by the black hole lowers its mass.

It works out, however, that the separation of pairs of particles in this way is at a very slow rate except around those black holes which produce the highest tidal forces in their vicinities. It may surprise you that the strongest tidal forces are not around the most massive, but around the *least* massive, black holes. Although the most massive ones have greater total gravitation, it increases more gradually toward the event horizon, so the difference in gravitational acceleration across a small distance is less than near a small black hole. Those little fellows have extremely strongly warped spacetime in their immediate vicinities, so they cause the highest rate of pair separation and lose mass at the relatively highest rate. As they get smaller, the rate of particles leaving them increases, and their mass loss accelerates. Finally, as the last bit of mass leaves the black hole, it goes off in a brilliant flash of gamma ray energy that could amount to 10^{29} erg or more.

Now suppose, a la Hawking, black holes of many different masses were produced in the big bang. On the basis of the above theory, it works out that those of very small mass would have evaporated and gone off long ago, while those of high mass evaporate so slowly that they have scarcely changed. In fact, a black hole of mass greater than the sun's will last virtually forever. However, those mini black holes that started off with about 10^{15} g—about the mass of a small minor planet—would just about now be finishing their lives. Can it be that this is actually happening? It would be very exciting to find evidence of exploding black holes.

Summary

Well, that about wraps up our story of the life of a star. The star will end, we think, as a black hole, a neutron star, or a white dwarf. If it becomes a black hole, it is lost from sight forever. If it is a neutron star, and is rotating, it might be observed for a while as a pulsar, but pulsars spin down eventually and stop producing radiation; most of those pulsars known already emit only at radio wavelengths. If the star ends up as a white dwarf, which we think is the usual way, the atomic nuclei inside it can only gradually cool down, with the result that over many thousands of millions of years the white dwarfs will simply fade out, like the slow fading of a red-hot poker withdrawn from the fire. Some stars of low mass short-circuit the main sequence altogether, as we have seen, and they contract to white dwarfs directly. All white dwarfs, eventually, cool off to become *black dwarfs*—cold blobs of degenerate gas drifting about in space.

---

10. Look elsewhere in this book for the necessary data and judge which of the following main-sequence stars are likely to end up as a) a black hole, b) a neutron star, c) a white dwarf.
1) spectral-type O star,
2) B star,
3) A star,
4) G star,
5) M star.

So what about the solar system? The sun, unless we are much mistaken, will end up as a white dwarf. If we could come back and observe it then from the present distance of the earth (the earth itself, as we saw, will likely be swallowed when the sun becomes a red giant), we would see it at first as an intensely bright-appearing star. Although hundreds of times fainter than now, it would still deliver far more light then than the full moon does now.

But this, too, is temporary. Gradually the sun will fade out to blackness. By then most neighboring stars that now shine brightly will also be gone. Occasionally a low-luminosity main-sequence star might chance to pass close enough for us to see it with the unaided eye, but the familiar brilliant constellations will not be there. Yet, far off in the Milky Way, star formation will still occur, and the life processes of new stars will still be beginning. Even this cycle will go on for a long time, but not forever. It too must stop when all the matter of the Galaxy has finally been cycled, in its final stars, into black dwarfs, neutron stars, and black holes.

But this dark future is far, far away. The sun has a life expectancy of another 5000 million years, and there are thousands of millions of other stars like it. There's a lot of energy yet stored up in those stars—great celestial batteries—which for an almost unimaginable future can make life of another sort possible for those wise enough to live it. Let us turn, then, from the life of a star to the life the stars support.

ACTV

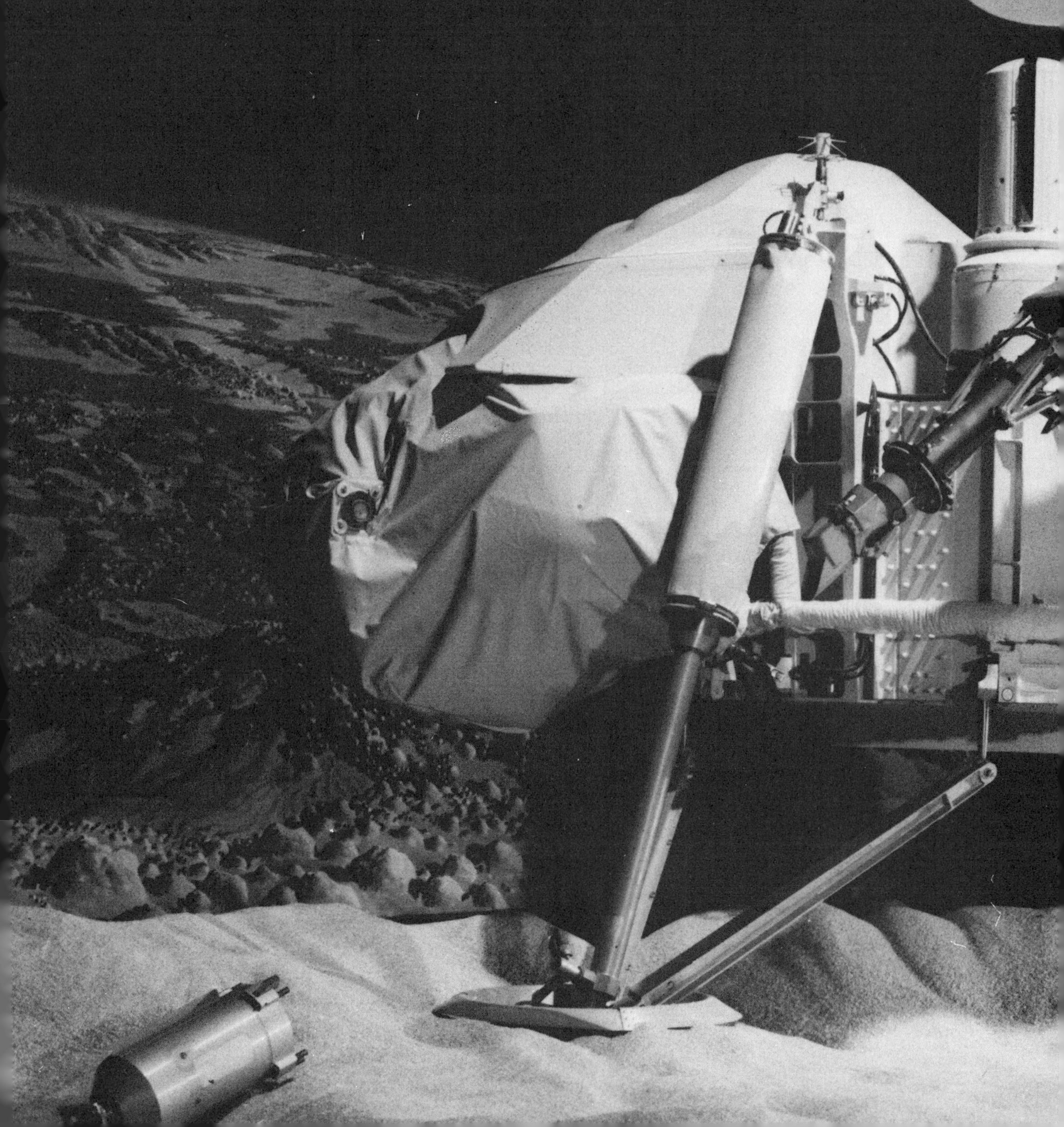

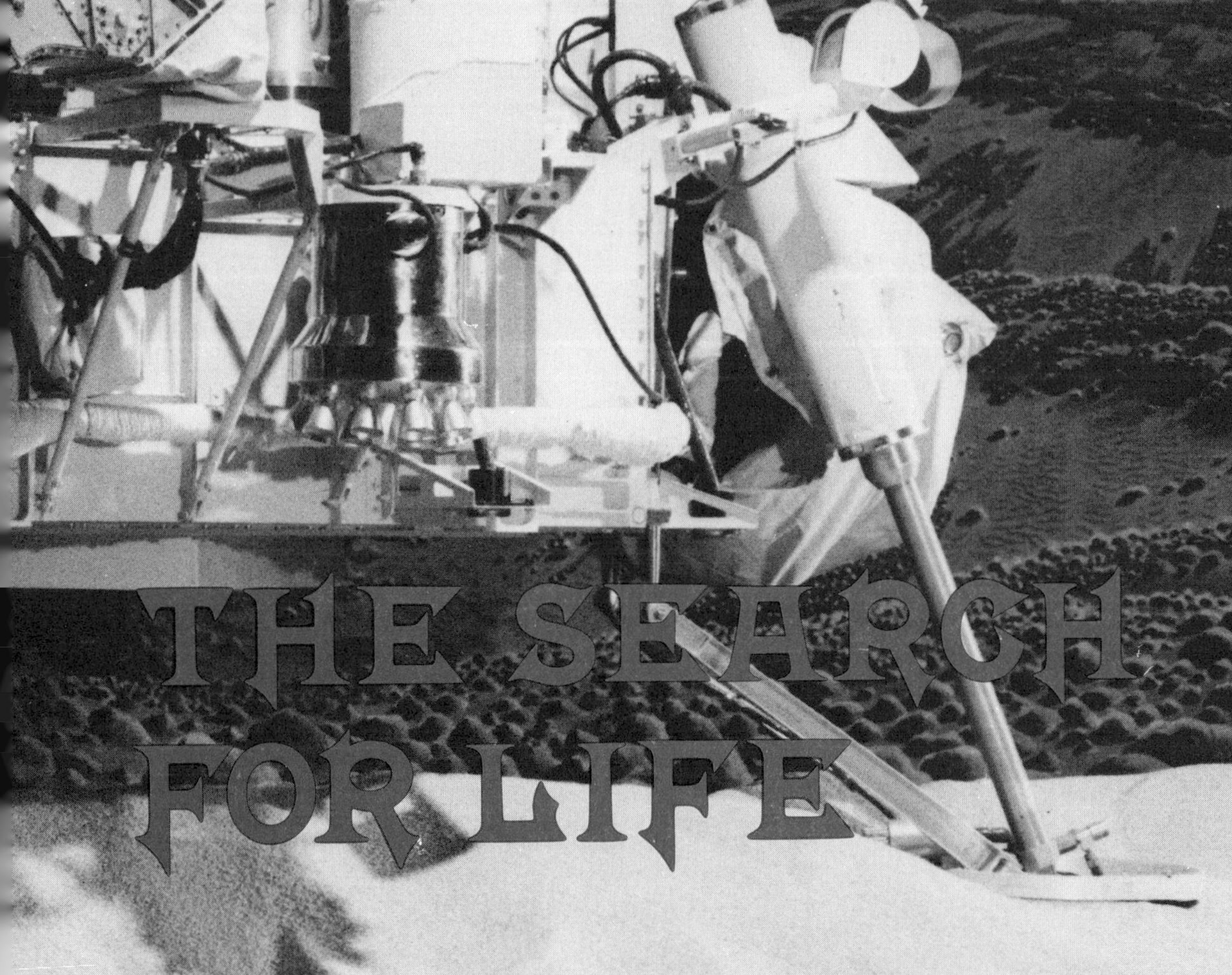

THE SEARCH FOR LIFE

The Living Earth

The earth, as we saw in Act IV Scene 5, formed by accretion of solid particles revolving about the protosun in the disk of the solar nebula. Being in the inner, hotter part of the disk, most of those particles were of nonvolatile metallic and rocky material, although volatile substances such as water, carbon dioxide, and nitrogen were contained in many of the particles in chemical combinations such as hydrates and carbonates.

As the young earth grew, and acquired its present mass and size, the weights of all those particles, bearing down on the interior, created a central pressure of some 4 million kg/cm^2. Much of the earth's interior, therefore, became compressed, hot, and molten. Most geophysicists believe that at this time the heavier atoms and molecules, especially of iron, sank inward, while the lighter "slag" drifted upward—a process called *differentiation*.

Moving to the upper portion of the earth's interior in the differentiation were materials containing aluminum, silicates, potassium, and even silicate and oxide crystals containing radioactive uranium and thorium. We have already seen how we measure the relative concentrations of those radioactive elements and their decay products to learn the age of the earth. Whereas, from such analyses the earth itself is found to be 4.6 thousand million years old, the age of the oldest crustal rocks in their present forms is about 3.7 thousand million years, thus showing that the rocks themselves were not present in the solar nebula.

Interior of the Earth

So we have the general picture of a dense *core* of mainly iron, surrounded by a rocky *mantle*. It is not surprising, then, that whereas the mean density of crustal rocks is only about 2.7, that of the earth as a whole is about 5.5. The best direct information on the internal structure of the earth, however, comes from analyses of seismic waves, vibrations released in earthquakes. Some of the waves travel along the surface of the earth; others are transmitted through its interior. When they encounter material of different density, the waves are refracted, somewhat as light waves are; some are also reflected at the bounda-

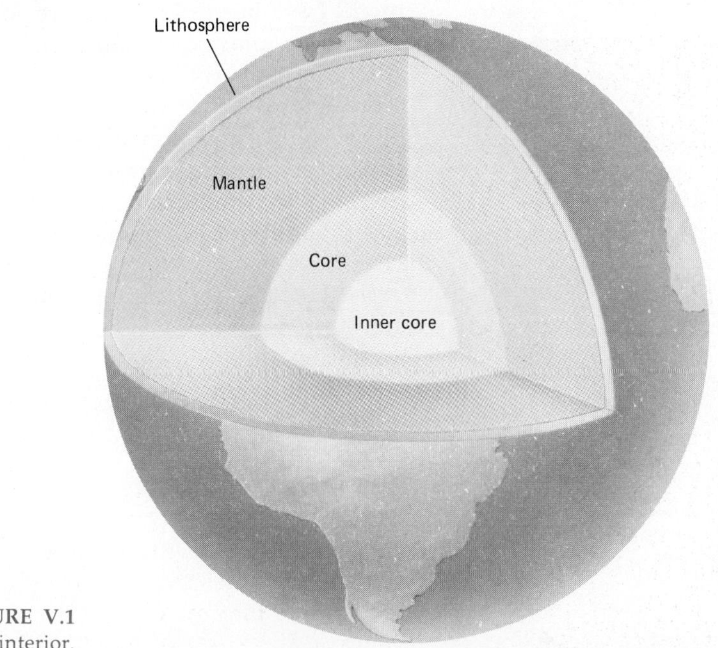

FIGURE V.1
The earth's interior.

ries of solid and liquid regions. Thus measurements of the speed, amplitude, and time of arrival of seismic waves from a remote earthquake enable seismologists to reconstruct a pretty detailed model of the earth's interior.

What is found is shown schematically in Figure V.1. The solid, rocky mantle, most of it with a rigidity of steel, extends down about 2900 km. Beneath the mantle is a liquid core, presumably of mostly molten iron. But there also seems to be an inner core about 2400 km in diameter that is highly compressed and hot, and probably of solid iron.

---

1. What fractions of the volume of the earth are occupied by the inner core, the entire core, and the mantle?

Atmosphere of the Earth

The earth's present atmosphere is not primordial, but came from the earth's rocks beneath the surface. There the heat produced by the radioactive decay of uranium, thorium, and the radioactive isotope of potassium (potassium-40, whose nuclei have 19 protons and 21 neutrons), decomposed hydrates, carbonates, and other compounds containing gases locked in chemical combination. The freed gases escaped to the earth's surface through a process called *outgassing*. Volcanos are the main mechanism of outgassing; when we realize that they are responsible for our vital atmosphere, we see that volcanos are our friends, not the frightful things they are often made out to be.

Water was the main gas to come to the surface; there it cooled and condensed and formed the oceans. The combined mass of the oceans (which in a sense are almost part of the atmosphere) is about one-ten thousandth that of the earth itself. Carbon dioxide was probably next in importance. Part of it dissolved in the ocean water, but being chemically active, most of it recombined with surface rocks. (It is interesting to contemplate what the earth might have been like if it were somewhat hotter so that the oceans had *not* condensed and the carbon dioxide had *not* reentered chemical combination. All that carbon dioxide left in the atmosphere would probably have produced a greenhouse effect like that which keeps the surface of Venue so hot and dusty.) Nitrogen appeared in the atmosphere, too, and although only about 1 percent as plentiful as the water, it remains the principal constituent of our present atmosphere. About 1 percent of the present atmosphere is argon; it is also thought to have outgassed. Argon is produced by the radioactive decay of potassium-40 in the crustal rocks.

The present atmosphere contains many other gases in trace amounts, the most important being water (in the gaseous state) and carbon dioxide. Today it also contains oxygen, but that came recently, as we shall see. There is another important respect in which the earth's primeval atmosphere is believed to have differed from our present one. Most experts are pretty sure it contained considerable amounts of methane and ammonia, and that these gases either outgassed or were formed by chemical activity on the surface. At any rate, from the time the earth had oceans of liquid water and an atmosphere, the process of erosion began, shaping and reshaping the faces of the continents many times over. The erosion also removed all traces of the great impact craters that must have been formed during the earth's final accretion stages.

---

2. In what respects are the earth's oceans like an atmosphere and in what respects are they different? How would your answer have differed if the earth's orbit lay between those of Mercury and Venus? What if it were beyond the orbit of Jupiter?

Terrestrial Magnetism

We have remarked several times that the earth has a general magnetic field. It has a strength (at the surface) of a little under one gauss, and rather resembles the field that is so easily revealed in the pattern of iron filings scattered on a paper over a bar magnet. The earth's magnetic field extends for tens of thousands of kilometers into space, but its strength drops rapidly with distance from the surface, and it gradually just fades into the general interplanetary magnetic field of the sun.

The earth's field is still strong enough, though, to trap electrons and ions at distances from 1000 to 20,000 km above the ground. These rapidly moving charged particles comprise a radiation discovered by early artificial earth satellites. Most of this radiation is in two broad belts surrounding the earth, called the *van Allen radiation belts* (Figure V.2), after James van Allen, the Iowa physicist who designed the satellite experiments that first detected them. Those particles causing the radiation are thought to originate from the solar wind and from collision products of cosmic rays with upper-atmospheric atoms.

It is remarkable that the earth's magnetic field is constantly changing its orientation. In fact, the earth's magnetic field has completely reversed itself 171 times in the past 76 million years. We detect these changes in polarity of the earth's field from the direction of magnetism in rocks. Molten rock, such

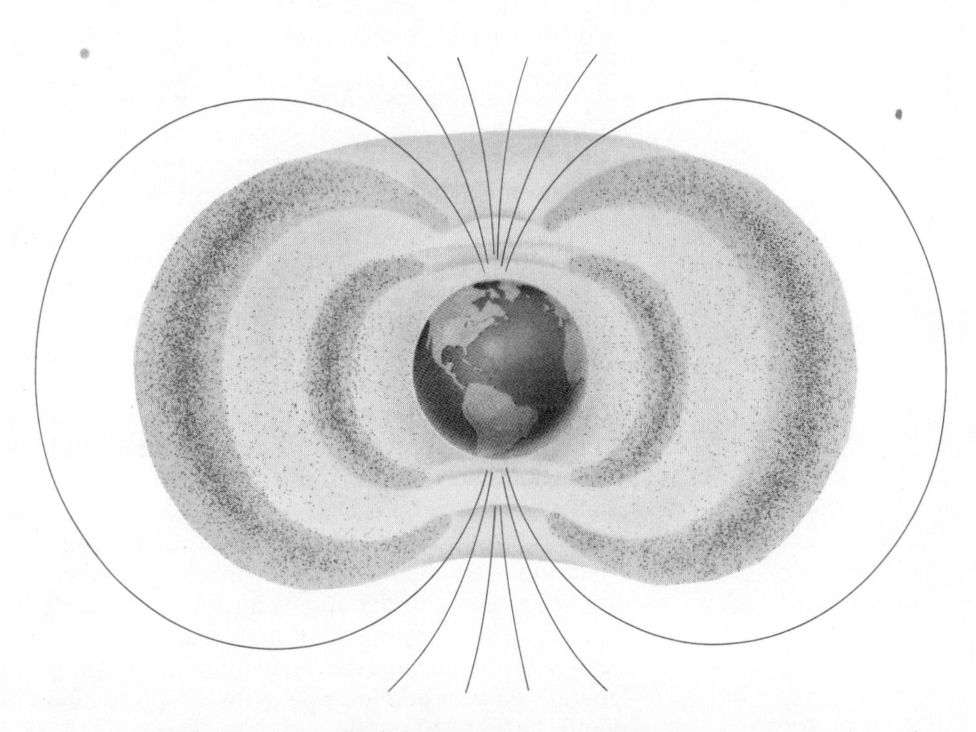

FIGURE V.2
Cross section of the earth's magnetic field and the Van Allen radiation belts.

as lava, contains iron compounds and is thus weakly magnetized in the earth's field. As it hardens, that magnetism is frozen into the rocks, serving as a permanent record of the earth's magnetic history. The earth's magnetic field is believed to be due to currents in the earth's molten core, but those circulations are unstable and change with time. It's a lucky break that these changes occur and are recorded in rocks, because they enable us to learn something extremely interesting about this active earth we live on, as we shall see shortly.

Plate Tectonics

Lots of school children, in studying maps or globes of the earth, notice that North and South America, with a little juggling, look as though they could almost be nestled up against Europe and Africa; it is as if those great land masses were once together but somehow tore apart. The same idea also occurred to the German meteorologist Alfred L. Wegener early in the twentieth century. However, he looked at the matter in considerable detail and made a pretty good case that the continents had, indeed, drifted apart. He based his arguments, in part, on mountain ranges and other geographical features that appeared on continents on both sides of the Atlantic and which looked as though they would be continuous if the continents were together. Well, nobody paid much attention to Wegener at the time, probably because too many school children had similar ideas.

Since the late 1950s, though, evidence has been pouring in that the continents really *are* drifting apart. Wegener had been right! The study of this phenomenon is called *plate tectonics,* and it has revolutionized geophysics just as the Copernican theory revolutionized astronomy.

The thin skin of surface rocks on the earth is called its *crust*. The crust extends to a depth of only 30 or so km, and even less under the oceans. The crust and upper mantle, to a depth of some 50 to 100 km, consists of a layer of the earth called the *lithosphere*. The lithosphere is not one solid piece, however, but consists of 10 major (and some minor) separate *plates*, fitting together something like jigsaw-puzzle pieces. These plates of the lithosphere float on a hot, plastic, and possibly partly molten layer just beneath them called the *asthenosphere;* it is from 100 to 200 km thick. The continents drift because the plates are moving around on top of the asthenosphere.

But how, you might ask, can a system of interlocking plates covering the entire surface of the earth move about? One might expect that they would not be free to circulate because of the presence of adjacent plates. The plates move about by jamming into their neighbors, and burrowing under them, right down into the mantle. Continents themselves don't actually burrow down under; when a continent reaches a plate boundary, it sort of stays there while the plate slides under it into the mantle. Far from a boundary a continent just rides along on top of its plate. It's a little like logs floating down a stream, and then jamming up against a low bridge over the water. When continents so crash into each other, though, they can build mountains.

But, you might wonder, if the plates go sliding around, bumping into each other and burying themselves, how can they cover the entire earth? Why don't they all dig under and disappear? Why isn't there a shortage of all that crockery? The answer is that the plates are being continually replenished, as quickly as they disappear. New material flows up from the mantle between separating plates and joins on to their trailing edges (or perhaps pushes them along). This is what happens at the *oceanic ridges* like the famous Mid-Atlantic Ridge. Sometimes all this activity occurs out of sight at the ocean floors, but sometimes some of the upflowing matter forms volcanic land masses. The whole nation of Iceland is sitting on the Mid-Atlantic Ridge and is currently being formed by new material from the mantle; that's why Iceland has all those volcanos and geothermal activity. Places where the plates burrow under are called *trenches*, a fine example being the Japan Trench, along the shore of Japan. Figure V.3 shows, schematically, a cross section of the earth

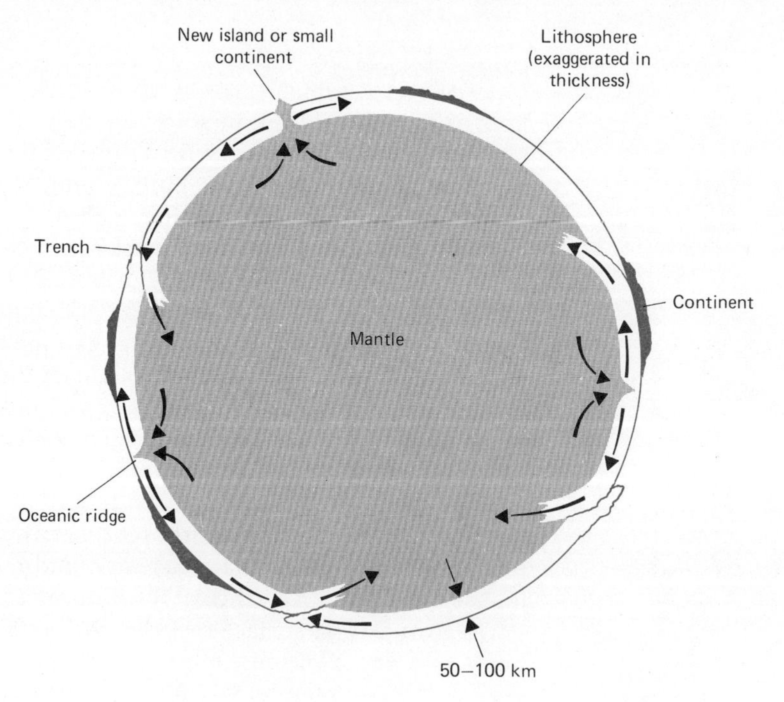

FIGURE V.3
The lithosphere (of much exaggerated thickness), showing the emergence and immersion of plates.

with the lithosphere exaggerated somewhat in thickness so you can see how this flowing out and digging in of material takes place.

Evidence for Continental Drift

Part of the evidence for the motions of the plates, giving rise to continental drift, comes from a reexamination of the coastlines of existing continents. Especially if the continental shelves are taken into account, the fits between continents that were once together are quite good. Geologically speaking, this tectonic motion of the plates occurs very rapidly and has profoundly rearranged the continents over the earth's surface in the last couple hundred million years, as can be inferred, for example, from fossil remains of tropical plants and animals in what are now polar regions of the earth.

The clinching evidence, though, is in the record of the changing magnetic field of the earth. Consider what happens when molten rock flows into the floor of the Atlantic along the Mid-Atlantic Ridge. It spreads out in both directions, hardens, and becomes part of the separating plates. As it hardens, it freezes in the magnetism of the earth at the time of its emergence from the mantle. We can measure the magnetism of the rocks on the ocean floor with delicate magnetometers towed behind ships, and see how their polarity switches back and forth many times as the ship moves away from the Mid-Atlantic Ridge. Since we have other records of the changing polarity of the earth's field, these Atlantic observations tell us just when the material in each part of the ocean floor oozed up from below and hardened, and how fast since then it has moved along with its plate.

Figure V.4 is a map of the world showing the outlines of the major plates. The many tiny dots along those boundaries (and the few elsewhere) are the sites of moderate and major earthquakes in recent years. As you might expect, most earthquakes occur at plate boundaries where the major disturbances due to the motions of the plates are occurring. The Nazca Plate,

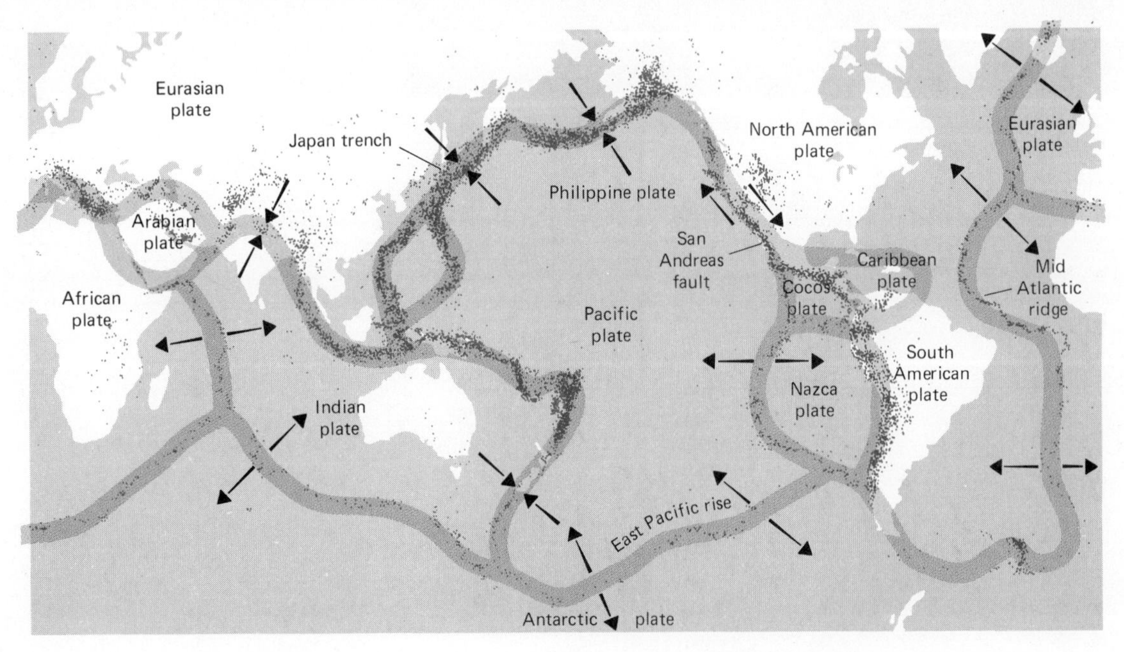

FIGURE V.4
Tectonic plates on the earth. The dots indicate regions of seismic activity, generally where the plate boundaries lie. The major plates are labeled, and the arrows indicate their direction of motion.

for example, burrowing under the west coast of South America, has produced all kinds of havoc while building the Andes. The Indian Plate, shoving into Asia, has built the Himalayas. A very seismically active area is Japan, where the Pacific Plate pushes under the Eurasian Plate at the Japan Trench.

---

3. Identify on the map in Figure V.4 other places besides Iceland where land masses may have been built up by the emergence of mantle material in an ocean ridge.
4. Identify on the map in Figure V.4 several places that you would expect to be very active seismically and several that you would expect to be very quiet.

If we extrapolate the plates' motions backward, we arrive, some 200 million years ago, to the supercontinent of *Pangaea*. At that time the continents of the earth were all together. The original Pangaea is shown in Figure V.5. The map may not be absolutely accurate, but will give you an idea how geophysicists think the continents lined up about the time early dinosaurs roamed the earth.

About 200 million years ago Pangaea broke up. Ever since, the moving plates have been separating the continents. The plates that consist mostly of ocean areas tend to move the fastest—typically about 10 cm/yr. At Easter Island, the Pacific Plate and Nazca Plate are separating at about 20 cm/yr. Those plates consisting mostly of continents are moving more slowly—typically 2 cm/yr, about the speed with which Europe and North America are separating. This may seem extremely insignificant, but consider that 2 cm/yr is 20 km/million years, or 2000 km/100 million years. Thus merely a few hundred million years separates Europe and North America.

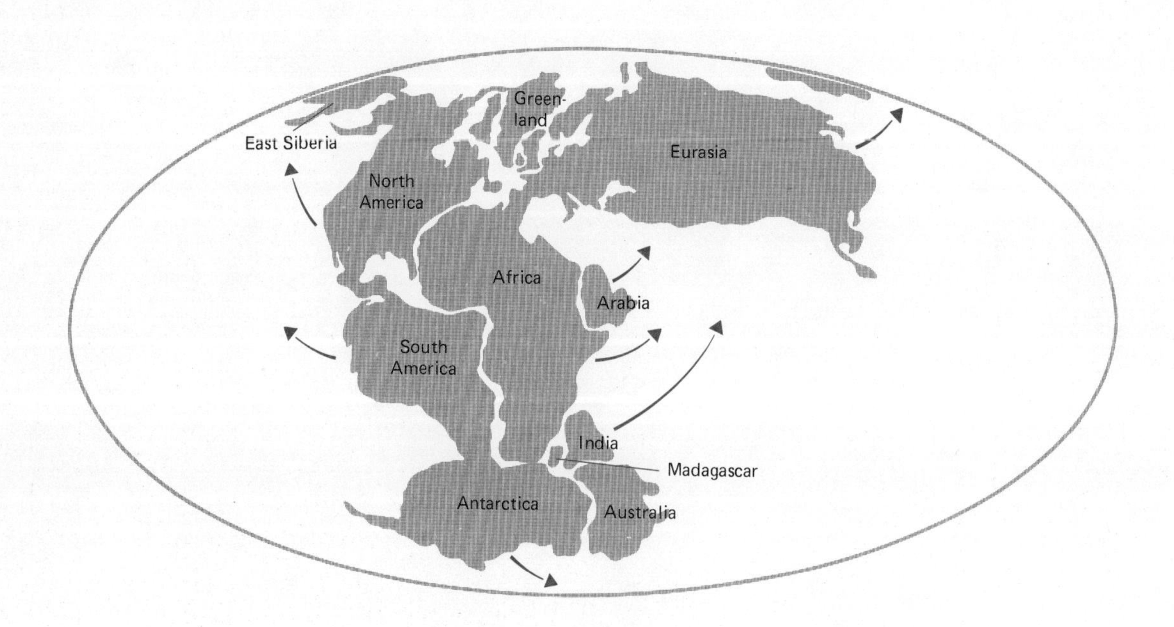

FIGURE V.5
A map of the earth as it appeared approximately 150 million years ago, when the continents are believed to have formed the supercontinent Pangaea.

5. If Europe and North America separate from each other at a constant rate of 2 cm/yr, how long a time has been required for them to reach their present separation of 4000 km?

Earthquakes

Because I live in Los Angeles, I have a special interest in the boundary between the Pacific Plate and the North American Plate. Naturally, where the solid crustal parts of the earth are pushing into each other or (in the case of California) are sliding against each other, fissures (*faults*, or cracks) in the earth's crust will result. One of the most famous of these is the *San Andreas Fault*, running from the gulf of California in the south to the Pacific Ocean just west of San Francisco in the north, and separating the Pacific Plate, containing southwestern California, from the North American Plate, containing the rest of the state. Displacements along it slowly carry Los Angeles northward; in several million years it will be on an island off the coast of San Francisco.

This creeping motion of the plates against each other builds up stresses in the ground around the fault, which are released in sudden slippages every 50 to 100 years, producing major earthquakes, including such famous ones as that in San Francisco in 1906. There is an impending one (at this writing) in the Los Angeles area.

FIGURE V.6
Photograph of the earth from space. *(NASA)*

Charles Richter (of the Richter scale of earthquake intensity) once said: "Some people have an irrational fear of cats, and others an irrational fear of earthquakes; the former should not have cats for pets, and the latter should not live in California." Dr. Richter (whom I met on an eclipse tour) has a fine sense of humor and was naturally partly kidding about living in California—but only partly. It is a region, like many on the earth, that is near or on a boudary between plates, and can expect many interesting seismic events over the years. Some may even be very damaging.

---

6. If you wish to move away from California to escape earthquakes, how about settling in Guatemala, Chile, Turkey, Italy, China, or Japan? Explain.

On the other hand, I relish the opportunity to live where I do. My time on this earth is short, cosmically speaking, and it is exciting to witness some of those grand changes that shape its face. If we could see a time-lapse motion picture of the earth's surface, with millions of years collapsed to a few seconds, we would find it a tremendously alive place, with its entire surface constantly rearranging itself; the motions of the plates would seem smooth and continuous. It is only on our tiny time scale, confined to barely more than a snapshot of the earth's cosmic existence, that things seem so abrupt and chaotic. In witnessing an earthquake, we are witnessing a tiny sliver of the evolution of the earth.

The Ice Ages

Other gradual changes on earth are the glacial periods, or so-called ice ages. Maybe they aren't so gradual at that; there is some evidence that during the onset of the last glacial epoch it started snowing, and didn't stop for weeks and weeks, piling up one meter of snow per week. In Russia, mammoths were found frozen in the ice with food still in their mouths and stomachs. Imagine the result of such an occurrence today in Europe or America. Unusually heavy snows (averaging about ⅓ m/week) in Buffalo, New York, in January 1977 left the city almost helpless, but that was relatively local and temporary (we hope!).

---

7. Consider what would happen in your town or city if snow fell at the rate of 1 m/week over the entire North American continent. Do you suppose the snow could be cleared fast enough to keep communications open? How? How do you imagine your community could survive?

It is too soon to be sure, but there is some hope now that we understand what brought on the previous ice ages. Because of the perturbative gravitational influences of the other planets, the earth's orbit is constantly changing, as is the tilt of its axis to the plane of its orbit (that is, the obliquity). Recent investigations show highly suggestive correlations between these periodic orbital changes in the earth's motion and glacial activity on earth over the past 450,000 years, as read from geological records. If the indications are correct, and investigators are not at all sure yet that they are, it means that we should now be approaching a new ice age, but that the cooling trend extends over the next 20,000 years. Thus we should have plenty of time to check out the theory, correct for its shortcomings, and plan for the future. But there is a catch: All this ignores what some authors call *anthropogenic effects,* which could well have a short-range and immediate effect. We shall return to this matter in Scene 5; meanwhile, I leave you to find out what "anthropogenic" means.

Origin of Life

It seems to me to be almost a miracle that throughout most of the earth's history, there has been life on it. Not very advanced life, to be sure; only in the past few hundred million years have there been complex animal forms; but at least single-cell, blue-green algae and similar organisms have been found in rocks as much as 3400 million years old. So far as we know, the earliest organisms required the existence of liquid water on earth to form, feed, and reproduce. Only on earth, of all known places, do we now find liquid water. On the other hand, under the present conditions on earth, we doubt that life could have ever begun.

Most biologists who have studied the origin of life believe that it required a *reducing* atmosphere, rather than the present *oxidizing* one. However, the earth almost certainly had no oxygen in its primordial atmosphere, and the best guess is that it had methane and ammonia, as well as water and carbon dioxide, to give it the right chemical makeup for synthesizing simple organic molecules necessary for the development of life. Biologists refer to the early conditions of an ocean of water and atmosphere of water vapor, carbon dioxide, methane, ammonia, and nitrogen as a "soup." Laboratory experiments show that if such a soup is irradiated with ultraviolet radiation, organic molecules such as sugars and amino acids are produced with large yield.

Three thousand to 4000 million years ago the earth, presumably, had no oxygen, and hence no ozone layer (ozone has molecules consisting of 3 atoms of oxygen) to filter out ultraviolet radiation from the young sun. Thus we would expect those prebiological organic molecules to be formed. That much seems almost inevitable. Somehow those amino acids and other molecules combined into proteins and eventually into the highly complex and long helical molecules we call DNA, those which form the chromosomes of living cells and which have the power of reproducing themselves. We cannot say whether their original formation was also almost inevitable or whether it was an incredible long shot. Anyway, it happened; then began the long road of biological evolution that took more than 2500 million years to produce animal organisms with different cells serving different functions, and another 1000 million years to produce the earliest manlike creature.

There is an interesting thing about the organic molecules in living organisms on the earth. The basic molecules are helical—corkscrew shaped—and the more complex ones are built up of these simpler ones. They are all with the right-hand thread; that is to say, in all living organisms, the helical molecules twist the same way. It's easy enough to understand why this should be so. In the early development of life, some molecule with the ability to reproduce itself got formed; it had to be of either right-hand or left-hand thread—it couldn't be both at once. It just happened to have a right-hand thread; it then fed and reproduced, and fed and reproduced, and had mutations, and natural selection, and all that, and eventually ended up with something holding a queer astronomy book in its hand. If life had sprung up independently all over the earth, we might expect by chance to find some forms with right-hand and others with left-hand helixes in their chromosomes. But all of us, from the lowliest bacterium to the biggest elephant, and even you, are made of organic molecules twisting the same way. Does this mean we all evolved from the *same* DNA molecule? And was its name Adam?

---

8. If a light beam is passed through a solution of natural sugar (cane or beet) in water, the light is always circularly polarized because the sugar molecules all twist the same way. Now, sugar is a simple molecule that can be synthesized in the laboratory. Synthetic sugar, however, has molecules with right- and left-hand twists in equal numbers. Light passed through a solution of synthetic sugar, therefore, is not polarized. However, if bacteria are introduced into the solution of synthetic sugar, after a while the solution polarizes light, but in the opposite sense from a solution of natural sugar. Can you suggest an explanation?
9. Helical organic molecules have been found in meteorites, but such molecules are found in equal numbers that twist to the right and to the left. Why do we conclude that these molecules are probably not of biological origin?

Oxygen in the Atmosphere

Eventually organisms formed that filled the oceans and absorbed carbon dioxide; they used the carbon contained therein to build themselves, and released the oxygen into the atmosphere. Thus oxygen gradually began to build up in the atmosphere. It was at first very gradual, and 1000 million years ago, oxygen was probably a very minor constituent of the air. In fact, 90 percent or more of the earth's atmospheric oxygen has likely been produced in the past several hundred million years from this photosynthesis by green plants.

The production of oxygen, however, is always competing with its absorption. Oxygen is absorbed in respiration of animals, of course, but even today that is a fairly minor effect. Mostly, it is absorbed in the oxidation of dead vegetable and animal material. That is, biological matter rots when it dies—which means that it eventually oxidizes, the dead plants reabsorbing the oxygen they released during life. The only reason there is any net production of oxygen at all is that a small fraction of the plant material, probably one part in ten thousand to one part in one hundred thousand, escapes decay by being conserved as carbon, buried in the earth. This material is our fossil-fuel supply—coal and oil. The present atmosphere of the earth contains about 21 percent oxygen, built up gradually by the slight imbalance of photosynthesis over the decay of dead organic matter. If the only source of oxygen is photosynthesis (we think it is the only important source) and if we could burn up (that is, oxidize) all fossil fuel on earth, we should break even with the bank and reduce the oxygen in the atmosphere to nil. At the present rate of our burning of these fuels, this should take some thousands of years; we shall see in Scene 5 that there are more immediate problems.

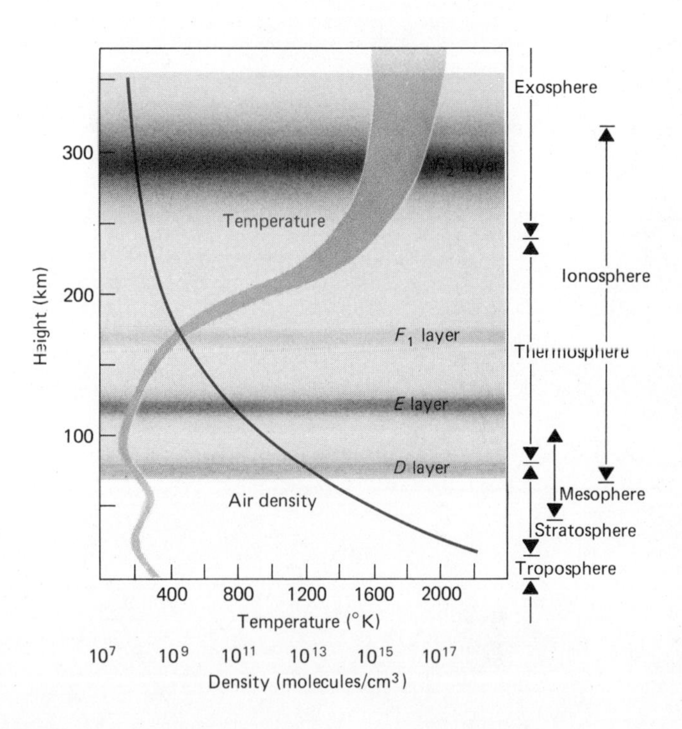

FIGURE V.7
The structure of the earth's atmosphere. The range of temperatures measured and estimated at various altitudes is indicated by the shaded band. The variation of density with altitude is shown by the solid line. The horizontal shaded zones are the regions of the atmosphere where the ionization of the air is especially high.

Levels of the Atmosphere

The present structure of the earth's atmosphere is summarized in Figure V.7. We see that the density of the air (shown by the solid line) drops rapidly with altitude, but that thin vestiges of the atmosphere extend to altitudes of hundreds of kilometers (actually, to about 1000, as shown by the altitudes of the highest aurora, or northern lights). The temperature of the air also drops with increasing altitude near the earth's surface, but it climbs a bit in a region called the *stratosphere*. Then it drops again up to an altitude of about 100 km. At higher altitudes it rises once more, and at 300 km the temperature is between 1000° and 2000°. There the air is heated by ionization of upper-atmospheric molecules by ultraviolet radiation from the sun. Layers of ozone in and above the stratosphere shield the ground from this rather toxic radiation. Other features of the atmosphere include layers where the air is ionized, and these ionized layers are effective in reflecting radio waves back to ground. Before the communications satellites, the ionized layers were important for long-range communication. Most weather occurs in the *troposphere*, the bottom 10 or 20 km of the atmosphere. Note that the ozone layer depends on oxygen, which in turn depends on living organisms; life itself has influenced the evolution of the atmosphere in a profound way.

Summary

Figure V.8 is a figurative railway track, built as the earth has aged. Each tie in the track corresponds to 2 million years. There are 2300 ties so far, and events in the history of the earth are marked at various ties along the way. Suppose typical railway ties are about 1 m apart and that the last tie is being laid as I write these words. The second to last one, 1 m back, was laid when manlike creatures first stood upright on earth. Only a few centimeters from

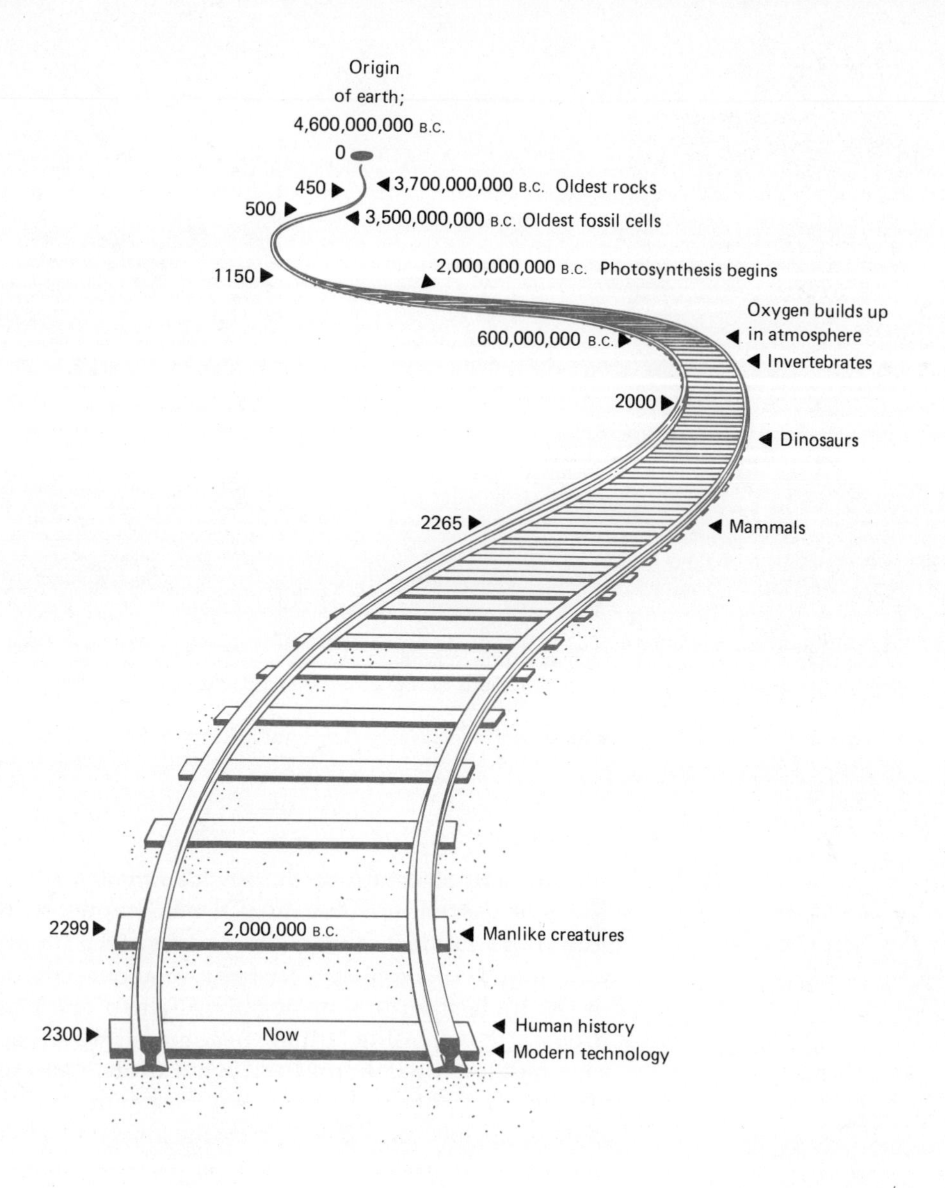

FIGURE V.8
A symbolic railway through time.

the front end of the last tie is where the history of man begins, as gleaned from his drawings and crude writing. Modern history begins a few millimeters back. Now bear in mind that the entire length of track is more than 2 km. Yet our technological society goes back only 200 microns—0.2 mm—just the thickness of that grain of dust on the front side of the tie that was just laid.

In the time represented by that grain of dust, man has drastically changed his environment. Only in the past few years have a few of us begun to realize that we may have permanently altered, to the disastrous worse, that fragile, thin, atmospheric envelope that protects us from the outside universe and makes our own existence possible.

But to the earth itself, chugging along with its plates burrowing under each other, switching its magnetic field around chaotically, having ice ages come and go, and whose existence is pretty much assured for 5000 million years, what man does to himself is a matter of no consequence whatsoever.

SCENE 2

The Moon: A Resident Alien

It is the nearest celestial world; yet it is an astronomical body—an alien world, far off in space. At least it seemed so to me when, as a boy, I first became interested in astronomy. How marvelous, I thought, if man could someday travel to the moon. Being a visionary of sorts I thought that probably one day he would, but of course not in *my* lifetime.

It seems incredible to me that most readers of this book have grown up to take for granted that we can go, and *have* gone, to the moon. Its over now, though; we have (in 1977) no further plans for the manned exploration of space. But at least we literally *saw* man set foot on the moon! I figure my share of the cost was a little over $100, and it was worth every penny. I probably spent more than that going to *movies* about man going to the moon.

If there are people living on earth 1000 years from now, and if they have history books to read, I suspect they will be relatively uninterested in the defects of our society, just as we are quite uninterested in the fact that the Egyptians and Greeks had slaves. We remember the Egyptians for their geometry and pyramids, the Greeks for their sculpture and Acropolis, and the Chinese for the Wall. I think we shall be remembered far less for the Vietnam War than for being the folk that first went to the moon.

But beyond all that, we learned something about another world. Only a very few percent of the Apollo effort went to science (about 5 percent, I understand), but even that small investment will provide us with material for study for years to come, and will reveal more and more about the nature of the solar system, and of our own origins. What a field day it has been for geologists, planetologists, and those relatively few astronomers who professionally study the moon! Not only have we brought back soil samples and rocks for detailed study, but we have set up spectrometers, magnetometers, solar wind detectors, ultraviolet cameras, seismometers, laser reflectors, and a host of other instruments, and have made the moon a laboratory for the study of the universe that could never have been created on earth.

We have learned that the moon was formed about the same time as the earth (4.6 thousand million years ago), and in about the same part of the solar system. It is made of the same basic stuff as the earth and has about the same

FIGURE V.9
The third quarter moon.
(Lick Observatory)

kinds of rocks. There are differences, of course; the moon is smaller than the earth, having only 1/81 its mass and 27 percent its radius. (The moon's mass is 7.35×10^{22}kg, and its diameter is 3476 km.) Being smaller and having less mass to weigh itself down, the moon is less dense than the earth—only 3.34 g/cm^3, compared to the earth's 5.5 g/cm^3.

Lack of Lunar Atmosphere

The small mass of the moon accounts for most of its differences from the earth. It has too little gravitation, for example, to hold an atmosphere. Being at the earth's distance from the sun, if the moon had an atmosphere, it would have about the same temperature as the earth's atmosphere, and the molecules that comprise it would be moving about the same speed as well. However, the earth has a velocity of escape of about 11 km/s, which means that a molecule (or anything) must be moving that fast to fly permanently away from the earth. On the moon, the velocity of escape is only 2.38 km/s.

Now the molecules in a gas are not all moving at the same speed; their average speed depends on the temperature of the gas, but some molecules

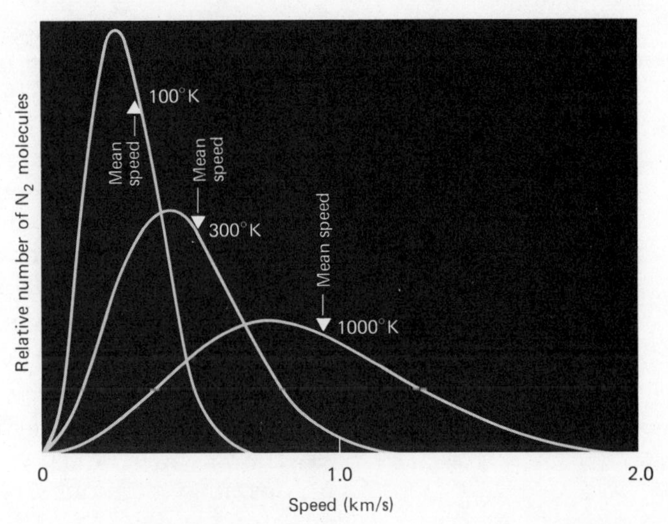

FIGURE V.10
Relative numbers of nitrogen molecules with various speeds for three different temperatures.

are moving faster and some slower than the average; and by virtue of frequent collisions between the molecules, each one changes its own speed many times a second. Figure V.10 shows the relative numbers of nitrogen molecules (the main constituent of the earth's atmosphere) moving at various speeds for three different temperatures.

From the kinetic theory of gases we can calculate how many molecules at any given time are moving at speeds greater than the velocity of escape from a planet. Most of those molecules do not actually escape, however, because many of them are moving downward toward the surface or are deflected downward by collisions with molecules above them in the atmosphere. However, some of those in the upper atmosphere moving in the right directions *do* escape. As other molecules, high in the atmosphere, obtain high enough speeds through recent collisions, they too escape. Gradually, the atmosphere of a planet can diffuse away if enough of its molecules are moving at a higher speed than the planet's speed of escape. Calculations show that in a thousand million years or so, a planet will lose its atmosphere if its molecules are moving at an average speed of at least one-fifth the escape speed.

At a temperature of 300°K, the mean velocity of nitrogen molecules is about ½ km/s. This is so much less than the velocity of escape from the earth that

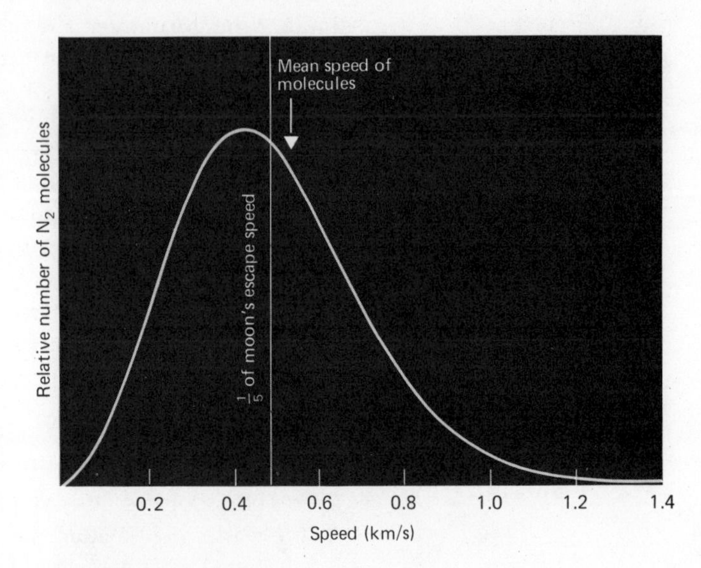

FIGURE V.11
On the moon, the mean speed of nitrogen molecules would exceed one-fifth the velocity of escape.

the earth can hold nitrogen almost indefinitely. The mean speed of nitrogen molecules, however, is greater than one-fifth the escape speed from the moon, so the moon could not hold that gas (see Figure V.11). In fact, the moon would not be able to hold any of the common gases in an atmosphere around it. Thus the moon can have no atmosphere nor any liquid water on its surface (water would evaporate in the vacuum, and escape as well). Thus there can be no weather on the moon, and no weather erosion. This is the reason why features thousands of millions of years old are still preserved on the face of the moon, revealing its geologic (or rather, *selenographic*) history.

---

1. According to the kinetic theory of gases, at a given temperature molecules of the smallest masses have the greatest average speeds. What specific gases would you expect the moon to have the best chance of retaining in an atmosphere?

The Moon's Motion

The moon revolves about the earth (with respect to the stars) in 27.32 days. However, because the earth moves in its orbit nearly 30° during a lunar revolution, carrying the moon along with it, the moon does not complete a revolution of the earth with respect to the sun—that is, a cycle of phases (say, from new moon to new moon)—until an average of 29.53 days. In other words, the period between new moons or full moons is a bit longer than a true revolution period of the moon; it is exactly analogous to the reason why a solar day is longer than a true rotation of the earth, which we discussed in Act III Scene 1. The true revolution period of the moon (with respect to the stars) is called the *sidereal month*, and its mean period of revolution with respect to the sun is called the *synodic month*. Figure V.12 explains it all.

The moon also rotates as it revolves about the earth, but because of the earth's tidal force acting on the moon for thousands of millions of years, the moon keeps the same side (approximately) turned toward the earth. That is, its period of revolution is equal to its rotation period. This means that any one place on the moon has sunshine for about two weeks and darkness for a similar time. Moreover, the moon has no air to blanket in the heat at night or to protect its surface from the heat by day. Consequently, the moon has a very great diurnal range of temperature—from about the boiling point of water (100°C) in the daytime to about −173°C (100°K) at night.

Thus the moon suffers large temperature changes each day. Also, it is subjected to the full force of cosmic rays, the solar wind, and the solar ultraviolet radiation. Furthermore, meteorites hit it from time to time, and such collisions were clearly far more frequent in the past. Thus there *is* erosion on the moon, but not the efficient kind caused by weather. Unlike the earth, the moon still has features visible on its surface that were formed during the first few hundred million years of its existence. On the other hand, because *some* erosion occurs there, we can tell which are the very old features.

---

2. It is commonly said that the dark side of the moon is turned away from the earth. Is this statement true? Why or why not?

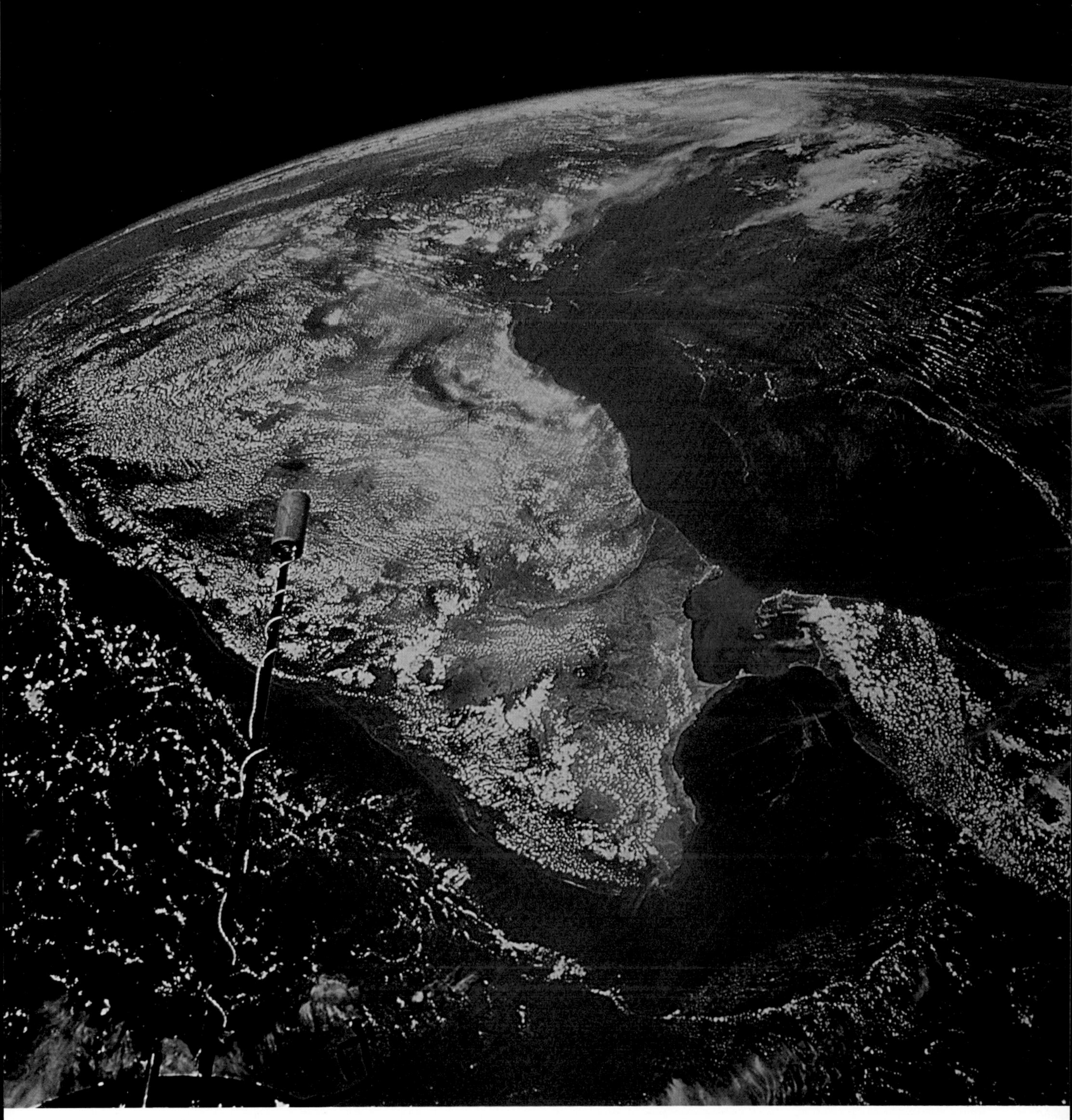

India and Ceylon, looking north with the Bay of Bengal on the right. Gemini XI photograph taken from an altitude of 410 nautical miles. (*NASA*)

New York City and Long Island as seen from Skylab III. (*NASA*)

Skylab II, 1972–1973. (*NASA*)

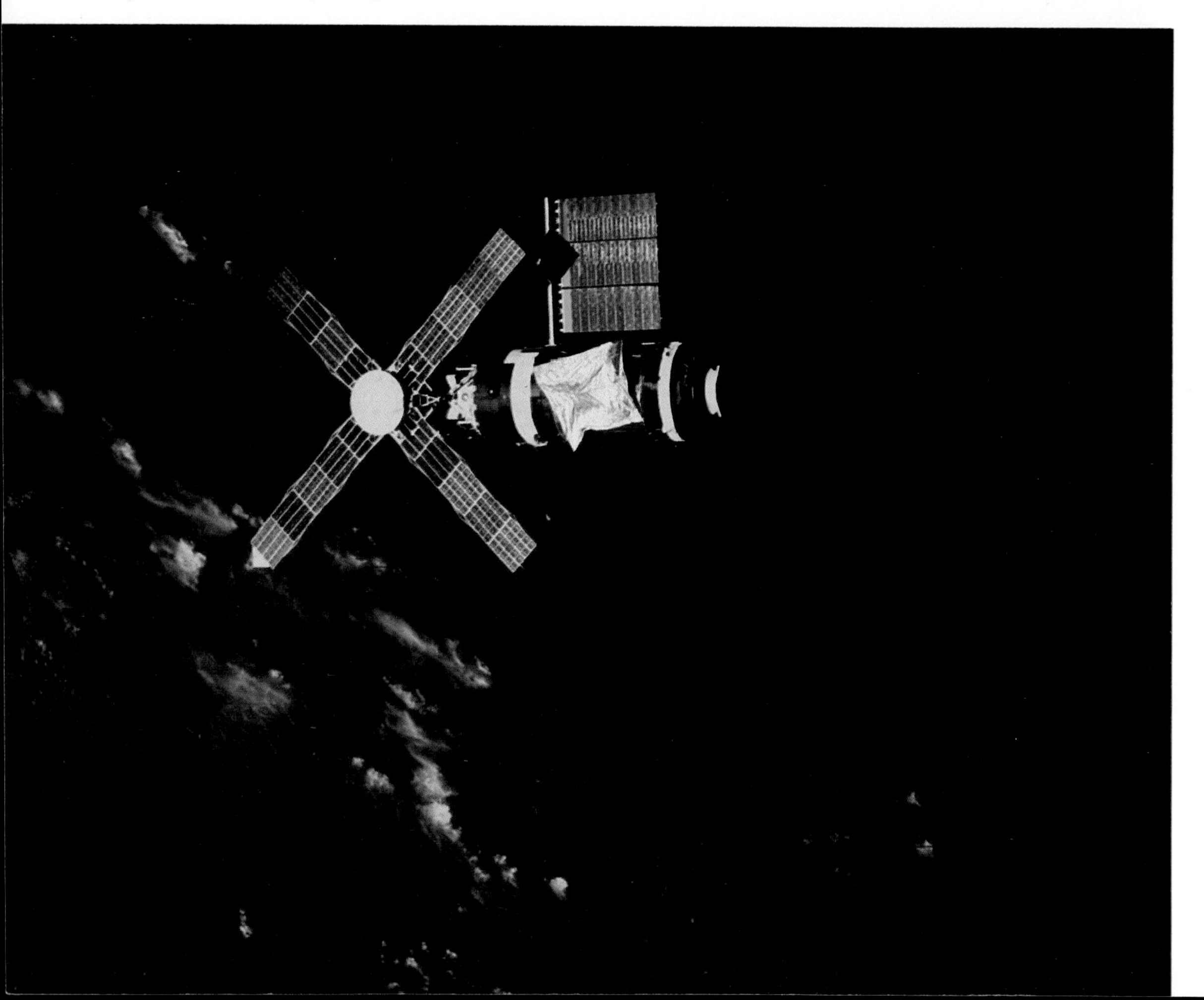

3. With what period would an equatorially mounted telescope (one with an equatorial mount) on the moon have to rotate about its polar axis in order to track the stars?
4. Describe the location and appearance of the earth in the sky as seen from the crater Copernicus on the moon. (Copernicus appears near the center of the moon's limb as seen from the earth.)
5. Describe the phases of the earth during the month as seen from Copernicus. At what phase of the moon does new earth occur?
6. Describe the phenomena observed from the crater Copernicus during a lunar eclipse as seen from earth. How about during a solar eclipse as seen from the earth?

FIGURE V.12
Synodic and sidereal months.

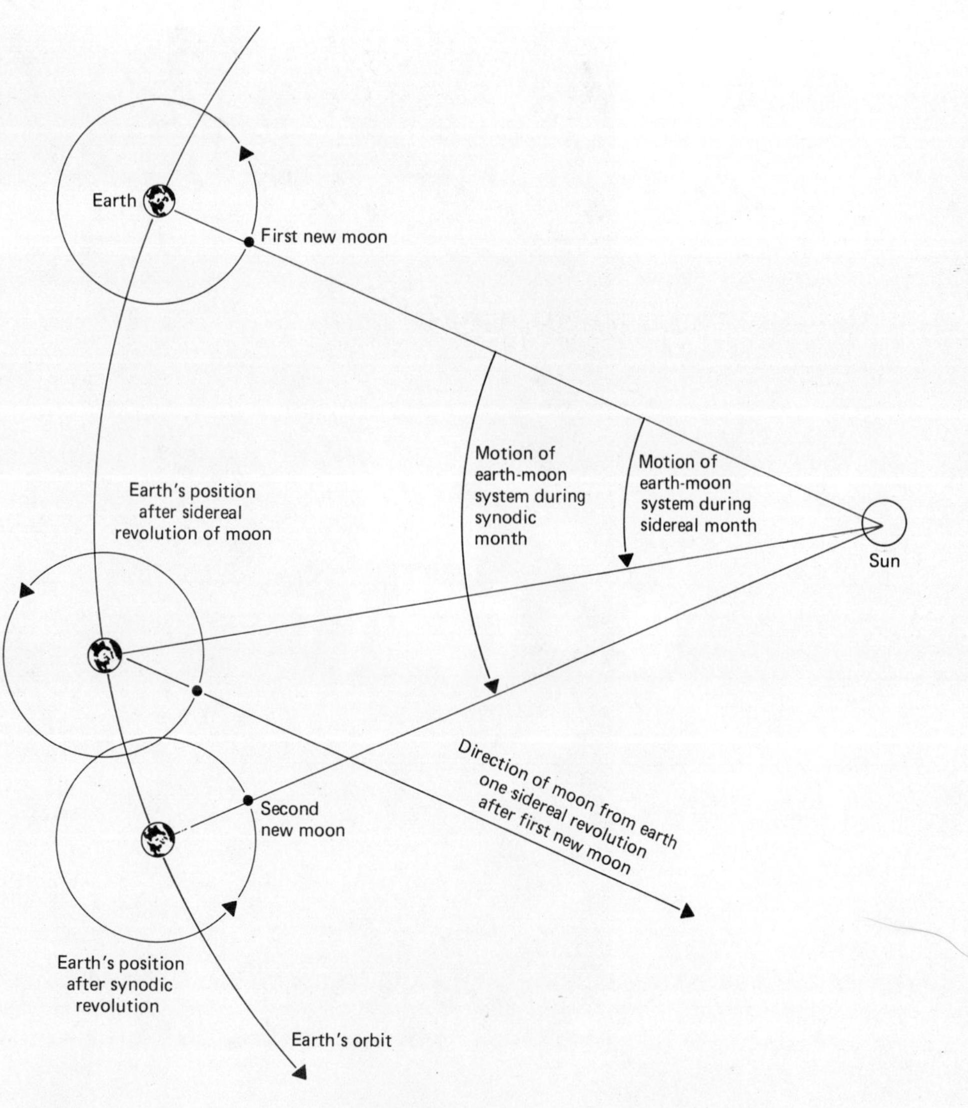

The Lunar Surface

The surface of the earth is highly evolved. It has oceans, resulting from vulcanism; mountains, resulting largely from plate tectonics; erosion features, due to weathering; and expansive evidence of vegetation. In fact the earth has preserved practically nothing of its original appearance. The moon, in contrast, is very primitive. It shows highly cratered regions—evidence of countless collisions with meteoritic bodies—and regions flooded with hardened lava, now covered with fine dust. And that's about it for the moon. There are the highlands—the mountainous cratered regions—and the lowlands (comprising about 17 percent of its surface)—the dried lava basins called *maria*.

The maria (latin for "seas") were so named by Galileo. He suspected that they were bodies of water—oceans, like those on earth. Other lunar observers quite naturally thought the moon was far more like the earth than it really is, and named many lunar features for terrestrial places. The maria, however, were given such names as the "Sea of Showers," the "Sea of Tranquility," the "Cloudy Sea," and the "Serene Sea." They're seas, all

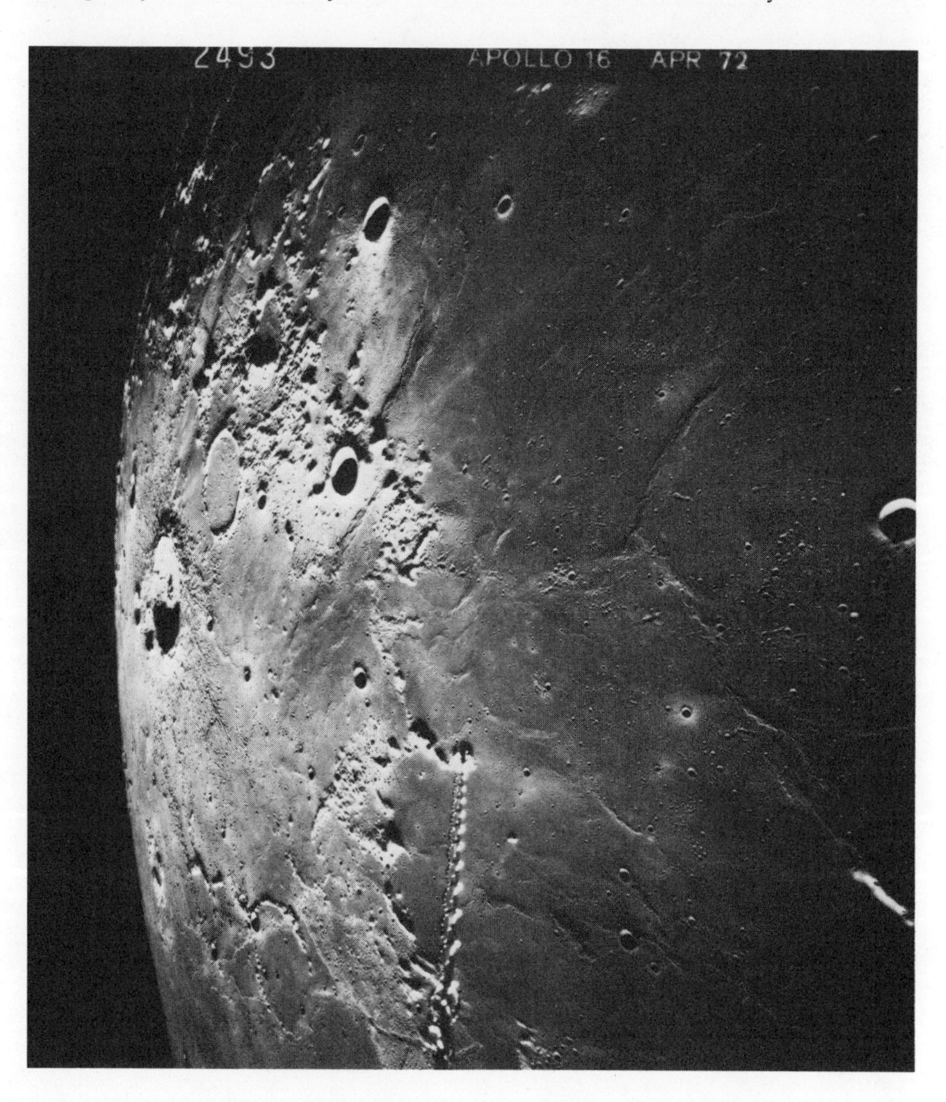

FIGURE V.13
A closeup of a part of a lunar mare photographed by the Apollo 16 crew. *(NASA)*

FIGURE V.14
The Alpine Valley, photographed by Lunar Orbiter V. *(NASA)*

right, but not of water; they are, rather, dust-covered seas of lava that flooded large areas of the moon's surface, probably when minor-planet-type bodies a hundred or more kilometers in diameter crashed into the moon more than 3000 million years ago. The maria also are the features of the "man in the moon."

There are ranges of great mountains on the moon, some of them nearly as high as the Himalayas in Nepal. But they also seem to have been formed by the impact of massive bodies with the moon, and not by crustal folding and lifting caused by plate tectonics, as on earth. The steep-walled valleys on the moon, such as the *Alpine Valley* in the lunar *Alps,* were not cut by running water, as were their terrestrial counterparts, but evidently by subsidence of blocks of the lunar crust.

Among the most conspicous features of the moon are its craters. Even those big enough to see from the earth number in the many thousands; the largest are about 1000 km in diameter. The craters of smaller and smaller size increase in frequency, down to potholes less than 1 m across. Most of the large craters are very ancient, and many are nearly obliterated by subsequent cratering on top of them by lava flows and by dust filling them in. A few large craters, though, are relatively young, possibly less than 10^9 years old. These

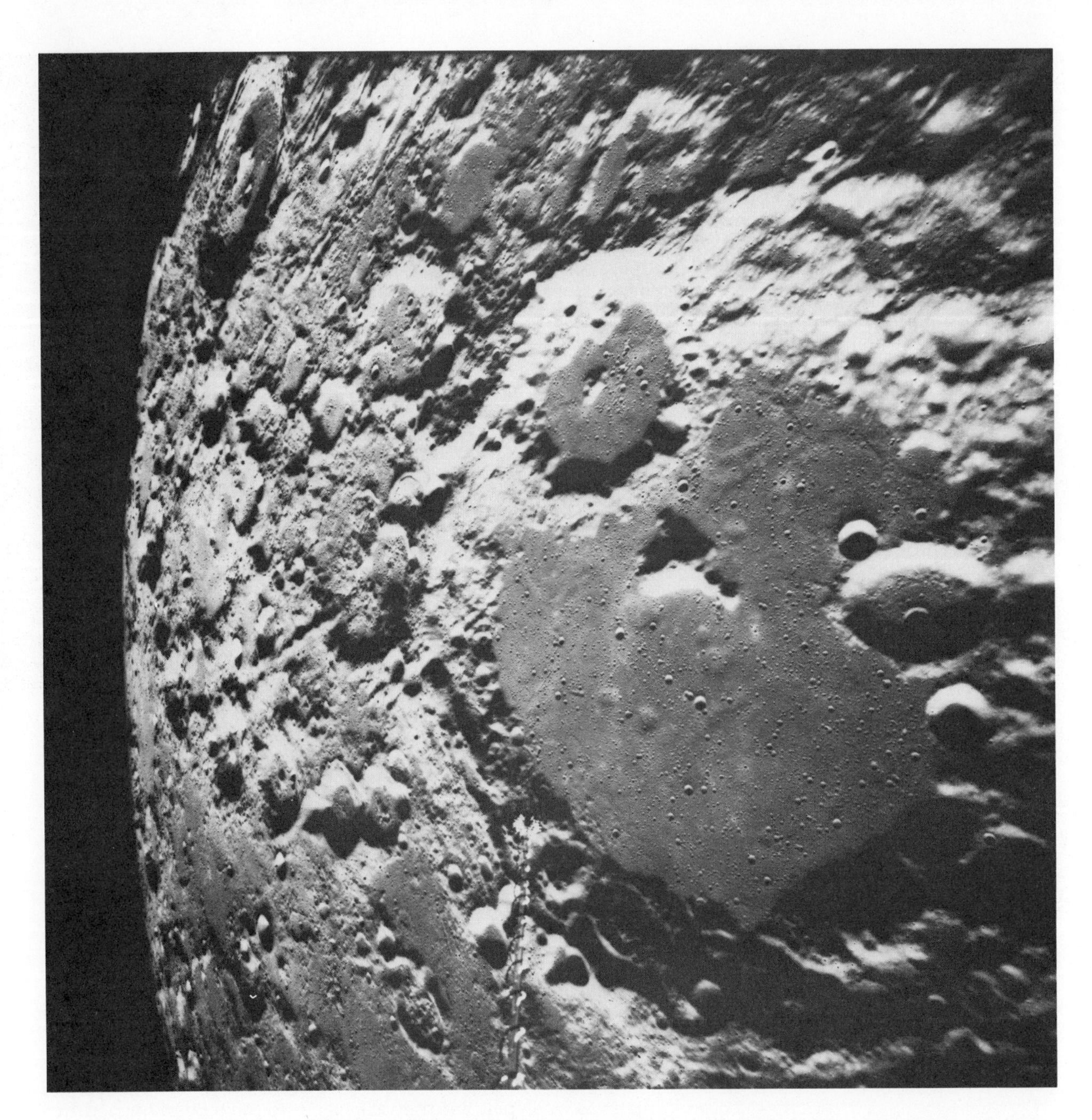

FIGURE V.15
A region of cratering on the moon, photographed by the crew of Apollo 16. *(NASA)*

include Tycho and Copernicus. In the impact explosions that formed them, large amounts of matter were splashed out, some of it leaving fine streaks, or *rays,* extending for many thousands of kilometers around the moon's surface. We shall see that the moon is not unique in being scarred with many impacts.

Nature of the Lunar Surface

The rocks on the moon are all igneous—at least judging from the many samples brought back by the Apollo astronauts and the Soviet unmanned landers—meaning that they formed by condensation from molten material. Thus they are not the primordial stuff of which the moon accreted; rather,

FIGURE V.16
A bleak lunar landscape from the window of Apollo 11 landing module. *(NASA)*

they formed from material that had been heated and melted. The rocks from the maria are basaltic—just cooled lava. On the other hand, most of those in the highlands, and those which cover the majority of the moon's surface, were formed by what is called crystal fractionization, which means that they crystallized from molten material and then floated to the top of the liquid. One would expect some rocks of this type in local lava pools, where the rock was heated by an impact. But the fact that they are quite general over the entire surface of the moon suggests that the entire lunar crust must have been molten early in its history, and that as it cooled, lighter material crystallizing out of it floated to the surface, where it remains today.

---

7. Describe a baseball game being played on the moon. Include discussions of the distance the ball might be batted, the strides of the base runners, the ease or difficulty of fielding the ball, the likelihood of rain checks and of calling the game because of darkness, and how the fans might boo the umpire.

Lunar Interior

As we have seen, we learn about the interior of the earth from the analysis of seismic waves. That is why the astronauts left seismometers (equipped with radio transmitters, of course) on the moon—to monitor seismic activity there and to learn about the lunar interior. To assure that there would be some seismic activity to monitor, we deliberately crashed some space vehicles (which had already served their purposes) onto the lunar surface. But the moon has also obliged by producing some moonquakes of its own. Some of these appear to be triggered by the tidal forces of the earth on the moon.

Anyway, we have learned a good deal about the insides of the moon. It has a crust that goes in some 60 km – on the side facing the earth. But on the other side the crust is deeper – about 150 km. The moon is lopsided! Worse than that: Its center of mass is about 2 km away from its geometrical center. Not much perhaps, but it shows that there are some density irregularities in the moon. In fact, the lunar-orbiting artificial satellites, including the Apollo command modules, gave some information about the interior of the moon and its mass distribution. Since the moon is not a perfect point mass, the motions of bodies revolving about it experience small perturbations. We analyze those perturbations and find what the mass distribution must be inside the moon. In the 1960s it was found that there were evidently concentrations of mass beneath the surface under some of the maria. These are called *mascons* (for "*mass concentrations*"). We don't know what they are, of course, but a guess is that they might be the remains of massive bodies that struck the moon, causing the lava flows that formed the maria, but stuck there in the outer and less dense mantle of the moon, which had by that time hardened.

FIGURE V.17
A large lunar boulder with many cracks. *(NASA)*

The present view is that the moon has that irregular crust we described, a rather deep mantle mostly of rigid rock, and probably an inner core of denser material about 1000 km in diameter. The moon appears to have cooled more than the earth, and much more of its mantle is rigid and solid. There is no evidence at all for tectonic activity—a circulation of crustal plates, as occurs on earth.

Origin of the Moon

Most of the various theories for the moon's origin are one or another of three general ideas: (1) that the moon separated from the earth by fission when the earth was just forming; (2) that the moon formed elsewhere in the solar system and was later captured; (3) that the moon accreted from particles in revolution about the young earth. Whereas no one is sure, most experts lean to the third of these ideas.

The idea of fission is that the forming earth had an extremely high rate of rotation. Under some circumstances a rapidly rotating body can be unstable and can divide into two bodies, which could then slowly separate as a result of their mutual tidal interactions. In fact, we have seen that the earth and moon do exert appreciable tides on each other, and that the friction of the tidal flow of water against itself and over coastal areas is gradually slowing the earth's rotation. In order that the earth-moon system conserve its angular momentum, the moon is, indeed, very slowly moving away from the earth, increasing the size of its orbit. But it is not known how the earth could have a rapid enough spin to fission into two bodies, and if it did, it should have a much faster spin today. In other words, it is hard to conceive a satisfactory fission model that works in detail.

---

8. Some people have noticed that the Pacific Ocean basin is roughly circular, and they have speculated that the Pacific Ocean is therefore a scar left when the moon pulled out of the earth. Criticize this hypothesis on as many relevant grounds as you can.

The second kind of theory—that the moon was formed elsewhere in the solar system and then captured—is more feasible than the fission model, but still presents problems. The idea of this theory is that the earth and moon both accreted from solid particles in roughly the same region of the solar nebula, but in completely different orbits. Later, perturbations by other planets changed the moon's orbit, so the moon came close to the earth and was slowed down enough at the same time to be captured. The problem here is that such a capture is improbable, and even if it did happen, we would expect the moon to have a very eccentric orbit, rather than its present nearly circular one lying almost in the ecliptic plane.

The preferred hypothesis is that the moon accreted in roughly its present orbit from particles revolving about the earth. It is much easier for the earth to have captured a disk of small particles than one large, fully formed moon. While the earth was forming in the solar nebula, lots of particles were colliding with each other near it, thereby slowing each other down to the point where many can end up in an orbit around the forming earth. Drag on the particles by the gas in the solar nebula can also slow them down. It is estimated that an accretion disk of particles could so form in orbit around the earth in a time as short as 1000 years, but easily in a few million years, which is still a snap of the fingers in the geologic time scale.

Origin of Lunar Craters

As the moon accreted, either in orbit around the earth or elsewhere in the solar system, it gradually collected matter and built up to its present size. In those very early years, though, there were still a great many particles left in the solar nebula, some of them very large. It must have taken some hundreds of millions of years to clean out the solar system of all that debris. During that

FIGURE V.18
The moon, photographed mostly from the far side by the crew of Apollo 16. *(NASA)*

time there were probably a number of periods when a new supply of rather big hunks became available, through collisions of large bodies—just as the minor planets seem to be colliding and fragmenting even today. During the first 500 or 600 million years there were many rains of particles on the moon (and, of course, on the earth and other inner planets), producing the extensive cratering we still observe there. Those collisions may have made enough heat to keep the moon's outer layer molten for a while; we have seen evidence that the moon's crust may have once been completely liquid.

Gradually the chunks were swept out of the interplanetary space, and the impacts became less and less frequent. By 4000 million years ago the moon's crust had hardened and most of the cratering had ceased, although occasional collisions causing overlapping lava flows must have continued for the first 1000 million years, thus leaving the maria. By that time the cratering must have been at a rate at least 1000 times less than in the beginning, and perhaps only a few tens as great as it is today. By 2000 million years ago, impacts became very rare, and only once in a great while did large particles, probably collision fragments of minor planets, chance to strike the moon and leave those craters of more recent origin. By now the moon is completely cooled off, to a depth of at least 1000 km. Seismic studies show that only an inner core is still soft and warm. To all intents and purposes that world, once seething with activity and molten rock, is now geologically dead.

Well, it's a nice place to visit, but I wouldn't want to live there.

SCENE 3

The Other Worlds

An interstellar wanderer, casually passing by the solar system, would probably be unaware that there are any planets here at all, because 99.86 percent of the system is the sun itself. The planets and their satellites make up the bulk of the rest, but they are pretty insignificant in comparison, as Figure V.19 shows.

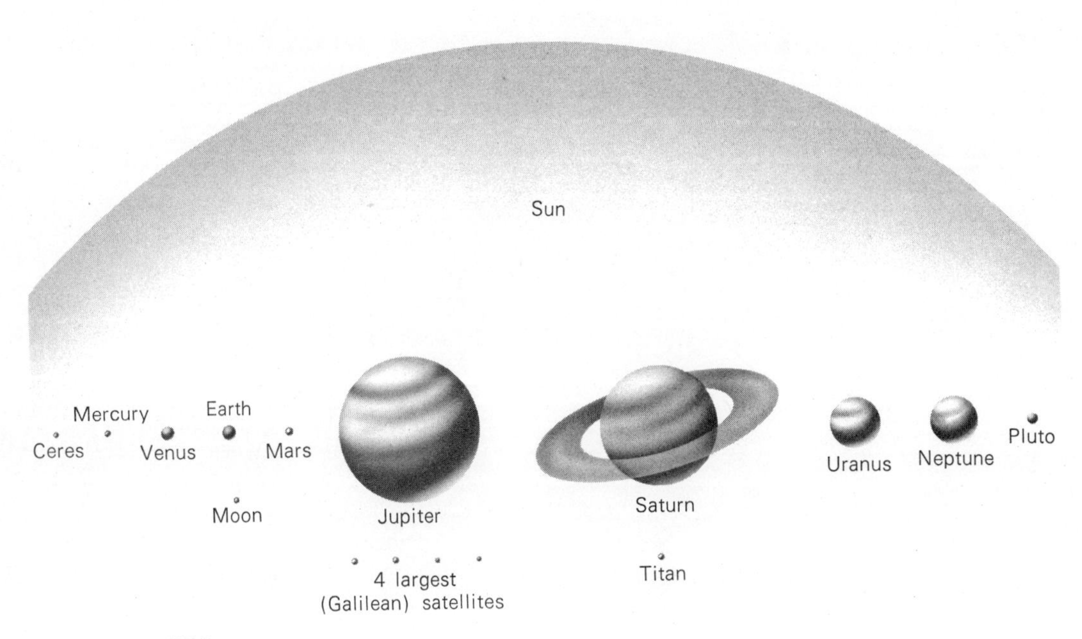

FIGURE V.19
The relative sizes of the sun and planets.

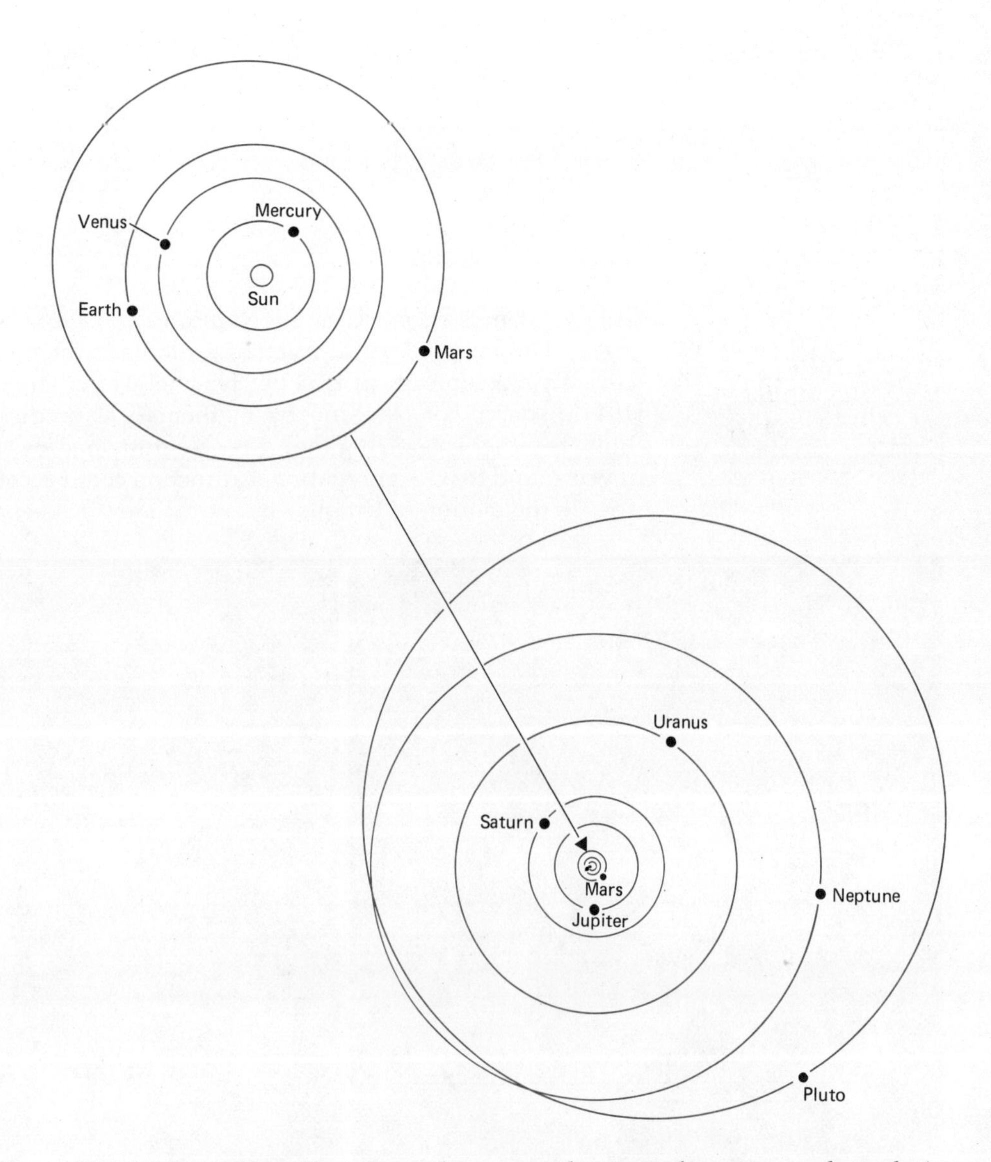

FIGURE V.20
The relative sizes of the orbits of the planets. Two different scales are used to show the outer and inner planets.

The planets all revolve about the sun in the same direction, and nearly in the same plane. The orbit of Pluto is most inclined (17°), and Mercury is next, with an orbital inclination of 7°. These same two planets have the most eccentric orbits, but even they look pretty circular, as can be seen in Figure V.20. Details of the planets and satellites are given in Appendices 6 through 8. For handy reference, though, I summarize in the following table some approximate figures, so you can see at a glance the major characteristics of the planets.

Planet	Distance from Sun (AUs)	Mass (earth units)	Diameter (earth units)	Revolution Period (years)	Rotation Rate	Escape Velocity (km/s)	Surface Gravity (Earth = 1)	Major Atmospheric Gases	Number of Known Satellites
Mercury	0.4	0.05	0.4	0.24	59 da	4	0.4	none	none
Venus	0.7	0.8	1	0.62	243 da	10	0.9	CO_2	none
Earth	1	1	1	1	24 hr	11	1	N_2; O_2	1
Mars	1.5	0.1	0.5	1.9	24.5 hr	5	0.4	CO_2	2
Jupiter	5	318	11	11.9	10 hr	60	2.6	H, He	13(14?)
Saturn	10	95	9	29.5	10 hr	35	1.1	H, He	10
Uranus	20	15	4	84	11 hr	20	1.1	H, He	5
Neptune	30	17	4	165	16 hr	22	1.4	H, He	2
Pluto	40	0.1(?)	0.5(?)	248	6.4 da	?	?	none?	none

Discovery of Pluto

All the planets except Uranus, Neptune, and Pluto were known to the ancients. Uranus is sometimes actually visible to the unaided eye, but was discovered telescopically in 1781 by Herschel. I have already told the story of the 1846 discovery of Neptune by mathematical prediction. The discovery of Pluto is rather interesting too. After Neptune had been observed for a while, it was found that its gravitational attraction could account for all the peculiarities in the motion of Uranus, except for a few discrepant early observations.

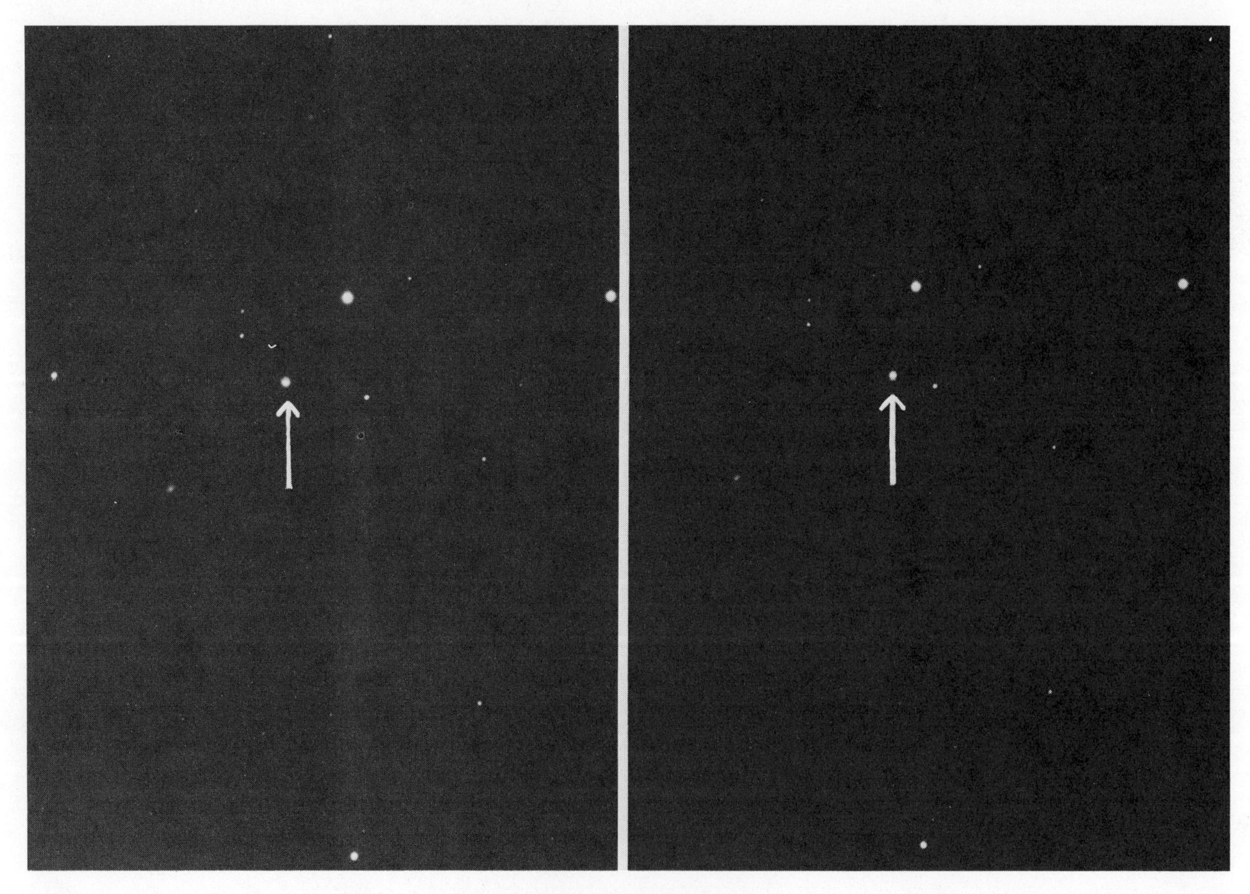

FIGURE V.21
Two views of Pluto, showing its motion relative to the stars in a 24-hour period. Photographed with the 5-m (200-inch) telescope. *(Hale Observatories)*

It turned out that those few early measures were just not very accurate, and actually there is no problem at all in understanding the motion of Uranus.

Before that realization, however, three different astronomers, following in the footsteps of Adams and Leverrier, set out to calculate where a ninth planet would have to be to do the apparent dirty work. The most persistent was Percival Lowell, who founded the Lowell Observatory in Arizona (unlike most astronomers, Lowell had money). Lowell searched for the unknown planet at his observatory from 1906 until his death in 1916. The search was continued at the same observatory by Clyde Tombaugh, who finally discovered a new planet in February 1930.

The orbit of the new planet is rather similar to that predicted by Lowell. Its mass is far too small, however, to have produced measurable (at that time) perturbations on the motion of Uranus. Although Lowell's calculations were formally correct, the discovery of the planet was completely accidental—a most interesting coincidence. Quite properly, the planet was named Pluto—sort of for Percival Lowell, whose initials appear in the planet's name.

Some General Properties of Planets

All the planets have atmospheres except Mercury, and probably Pluto; on Mercury it's too hot, on Pluto it's too cold. The velocity of escape from Mercury is about the same as that from Mars; yet Mars has a thin atmosphere, because it is far enough from the sun to have a low enough prevailing temperature that the molecules in its air are not moving fast enough to escape. Pluto is so far from the sun that the common atmospheric gases it might be expected to have would be frozen.

The rotation periods of Mercury and Venus have been known only since the development of radar. Radar waves beamed to those bodies in a very narrow range of wavelengths are reflected back over a broader range, because the waves reflected from the approaching and receding limbs of a planet are Doppler-shifted by different amounts. The amount of spread of wavelength or frequency tells us the rotation rate. We observe the rotations of Jupiter, Saturn, Uranus, and Neptune by comparing the Doppler shifts in the optical spectra of light reflected from their opposite limbs. We can time successive appearances of visible surface features on Mars to find its rotation rate. Pluto, though, looks like a point of light, and we infer its rotation only by the fact that it varies in light by about 20 percent with a very regular period of 6.387 days; evidently it has an uneven surface that reflects light differently from different hemispheres.

---

1. How might you attempt to learn the rotation periods of some of the brighter minor planets?

If you compare the gross characteristics of the planets, you will see that all but Pluto fall nicely into two different groups. The *Jovian* planets—Jupiter, Saturn, Uranus, and Neptune, far from the sun—are all large, formed of relatively lighter elements, have lots of satellites, and are rapid rotators. The *terrestrial* planets—Mercury, Venus, Earth, and Mars, close to the sun—are all small, made of rocks and metals, have few or no satellites, and are slow rotators. Pluto is more like the terrestrial planets in size and mass, but is remote and cold. It must have had an origin different from those of the Jovian planets. Its orbit is eccentric enough that at times it comes closer to the sun than does Neptune (as you can see in Figure V.20). Because of the large inclination of Pluto's orbit to those of the other planets, it cannot actually collide with Neptune. However, over hundreds of millions of years the orbits of the planets are constantly changing because of their perturbations on each others' motions, and some people think that Pluto may once have been a satellite of Neptune that somehow escaped—perhaps because of perturbations of its other satellites. Anyway, Pluto doesn't fit into the scheme very well, so we'll just not say any more about it.

---

2. Opposition of a planet occurs when that planet is opposite the sun in the sky. Inferior conjunction occurs when a planet is between the earth and sun. Superior conjunction occurs when a planet is on the far side of the sun, so that the sun is between it and us. Which planets can and cannot pass through
a) opposition,
b) inferior conjunction,
c) superior conjunction?

3. Explain why Mercury can be seen only at certain times when it is either low in the sky above the eastern horizon shortly before sunrise, or low in the sky above the western horizon shortly after sunset.

The Jovian Planets

The Jovian planets themselves fall into two subgroups. Jupiter and Saturn are pretty much alike, and so are Uranus and Neptune. We think that they all got their start by accretion of solid particles, as did the terrestrial planets. Since we have every reason to expect that metallic and rocky particles were out there as well as near the sun, we suspect that the Jovian planets all have small rocky and metallic cores, perhaps each comparable in mass to the earth. But a lot of ices should have condensed from the solar nebula far from the sun, and the Jovian planets should have gotten a lot of them too. We think that Uranus and Neptune, in fact, are probably made up mostly of icy stuff.

But those large planets also attracted and held on to a lot of the gas in the solar nebula, most of it, of course, being hydrogen and helium. That gas, falling in on the planets from a considerable distance, carried a lot of angular momentum down with it, which is why, presumably, the large planets formed their own disks, rather like miniature solar nebulae, and developed satellite systems in their equatorial planes (in contrast, note that the moon's orbit about the earth is nearly in the *ecliptic* plane, and not in the earth's equator plane). It also accounts for the rapid rotations of the Jovian planets.

Jupiter seems to have just about the same chemical composition as the sun—mostly hydrogen and helium, and in about the same relative proportion. But Jupiter is nowhere near massive enough to have been able to heat up that hydrogen to the point where it could undergo fusion into helium. The hydrogen in the inner 46,000 km of the 71,400-km radius of Jupiter is degenerate. But for a body of the mass and size of Jupiter, that degenerate hydrogen acts like a highly conducting liquid metal rather than like a gas. Jupiter, in fact, is just about as large (in diameter) as a body of degenerate hydrogen can be. If it were somewhat more massive, it would have gotten hotter inside, the hydrogen would have been a degenerate *gas*, and Jupiter would have ended up more compressed and smaller than it is.

Outside Jupiter's very large core of liquid metallic hydrogen, the hydrogen is molecular, but is still compressed to a liquid state. Only in the outer 1 percent of its radius is there a gaseous atmosphere. That gas just gets denser and slushier lower down until before you know it, it's liquid. It would be a very easy planet on which to make a soft landing; the spaceship could dive into the atmosphere, and gradually slow down as it sinks into the slush. Getting out again might pose some problems to the astronauts.

Jupiter still has a lot of internal heat stored up inside, and it is gradually radiating that heat away as infrared radiation. In fact, it radiates more than twice as much energy as it receives from the sun. Its luminosity is three ten-millionths (3×10^{-7}) that of the sun.

Saturn is rather similar to Jupiter, but is less massive and less dense. The mean density of Saturn, 0.7, is actually less than that of water; Saturn would float if we could find a suitable pond to launch it in. It is also mostly liquid hydrogen, but the core where the hydrogen is metallic is very much smaller than in Jupiter. Saturn, like Jupiter, radiates away its internal heat, but has a self-luminosity only about one-fifth that of Jupiter's. Uranus and Neptune probably have no metallic hydrogen at all.

---

4. Describe carefully the distinction between the sunlight Jupiter and Saturn reflect into space and the energy they radiate into space. By which of these radiations do we observe those planets when we look at them through a telescope? How might the other radiation be observed?

The rapid rotation of Jupiter and Saturn make them noticeably oblate. Jupiter's equatorial radius, as I have just mentioned, is 71,400 km, but its poles are only 67,000 km from its center. Its rapid rotation and the high conductivity of its metallic hydrogen are probably responsible for Jupiter's very strong and extensive magnetic field. Since the 1950s we have been observing synchrotron radio radiation from the region surrounding Jupiter, showing that it has a magnetic field in which electrons are spiraling at very high velocities. It's sort of like the van Allen radiation belts of the earth, but on a tremendously larger scale. There are also sudden bursts of radio energy from Jupiter from time to time. Some of these are found to be correlated with the revolution of its innermost satellite, Io. Evidently Io interacts with the magnetic field around it in such a way as to trigger those bursts of radiant energy.

Pioneer Probes of Jupiter

The atmosphere and surrounding *magnetosphere* of Jupiter were studied in detail by the spacecraft Pioneer 10, launched in March 1972, and by Pioneer 11, launched in April 1973. They passed Jupiter in 1973 and 1974, respectively, the first 132,000 km and the second 42,000 km above the Jovain clouds. The path of Pioneer 11 was so arranged that the gravitational pull of Jupiter swung its direction around so that it is going right back across the solar system, to pass near Saturn in 1979. Eventually both Pioneers will move off into interstellar space, escaping the solar system forever. Two Voyager spacecraft, launched in 1977, are scheduled for arrival at Jupiter to further explore that world in 1979. Like Pioneer 11, the Voyagers will take advantage of Jupiter's gravitation to deflect their trajectories and speed them out to still more remote planets.

The Pioneer spacecraft found that the magnetic field strength at the surface of Jupiter is from 11 to 14 gauss, which is a lot of magnetism when you consider the volume of space it occupies. And both spacecraft encountered enormous amounts of radiation from high-speed charged particles – in fact, many hundreds of times the dose that is lethal to humans. Astronauts had better not venture there!

The Jupiter Atmosphere

Five gases have been observed spectrographically in the atmosphere of Jupiter: hydrogen, helium, methane, ammonia, and water; there must be many others as well. The planet is shrouded in an opaque cloud layer with a temperature of about 125°K at the top, and consisting of bright and dark bands parallel to its equator. The white bands are thought to be upcurrents of gases, and the dark ones places where the gases cool and circulate down again. In the process there is evidently a considerable amount of chemical activity, for the dark bands have a lot of red or red-brown color. Astronomer Carl Sagan and biochemist Cyril Ponnamperuma (and others as well) have experimented in the laboratory to see what happens, under the conditions present in Jupiter, to an admixture of elements known and presumed to exist in Jupiter's atmosphere. They find that among the many chemical compounds that can form are red organic polymers, and suggest that such might account for the red color in the Jovian clouds.

One interesting feature of the Jupiter atmosphere is the famous Great Red Spot. The spot was first noticed about 300 years ago, and has been there continuously since, although sometimes with diminished vividness of color. Its width is a pretty constant 14,000 km, and its length varies from 30,000 to 40,000 km. It is easily seen on some of the color photographs of Jupiter made by the Pioneer spacecraft television cameras and reproduced in this book. The Pioneers showed that the spot is a great storm, rather like a hurricane. Since it doesn't have any land to move over and dissipate, like a self-respecting Florida-type hurricane, it has persisted for a long time. The spacecraft also observed other similar but smaller spots, which are presumably less permanent.

Satellite and Ring Systems

We saw in Act IV Scene 5 how satellite systems of the large planets are believed to have come about, and I mentioned it again briefly earlier in this scene. We must really *expect* these great planets to have satellites in their equatorial planes. Uranus, in fact, has its polar axis practically in its orbital plane, with its equator roughly at right angles to the ecliptic. (Actually, its obliquity is 98°, and technically Uranus rotates backward.) The satellite system of Uranus, similarly, sticks out of the main plane of the solar system at that same crazy angle!

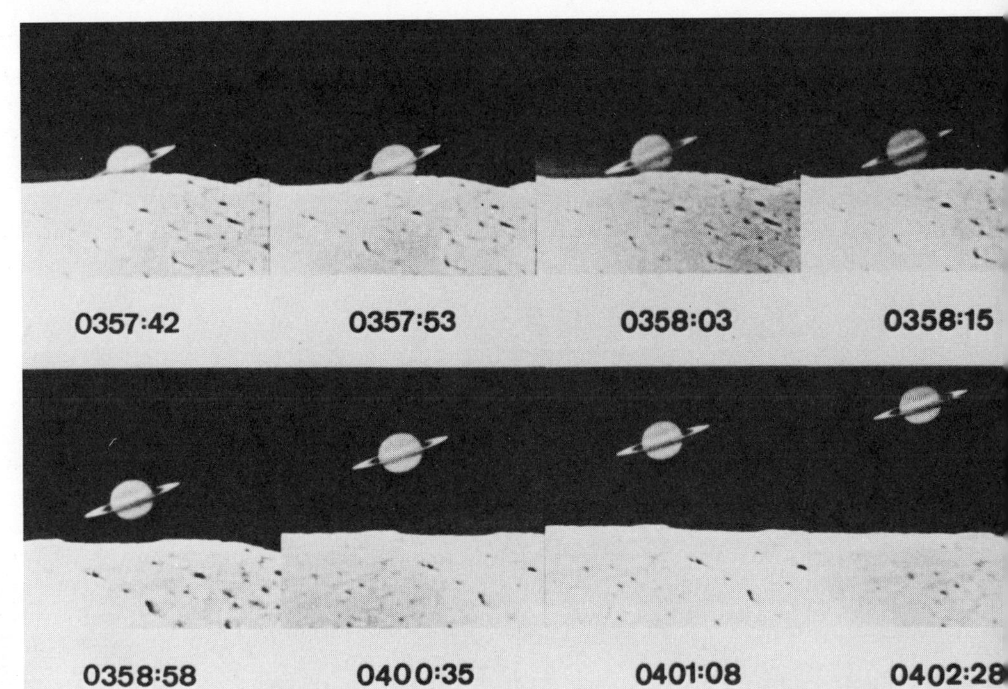

FIGURE V.22
A series of photographs showing the emergence of Saturn from behind the moon following the occultation of 13 November 1967. *(Photograph by B. A. Smith and C. J. Barnes, New Mexico State University Observatory)*

5. Describe the seasons on Uranus.

Until recently, only *one* planet, Saturn, was known to have rings. In 1977, however, a small ring system about Uranus was discovered. Both ring systems are made of countless tiny particles, the largest only a few meters across, but most very much smaller. In both ring systems the particles revolve in their planet's equatorial planes. The sunlight reflected off the myriad particles in Saturn's rings gives them the illusion of being solid. The five small rings of Uranus reflect too little sunlight to be observed directly; they were discovered when Uranus passed in front of a star on 10 March 1977. The rings are not solid and did not occult the star completely, but the star did dim somewhat as each ring portion passed by. The outer diameter of Saturn's rings is 275,000 km, 2.31 times that planet's diameter. The outer ring of Uranus has a maximum diameter of 102,000 km, 1.83 times that of Uranus itself. The rings of both Saturn and Uranus consist of material too close to those planets to accrete into a satellite.

6. How often are the rings of Saturn and of Uranus turned edge-on to our line of sight? Assume, for this question, that the orbits of both planets are precisely in the ecliptic plane and that we are observing the planets from the sun. See the table at the beginning of the scene or in Appendix 6 for necessary data. [Hint: The plane of each ring system remains at a constant orientation in space while the planet revolves about the sun. Why?]

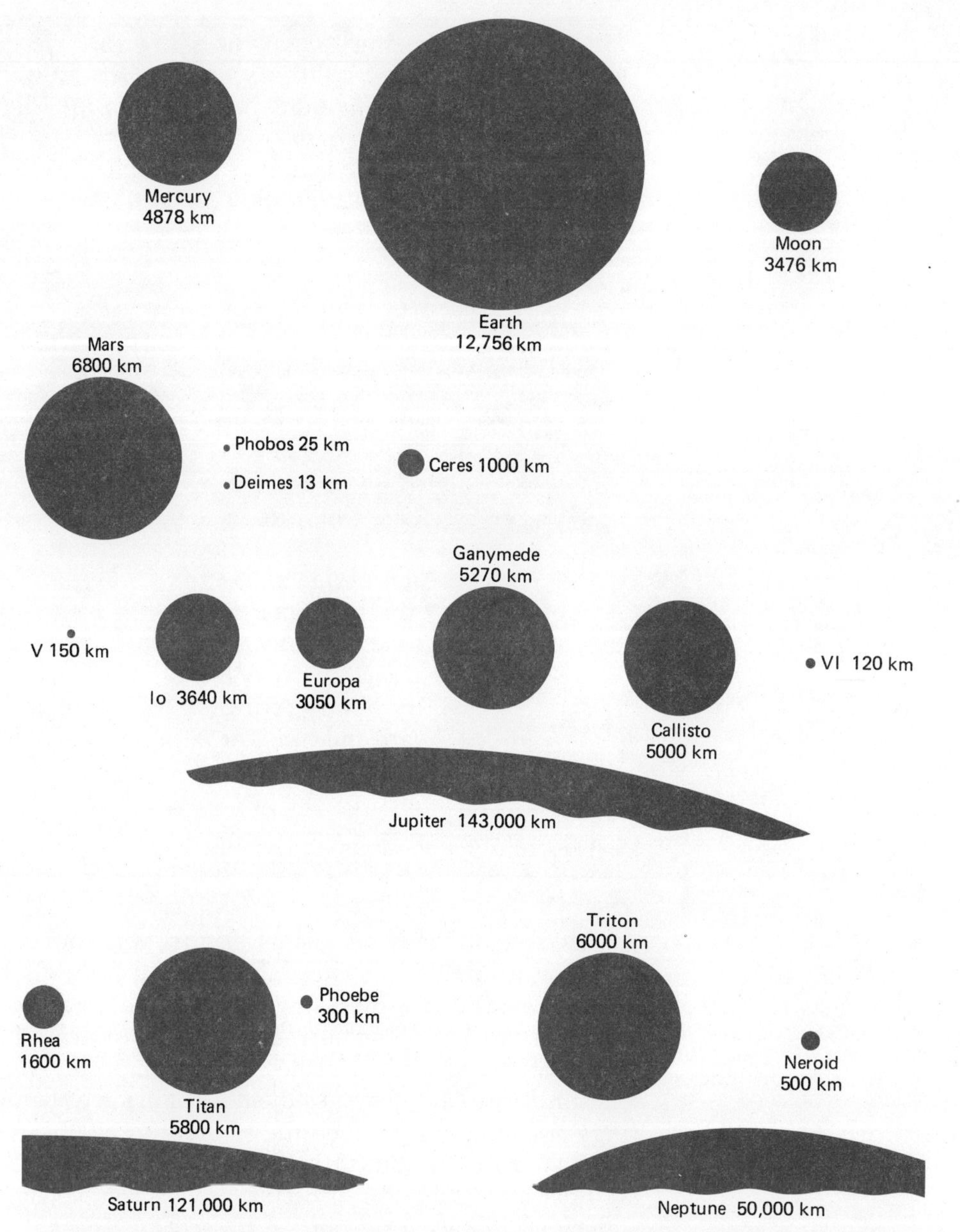

FIGURE V.23
The relative sizes of some of the planets and several of the smaller members of the solar system.

Some of the satellites of the Jovian planets are pretty big. Five are larger than the earth's moon, and three are about the size of the planet Mercury. Titan, Saturn's largest satellite, has long been known to have an atmosphere of its own, and the Pioneers found that Io, innermost of Jupiter's moons, has an atmosphere. Figure V.23 shows the relative sizes of some selected small bodies in the solar system.

Jupiter, by the way, has two kinds of satellites. The inner five, which are the largest, revolve about it in its equatorial plane, and form the regular satellite system. At least eight more distant satellites, however, are in the ecliptic plane, and have far more eccentric orbits. The outer four even revolve from east to west, opposite to the sense of almost everything else in the solar system. They are almost certainly not primordial to Jupiter's family of satellites, but likely are captured minor planets that wandered too far from their own fold. Small outer satellites of Jupiter are still being discovered from time to time, especially by such energetic fellows as Charles Kowal at Caltech. In 1974 Kowel found the last confirmed satellite of Jupiter, its thirteenth.

So that's the way it is in the outer part of the solar system. Giant worlds exist out there, but they are cold, without surfaces that you could walk on, and at least one of those worlds is surrounded by absolutely lethal radiation. Not very salubrious vacation spots for earthlings. Let's see if we can do better in the inner solar system.

The Terrestrial Planets

The terrestrial planets are, indeed, very different, all being accreted from rocky and metallic material. All are now believed to have large iron cores, which in Mars, the earth, and Venus extend about halfway to the surface. Mercury, though, having formed very near the sun where only the most refractory material could condense out of the gas of the solar nebula, has a much larger proportion of iron; its core extends through three-quarters of its radius.

Mercury's rotation period of 58.6 days is exactly two-thirds of its period of revolution about the sun. Solar tidal forces have locked it into that period. Evidently it has a very slightly greater diameter in one equatorial direction than the others (an asymmetry of only a few kilometers or less would suffice). Every time Mercury passes perihelion (is nearest the sun), its longest diameter is pointed toward the sun. The period of Venus is 243 days, but is retrograde—that is, from east to west, the reverse of its direction of revolution.

---

7. Draw a diagram showing how alternate ends of the long diameter of Mercury are directed to the sun at successive perihelion passages. How is the long diameter oriented at aphelion?

None of the terrestrial planets could have captured gases directly from the solar nebula, so their atmospheres have come from outgassing, as in the earth. We expect the principal gases to so exude to be water, carbon dioxide, and nitrogen, in that order. All are accounted for on earth and, as we shall see shortly, on Mars. Venus' present atmosphere is mostly carbon dioxide, and nitrogen can well be there. Water, however, exists only in trace amounts on Venus, along with carbon monoxide and hydrogen chloride. Where did its water go? Perhaps it dissociated under the action of ultraviolet solar radiation, releasing hydrogen, which escaped, and oxygen, which combined with more carbon to make even more carbon dioxide. Or, perhaps for some unknown reason Venus never had appreciable water locked up in chemical combination in the particles from which it accreted.

On the earth most of the carbon dioxide to outgas is now combined in carbonates on the surface. If it were hot enough here, that carbon dioxide would be released and would be a major constituent of our atmosphere. Then we'd be in real trouble, because that gas would blanket in the infrared radiation the earth radiates, making it very hot around here. (Heating by this mechanism is called the *greenhouse effect*.)

Well, that's just the case on Venus. All the carbon dioxide is in the atmosphere, producing a surface pressure there about 100 times that of the sea-level atmospheric pressure on earth. The ground temperature on Venus is

FIGURE V.24
The surface of Venus, showing fairly rough rocks, photographed by the landing apparatus of Venera 9. *(Novoste from SOVFOTO)*

FIGURE V.25
The surface of Venus, photographed by the landing apparatus of Venera 10. The vertical stripes in this figure and in Figure V.24 occur where the transmission carried non-picture information for other scientific experiments. *(TASS from SOVFOTO)*

consequently about 750°K (900°F). We don't see the surface of Venus; the optical radiation we observe from earth comes from above opaque clouds that perpetually hide the planet. The temperature at the tops of those clouds, high in the atmosphere, is only 250°K, well below freezing. We first learned of the high surface temperature by means of radio observations of the planet; the clouds are transparent to radio wavelengths. The surface temperature is also verified by instruments landed on the planet by the Soviet Venera space probes.

The clouds themselves are almost formless. They do show indistinct features in the ultraviolet, indicating what appear to be streaks swept by high-altitude winds of speeds up to 300 km/hr. What the clouds are is not known for sure. Some experts think they are largely sulfuric acid, with some hydrogen chloride, hydrogen fluoride, and sulfur mixed in.

Whereas the Venerian clouds hide the planet's surface from us, they do not keep all light from getting in. The Soviet Venera 9 and 10 spacecraft, which landed on the surface of Venus in 1976, obtained clear photographs and televised them back to earth during the hour or so before each spacecraft was destroyed by the high temperature and atmospheric pressure, and possibly by corrosive chemical activity. In one location the rocks were found to be rough and jagged, and in the other somewhat rounded. So some erosion oc-

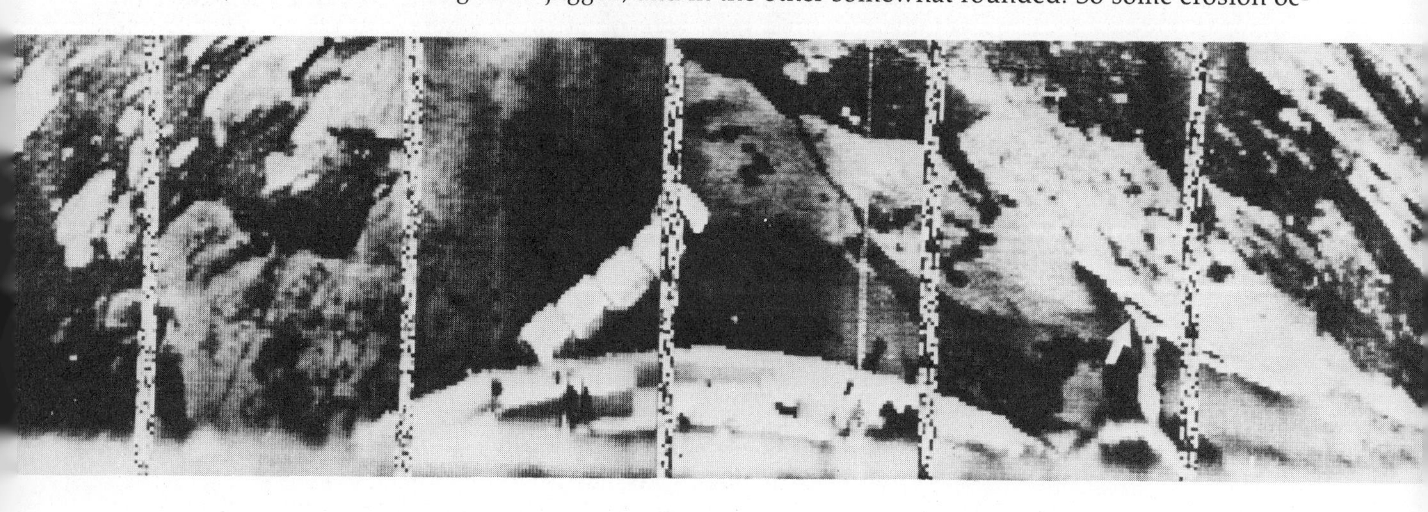

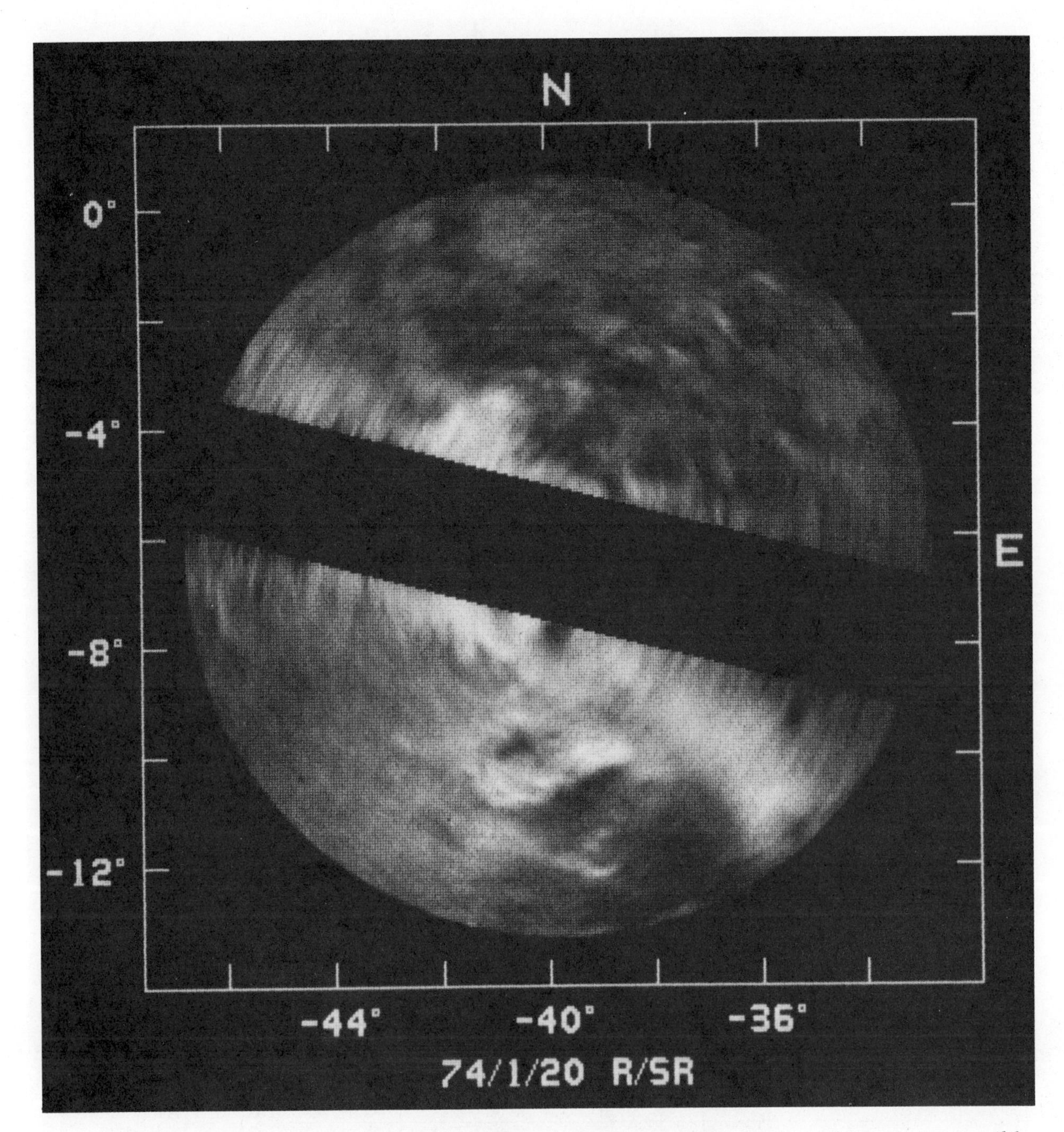

FIGURE V.26
Many large craters are revealed in this radar map of Venus. *(NASA/JPL)*

curs there, but not due to liquid water. The surface of Venus is a terribly hot, oppressive, stifling, dusty place.

All the terrestrial planets, like the moon, were very heavily cratered during their final accretion phases, 4000 million to 4500 million years ago. A large fraction of the surface of Mercury has been photographed with a resolution of 150 m by Mariner 10, which obtained more than 8000 television pictures of that planet and of Venus. Mercury's surface is still almost completely covered with craters. On Mars, most small craters were long ago eroded away, and over about half the planet vulcanism has obliterated even the large craters. Radar observations of the surface of Venus show many scattered craters from

FIGURE V.27
A mosaic of photographs of Mercury transmitted to earth on 29 March, 1974 by Mariner 10. *(NASA/JPL)*

30 to 1000 km in diameter, and those 200 km or more across seem to be about as abundant as on Mars. Radar has also revealed on Venus a large rift canyon some 1500 km long.

On the earth the old craters are now gone, although a few fossil craters are known that seem to be as old as 1000 million years. The best-known site of old craters is in the Canadian Shield near Hudson Bay. Other old features can be recognized in mountainous areas in the United States. The very well preserved Meteor Crater in Arizona is, of course, of very recent origin. It is only about 1.3 km across. It is estimated that it takes a body 100 km or so in diameter to make a crater 1000 km across, like the largest observed on Mercury, Venus, and Mars. It's a good thing for us that there aren't many of such chunks around any more.

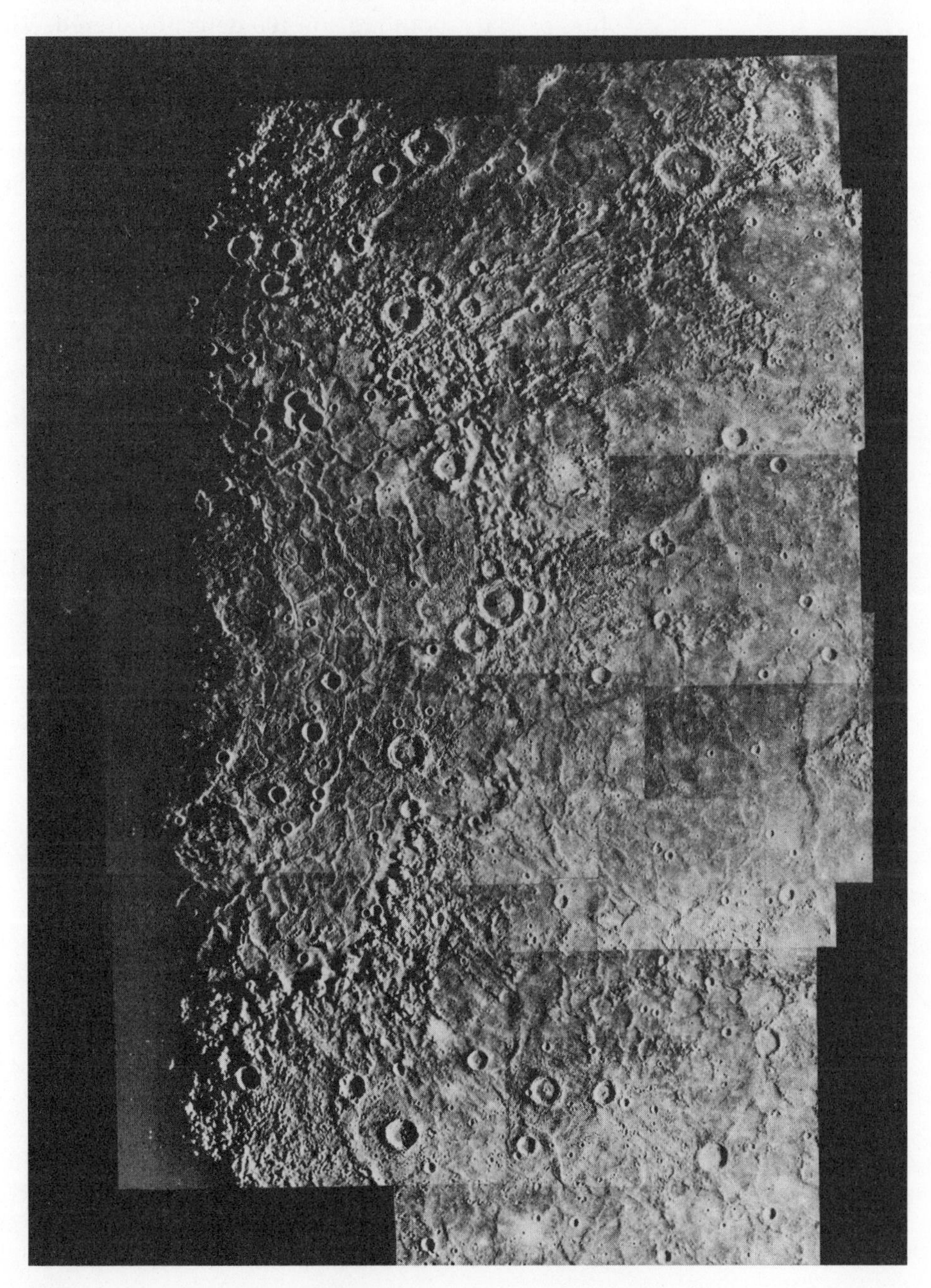

FIGURE V.28
In the left half of this mosaic of Mariner 10 photographs is the ring basin Caloris, 1300 km in diameter *(NASA/JPL)*

Mercury and the moon are quite similar, except that Mercury was hotter and cooled more slowly, and careful analysis shows that its surface was more affected by the heat. There are long compression faults on Mercury, showing large-scale subsidence; the circumference of the planet has evidently shrunk about 2 km since its formation. The daily temperature range on Mercury (albeit, a 59-day "daily") is greater than on any other planet. The hottest spot is probably the *Caloris* basin, so named because it has the sun overhead at noon when Mercury is at perihelion. Daytime highs get up to about 700°K, not quite as hot as at the ground on Venus; but Mercury nighttime lows are around 100°K.

We can order the terrestrial planets and their satellites according to how advanced they are in erosion. At the bottom of the list are the two tiny satellites of Mars; both are still 100 percent cratered. Virtually all features on the moon and on Mercury are older than 2500 million years. Venus and Mars are intermediate. We know the surface of Mars much better of course; there about 50 percent of the features are younger than 2500 million years, and about 15 percent are younger than 600 million years. The earth is by far the most evolved. At least 98 percent of its surface is younger than 2500 million years, and 90 percent is younger than 600 million years.

---

8. The atmosphere of Venus is very much denser than that of the earth, yet the earth's surface is more eroded than that of Venus. Why is this so?

The Martian Dream

We end this survey of our neighboring worlds with a more careful look at Mars, which has excited the interest and imagination of man for centuries. In some ways it is like the earth. Its day lasts 24 hours 37 minutes. Its obliquity is 24°, compared to 23½° for the earth; thus Mars has seasons similar to ours, although each season lasts almost six of our months. For more than 100 years we have been watching seasonal changes on Mars, the most conspicuous being the growth and recession of its polar caps. Seasonal color changes were also reported in the temperate zones of the planet; some observers thought those regions turned green in spring and summer and brown in fall. It is actually less a change in color, however, than a change in the visibility of various surface markings, as viewed through earthbound telescopes.

Mars also was known for a long time to have an atmosphere. Vast yellow dust storms were sometimes seen to cover the planet. In fact, when Mariner 9 reached Mars and went into orbit around it in 1971, for the first weeks of observation the entire planet was hidden by such blowing dust. White clouds on Mars have long been observed from earth, too. We know now that some of these are quite high – about 45 km, and are probably carbon dioxide crystals. Other clouds are lower, 15 to 30 km, and are believed to be of ice crystals, like our cirrus clouds.

The Martian atmosphere, however, is now very thin; the surface pressure is somewhat less than 1 percent sea-level pressure on earth. The best determination of the abundances of the various gases present there is from data gathered by the Viking landers in 1976:

Gas	Percentage
Carbon dioxide (CO_2)	95%
Nitrogen (N_2)	2 to 3%
Argon (A)	1 to 2%
Oxygen (O_2)	0.1 to 0.4%
Water (H_2O)	0.01 to 0.1%
Krypton (Kr)	less than 0.0001%
Xenon (Xe)	less than 0.0001%

The abundance of the argon is a good indication of how much total gas
The abundance of the argon is a good indication of how much total gas
outgassed, and hence what Mars' atmosphere may have been like in the past. A best guess at the time of this writing is that the Martian atmosphere may have once had a pressure of about 10 percent that of the earth's.

Before the space age, all astronomers (amateurs, too) who spent much time looking at Mars through a telescope agreed that it abounds in surface detail. The problem is making out what that detail is. Under the best conditions, Mars shows telescopically about the same amount of resolution that the full moon does when seen with the unaided eye. Those of you who have looked at Mars through a telescope probably saw mainly a shimmering red or orange ball, with perhaps a slight spot of white at one or the other pole. The shimmering is, of course, caused by the earth's atmosphere, a phenomenon called *seeing*. Seasoned observers, though, waiting for the steadiest (terrestrial) atmospheric conditions, occasionally see much more, especially if they wait for rare moments when sharp features seem to flash into view for a brief instant. I have had the experience myself several times, and can fully understand how the professional Mars observers can have become terribly excited at times.

The trouble is that the eye tends to simplify complex patterns that are seen indistinctly. What some observers called many fine specks, others saw as connected blobs. So it was that in 1877 the Italian astronomer Giovanni Schiaparelli thought he saw fine straight lines on Mars. He called them *canali*, or "channels." Then lots of other people saw them too. Most famous among them was Percival Lowell, who devoted most of his life (when he wasn't looking for Pluto) to studying Mars. In fact, the main reason he established his Flagstaff, Arizona, observatory was to study that mysterious red planet.

Lowell mapped hundreds of canals. He saw them connected at spots he called *oases*. He believed they were real water canals, built by intelligent Martians to carry the waters from the melting polar caps across the desert to irrigate their fields. So grew the Martian dream.

It wasn't that observers like Schiaparelli and Lowell were seeing things. The features were there all right. It's just that they were misinterpreted, because of the limited resolution permitted by earthbound telescopes, due especially to the turbulence of the earth's atmosphere. The first close-up view of Mars came in 1965, when Mariner 4 flew past the planet and televised 22 pictures back to earth. The photographs showed a bleak and barren planet, covered with craters, more like the moon than the earth. But then Mariners 6 and 7 flew by in 1969, and sent back a couple of hundred more pictures, including 33 of the south polar cap.

The real coup came in 1971, when Mariner 9 went into orbit around the planet and sent back 7329 photographs that have mapped Mars far better than we knew the face of the earth 200 years ago. Mariner 9 even photographed those little satellites, Phobos and Deimos (which mean "Fear" and "Panic," companions of the god of war), 25 and 13 km in diameter, respectively. What some people hoped would be the grand finale came in 1976, when the Viking 1 and 2 landers settled down on Mars and proceeded to send back pictures and chemically analyze its surface. The Vikings did not, as we shall see, provide the final answers. Nor did they (yet) fulfill the Martian dream. But Mars is so exciting a place that the dream has also not been shattered.

The Viking Landers

Let's skim off some of the cream of what we've learned about Mars with all of this space activity. First, how about those polar caps? They do, indeed, come and go. The south cap, in particular, reaches halfway to the Martian equator in midwinter, and often disappears entirely in summer. The main stuff of those white caps is frozen carbon dioxide (like Dry Ice). But not all! Part of the caps is water. At least in the north, there is a permanent cap of water ice. It doesn't go away in summer, and it is probably hundreds of meters thick. If it could all be melted, it is estimated that it could cover the planet with water to a depth of one-half meter. Furthermore, a lot of water could be frozen as permafrost under the Martian surface; we don't know, of course, for sure.

There is also vulcanism, and on a large scale. At least 12 huge volcanos exist on Mars, and their eruptions have flooded a large part of the planet with newly formed lava rock. Most spectacular among them is Olympus Mons, 25 km high and 600 km across and displaying a giant caldera 70 km across at its summit. It is by far the largest volcanic pile ever seen by man. Take a look at Figure V.31.

Canyons also are present, especially a great canyon (Valles Marineris) 5000 km long, 75 km wide (on the average), and 6 km deep. Although there is what appears to have been lifting and subsiding, there is no evidence for *horizontal* plate tectonics, as on earth. This can explain why certain Martian volcanos,

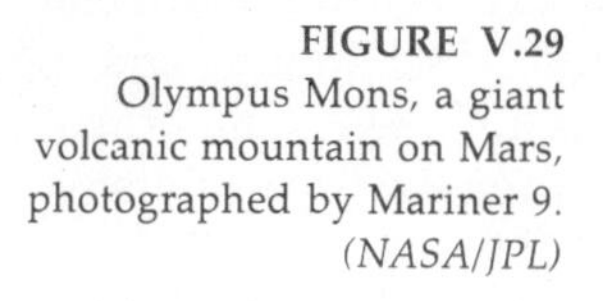

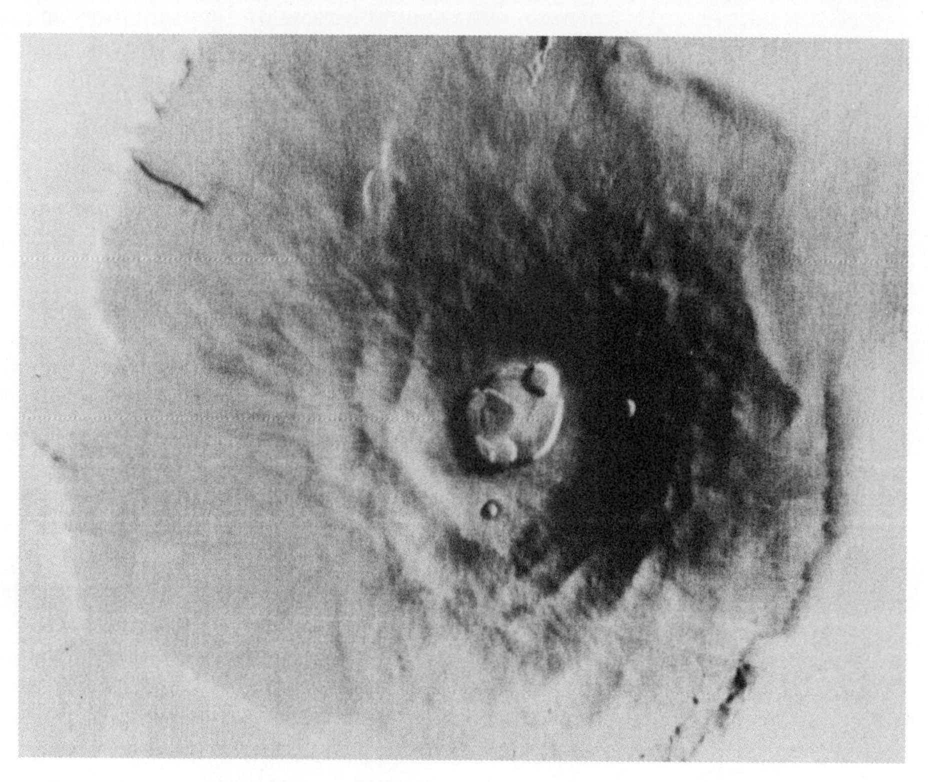

FIGURE V.29
Olympus Mons, a giant volcanic mountain on Mars, photographed by Mariner 9. *(NASA/JPL)*

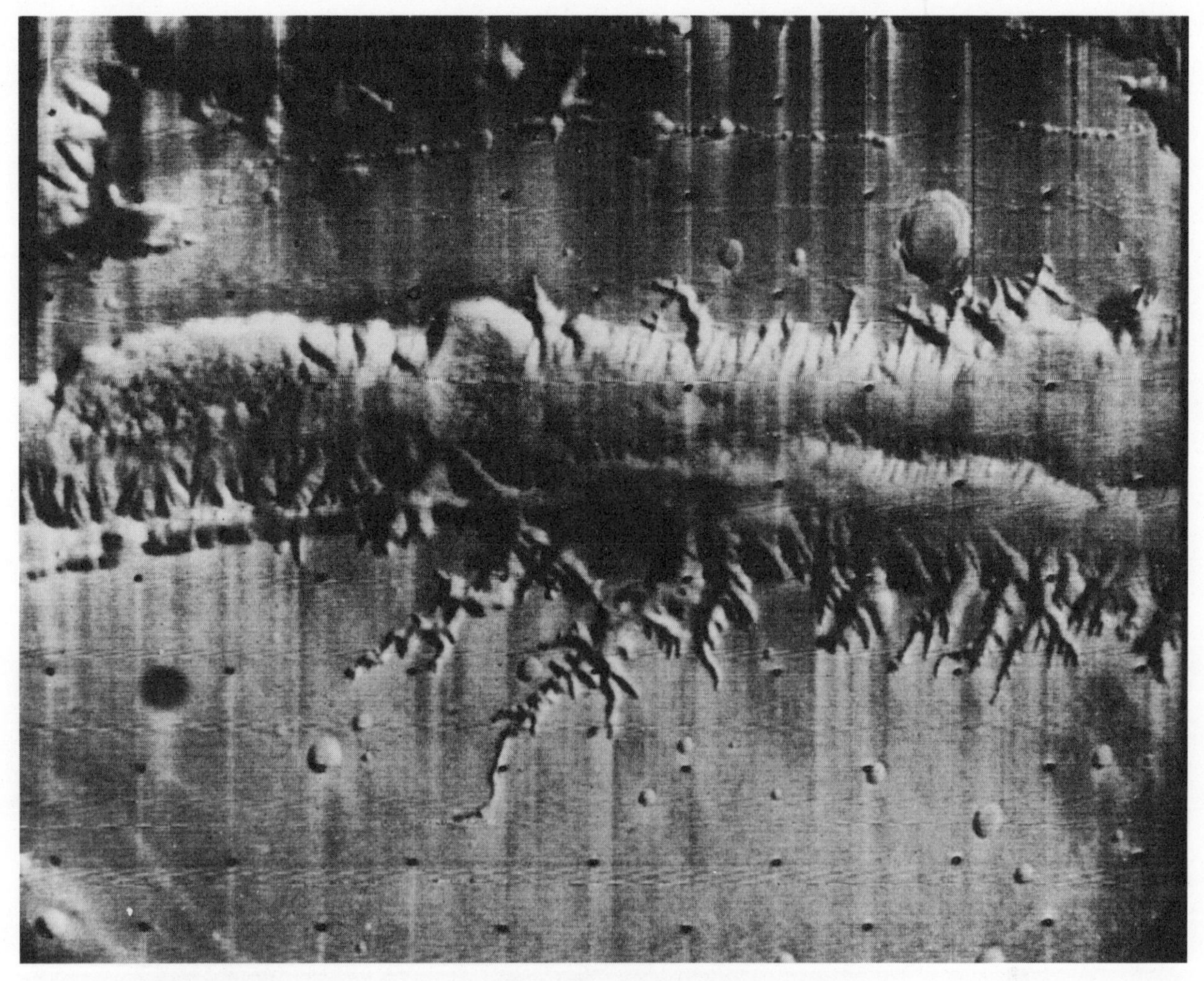

FIGURE V.30
Valles Marineris, a huge Martian canyon photographed by Mariner 9. *(NASA/JPL)*

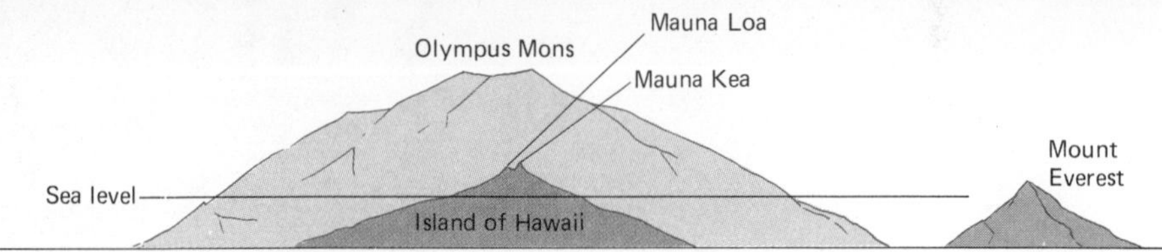

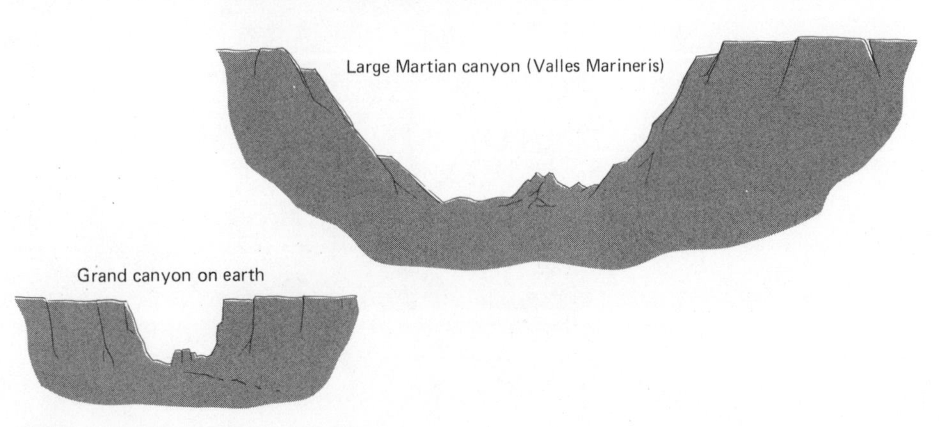

FIGURE V.31
The relative sizes of Olympus Mons, Mauna Loa, Valles Marineris, and the Grand Canyon, The vertical scale is enlarged 5 times.

such as Olympus Mons, have grown so huge. On the earth, a volcano grows over a "hot spot" in the mantle beneath it. Now if a crustal plate is slowly moving over that hot spot, the volcano emerges at different places on the surface of the crustal plate as time goes on. Thus island chains, like that of Hawaii, are built up over the tens of millions of years, while a moving plate (in the case of Hawaii, the Pacific plate) carries new parts of the earth's surface over the exuding magma.

---

9. Draw a diagram to show how an island chain can be built up on a tectonic plate moving over a hot spot in the earth's mantle.

I think the most interesting things found on Mars are dry river beds. Apparently all experts now agree that there is no question that running water was once present on Mars. Braided streams have been found, and even ancient islands in dry stream beds. Certainly there has been erosion by some liquid, and water seems to be the only candidate. Where is that water now?

FIGURE V.32
A dry Martian river bed, showing the braided pattern characteristic of many terrestrial stream beds. *(NASA/JPL)*

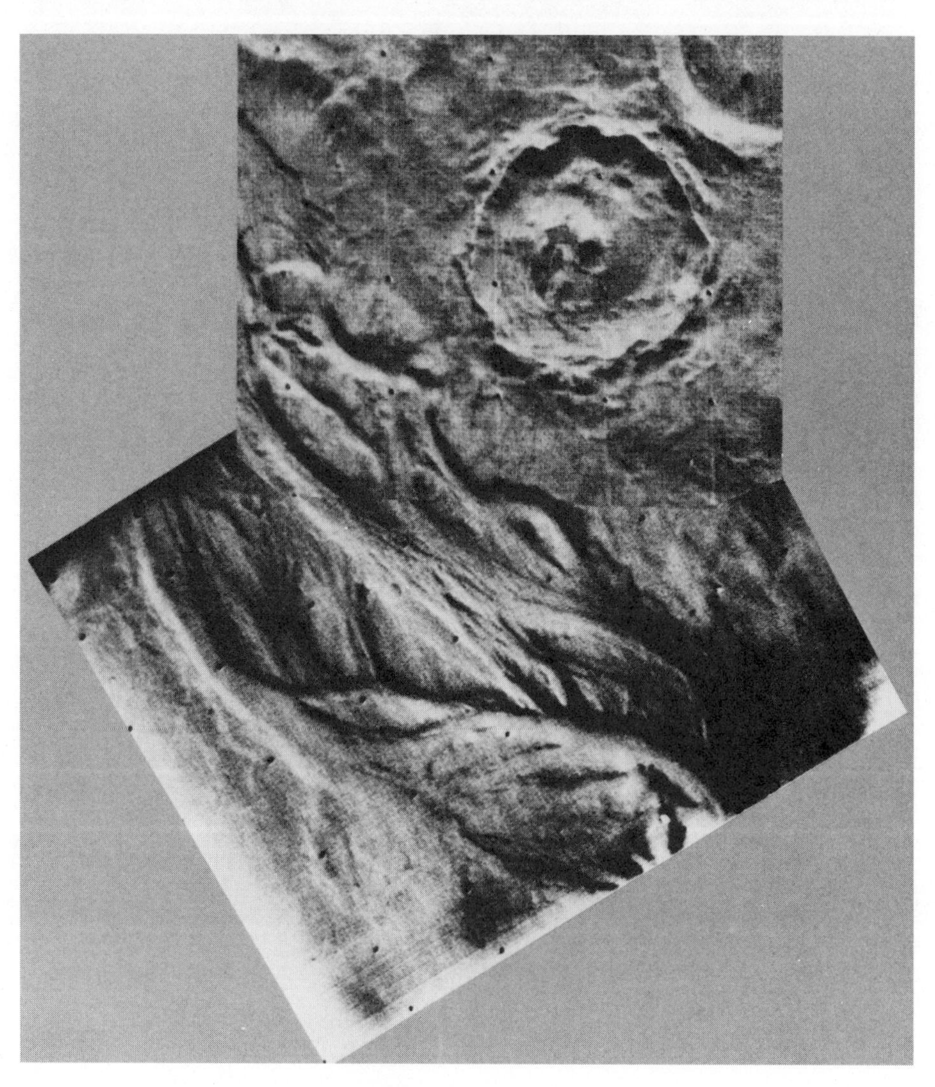

Evidently frozen, either as permafrost or in those polar caps, or both. Or perhaps most of it has evaporated into space, although a considerable amount of frozen water appears to be left.

Then why does the water not flow now? In part, perhaps because Mars may be in a glacial period, like the ice ages on earth. But there is another problem. Geophysicists have pointed out that liquid water can simply not survive in the low-pressure atmosphere Mars has today. Various experts estimate that the atmospheric pressure would have to be from 5 to 50 times as great as at present to allow liquid water to flow over the surface of the planet, and not immediately evaporate. Yet, as I said above, the abundance of argon suggests that the requisite atmospheric pressure may have existed in the past, perhaps more than 1000 million years ago.

Does this mean that Mars may once have been a planet of flowing water, like the earth? If so, has that water evaporated and dispersed into space, along with Mars' gradually dispersing atmosphere, or perhaps frozen out, never to flow again? There is, alas, some reason to think that such a fate may, indeed, have fallen upon Mars. The authorities disagree on the ages of the youngest great volcanos; some say a few hundred million years, and others say over a thousand million years. In any case, those youngest looking volcanic lava slopes do *not* have water channels eroded into them, which strongly suggest that they came after the last running water. Mars may well be in need of a celestial plumber.

But the Martian dream is not completely dead. If running water was there in the past, maybe life existed as well. Maybe life, in some form, or its remains, still exists in the Martian soil. The idea may be improbable. Most scientists think it is very unlikely. Yet the chance is there.

The Viking landers—in addition to carrying experiments to measure weather, seismic events, and all kinds of other things, not to mention television cameras to see any Martians who might be walking up—had long arms that went out a few meters to scoop up soil and bring it back into the laboratory for analysis. There were four biological experiments. Three were to test for respiration of living animals, absorption of nutrients by any living organisms, and exchange of gases between the Martian soil and the laboratory environment for any reason whatsoever. The soil samples were isolated and incubated in contact with various gases, radioactive isotopes, and nutrients in the various experiments to see what would happen. The fourth experiment pulverized the soil sample, and analyzed it carefully to see what organic material it contains.

I wish I could report definitive results to you. But then, if I could, you would have read about them in the newspaper; what more exciting scientific discovery could there be than that of organisms, living or dead, on another world?

But for today, at least, the answer is, "Well, gosh . . . ???" The Viking experiments that tested for absorption of nutrients and gas exchange clicked away like mad. It was as if life were rampant on Mars. Too rampant, in fact, for anyone to believe. Probably it was the result of a very chemically active soil on Mars, not living or dead organisms. The reason is that the organic chemistry experiments showed no trace whatsoever of any organic material present. At this writing, the possibility of biological organisms has not been entirely ruled out, but most experts consider it extremely unlikely. To be up to date, follow current periodicals. (But be sure to read a reputable journal, not some sensational rag like the National "whateveritis.")

FIGURE V.33
The Martian landscape, photographed by Viking 1. *(NASA)*

Summary

So in our survey of the solar system, we have found no place—not even Mars—where we think life can flourish. Anyway, there is no place other than the earth where *we* could live. Oh, of course, we could survive for limited times, as we have on the moon, and in the skylab, by carrying our environment with us. But that's only a temporary visit, at best. I sure wouldn't want to ask my kids to raise my grandchildren on Mars, or on the moon, or in skylab, with no lakes to swim in, no mountains to hike in, no wildlife to watch, and no oxygen to breathe outside a hermetically sealed enclosure.

And elsewhere in the solar system is even worse! You couldn't even stand on a Jovian planet, even if you could stand the pressures, the cold, the poisonous atmosphere, and the radiation. You would melt (literally) on Mercury and would crush under the atmospheric pressure on Venus, even if a rain shower of sulfuric acid didn't do you in.

Life is teeming—almost everywhere—on the earth. But so far as we know, in the solar system it is unique here. So far as we know it *never* existed elsewhere in the solar system. We have found precious few places in it that we'd even want to *visit,* and then not for long.

Does all this mean, then, that we are alone in the universe? I promise you no definitive answer to that question, and indeed, neither can any other honest person right now. But, if you are willing to read on, we shall at least explore the other possibilities in the next scene.

SCENE 4

Is There Intelligent Life Elsewhere in the Galaxy?

There are probably about 1000 million other galaxies near enough or bright enough to be potentially observable with present telescopes on the earth. Among them there could be 10^{20} or more stars, and a significant fraction of them could have planets. Many of those possible planets could have life. We may never know. Most of those galaxies are far off in space, and we see them as they were when light left them far in the past, when they were at much more primitive stages of evolution than they or we are now. To assess the chances of life in those systems, let us first explore our own Galaxy.

1. What are the problems of two-way communication by, say, radio with a hypothetical civilization on a planet revolving about a star in another galaxy?

Number of Habitable Planets in the Galaxy

Our Galaxy may have about 4×10^{11} stars; the actual number could well be half or twice that number, but that's as good a guess as I can make now, so let's assume it for the sake of discussion. About half those stars are probably main-sequence stars of the first generation in the Galaxy; if they are, they may not have formed from material with enough heavy elements to produce planetary systems. Judging from the incidence of duplicity of stars in the solar neighborhood, we estimate that at least half of those left would belong in double- or multiple-star systems. Planets cannot exist in stable orbits around stars in typical binary systems, because the gravitational accelerations of two stars acting together would result in highly chaotic motions of a planet; anyway, we don't think planets could even form in a double-star system. Some recent studies suggest that binaries may be even more common than generally thought, making up 90 or even 99 percent of solar-type stars. But let's be optimistic and suppose that there are nevertheless 10^{11} stars in the Galaxy that can have, and *do* have, planetary systems.

Astronaut Harrison Schmitt working beside a large boulder on the moon. The lunar roving vehicle is at the left. (*NASA*)

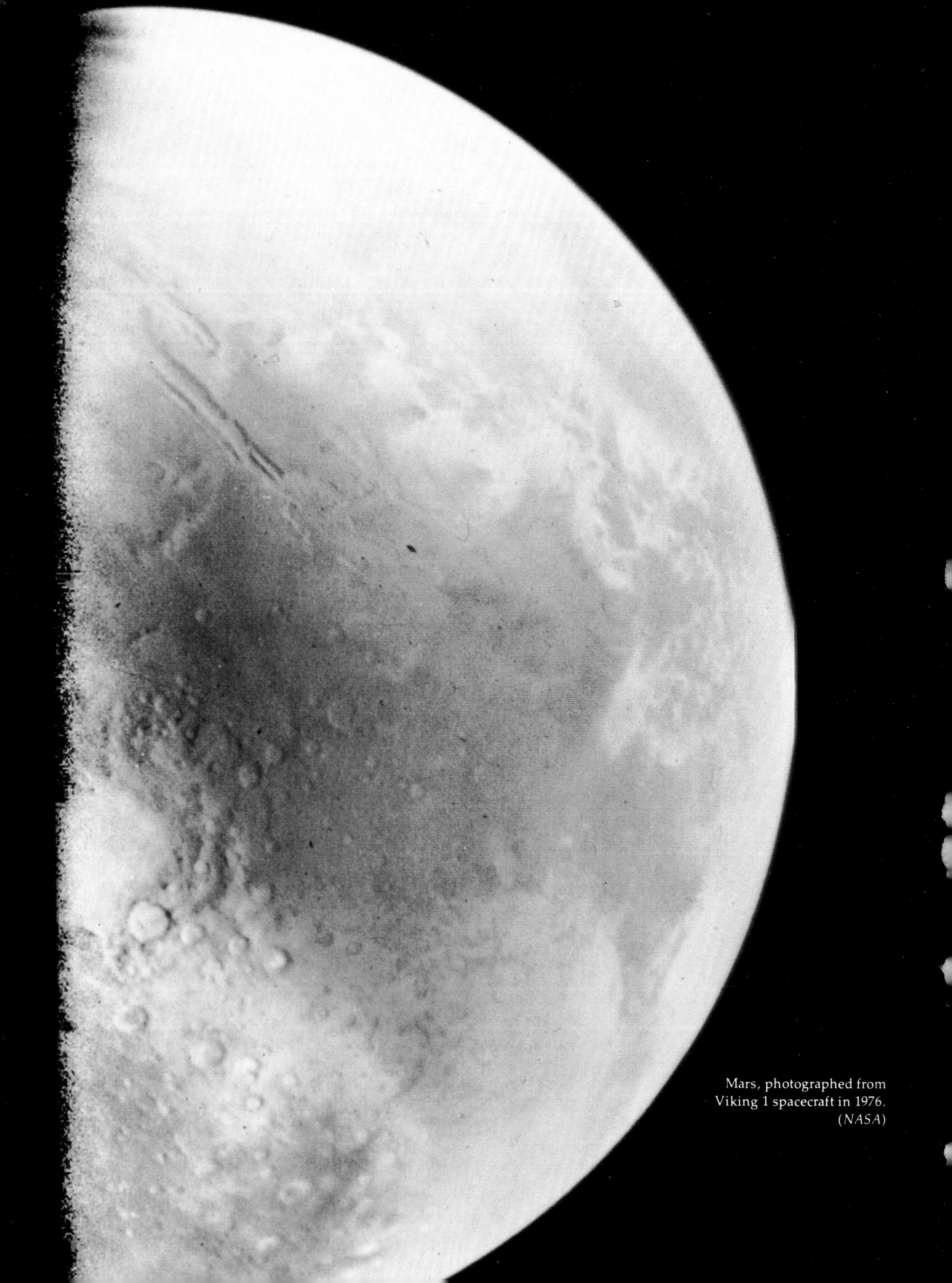

Mars, photographed from Viking 1 spacecraft in 1976. *(NASA)*

A region near the north polar cap of Mars, photographed by the Viking 2 orbiter. (*NASA*)

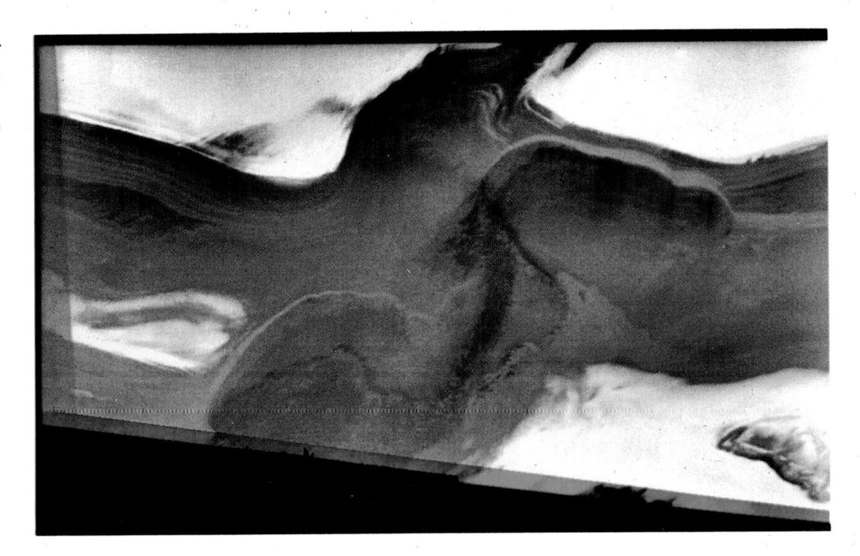

Looking south from the Viking 2 lander on 6 September 1976. (*NASA*)

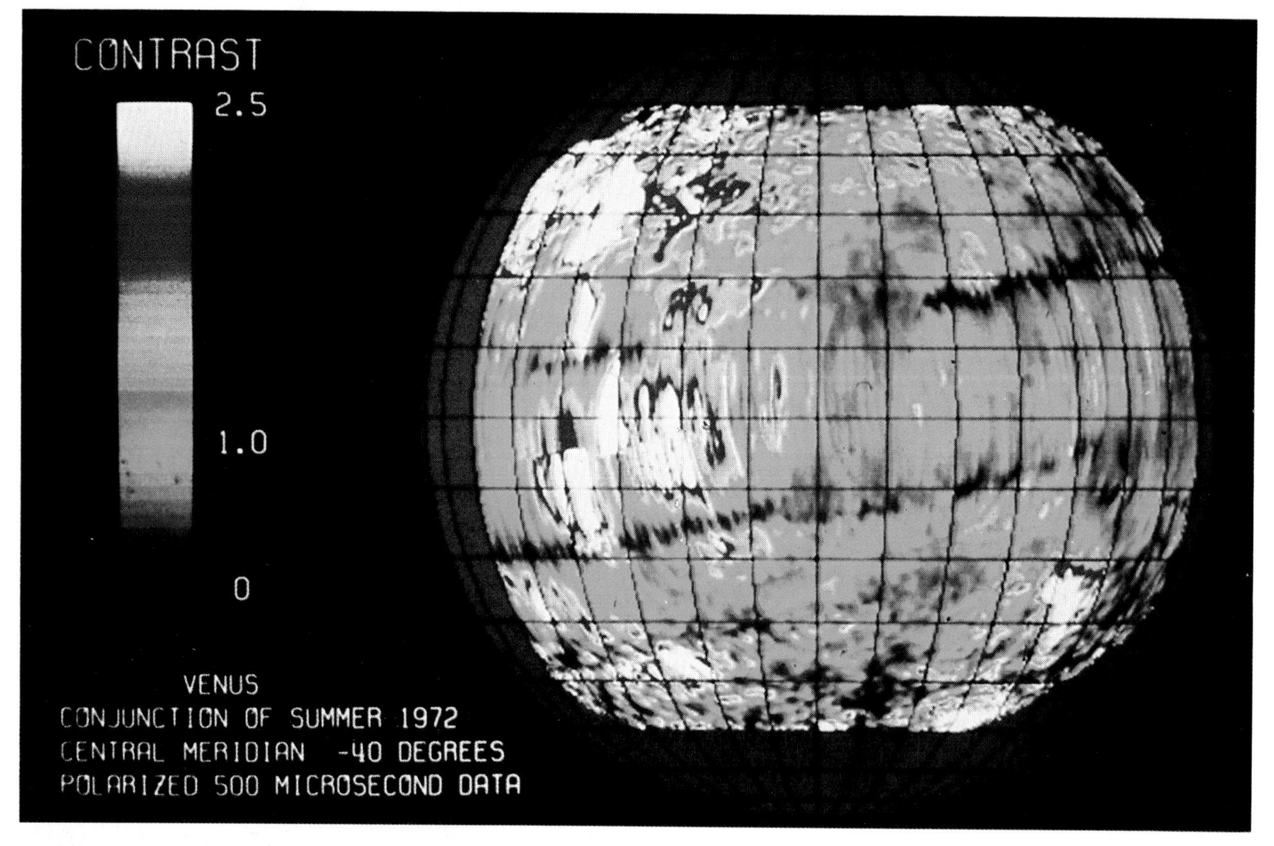

A radar map of Venus, prepared by the National Astronomy and Ionospheric Center, Arecibo, P. R. (*Cornell University*)

An ultraviolet photograph of Venus made with Mariner 10. (*NASA*)

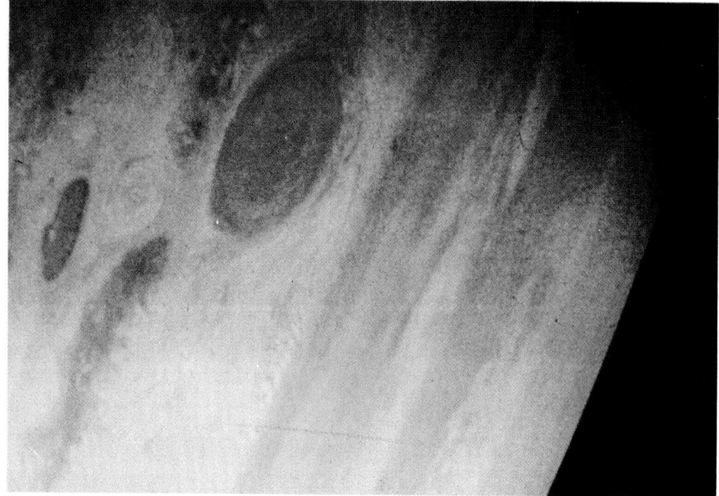

The great red spot on Jupiter, photographed by Pioneer 11 at a distance of 545,000 km. (*NASA*)

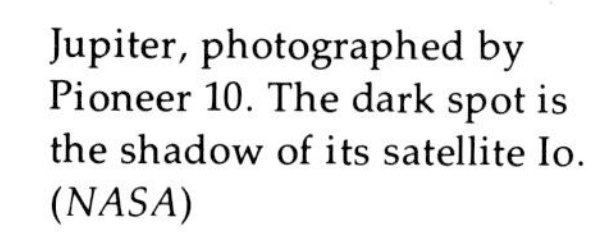

Jupiter, photographed by Pioneer 10. The dark spot is the shadow of its satellite Io. (*NASA*)

FIGURE V.34
Many of the thousands of millions of stars in our Galaxy may have habitable planets. This is a part of the Milky Way in Cygnus. *(Hale Observatories)*

We do not, however, expect life in all such planetary systems. On the earth, life evidently took about 1000 million years to get started, and we have seen that hot and luminous main-sequence stars don't live that long.

Faint main-sequence stars also have to be ruled out as good prospects for having planetary systems with life, because they give off so little radiation that a planet would have to be very close to heat up enough to have liquid water—much closer, in fact, than Mercury is to our sun. In such close proximity to its parent star, such a planet would almost certainly be locked into synchronous rotation, so that one side always faced the star. That side would be very hot and the other very cold, so any atmosphere would migrate to the cold side and condense.

Thus we are left, for good candidates for stars with planetary systems capable of supporting life, with stars not too different from the sun—those of spectral-type G and the cooler ones of spectral-type F. Only about 10 percent of the stars fill this bill, so we are left with (optimistically) 10^{10} planetary systems in the Galaxy.

In our planetary system, only the earth is near enough the right distance from the sun to have the necessary conditions to support life. Perhaps some systems could have two such planets, and others may have none. We are probably not too far off in guessing that there should be about one suitable planet per planetary system. So that leaves us (again, for the purposes of discussion) with 10^{10} planets in the Galaxy (that's 10 thousand million) capable of supporting life.

---

2. How many of the stars listed in Appendices 10 and 11 might you expect to have habitable planets? Explain.

Number of Civilizations in the Galaxy

Laboratory experiments suggest that if those hypothetical planets developed at all in the way the earth did, it should be almost inevitable that simple organic molecules would form on their surfaces. At present, however, no one knows whether it is inevitable that those simple organic molecules will somehow find their way into self-replicating molecules, such as DNA or some alien analogue. If there is liquid water around, chances are probably better, but even then they are by no means certain. Some biochemists believe that, given the right conditions and enough time, life forms *are* inevitable. Most, however, allow that the odds could be only one in ten, and some believe it is a long shot indeed. For this discussion, though, let's be optimistic again, and assume that the chances are somewhere between 10 percent and 100 percent that life will eventually form on those 10^{10} habitable planets. That means between 10^9 and 10^{10} planets with life.

Some biologists subscribe to the theory that given life, natural selection will lead, in time, to an intelligent, communicative species, with some kind of civilization. Well, in the one case we can test (so far) it kind of worked out that way; that is, we're here. On the other hand we took 4.6×10^9 years to make it, and no one knows if we were retarded in our progress or were unusually fast at it. Moreover, there is no assurance that every intelligent species will have an interest in trying to communicate with other intelligent civilizations in the universe. In fact, one could easily make the case that a truly intelligent society would know better. Well, it *may* be that every planet with life on it will, in the 10^{10}-year age of the Galaxy, sometime develop a communicative civilization. But it could also be extremely unlikely. I think we are being optimistic again by supposing the odds in favor of the evolution to such a civilization to be in the range from 10 percent to 100 percent. Nevertheless, we shall assume it, which leaves us with between 10^8 and 10^{10} planets in the Galaxy that at *some time* will have intelligent, communicative civilizations, ready and willing for contact with other such civilizations.

---

3. What circumstances must be realized in order that the emergence of life on a planet eventually leads to a civilization that wishes to communicate with other civilizations?

Already you may feel that with "reasonable" guesses you can get almost any answer you want; nevertheless we shall now push this numbers game further. The big question is, given such a communicative civilization, how long does it last, in comparison to the total lifetime of the star its planet revolves about (which we shall take to be 10^{10} years)? If a civilization can last in the technological state for a million years, that's one-ten thousandth the time its planet can support it. But that seems to me to be extremely optimistic for an *average* civilization—at least if we have any right to judge by ourselves. We have had a technological society with the ability to communicate by radio waves for less than 50 years, and unless I am unduly pessimistic (see the next scene), we will be very lucky indeed to survive another 50 years. If we suppose that the typical lifetime of a technological, communicative civilization is somewhere in the range 100 years to 1,000,000 years, we find that at the *present time* we should expect from one to one million such civilizations in the Galaxy. We could choose numbers to make the number higher, but only by adopting odds that would stretch the credulity of most investigators. If we reduce the estimate below the lower limit, of course, we have a problem explaining what *we're* doing here!

Although I have treated this matter with some levity—I think we have very little basis for making such estimates—this is nevertheless the kind of exercise serious investigators must go through in trying to estimate the prevalence of other civilizations in our Galaxy. They haven't much choice but to make intelligent guesses, because we have so little factual data to go on. If only the Vikings had given us positive information about life on Mars, at least it would have greatly strengthened our confidence that given the right conditions, life itself is almost inevitable. Absence of such information is not particularly discouraging, but it does still keep us guessing. Anyway, for the sake of further discussion, I shall choose my optimistic estimate—that there are one million civilizations currently existing in the Galaxy. I think few of the serious professional guessers would argue that the figure should be very much higher.

Now let us try to estimate the most likely distance to the nearest civilization. In our part of the Galaxy, there is about 1 star for every 10 cubic parsecs. If there are one million civilizations in the Galaxy and a total of 4×10^{11} stars, about 1 star in 400,000 has a planet with a civilization. With these numbers, it works out that there is an even chance of the nearest other civilization being within about 80 parsecs, or about 250 light years. If there were only 1000 civilizations in the Galaxy, the nearest one would have an even chance of being within 2500 LY. Now how do we go about visiting those fellows?

Interstellar Travel

Interstellar travel is not impossible. In fact, the United States has already launched several unmanned interstellar vehicles—Pioneers 10 and 11 and the Voyager spacecraft to Jupiter and other outer planets. After completing its assigned tasks of exploring the giant planets, each will wander endlessly through interstellar space—unless, of course, some alien astronaut should, by some extraordinary long shot, chance to come across one of them many millennia from now. We are prepared for this contingency; each Pioneer carries a plaque with symbols engraved on it in a cryptic code, telling about the world from which the space vehicle came. The plaque also has a drawing of a naked man and woman, to show what *we* are like. (The drawings were made by Mrs. Carl Sagan, wife of the Cornell astronomer, who is a fine artist. Carl denies that he posed for the male picture.)

The plaques, designed by Carl Sagan and his wife, are not just a stunt. It is, admittedly, extremely unlikely that they will ever be seen or appreciated, and even if they are, the space vehicles themselves will tell their discoverers far more about the race that built them. The Sagans designed the plaques mainly

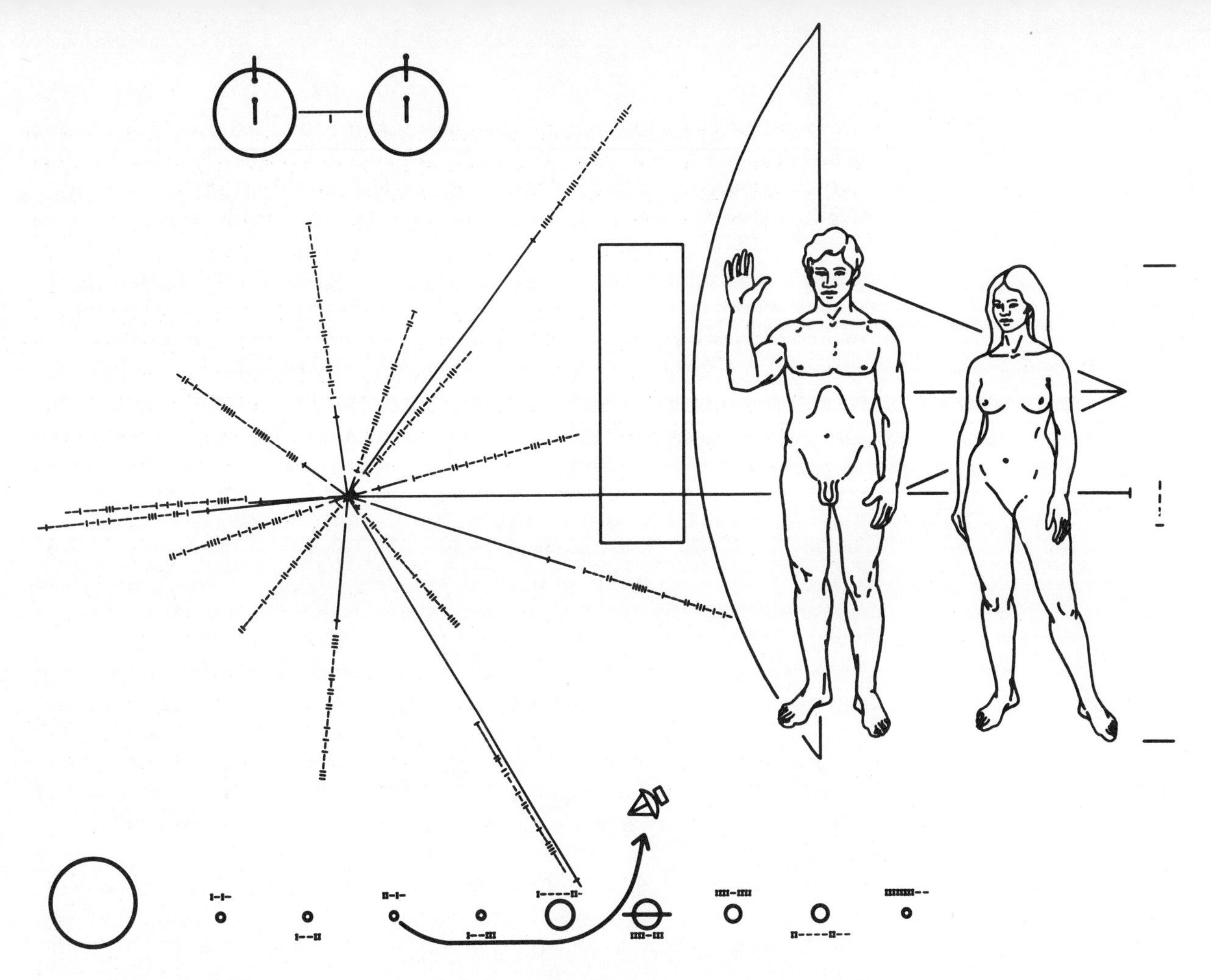

FIGURE V.35
The plaque carried by the Pioneer and Voyager spacecraft. *(NASA)*

for *us*, to emphasize that we may indeed not be alone in the universe, and now that we have the ability to explore remote worlds, we should be fully aware of the implications and the profound opportunities that our ventures may entail.

But the Pioneers and the Voyagers will be traveling very slowly in interstellar space—just 10 or so kilometers per second. Even if we could send *men* into space as fast as one-tenth the speed of light, it would still take 2500 years to reach the nearest civilization, and only if we knew where it was in advance. The spaceship would have to have a huge life-support system to provide for the astronauts and their descendants—destined to a voyage of many generations without knowing what, if anything, they will find at the end of their journey. Perhaps, we could freeze the astronauts somehow, to slow down their life processes, thus enabling them to hibernate for thousands of years, and then wake them up in time for their arrival. I know of no reason why such travel is necessarily impossible, and cannot rule out that we may someday try it, but I don't think it is a high-priority project for the immediate future.

---

4. At a speed of one-tenth the speed of light how long would it take a spaceship to reach Alpha Centauri, the nearest star beyond the sun? Would you expect to find a civilization on a planet revolving about that star? Why?

On the other hand, what if astronauts could travel at a very much higher speed? We recall from Act II Scene 4 that if one can travel at nearly the speed of light, his time passes more slowly than time for those at home, and he ages more slowly than those remaining at home. In principle, an astronaut could travel at nearly the speed of light to a star hundreds of light years away and return to earth still a young man, and meet his descendents of many generations. Let's see how much time we could save this way in traveling to a star, say, 100 LY away and back (200 LY round trip). The following table shows the travel times required for various speeds. The first column gives the astronaut's speed in percent of the speed of light. The next column shows the factor by which his time slows down. The third column shows how long his trip would take in years, according to people on earth, and the final column the number of years of his own time.

Speed (percent of light)	*Factor by Which Time Slows*	*Years for 200 LY Round Trip, According to Earth*	*Years of Astronaut's Time for Same Trip*
10	1.005	2000	1990
50	1.155	400	346
75	1.512	267	176
90	2.294	222	97
98	5.025	204	41
99	7.089	202	28
99.9	22.37	202	9
99.99	70.71	200	2.8

You can see from the table that you have to get pretty near to the speed of light before the time dilatation buys you much advantage. Unfortunately, when the speed of the vehicle gets close to that of light, it takes an enormous amount of energy to provide further increases in its speed. Let's take a rather modest example: Suppose you decide to travel at 98 percent the speed of light, so while your round trip, as measured by people at home, takes just over 200 years, you have aged only 41 years. Let's allow ten tons for the payload (three to five automobiles) – not much to contain you and your life-support systems, in fact your whole world, for 41 years of your life – and we'll add another ten tons for power plants. If you want to do your accelerating at a comfortable 1 g (the earth's surface gravity), it would take 2.3 years (at each end, of course). Astronomer Sebastian von Hoerner has pointed out that the energy would have to come from the complete annihilation of matter, and we don't yet know how to do that on a large scale. But even if we solved that problem, it would still require the equivalent of 40 million annihilation planets of 15 million watts each to supply the energy, which would then have to be transmitted backward (for propulsion). That transmission would need the equivalent of 6000 million transmitters of 100 thousand watts each, and all the annihilators and transmitters would have to operate with perfect efficiency. And the whole works would have to be contained within a rest mass of 10 tons.

Just to accelerate you and your spaceship to 98 percent of light's speed *once* – not counting slowing down and stopping at the other end, turning around, speeding up again for your return trip, and slowing for your final stop at earth – takes a total of 4×10^{29} erg. This is roughly the amount of energy the entire human race consumes, at the present rate, over 200 years.

The above figures on energy need are based on the most fundamental physical principles, involving conservation of energy. It is because of these very basic considerations that most of us do not regard interstellar travel by man as imminent.

---

5. Actually, even with complete annihilation of matter, a 20-ton spaceship could not carry enough fuel (within that weight limit) to produce 4×10^{29} erg. Can you imagine a source of fuel en route? What might be the problems of using that fuel?

UFOs

Well, if we can't visit other civilizations, maybe they can visit us. Indeed, many people think they have visited us. There are thousands of reports of *unidentified flying objects—UFOs*—which are widely believed to be spaceships from other planets from beyond the solar system. In fact, a Gallup poll some years ago indicated that some 90 percent of the American people believed that UFOs are evidence of visits by alien civilizations.

There are close to a thousand UFO sightings reported each year. To be sure, most are easily explainable in terms of natural or known phenomena, but a residual of reports—let's say, 10 percent— remain unexplained (perhaps because of insufficient evidence). On the other hand, most "UFOlogists" point out that only one-tenth or so of the people seeing strange things in the sky report them. Thus there could well be 1000 or more potentially inexplicable UFO reports each year.

Can these UFO reports be evidence of extraterrestrial visitors? As a first step toward an answer to this question, we should, I think, ask whether it's theoretically possible. Well, at the outset, I'll grant that it probably is. But now let's see how we might assess the likelihood of the extraterrestrial hypothesis being the correct one for explaining the UFOs. Suppose there are, in fact, one million civilizations in our Galaxy. Then one star in 4×10^5 should have such a civilization. In that case, within 1000 pc (or 3000 LY) we should expect among the 400 million stars about 1000 civilizations.

Now we can imagine no way that any of those 1000 potential civilizations could be aware of us. We have sent out no signals of our presence save our inadvertent radio and television programs and commercials, beamed through space for the past half-century. But even those early *Amos 'n Andy* radio programs are now only on the surface of a sphere of radiation less than 50 LY in radius, centered on the solar system but moving outward with the speed of light. Since we don't expect a civilization within 50 LY, chances are that no one on any other planet knows we exist. Even if a wandering spaceship nearby should "pick us up," its message with the news could not get back to its parent planet any faster. Thus the only visiting spaceships we would expect around here are those which were just out scouting at random.

Now within that 1000-parsec radius, about 10^7 (10 million) solar-type stars could conceivably support life. That's 10 thousand stars for each civilization—granting the most optimistic estimates on the number of civilizations. Thus if each civilization launched a spaceship per year toward a star that it thought could have another civilization (say, us), the chance would be only 1 in 10 thousand that one of those spaceships was directed our way. That means we should expect one interstellar visit every 10 thousand years or so.

On the other hand, to have 1000 alien visits each year, each civilization would have to be launching 10 million spaceships annually, sort of at random, to the various stars about them that might be expected to have planets. That's 25,000 launchings each day, or about one every 3 seconds from every possible civilization (by our optimistic assumptions); and this assumes that every one of those civilizations had mastered the art of interstellar travel, which we have not and cannot imagine how to. It seems unlikely that we are at the very bottom of the ladder of advancement of *all* possible civilizations that we would predict to exist, with liberal assumptions, in our part of the Galaxy; to account for the UFOs with spaceships, however, we have to do so.

Perhaps I have painted too one-sided a picture. Perhaps, for example, all 1000 hypothetical sightings each year are of the *same* alien ship. But the descriptions are not all the same. Some people report formations of lights; others a single light; some large colored lights changing in various ways; some disk-shaped things; others cigar-shaped vehicles, with or without portholes; and some report landings of vehicles, with strange beings walking about. By the way, almost without exception, these beings, as reported, are anthropomorphic—each having a head, a neck, two arms, two legs, and so on. (Which may say something about the pilots of the alien spaceships or something about the imagination of the reporters.)

But, some say, those strange apparitions I am talking about are mostly explainable in terms of natural phenomena; it is only a rare few that are really alien spaceships. On the other hand, if all but a small handful of the allegedly "unexplainable" reports can really be explained away, one might wonder whether the last few might be as well, if we had better information about them.

Evaluation of UFO Reports

Well, arguments such as these have made most scientists, including me, highly skeptical that UFOs are intelligently controlled spaceships from planets revolving about stars beyond the solar system. Such a theory may not seem farfetched to a layman, but it seems extremely farfetched to those of us who know something about astronomy.

Still and all, when we account for all the obvious hoaxes and all the likely natural effects that can account for the phenomena described, and can eliminate all the stories by those who are overimaginative or likely to be unbalanced, even then there remain a few "tough cases"—stories by evidently highly reputable and experienced people that are hard to explain. What about them? Well, most professional scientists are so conditioned by the public's sincere desire to believe in the most bizarre or sensational hypothesis that they lean the other way—in order to balance somewhat a very unbalanced treatment of the subject by the press and in the popular literature. Most of my colleagues publicly express the opinion that the whole UFO business is nonsense. But of course no one can ever be certain—he can only guess the odds.

The fact is, there is no hard, convincing evidence for extraterrestrial visitors—just strange stories and hearsay. It's terribly hard to check up on anything about the subject. As MIT physicist Phillip Morrison puts it, when one (that is, a UFO) lands in Cincinnati and is taken intact to the Smithsonian Institution, dissected, examined, and made available to competent and interested scientists from anywhere to come and see for themselves, then we'll have some hard evidence.

The problem is that most of us want very much to be able to believe in extraterrestrial life. Certain evidence of its discovery would probably be the most important news of the century. But especially as professionals, we have a particular responsibility not to go off half-cocked. If we even hint that maybe there's some evidence for life elsewhere, the press and the public tend to shout, "It's proven!!" Many scientists have learned to be doubly cautious in dealing with subjects of particular newsworthy interest. You can find so-called newspapers in your supermarkets which carry all kinds of sensational "scientific news releases" that are pure bunk. So you must be patient with us if we scientists sometimes tend to be overly conservative.

Yet, it would be equally mistaken to assume that there could be nothing to the UFO phenomenon just because we think there is *probably* nothing to it. Consequently, I have cooperated with a small group of scientists who have agreed to look seriously at any evidence that has promise of being hard and testable. I have been personally involved in two cases, and peripherally in a third. One was completely explained; one was probably explained; the third turned out not to be a significant mystery after all.

The reason for this discussion is to make the point that there is a difference between being open-minded and being gullible. Some things have been proved impossible years ago (like the hollow, inhabited sun); we should not waste our time on them. Some are very improbable but not positively impossible; in those cases, we should look at any real, honest evidence that pertains to the case. Some are unlikely, but could be; those we should examine especially carefully—we might discover something new and exciting. But we must be very careful not to mold our judgment to what we *want* to believe.

Communication with Other Civilizations

So now where do we stand? There may be other intelligent civilizations in the Galaxy. We can't imagine how to visit them, even if we knew where they were (if, indeed, they are). We think it to be exceedingly unlikely that any of them have visited us. How then are we to ever know? How about trying to communicate?

Communication means that we use electromagnetic radiation to carry signals. There's not much point in trying to send light signals; the earth, as seen from any other star, is sitting right on top of the sun, whose great luminosity would completely swamp any light signal we could hope to transmit. The best bet would seem to be at radio frequencies. We can beam a lot of energy in a small wavelength band, and in that band of frequencies far outshine the sun. In fact, we have the technology right now to send radio signals that could easily be detected by a civilization on the far side of the Galaxy with the same technology as we have, if they were looking for those signals. Why, in a few centuries, the nearest civilizations (if they exist) could probably even eavesdrop on our radio and TV shows—which would doubtless encourage them to turn off their radio receivers forever!

But if the nearest probable civilization is 200 to 300 LY away (by our optimistic estimate), we'd have to wait centuries for a reply—assuming that someone could figure out what we were trying to say. It might take a long time for a floating blob of gas to decode what we were getting at in describing how our hearts pump blood through our arteries. If Congress appropriated funds for the project, I suspect it would not be many years before someone—Senator William Proxmire, for example—was asking questions about that item in the budget. (As a matter of fact, a radio message was beamed toward the globular cluster, M13, from the Arecibo radio telescope in Puerto Rico. But it was more of a demonstration of what was possible than an attempt to communicate. I think M13, having all early-generation stars, is highly unlikely to have any planets, let alone life.)

---

6. Suppose we could carry on a two-way radio communication with another civilization. After a question is transmitted, what is the time we would have to wait before we could expect to receive an answer if that civilization is

a) on the moon,
b) on Jupiter (assume Jupiter is at its nearest to the earth),
c) on a planet revolving about Alpha Centauri,
d) on a planet revolving about the star Tau Ceti (see Appendix 10 for the distance to that star),
e) at the center of our Galaxy?

On the other hand, we could try listening. In fact, we have already. The first such project, the idea of Frank Drake, now director of the Arecibo installation, was to monitor a few neighboring solar-type stars just to see if any radio messages could be detected that looked like intelligently coded signals. Drake's idea was to observe at a radio wavelength of 21 cm, that of the emission line of neutral hydrogen, so prevalent in the Galaxy. Drake felt that any intelligent technological civilization would know about that 21-cm radiation and figure that radio astronomers elsewhere (here, for example) would be trying to observe neutral hydrogen in the Galaxy at 21 cm, and hence might chance to stumble on coded signals from their civilization, broadcast at that wavelength. The experiment was called *Project Ozma* (for the good princess in Frank Baum's *Wizard of Oz* tales). Several other experiments, involving surveillance of some hundreds of stars, have been carried out both in the United States and in the Soviet Union. Nothing has been turned up, but that is not particularly discouraging, since one would expect to have to survey many thousands of stars to hope to achieve success, even with the most optimistic estimates on the number of other civilizations.

---

7. Suggest ways in which you might code a radio message sent into space so that a civilization that happened to receive it would have a good chance of recognizing it as intelligently contrived.

Project Cyclops

Then how could we hope ever to know if we are alone? The matter was studied in 1971 by a NASA team under the direction of Bernard Oliver. The Oliver Committee designed a system, called *Project Cyclops,* that should be able to detect any civilization within a few thousand light years trying to contact other civilizations in the Galaxy by radio. The project consists of a vast array of radio telescopes, covering an area more than 5 km across. It would cost thousands of millions of dollars, but would almost certainly pay for itself in technological "fallout" even if no other civilization were ever detected. In any case, it would be a marvelous radio telescope, and would certainly tell us a great deal about the universe. I can imagine few better ways to spend our resources.

Blasphemy! you say. With all those people starving and cancer yet to be cured! But I think we are more likely to discover something that may help us find a cure for cancer by building Cyclops (or some other such venture) than by *not* building it. And I am quite certain that whether or not we build Cyclops is completely irrelevant to the feeding of the starving multitudes.

On the other hand, what if—just *what if* (as I doubt)—there *were* many other civilizations in the Galaxy, sending radio messages throughout space to be received and (with ingenuity) decoded by other races, and then *they* were to pass on more messages to *other* Galactic observers. No one civilization would ever expect a specific reply—quite likely it would no longer exist by the time its messages were received. But all the same the word would still be passed around. The idea has been called *The Galactic Club* by Stanford's Ronald Bracewell. If that club *does* exist, it would be great to belong—especially when the dues are so low.

FIGURE V.36
Artist's conception of the array of antennas proposed by Project Cyclops in 1971. Each of the large radio telescopes in the system would have an aperture of at least 100 m, and would be designed to receive signals in the 1420 to 1660 megahertz frequency range that might originate from a distance as great as 1000 light years. *(NASA/Ames Research Center)*

FIGURE V.37
Artist's conception of a closeup of several of the antennas in the proposed Project Cyclops. The building at the right, near the center of the array, is the proposed data collection and processing facility. At present (1978) there are no plans to actually build the system. *(NASA/Ames Research Center)*

But what if it does *not* exist, or if we fail to find out, for lack of imagination or vision? Then, if our own civilization is about to snuff itself out, as I very much fear it is, so far as we shall ever know, the universe can be sterile—devoid of life forever.

Consider a puppy, born alone in the wilderness. By nature he is gregarious—as we are. His natural inclination is to trot out into the world in search of friends. He may be naive and even foolish to do so, but what is his alternative? To curl up in his cave until he dies?

SCENE 5

Is There Intelligent Life on Earth?

In part, this scene is an editorial. But I think I can back up my opinions with cold mathematical realities.

The world is full of people. In my opinion, it is too full. It is very difficult to get into any state park in summer. Yellowstone is terribly overcrowded in season. You can't get through the streets in Rome. Children, I am told, are begging on the streets of Calcutta. You can't park an automobile in downtown Edinburgh. The welfare rolls are growing in our great cities (and elsewhere). The Muir Trail in the Sierras (between Sequoia and Yosemite) has become so crowded that its waters are polluted. Most of our great animals, both on land and in the sea, are being crowded to extinction; tigers are endangered, many species of whales are endangered, elephants in Africa are thinning out, the California cougar is almost extinct, and on and on. Our wilderness is vanishing, and I shall miss it. My grandchildren, if I have any, will never know it. I think they will be the poorer for it.

Meanwhile, the world population is growing. It is growing, according to United Nations figures for 1976, at 1.9 percent per year. Do you know what this means? It means that at the present rate the population of the world *doubles every 37 years.* Again according to U.N. data, the population is growing fastest in Africa, where 40 out of 47 countries double their population in less than 35 years and 10 nations in less than 23 years. Mexico and some other Latin American nations are also doubling their populations in 20 years or so. If you have traveled in Mexico, I am sure you have enjoyed the goodwill and hospitality of our neighbors to the south, but you must have been disturbed by the poverty of so many of the people there. Yet, Mexico must *double* its roads, hospitals, schools, homes, factories, in fact everything it has, in just 20 years *to maintain that present level of poverty.*

Geometrical Progressions

Do you *really* understand what a doubling every 37 years means? It is called a geometrical progression. Consider another example of such a geometrical progression. Take a sheet of ordinary typing paper about 0.1 mm thick. Suppose you cut that sheet of paper exactly in half. Now lay these two pieces on top of each other, and cut the two of them exactly in half. Now lay the four sheets on top of each other, and again make a cut, this time producing 8 sheets of paper, and pile them up. Continue the cutting and stacking process. Admittedly, pretty soon the papers begin to slide a bit, and your scissors slip, but suppose that you had infinitely sharp shears, and could continue cutting and stacking, until you had finally stacked up the pieces resulting from your one-hundredth cut. Guess how thick the stack of paper is. The answer is 1.34×10^{10} LY—probably near the distance to the most remote quasar ever discovered (next act).

---

1. Take an ordinary sheet of paper, a straightedge, and a pen or pencil. Draw a line (representing a cut) that roughly bisects the paper. Now, perpendicular to your first line draw a second line bisecting one of the halves. Keep this up, counting the number of times you are able to continue to bisect the smaller and smaller rectangles. Try to imagine continuing to 100 bisections.
2. A lad answered a help-wanted advertisement for a delivery boy. When asked what salary he would work for, the youth agreed to work for one cent for the first day, provided that the employer would double his salary each day. How much did the boy earn after 30 working days?

FIGURE V.38
Space is growing short.
(John Littlewood, Editorial Photocolor Archives)

Once when I was flying from Los Angeles to New York, and the plane was over the Arizona and New Mexico area, the woman sitting at the window beside me looked out and remarked how wide open our nation is—how much room we have left. A few hours later, we chanced to share a cab into Manhattan, and were caught in the rush-hour traffic. She made no such utterance then.

Now Manhattan Island has just over 100 m^2 for each man, woman, and child living within it. Most people regard that area as rather crowded; but let us suppose it could be more than 100 times as crowded as now, so that there was only 1 m^2 (about 1 square yard) per person. Moreover, let us suppose that the entire 1.358×10^8 km^2 of land area of the world could be equally crowded. At the present rate that the world's population is increasing, this would happen in just over five and one-half centuries.

One could argue, I suppose, that the world can support more people even than one per square meter. You can imagine people living their lives in multilayer bunks, with tubes in everyone's mouth, and feeding on sea plankton. Yet, remember, our bodies are made mostly of water, and the oceans of the world contain most of the earth's water supply. Thus if we people multiply without limit, it must be at the expense of ocean water (you can't make something from nothing). At the present rate of the world population growth, the oceans would be converted to people in less than 1150 years. If we could find a way to convert the solid material of the earth into people, the entire earth would be converted to people in about 1640 years. In just over 2300 years we would have to transform the entire mass of the solar system into people to accommodate the present rate of increase in their numbers. If you could create people out of nothing, in about 5700 years there would be a great solid sphere of humanity 166 LY in radius, growing so fast that its surface would be moving at the speed of light!

As a matter of fact, the world population has *not* been increasing at a constant rate; it has been increasing at an *increasing* rate—indeed, at a rate proportional to the population itself! To show this, I plotted Figure V.39 from data available from the United Nations on the world population since 1850. For each of several dates I simply divided the total world population into the total number of square kilometers of land area of the earth, to see how much space, on the average, there is for each human. Notice how well the plotted points (representing the actual population figures of the world) follow the straight-line prediction. By this reckoning, the population becomes infinite, and the space available to each of us goes to zero on 20 August 2023.

Note that the present trend indicates that Doomsday comes in just over four decades! The rate of the world population growth is still *increasing*. When will it taper off? If you have a child that was born within the past few years, don't you see that something has to change drastically within the next few years for your child to have a chance to survive to middle age?

---

3. Why does Figure V.39 imply that the time required for the world population to double (the doubling time) has been decreasing over the past 200 years, instead of remaining constant?

The Future

My point in the above discussion is not to predict that the world population will become infinite, or that the oceans will be converted to people, or that we will live in a density of one person per square meter. That cannot happen. My point is to demonstrate that there is a limit to the population that the world can support, and that that limit will be reached *very soon*—at most in a few decades. The ultimate limit is not a matter of technology or our ability to grow food—it's a simple matter of conservation of mass—you can't make something from nothing, so eventually people will have to stop reproducing faster than they die. In no possible way can the earth feed an ever-increasing number of people, let alone clothe them, house them, provide oxygen for them to breathe, or, ultimately, even provide standing room for them.

Yet, I have heard it seriously argued that there is no limit to the population of humans, because we can move out and colonize the planets. But consider this: The entire combined area of the moon and Mars is only a little greater than the land area of the earth. Even if you *wanted* your children and their children condemned to lives on those barren and hostile worlds, we could put off the problem only for about one more generation, because in one doubling time of the population we would have filled those worlds as full as the earth is now! What about planets revolving about other stars? Even if we could get there, and again, even if we *wanted* our descendants to grow up there, and even if we could populate every available planet that could support life (according to those optimistic estimates we made in the last scene), it would take us less than 400 years to fill up all the planets within 160 LY with people at the density of the present earth population. And to accommodate our ever-increasing population by extending it even further, we would have to transport people at speeds faster than light—and this is impossible.

Long ago, Malthus warned of the same problem. He pointed out that the population increases geometrically, while our ability to grow food increases only at a rate proportional to time. His predictions of gloom nearly two centuries ago were premature, and consequently many people have been lulled into believing that the population problem is only a theory that has not been proved right. Wrong. It is not a theory; the limit to population growth is a mathematical fact. Our capacity to produce food has limits—ultimately (but probably much sooner) by the conservation of matter and energy itself. Malthus was absolutely correct in predicting that the population of the world will cease to grow. The question is *why* will it cease to grow?

I *wish* the reason were that we would realize there is a problem, and that there would be worldwide cooperation in stopping the growth of our numbers, and even reducing them to a level we might decide is desirable. The problem is to find agreement on how to go about it. There are nationalism, ignorance, and ways of life that have become habits hard to break. The governments of the world have not, by and large, shown interest in stabilizing population. Richard Nixon, when he was president, rejected a report of a committee he appointed to study population problems on the grounds that it was not the government's business to interfere with people's family planning. In the early 1970s the United Nations held a worldwide conference in the Balkans which resulted in a resolution to the effect that the problem of the world is not overpopulation but distribution of wealth.

I suspect the severity of the problem will be obvious only when it is too late to do anything about it. Notice how suddenly things happen in Figure

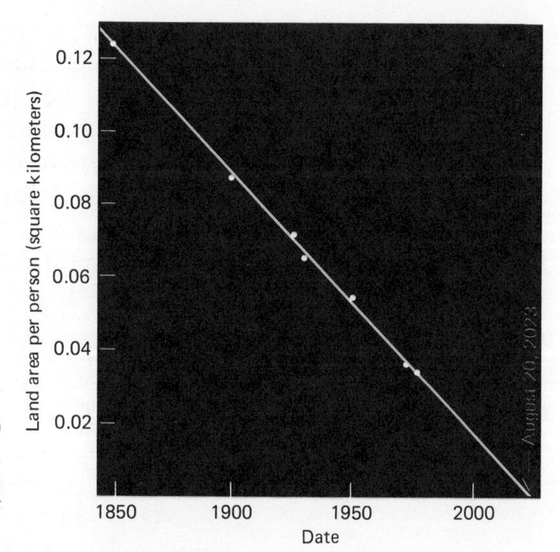

FIGURE V.39
The land area of the earth available to each living human being, as a function of time.

V.39. Exactly how disaster will strike, however, is harder to predict, because the problem involves other things, such as energy supplies, weather and food production, wars, and the general collapse of order in the society.

The World's Resources

Consider, for example, our use of energy. For the most part, we have been relying on fossil fuels for our energy—at least in North America and Europe. At the present rate of usage, the supply, including coal and undiscovered petroleum, might last for a century or two. But there are other problems with the combustion of fossil fuels, as we shall see. Moreover, if the underdeveloped and developing nations were to use *their* share as fast as we in the United States and Western Europe burn ours up, the fuel supply would surely be exhausted in a few decades. Worse yet, our use of fossil fuel has been *increasing*— and at a faster rate than our population.

In early 1974 a tightening of oil supplies led to a minor crisis in the United States and Europe, and there were horror stories of violence in queues at gasoline pumps. For a time, it looked as though economical automobiles might become popular, but when the gasoline started flowing again (at greatly increased prices) its use began climbing at the normal rate, and enormously fuel-hungry motor cars regained their popularity. Such is the way people worry about the future of our energy sources.

If most of the people of the world are eating today, a good fraction are just *barely* eating. As the population keeps growing, think how critical it is that food production be maintained throughout the world. What happens if the monsoons fail to arrive in southern Asia one year or if severe northern winters, like that of 1977, should occur several consecutive years? The survival of the world—or at least our present society—is terribly dependent on relatively stable worldwide weather conditions.

The Role of Weather

Of course we can never count on the weather. There have always been short-term fluctuations for unexplained reasons and also the long-range changes marked by the ice ages, which we think we are beginning to understand. We don't expect the onset of the next ice age for thousands of years yet—at least if Nature has her way. But consider for a moment what would happen if a premature ice age *did* come. What if, as we think happened last time, it just suddenly started snowing in northern latitudes, and snowed day after day, a meter or so a week, and kept on snowing for months. In 1977 the city of Buffalo, in New York State, had a mild taste of this sort of thing; it snowed pretty constantly for several weeks, and enough snow accumulated to nearly cripple that city. But that was a temporary and local event. What would happen if two or three times as much snow as fell on Buffalo covered most of our breadbasket land in northern Europe and America, and the fall was permanent? It is doubtful that much of our present society would survive it. Can *we* bring around such profound and devastating changes in weather patterns?

One thing we are doing is accelerating the combustion of fossil fuels. Recall that our oxygen supply has been built up over several hundreds of millions of years just because that coal and oil did not oxidize when the organisms that formed it died. In principle, if we burn it all up, we should use up that oxygen and turn it back into carbon dioxide. The depletion of oxygen at the present rate, however, would require thousands of years. Far more important is the concentration of carbon dioxide in the atmosphere. On the average it is about 0.03 percent by volume, but it has long been known to be in higher concentrations over industrialized areas. Since the total concentration is so low, to raise its concentration very significantly we need pour relatively little into the atmosphere, at a rate too great for it to be removed by absorption in the sea and in surface rocks. Measurable increases of the mean content of carbon dioxide in the atmosphere over the past few decades have already been reported. That gas blocks some of the infrared energy the earth tries to radiate into space—the greenhouse effect, which we discussed in connection with Venus. It might take a long time to turn the earth into a Venus hellhole, but we don't have to heat up the surface of the earth very much to foul up our food production and starve most of us.

On the other hand, we are also polluting the atmosphere with particulate matter—solid crystals of carbon and other junk. This stuff serves to shield us from sunlight, reducing the energy that strikes the ground. If that effect won over, the result would be premature cooling. This kind of fiddling around with the incident sunlight on the earth worldwide—the solar *insolation*—could constitute an anthropogenic effect on ice ages that we hinted at in Scene 1. What would happen if it cooled enough that the oceans froze? Since the ice reflects sunlight far better than water, the absorbed sunlight would be further reduced and the oceans might never thaw out again. Mars appears to have lost most of its atmosphere and may be permanently glaciated. No one knows what we're doing with our polluting of the air with particles and with our increasing its carbon dioxide content, but it'll be darned lucky indeed if all our tampering just happens to cancel out and have no effect at all. Nature can be capricious enough with her crises without our helping her along!

I am not competent to offer an opinion on whether the aerosols we are dumping into the atmosphere will actually disrupt the protective ozone layer (without which all biological life is endangered), but a number of people who *are* qualified are worried about it. Our official reaction, though, is that since we are not yet *sure* it's hurting anything, we'll just go on spraying around. When and if we have established that the aerosols are, in fact, destroying the ozone, then of course we'll stop, but then there won't be much need to!

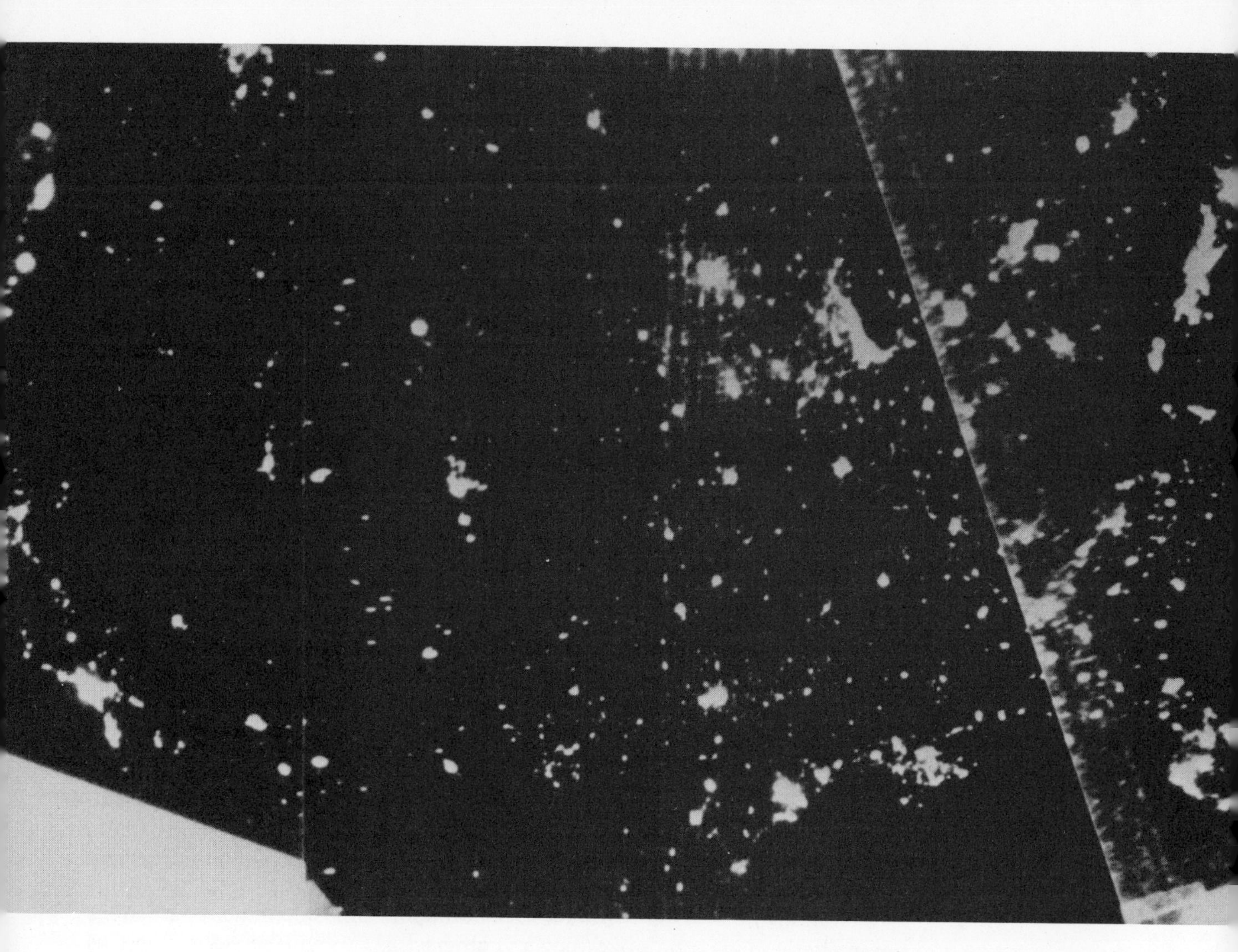

FIGURE V.40
Light pollution is a problem to astronomers searching for new observatory sites. This mosaic of satellite photographs of the United States at night shows clearly where the major population centers are. *(Kitt Peak National Observatory)*

Just our continued accelerated use of energy, whether from burning fossil fuels or in nuclear reactors, will eventually raise the mean temperature of the earth's surface. At the present rate, it is quite conceivable that within a hundred years or so the rise in temperature could be enough to affect production of food adversely. I suspect we'll never know whether *that* possibility is serious enough to worry about.

Other Crises

If, in a decade or so, you go to the supermarket and find that it has closed because the food is gone or, even if it is open, that great crowds of people are struggling for the few packages and cans on the nearly empty shelves, you can well expect violence. Even today it is becoming unsafe to walk the streets alone at night in many of our large cities. What can we expect when decent citizens cannot buy bread for their families because there is none to buy, except possibly on the black market? Is this scenario really so farfetched?

Perhaps it is, because probably before that point is reached, the competition among nations for diminishing resources and food will result in wars that will end our society. At least until recently (and maybe still) the stockpile of nuclear weapons was increasing faster than the population; even back in 1970 the total supply of nuclear bombs in various nations' arsenals was enough to supply the equivalent of 10 tons of TNT for every man, woman, and child on the earth. Ten tons of TNT provides enough energy to raise a 10,000-ton apartment building 500 m into the sky; man has the standby weaponry power to raise and drop such a building on every living person!

I have touched on a number of crises that either are or might well be facing mankind: energy, food, weather, violence, war, and, most fundamental and basic to all the other crises, overpopulation. Even if these problems could be faced squarely and solved, however, there are still challenges to the long-term survival of man's civilization. For example, there is the problem of genetic degeneracy. Improved medical care increases life expectancy, but encourages survival of genetically unfavorable mutations. And there is the threat of horrible boredom. Is man, by nature, able to cope with a permanent life on a stable planet, completely filled and developed, or would the absence of challenges itself lead to his stagnation and downfall?

I do not know whether these last problems are serious ones to man or not, but I see little hope that he will survive to ever find out. Can you conceive another 100 years at our present rate? another three doublings of our numbers, bringing the world population to over 30 thousand million people—eight times its present number? I figure we'll be lucky to last 40 years. That's longer than I can expect to be around, but I worry about my children. Is the future *really* this gloomy, or have I painted an overly dismal picture? Think about it yourself. What alternative scenario do *you* envision?

---

4. Would you recommend that the NASA budget be canceled and that the money so saved be spent instead to find ways to avoid the crises discussed in this scene? Why or why not? What alternative might be appropriate?

Of course it's obvious the problems *could* be solved. It's not a matter of technology. We *know* what causes overpopulation. We *know* which sources of energy are safe and which are not, and which should be developed. We don't know what we are doing to the weather, but we *know* how to *stop* doing it to it. We can at least imagine ways in which we could live in peace without violence and war. It isn't scientific principles that need discovering or gadgets that need inventing. It's *people* who need convincing. People are basically decent. But by our nature we have prejudices, nationalistic ideals, family habits, superstitions, hopes, beliefs, hates, ways of life, and a host of attitudes and notions that make it difficult for us all to think in unison and work together for our common survival.

In the past it wasn't necessary. Our nation was built by individual enterprise. I believe in it today. But the time for unbridled expansion has passed. Times have changed faster than we have, as have the conditions for our survival. Concerted action must be taken at once; I believe it will be too late in 20 years. It will be hard to convince the people of emerging nations that it is folly to follow in our footsteps. It will be equally hard to convince ourselves to change our aspirations, and to exploit only the resources of the earth that are replenishable.

FIGURE V.41
An ominous exhibit at the Bishop Museum, Honolulu, Hawaii. *(Bishop Museum)*

I do not expect it to happen. Not soon enough, anyway. In about 30 years, perhaps, depending on how the crisis first appears, it may become horrifyingly obvious that we have goofed. Then, I suppose, the good people will say, "Why didn't someone warn us? Where were the scientists? Why didn't they save us?"

Have other races of beings in the Galaxy, if they exist, similarly rushed headlong to their own destruction? Or have they learned to appreciate their planets—to treat them with respect, and to live within the means that their worlds can provide?

Perhaps if we could find an affirmative answer to that last question, we would find hope for ourselves. Perhaps the discovery of a successful civilization, anywhere, would not only be the news of the century, but would quite possibly be our own salvation as well.

ACT VI

THE GRAND QUESTIONS

SCENE 1

The Galaxian Zoo

Existing telescopic photographs reveal images of tens of millions of galaxies. Nearly all are either *elliptical* or *spiral* galaxies. Ellipticals are symmetrical spheroidal or somewhat flattened spheroidal systems of stars, which do not generally have conspicuous bright or dark nebulae, nor do they have any young highly luminous stars.

FIGURE VI.2
The Small Cloud of Magellan, an irregular galaxy. *(Courtesy of the UK Schmidt Telescope Unit, Royal Observatory, Edinburgh)*

Spiral galaxies, by contrast, have flat rotating disks in which are embedded the spiral arms of gas and dust clouds, and young clusters and associations of stars, including many main-sequence spectral-type O and B stars that make the arms stand out like those of a pinwheel. At the center of a spiral there is generally a large nuclear bulge of stars. Some spiral galaxies have thin tightly wound arms; in these the bulges tend to be very large. In other galaxies, the arms are more prominent, less tightly wound, and show high resolution into clusters, nebulae, and many bright stars; in these the nuclear bulges are smaller, or sometimes nearly absent. Relatively less common are *irregular* galaxies, which are sort of extreme examples of these spirals with wide-open arms, but with less or no rotational symmetry. The two nearest external galaxies, the Large and the Small Magellanic Clouds (Act III Scene 5), are usually classed as "irregular."

FIGURE VI.3
The spiral galaxy NGC 4622 in Centaurus. *(Kitt Peak National Observatory)*

FIGURE VI.4
The barred spiral NGC 1530 in Camelopardalis. *(Kitt Peak National Observatory)*

The spiral galaxies themselves fall into two main families—*ordinary* spirals (like our Galaxy and the Andromeda galaxy) and *barred* spirals. The latter have straight barlike arms running through their nuclear bulges, with the spiraling arms beginning either at the ends of the bars or from the circumferences of circular rings attached to the ends of the bars. The bars and rings (if present) are themselves like the spiral arms in that they contain gas and dust clouds, young clusters, and highly luminous stars.

These pages contain many photographs which should give you a pretty good idea what spiral galaxies are like. There are a number of systems for further classifying spirals, and the subject is described fully in a number of references cited in the Bibliography in Appendix 1. All spirals rotate in directions such that their arms trail.

A lot of people have speculated about evolution of galaxies from one type to another. In fact, even Hubble had considered the possibility that the continuous sequence of galaxies from spherical ellipticals to highly flattened spirals might represent an evolutionary sequence. Our current view, though, is that the nature of a galaxy depends mostly on two initial properties of the pregalaxian cloud of matter from which the galaxy presumably formed: the mass and its angular momentum of that cloud. If an original cloud of a given mass did not have too much initial rotation, the gas, as it condensed, could get dense enough to break up completely into smaller condensations that became stars (or star clusters). On the other hand, if it had a great enough original rotation, as it condensed it sped up its rotation, to conserve angular momentum, and flattened into a disk—more or less as occurred in the solar nebula. In the disk the density never became great enough for complete formation into stars, and a lot of gas was left over. This gas, along with matter ejected from stars and recycled into it, eventually ended up in spiral arms, where star formation continues today.

Stellar Content of Galaxies

The significant thing about elliptical galaxies is that they show no evidence of new or recent star formation. Their brightest stars are like those red giants in globular clusters. Although those giants are roughly a thousand times as luminous as the sun, they are nearly a thousand times *less* luminous than the brightest young stars in spirals. In other words the *stellar populations* of spiral and elliptical galaxies are quite different from one another.

It was Walter Baade who first clearly called attention to the obvious differences between different stellar populations. I mentioned Baade in Act IV Scene 7. In my days as observer on the Palomar Sky Survey, I spent many cloudy nights talking with him on Palomar Mountain. He immigrated from Germany early in his career, and spent most of his productive life at what was then called the Mount Wilson Observatory, and later the Mount Wilson and Palomar Observatories. (Now they are the *Hale Observatories,* named for the American astronomer George Ellery Hale, whose vision and energies resulted in the founding of both observatories.) I will always particularly remember Baade for his special interest in fine cheeses and beers.

Baade was highly regarded as a superb astronomical observer. During the years of World War II the city of Los Angeles (like all western U.S. cities) was dimmed out so that potential enemy aircraft would have difficulty spotting it for attack. The dimout was a great boost to astronomy, because it removed the bright glare in the sky above the observatory on Mount Wilson, which

FIGURE VI.5
Walter Baade. *(Courtesy of the Archives, California Institute of Technology)*

overlooks Los Angeles. That fact, the availability of new photographic emulsions sensitive to red light, and especially the patience and perseverance of Baade at the 100-inch telescope, enabled him to resolve for the first time the nuclear bulge of the Andromeda galaxy into red giant stars. This was what led him to the realization of different stellar populations.

Baade designated as belonging to *population I* those stars which, like the sun, are found in or near the spiral arms of galaxies, and thus in the vicinity of interstellar gas and dust clouds. Population I includes all recently formed stars, and hence all stars that are highly luminous and much more massive than the sun (which would have burned out by now if they were older than a few thousand million years).

Baade's *population II* consists of *old* stars — all at least 5000 million years old — those stars, for example, of the first generation or of early generations in our Galaxy. They are the kinds of stars found in globular clusters and elliptical galaxies. The most massive stars of population II are those just a bit more massive than the sun (about 1.2 solar masses); the originally more massive stars in those systems have already spent their fuel and died. Population II stars are in regions of galaxies devoid of dense interstellar matter, and thus in which star formation has not occurred for a very long time.

Whereas elliptical galaxies contain stars of pure population II — all old stars — we are not sure that there are any galaxies containing only young stars. Some irregular galaxies seem to contain a large fraction of young stars, but spirals, in particular, are a complete mixture. The nuclear bulge of a spiral galaxy consists of population II stars, as does the sparse distribution of stars on either side of its disk and surrounding it in a great halo or corona. In fact, the nuclear bulge and corona of a spiral are almost like a large spheroidal or elliptical galaxy in which the disk with its spiral arms is superposed. The spiral arms themselves contain young stars of population I, but as time goes on, those stars become more and more dispersed throughout the disk. The disk of a spiral galaxy thus contains a mixture of stars of many ages.

1. Where might any interstellar gas in an elliptical galaxy come from?
2. Draw hypothetical Hertzsprung-Russell diagrams for a spiral galaxy and for an elliptical galaxy. Identify and interpret the various sequences on your diagrams.

"Weighing" the Galaxies

Now we saw in Act III Scene 5 how we are able to obtain the distances to other galaxies (albeit with considerable uncertainty). Once the distance to a galaxy is known, its total luminosity can be found from its magnitude and the inverse-square law of light, just as for stars. It is easy to calculate the diameter of a galaxy of known distance from its angular size, just as Shapley found the diameters of globular clusters (Act III Scene 4). However, before I can give you a final summary of the properties of galaxies, I must explain how sometimes we can estimate their masses.

One way to find a galaxy's mass is from its rotation. Of course we don't *see* it rotating; van Maanen (Act III Scene 5) *thought* he did, but we have seen that he was wrong. We *can*, however, see how fast a galaxy is rotating by observing the Doppler shift of the light from different parts of it – at least we can for the relatively nearby galaxies in which we can inspect individual portions.

Suppose, for example, we observe the spectrum of a spiral galaxy that is oriented almost edge-on to us, as shown schematically in Figure VI.6. Suppose the nuclear bulge at *B* shows a Doppler shift of its spectral lines, indicating that it is moving away from us at 1000 km/s. That is, presumably, the speed with which the galaxy as a unit is moving in our line of sight, since the

FIGURE VI.6 The rotation of a spiral galaxy.

bulge must be at the center of mass of the galaxy. But what if points A and C, on the other hand, show Doppler shifts indicating speeds away from us of 1300 km/s and 700 km/s, respectively. With respect to the center of that galaxy, point A is turning away from us and point C toward us, each at 300 km/s. From the foreshortening of the image of the presumably round galaxy we can rather easily calculate its tilt to our line of sight and thus convert the observed speed of 300 km/s to true orbital speeds of A and C around the center.

The analysis from here on gets rather difficult if done rigorously, but we can illustrate the idea very simply. We observe how far A (or C) is from the center of the galaxy, so we know the radius of its orbit. We know how fast A is moving, so we can easily calculate how long it takes to complete one revolution. Then we apply Kepler's third law to calculate the mass of the galaxy, just as we can calculate the mass of Jupiter from the distance and period of any of its satellites. Although this oversimple analysis ignores the part of the galaxy's mass outside the orbit of A (or C), it gives the general idea of how from studies of their rotations, we have obtained estimates of the masses of several dozen spiral galaxies and at least one fairly flat elliptical.

---

3. Suppose spectra are observed of the light coming from many places over the image of a spiral galaxy, and the radial velocity is measured from each spectrum. If all these radial velocities turn out to be the same, what can you conclude about the orientation of the galaxy in space?

We can also estimate the masses of pairs of galaxies that are almost certainly in mutual revolution about each other, more or less as we do for double stars. We can see the separation of the two galaxies in a pair, and we can measure the Doppler shift in the spectrum of each and thus find the difference in their radial velocities. The only trouble is that we don't see the galaxies actually moving (their revolution periods are hundreds of millions of years) and we don't know the orientation of their orbit in the sky nor where they are in their orbit. Thus there is no way to find the mass of an *individual* pair of mutually revolving galaxies.

On the other hand, we can analyze *many* such pairs of double-galaxy systems. Then we can assume that their orbits are oriented at random and that the galaxies occupy random places in their orbits, and we can calculate, on the average, what the various unknown orientation effects are. This means that we can calculate the *average* mass of the galaxies in a large number of such pairs. It turns out that this procedure is very important.

We can also estimate the mass for a cluster of stars, galaxies, or other objects held together by their mutual gravitation. Each member of that cluster moves under the gravitational influence of all the other members combined. It's speed, therefore, is constantly changing, and in a complicated way. However, the average of the speeds of all the members of the cluster depends on the total gravitation of the system, which in turn depends on its mass and size. Thus if we observe the average of the speeds of many members of a cluster and also the size of the cluster, we can calculate what its total mass must be to exert the gravitational accelerations necessary to produce those speeds.

We can photograph the spectra of individual stars in some clusters, measure the Doppler shifts of their spectral lines, and thus calculate their average speeds—compared to the centers of mass of the clusters. In this way, we find the masses of several star clusters. Elliptical galaxies are somewhat like enormous star clusters. However, we cannot measure spectra (and hence Doppler shifts) of individual stars in other galaxies. Rather, the spectrum we observe for a galaxy is contributed by the light of millions of stars, all moving at different speeds and each producing its own private Doppler shift. But this makes each line in the composite spectrum of a galaxy very broad, because it is Doppler-shifted many different ways at once. Thus we can tell roughly how fast the stars are moving in an elliptical galaxy, and can estimate its mass from the widths of its spectral lines. The observations are difficult, however, and the analysis has been carried out for only a few dozen individual galaxies.

A similar analysis, however, can provide an estimate of the mass of an entire cluster of galaxies. There are several large clusters for which radial velocities have been determined from the Doppler shifts in the spectra of many of their member galaxies. Although the results are somewhat uncertain, at least approximate masses are found, with various degrees of confidence, for between one and two dozen clusters. Thus, again, we have ways of estimating, on the average, masses of galaxies that are members of rich clusters ("rich" means the cluster has lots of members).

Now let's see where we stand. For a few dozen spirals, we have masses determined from their rotations. For perhaps a couple of dozen ellipticals, we have estimated masses from the widths of the lines in their spectra. We have also estimated average masses for about a hundred double-galaxy pairs and total masses for about two dozen clusters. I must admit, though, that sometimes these different methods give results that are not in very good agreement with each other. Masses found for clusters of galaxies tend to be somewhat larger than we would expect from the numbers of galaxies they contain, if individual galaxian masses are correctly determined from their rotations or the widths of their spectral lines.

In fact, some years ago, astronomers were pretty worked up over this discrepancy. It was hypothesized that there may be "missing mass" in clusters (perhaps in the form of invisible gas) to give them enough gravitation to account for the high speeds of their member galaxies. As better data become available, though, the discrepancy seems to be diminishing, and I'm rather optimistic that eventually we'll reconcile all the data pretty well. But everything is not yet nicely and completely cleaned up the way we might like. I have taken you to the front line in this area of research on galaxies and have shown you some of our dirty linen. Shortly, in this scene, I shall show you some linen that's a lot dirtier!

---

4. It has been suggested that the so-called "missing mass" in a cluster might be due to gas or dust between the galaxies in the cluster. On the other hand, we can rule out enormous quantities of dust and also a large amount of neutral hydrogen gas at fairly high density. What kinds of observations do you think can rule out these possibilities? [Hints: What does dust do to starlight? How is neutral hydrogen observed in our own Galaxy?]

Gross Properties of Galaxies

But first, let me summarize the properties of these different kinds of galaxies.

Spirals are typically 30,000 parsecs (about 100,000 LY) in diameter, but probably range from 10,000 to 50,000 parsecs across. We think our own Galaxy is more or less typical, but it may be a little larger and more massive than average. The smallest spirals probably contain a few thousand million and the largest a few hundred thousand million stars.

Irregulars are, on the average, smaller than spirals, although there is some overlap; the largest galaxies usually called irregular, like the Large Magellanic Cloud, have more stars than some of the smallest neater looking galaxies classed as spirals. The irregulars range from 2 or 3 thousand to 10 or 15 thousand parsecs across and contain from 100 million to a few thousand million stars.

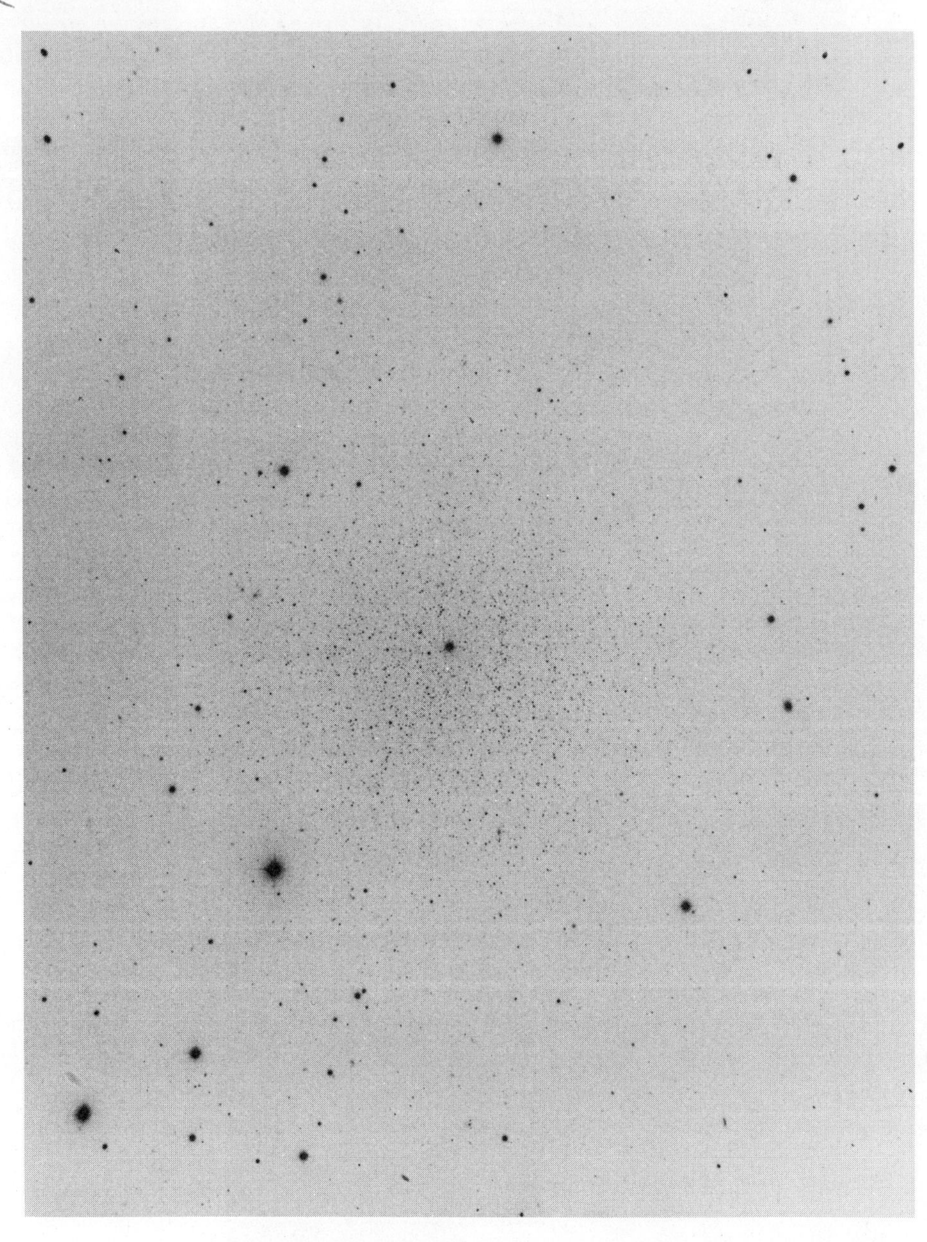

FIGURE VI.7 The dwarf elliptical galaxy Leo II (negative print). *(Hale Observatories)*

The spiral galaxy M51 in Canes Venatici. (*U.S. Naval Observatory*)

The edge-on spiral galaxy NGC 4565. (*U.S. Naval Observatory*)

The irregular galaxy M82 in Ursa Major. (*U.S. Naval Observatory*)

The spiral galaxy NGC 253 in Sculptor. (*Hale Observatories*)

The Large Magellanic Cloud, the nearest external galaxy. (*U.K. Schmit Unit, Royal Observatory Edinburgh*)

M32, a small Local Group elliptical galaxy in Andromeda. (*U.S. Naval Observatory*)

The peculiar elliptical galaxy NGC 5128 in Centaurus. (© *Association of Universities for Research in Astronomy, Inc. The Cerro Tololo Inter-American Observatory*)

Ellipticals range continuously from supergiant galaxies down to dwarfs. A respectable giant elliptical, itself a rare breed, has the diameter of a typical spiral, but from ten to a hundred times as many stars. The very rare supergiant ellipticals, also called *cD galaxies,* are found, so far as we know, only in clusters of galaxies and range up to a million million (10^{12}) times the total luminosity of the sun. By comparison they dwarf our Galaxy, extending over diameters of hundreds of thousands of parsecs and containing up to a hundred million million (10^{14}) stars. Medium sized ellipticals, much smaller and less massive than typical spirals, are very common. Most common of all are the *dwarf* ellipticals. Only a few hundred are actually known, because they are so faint they cannot be seen very far away, but we have reason to think they are by far the most common kind of galaxy. The smallest we know probably contain only a few million stars, and are spread over a diameter of only about 1000 parsecs. They are so diffuse and faint in surface brightness that you can see distant background galaxies shining right through them.

So that's a Cook's tour through the zoo of "normal" galaxies. Now let's wander over through the snake house and into the enclosures beyond to see some of those very esoteric beasts, many of which we don't at all understand.

---

5. If extragalactic globular clusters exist as "galaxies," as what kind of galaxies would you classify them?

Interacting Galaxies

There are some pathological-looking galaxies that we think we can now understand, thanks to studies of Alar Toomre, at MIT, and his brother Juri at the University of Colorado. The Toomres have considered how two galaxies that pass close together in space can distort each other because of their mutual tidal forces.

The Toomres have carried out their very extensive calculations on a large computer, and in each experiment have had the computer prepare pictures of two hypothetical interacting galaxies at many points along the encounter. Figure VI.8 shows five such plots selected from a very large sequence of plots that describes just one such hypothetical encounter. The successive plots in the figure are separated in time by about 200 million years. Each galaxy is shown in a different color so you can keep them and the material they lose separate from each other. The projected rectangle is in the plane of the relative orbit of the galaxies. Now imagine how the fifth frame would look it it were viewed a little more nearly edge-on to the orbital plane; then look at Figure VI.9, which shows the two galaxies NGC 4038 and NGC 4039. The Toomres have similarly accounted for the strange and wonderful appearances of a number of other pairs of galaxies in terms of tidal distortions.

The Hubble Law

At this point I must interrupt our tale to remind you about the *Hubble law,* first announced by Hubble in 1929. We have already mentioned it in Act III Scene 5, and shall return to it to discuss its interpretation and ramifications in Scene 3. But what Hubble discovered was that the lines in the spectra of remote galaxies are always shifted to longer wavelengths—toward the *red* in

FIGURE VI.8
A sequence of frames from a computer-generated motion picture that simulates the tidal distortion of two interacting galaxies. In this computer run the initial conditions were chosen to see if the peculiar pair of galaxies NGC 4038 and NGC 4039 could be accounted for in terms of tidal effects. Compare the last two frames with the photograph in Figure VI.9. The time interval between successive frames shown here is 200 million years. *(Courtesy of Alar and Juri Toomre)*

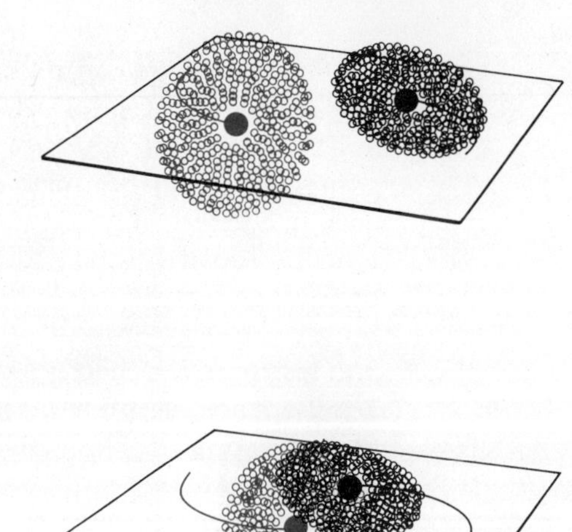

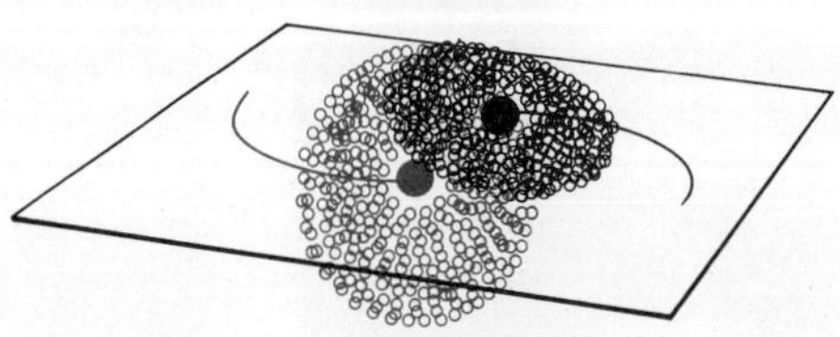

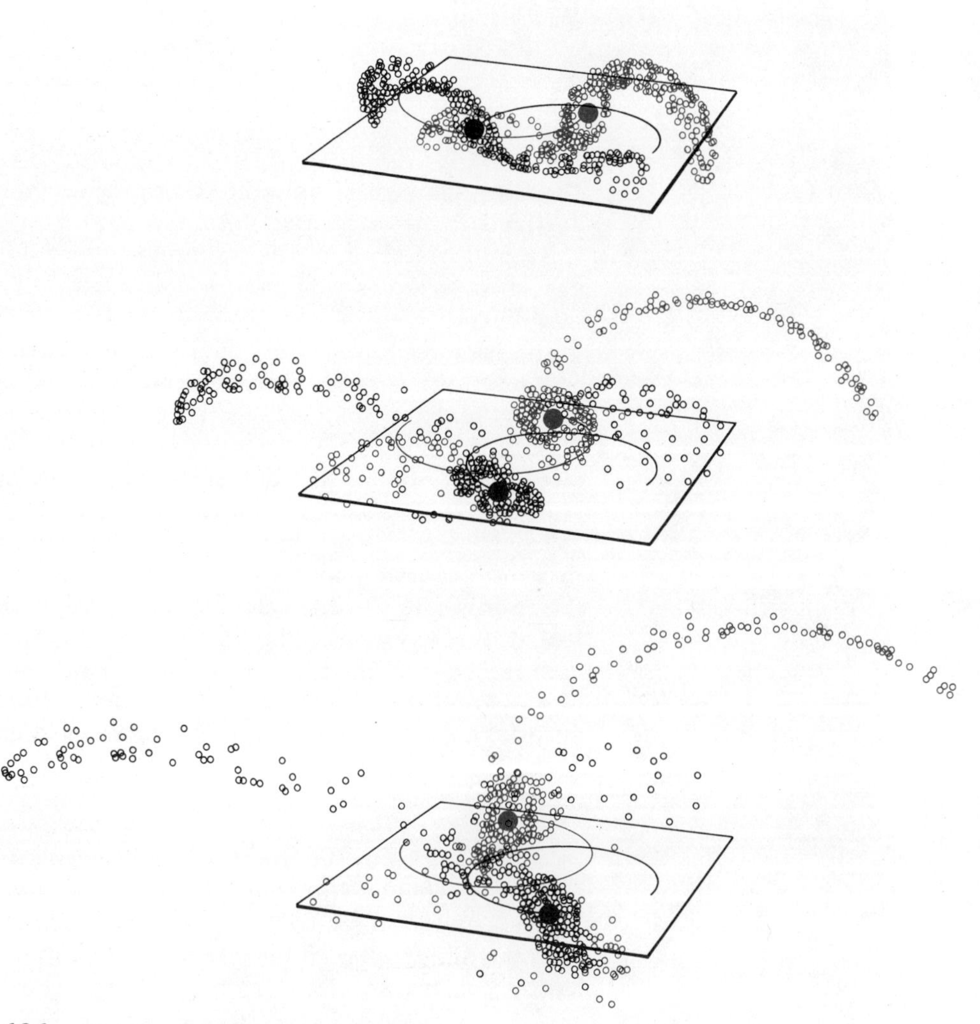

FIGURE VI.9
NGC 4038 and NGC 4039. Compare this figure with Figure VI.8. *(Hale Observatories)*

the visible spectrum. Moreover, he found that this *redshift* of the spectral lines is in direct proportion to the distance of the galaxy observed.

Now we know of only two effects that can produce large redshifts: very strong gravitational fields *(gravitational redshifts)* and Doppler effects (indicating velocity away from us). But the redshifts in the spectra of the galaxies cannot be gravitational, or the spectra (as well as the visible appearances of the galaxies) would show other effects that are not present. Thus we assume that the redshifts indicate that galaxies are all moving away from us, and the greater their distances, the faster they move. (We shall see in Scene 3 that the Hubble law is *expected theoretically*, and shows that the universe is expanding.)

There are actually two contributions to the radial velocity of a galaxy. One is due to its own motion within a cluster or group, or double-galaxy system, to which it may belong, simply because of the gravitational accelerations produced by the other members. The other contribution is the velocity away from us because of the general expansion of the universe—that is, the Hubble law. To keep the two contributions separate, we call the latter the *cosmological redshift*. Both the redshifts due to the individual (usually random) motion of a galaxy and the cosmological redshift are, of course, Doppler shifts.

The individual radial velocities of galaxies in their groups or clusters are typically a few hundreds of kilometers per second, and are only rarely as great as 2000 km/s. On the other hand, the observed cosmological redshifts of galaxies indicate radial velocities that range up to about half the speed of light. So although the two contributions to the redshift are, in general, superimposed on each other, except for the relatively nearby galaxies, the cosmological redshift dominates and the individual random motions of galaxies produce just a small scatter about the uniform Hubble law.

Also in Scene 3 we shall see that the proportionality of velocity with distance cannot apply to extremely remote galaxies, and just how the Hubble law deviates from the simple proportion depends on which theory of cosmology turns out to be correct. Nevertheless, we *can* say that more remote objects always have *greater* cosmological redshifts than nearer ones.

---

6. Why is it that the cosmological redshift is the dominant contribution to the Doppler shift in the light from a distant galaxy, but not in the light from a nearby galaxy?

Radio Galaxies

All galaxies emit radio radiation, along with other electromagnetic radiation. Even stars emit radio energy (the sun does). However, ordinary stars do not emit enough energy at radio wavelengths for us to observe. We observe radio energy from the sun only because it is so near to us; if the sun had the distance of even the nearest other stars, we would not be able to detect its radio wavelengths.

The observed radio sources in galaxies, therefore, are other than stars. Usually they are interstellar gas clouds, and most often the radio energy itself is synchrotron radiation, emitted by electrons moving at nearly the speed of light in interstellar magnetic fields. In our own Galaxy radio energy is also emitted by such specific objects as the Crab nebula and other supernova remnants, as well as pulsars, some planetary nebulae, and certain stars that occasionally emit flares of radio energy. In addition radio energy is found to be coming from the nucleus of our Galaxy, and we shall return to that point shortly. Our Galaxy appears to be more or less typical among spirals as a radio emitter, and is called a "normal" galaxy. Typically, the total radio luminosity emitted from a normal galaxy is about 10^{38} erg/s—a tiny fraction of 1 percent of the total amount of electromagnetic energy it emits at all wavelengths—including that which it emits in the visible spectrum.

Normal spiral galaxies are also often observed in the radio energy of the 21-cm line of neutral interstellar hydrogen. In such cases we can measure the Doppler shifts of that 21-cm line and find the radial velocities of the galaxies; the 21-cm radial velocities always agree with those obtained from the redshifts of the lines in the visible spectrum, as expected.

Unlike the normal galaxies, however, some galaxies are peculiar in that they emit very intense synchrotron radiation at radio wavelengths. These are called *radio* galaxies; they emit at radio wavelengths as much energy as they do in the visible spectrum, and sometimes far more—up to hundreds of times as much. One of the first cosmic radio sources to be identified is in the constellation Cygnus, and is called *Cygnus A.* The source of the radiation is a

FIGURE VI.10
The peculiar galaxy identified with the radio source Cygnus A. *(Hale Observatories)*

galaxy in a remote cluster of galaxies. Cygnus A is one of the strongest radio sources in the sky, as observed from here on the earth, despite the fact that the cluster of galaxies is about 1000 million LY distant. Extreme radio galaxies, such as Cygnus A, emit more than 10^{45} erg/s at radio frequencies. There are now thousands of discrete radio sources cataloged, each occupying a small region of the sky. Hundreds have been identified with galaxies and clusters of galaxies, and most of those which have not been identified with visible objects are thought to be radio galaxies too remote to observe in the visible spectrum.

---

7. Suppose a radio galaxy such as Cygnus A has a distance of 10^9 pc. What is its distance in astronomical units (see Appendix 5)? Suppose that, as seen from earth, the radio galaxy appears about as bright (in radio wavelengths) as does the sun. How many times as intense as the sun's is the galaxy's total emission of radio energy?

Sometimes a radio galaxy has a characteristic feature in its optical image. Consider M87, for example. On an ordinary photograph (Figure VI.11) it looks like a typical, well-behaved giant elliptical galaxy. On a very short exposure, however, where the photographic emulsion is not all burned out by the light from its million million or more stars, there appears a bright luminous jet, seeming to radiate right away from the very center of the galaxy

FIGURE VI.11
The elliptical galaxy M87. *(Kitt Peak National Observatory)*

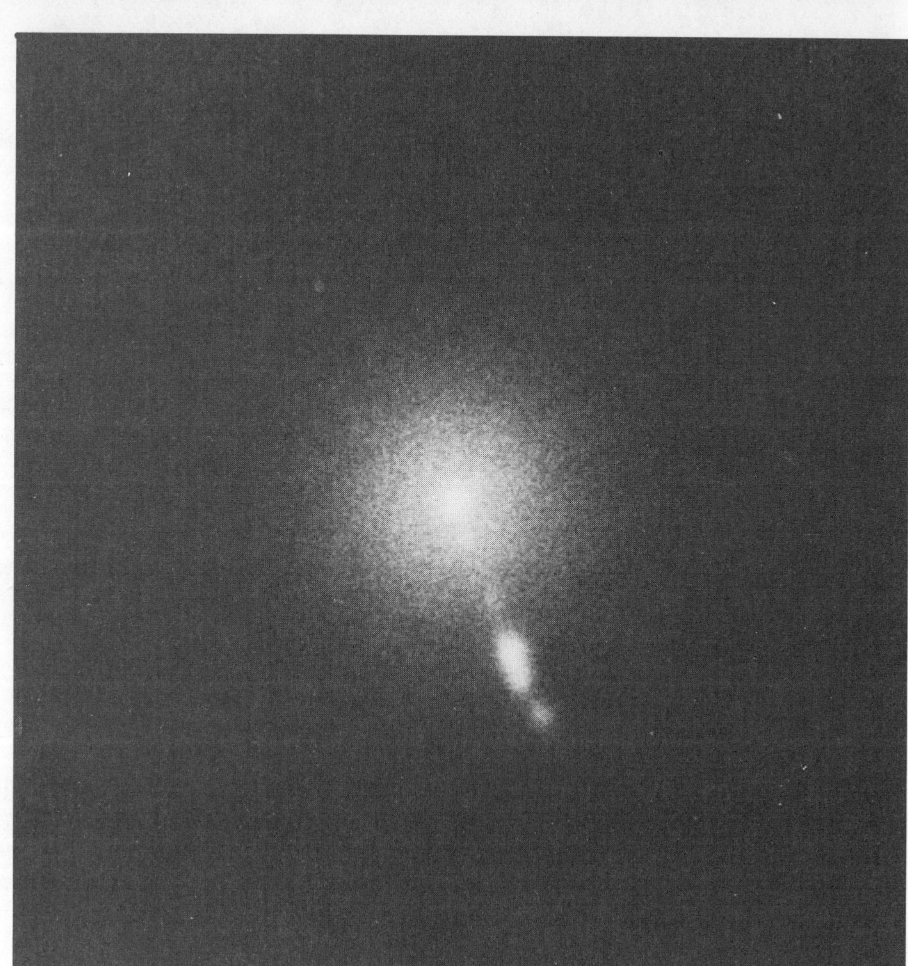

FIGURE VI.12
A short exposure of M87, showing the jet. *(Lick Observatory)*

(Figure VI.12). That jet is the source of the strong synchrotron radio radiation from M87. It *looks* as though it might have been shot out of the nucleus of the galaxy in an explosion, but there is no real evidence for any such ejection.

A number of other radio galaxies look ordinary on photographs, but have small intense sources of radio waves within them. None is known to have a jet, like M87, but most are so far away it would be difficult to observe such a feature. Other normal-appearing radio galaxies have central radio sources surrounded by larger extended regions of radio emission; these are called *core-halo* sources.

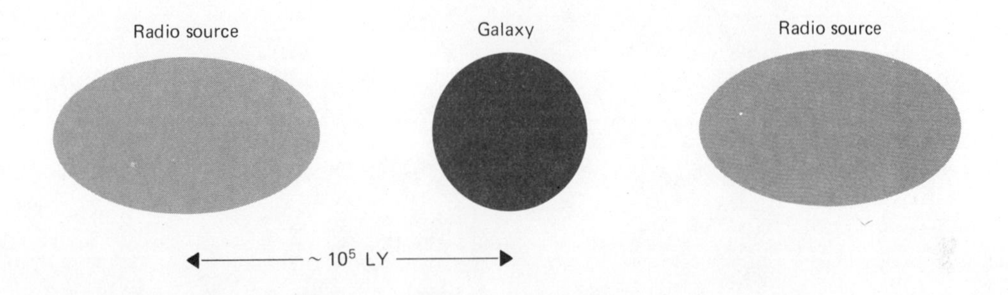

FIGURE VI.13 A schematic representation of the pair of radio sources associated with a typical radio galaxy.

The most common kind of radio galaxy is a giant elliptical, or maybe supergiant cD galaxy in a cluster; such a galaxy *looks* normal, but is centered on a *pair* of large radio sources. Figure VI.13 shows a typical layout. The galaxy is in the center between two large round regions of apparently empty space from which the radio radiation comes, each about 60,000 parsecs in diameter, and the two are separated by about 200,000 parsecs. NGC 5128 is an unusual radio galaxy and is strange in two respects: First, it is almost unique in being an elliptical galaxy with a heavy absorbing dust lane (Figure VI.14); second, it is centered on *two* pairs of radio sources, an inner intense pair and an outer more diffuse pair (Figure VI.15).

The energy involved in the typical double-source radio galaxy is not only immense; its origin is not yet understood. Those 10^{44} to 10^{45} erg of radiation leaving its two sources each second are being emitted by electrons at speeds near that of light moving in magnetic fields. The field strength required may sound low, only about 10 microgauss, but it is spread over an incredible region. The total energy that must be stored in those magnetic fields and moving electrons must be at least 10^{60} erg. It is equivalent to the energy obtained by the complete annihilation of a million suns. We have as yet no idea where the extensive magnetic fields come from, how the electrons get out there (and there must be protons and other atomic nuclei as well), and where the energy for it all comes from. We cannot imagine that radio galaxies last as such indefinitely, but suspect that their astonishing radio properties must be temporary phenomena.

FIGURE VI.14
The peculiar elliptical galaxy NGC 5128 in Centaurus. *(Kitt Peak National Observatory)*

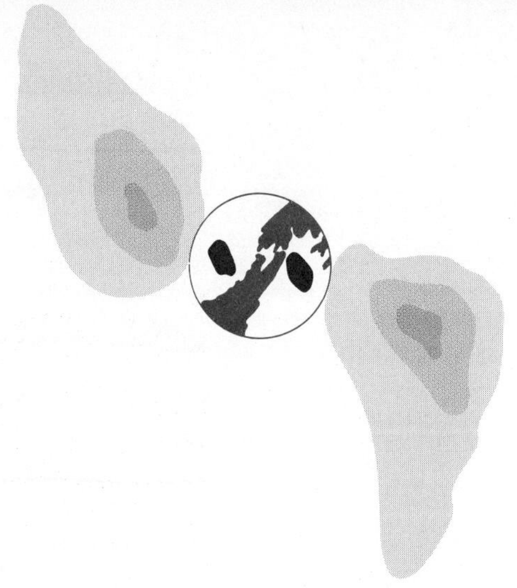

FIGURE VI.15
Optical and radio images of NGC 5128.

---

8. Some theorists think that the matter emitting radio waves in a typical double-source radio galaxy is being ejected from the center of that galaxy. If so, and if the two sources are 200,000 pc apart, what is the minimum possible age for the system since the onset of its energetic activity?

Quasars

There is another class of radio-emitting objects that is even more remarkable. In fact, their discovery is an interesting story in itself. The radio astronomers scan the sky and catalog the directions from which radio waves are coming. Then optical astronomers search telescopic photographs for images of objects in those directions that might be the sources of the radio waves. (Quite often the optical and radio astronomer is the same person.) Well, in 1960 especially accurate positions were obtained for two radio sources. The objects on the photographs at those positions were *stars*. This was totally unexpected, because, as I have said, stars (except in occasional flares) do not emit strongly enough to be observed at radio wavelengths. But these two stars were suspicious, because they had very strange spectra with unidentified spectral lines. By 1963 the number of such radio "stars" had grown to four. Several theoreticians even began advancing theories to explain the unusual radio emission. Then Caltech's Maarten Schmidt dropped a bombshell!

Schmidt is another of those fine Dutch astronomers. He is a terribly nice person, generous of his time, and active in the education of young people as well as of advanced astronomy students, despite the fact that he was appointed Director of Hale Observatories as of July 1978. He also plays an enthusiastic game of softball. Schmidt's original field of research was the structure of our Galaxy. However, he had become interested, among other things, in those strange radio stars. So he took a new spectrogram of one with the 200-inch telescope.

FIGURE VI.16
Maarten Schmidt.
(Photograph by the Author)

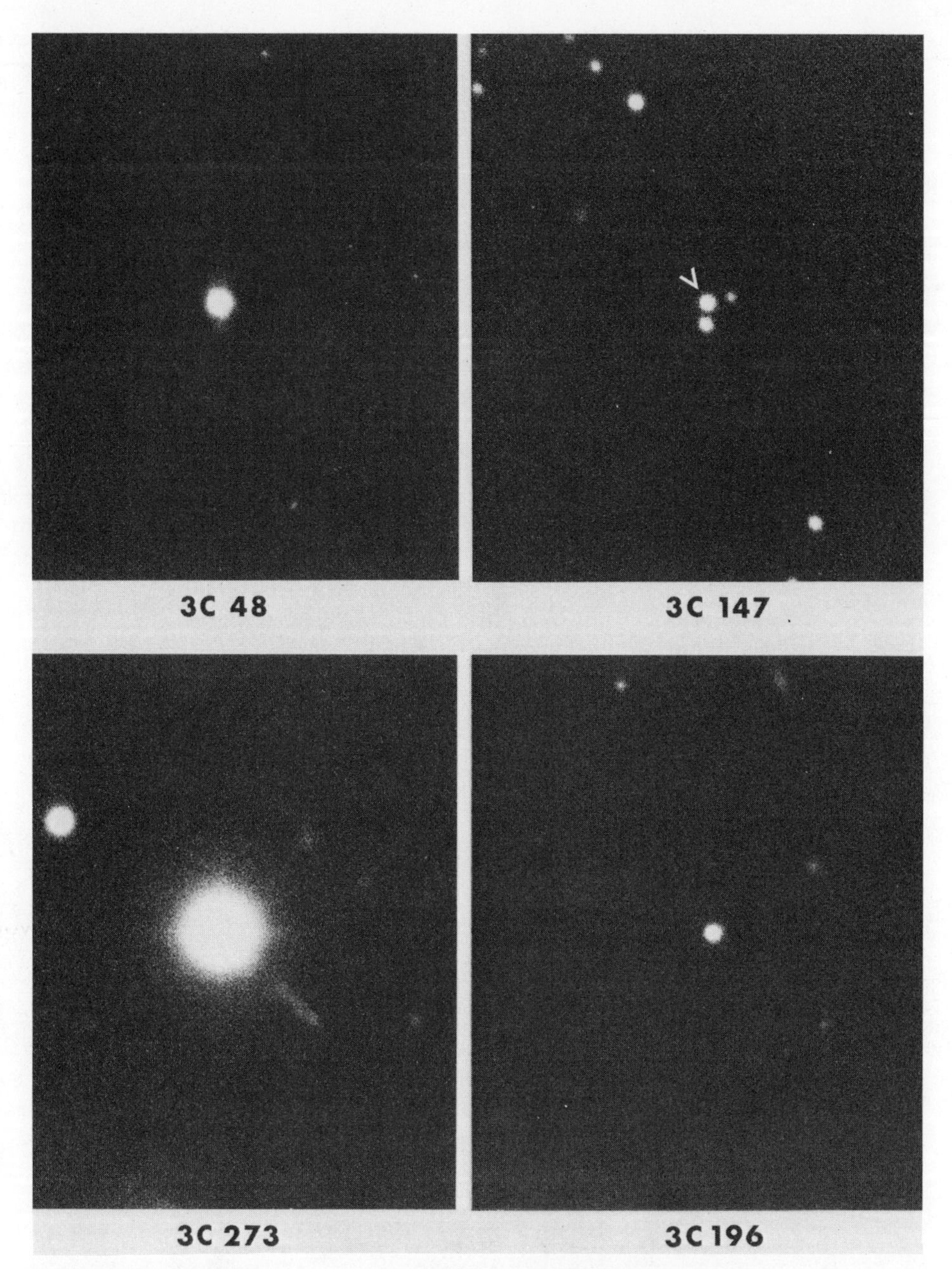

FIGURE VI.17
Four quasars photographed with the 5-m (200-inch) telescope. *(Hale Observatories)*

When he examined his spectrogram, wondering about the strange pattern of spectral lines he saw, it suddenly hit him that they were the familiar Balmer lines of hydrogen. What was funny about them is that they were in the wrong place for a stellar spectrum; they were strongly redshifted, indicating a velocity of recession of the star of about 15 percent the speed of light! Only galaxies were known to have such redshifts; but this object was a *star*—or was it?

Then Schmidt looked at the spectra of the other strange stars. Once he realized that the spectral lines could be greatly shifted from their normal positions in the spectrum, he was able to identify them too. The other "stars" were also moving away at great speeds—in fact, at far greater speeds than the first one. Obviously these objects could not be stars. They were either highly unusual galaxies, or things inside galaxies, or at least things out there among the most remote galaxies. They only *look* like stars, and hence were at first called *quasi-stellar radio sources*, or *QSS*s; now the name has been shortened to *quasars*.

At the time of this writing, more than 400 quasars are cataloged. They all have very large redshifts, and the overwhelming majority have redshifts greater than those of any ordinarily recognizable galaxies. The largest redshift yet observed, if a Doppler shift, implies a radial velocity of more than 90 percent the speed of light. If the redshifts are *cosmological*, as most of us think they must be, the quasars are the most remote individual objects seen in the universe.

Now, as far as their radio properties go, quasars are not particularly unusual; they are, more or less, like typical radio galaxies. Their radio radiation often comes from two separate sources, with the optical quasar image centered between them, as is the case for typical radio galaxies. In quasars the sources appear smaller, in angular size, than they do in radio galaxies, and at a smaller angular separation, just as you would expect if they were farther away than most galaxies. So if you were to judge by their radio properties, you would say they were very remote radio galaxies.

The perplexing thing about quasars is their array of *optical* properties. They look like faint stars—unresolved points of light. However, to be visible on telescopic photographs at all, at their presumed distances, they must really be enormously luminous—often ten to one hundred times as bright as normal bright galaxies. Many are variable in their light as well; they don't pulse in regular periods, but change erratically, sometimes in intervals of only a few months. For an object to change in light output in only a few months, it must be smaller than a few light months across; otherwise it couldn't change light as a unit, because light from different parts of it would take different times to reach us and the variations would appear washed out. So at least the parts of quasars emitting light are extremely tiny—maybe not compared to stars, but certainly compared to galaxies. Moreover, the quasars emit ultraviolet radiation especially strongly.

This last property of quasars led Allan Sandage to inspect a number of exceptionally blue stars (with excess ultraviolet radiation) lying in directions far from the Milky Way, to see if any of them might be quasars that did not emit radio waves strongly enough to be detected. Most of those blue stars turned out to be blue stars, but some, to Sandage's surprise, have large redshifts, like the quasars, but are *radio quiet*. If we take account of the large number of blue stars that are possible quasars, we find that the sky could contain many thousands of potentially observable quasars.

The total emission of energy by a typical quasar is so great, at least if its distance is as great as would be required if its redshift is cosmological, that many people have wondered whether quasars might really be much closer—with distances typical of relatively nearby galaxies. Well, if the quasars are not extremely far away, it is true that they don't have to be so extraordinarily luminous to account for their observed magnitudes. This condition would pretty well remove the difficulty of explaining where their energy comes from, but it poses an even greater problem: What unknown physical effect causes the redshifts in their light?

Moreover, many of the properties of quasars are just like those of radio galaxies, and no one questions that the redshifts of radio galaxies are cosmological. Quasars are a lot brighter in visible and ultraviolet radiation, to be sure, but if we could understand the incredible source of energy that keeps radio galaxies emitting, perhaps we would also understand the somewhat more incredible amount of energy that fires up quasars.

---

9. Suppose the quasar phenomenon is something that can occur only in a very young galaxy, say less than 10^9 years old, but not in older galaxies. How would this affect the observed distribution of quasars of various redshifts?

Most astronomers now think that quasars are bright spots inside galaxies that would otherwise appear more or less normal if they were near enough to be observed. At the large distances of quasars, the light of a normal galaxy would be hard to detect, especially against the very intense light of a quasar itself. In fact other kinds of peculiar galaxies give us a strong hint that quasars are active goings-on in the central *nuclei* of galaxies. But to understand those hints, let us take a moment to look at the nucleus of our *own* Galaxy.

The Galactic Nucleus

The *nucleus* of our Galaxy is the centermost part of its nuclear bulge. We don't see it in visible or ultraviolet light, because it is hidden in those wavelengths by the absorbing dust in our Galaxy. On the other hand, radio waves and infrared radiation, of wavelengths long compared to the sizes of the dust grains, pass on through to the earth, and X rays, being of such high energy, plow on through. Observations at these wavelengths in recent years have revealed that there is something very special about the nucleus.

At the shorter infrared wavelengths (about 20,000 Angstroms) the galactic nucleus looks about the same as does that of the neighboring Andromeda galaxy. We think that this is the normal infrared radiation from stars, and from its extent we derive the approximate distribution of stars in the central part of our Galaxy. We find them to be packed quite densely in there; within 1 parsec of the center there are about 4 million stars per cubic parsec, and within 0.1 parsec their density is $300{,}000/\text{pc}^3$. The density of stars at the very center is estimated to be about 30 million per cubic parsec; even there, though, they are hundreds of astronomical units apart, and actual collisions between stars are very improbable.

At longer wavelengths the nature of the infrared radiation indicates that it is emitted by a pretty fair amount of dust in the nucleus. Some of the dust is dispersed sparsely over a considerable volume, but much of it seems to be collected in relatively dense clouds, probably around stars. The total infrared energy emitted from the region of the nucleus is about 10^{42} erg/s, which could be as much as 10 percent of the total luminosity of our Galaxy.

There is a strong radio source in the nucleus of the Galaxy as well; it is known as Sagittarius A. At least some of that radio radiation is synchrotron radiation, showing there are energetic electrons and magnetic fields around. And there in the nucleus, protected from ultraviolet radiation of hot stars by all that dust, is where many of the complex molecules form that we have been finding in the past decade, by means of their radio emission lines (I described them at the end of Scene 4 Act IV). On top of all this, 21-cm observations show that hydrogen is streaming out of the nucleus at about 50 km/s.

Thus the nucleus of our Galaxy is obviously a very busy place. There is that vast amount of infrared radiation, dust, hot gas, molecules (including complex organic ones), magnetic fields, high-speed electrons, gas flowing out, and even X rays.

Seyfert Galaxies and Possibly Related Objects

We also see small bright nuclei in the centers of a few other galaxies, but if most galaxies are more or less like ours, we would expect to be able to see the nuclei in only the nearest of them. There is, however, a class of galaxies, called *Seyfert galaxies,* that are observed to have especially active nuclei. They are named for Carl Seyfert, the astronomer who described the first dozen or so known. Seyfert was tragically killed in an automobile accident. I think he would have been pleased, though, at the tremendous interest expressed today in "his" galaxies.

A Seyfert galaxy usually looks like any other on an ordinary photograph. However, its spectrum shows strong broad emission lines coming from its nucleus, indicating that there is a lot of very hot nuclear gas. Seyfert galaxies

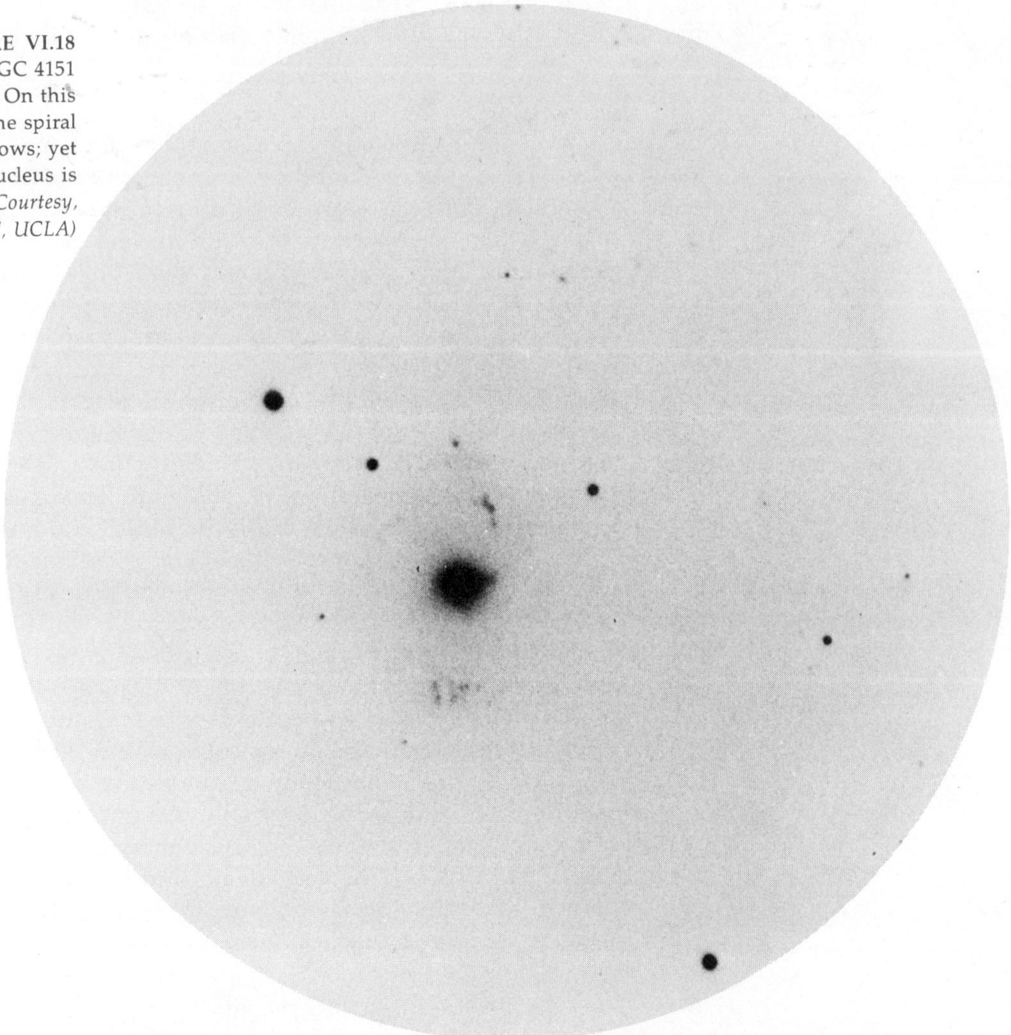

FIGURE VI.18
The Seyfert galaxy NGC 4151 (negative print). On this short exposure the spiral structure barely shows; yet the brilliant nucleus is already burned out. *(Courtesy, H. Ford, UCLA)*

also emit enormous amounts of infrared radiation—sometimes up to a hundred times as much energy as they emit in visible radiation. About half of them are known radio sources, and at least one Seyfert galaxy, NGC 4151, is a known X-ray source.

It may seem to you that, in their nuclear properties, Seyfert galaxies are really rather like our own Galaxy, only much more so. They are indeed. Although there are now only about a hundred known Seyfert galaxies, their properties are easy to recognize only in comparatively nearby galaxies. It could be that from 1 to 2 percent of all spiral galaxies are Seyferts. Or, alternatively, it could be that *all* spiral galaxies display the Seyfert characteristics 1 or 2 percent of the time.

Whereas we cannot easily recognize Seyfert galaxies at large distances, there are a great many more remote galaxies that are observed to have small bright nuclei. They are classed as *N galaxies*. Photographically, an N galaxy appears as an almost stellar nucleus superposed on a faint image of the rest of the galaxy. Those nuclei sometimes have very high luminosities.

In fact, there seem to be galaxies displaying a sort of continuum of the properties of active nuclei—ranging from the relatively sedate (but still pretty interesting) nucleus of our own Galaxy, through the Seyfert galaxies, and the N galaxies to . . . yes . . . the quasars. Most (but *not all*) astronomers regard quasars as most likely to be extreme cases of galaxies with active nuclei.

Perplexing Mysteries

Of course, that doesn't tell us what quasars are. Perhaps, when we understand Seyfert and N galaxies (and the nucleus of our own Galaxy), we will understand the main features of quasars. Or perhaps not. Maybe a quasar is a region of active star formation within the nucleus of a galaxy—or even a young galaxy in the process of formation. Or maybe it is matter ejected from evolved stars in a galaxy falling into a black hole, or matter in some highly condensed state. Or maybe it is a young galaxy, or a young region within a galaxy, with lots of stellar evolution going on, and lots of such things as supernovae. All these ideas, and many more (except, probably, the right one), have been suggested.

O.K., so you've seen some more dirty linen. Actually, astronomers love the challenge because it means we have yet some problems to solve, and maybe we can keep busy at our jobs for a few more years. But I must say, the situation isn't helped any by certain catalogs of crazy-looking objects. One nasty person who has given us such a catalog, and who keeps adding to it, is Halton C. Arp. I don't know what the "C." stands for, but all his friends call him "Chip." He's in danger of losing all those friends, if he keeps up what he's doing now!

Now, mind you, I am kidding, because I am very fond of Chip Arp. In fact, we were graduate students at Caltech together. But in those years he was nice. He found the period luminosity relation for W Virginis stars and studied globular clusters and nice respectable things like that. But lately, he's been turning up all those unsightly galaxies that one can find if he peruses the photographs of the sky as carefully as Chip does. I don't know how to explain some of the things he finds. Neither does he. There are interacting galaxies, maybe exploding galaxies; there are what seem to be near and distant galaxies connected by "bridges" of matter, and all sorts of esoterica. The universe is becoming a very complicated place, and this is unsettling. Chip suggests that there may even be new physics to be discovered—even new reasons for the redshift. I personally doubt this—at least on the basis of the evidence so far. So do most astronomers. There does not yet seem to be a consistent pattern to all the anomalies, as there was when Newton, and later Einstein, broke forth with new physics.

On the other hand, it is always the *un*explained phenomena and the observations we *do not* understand that lead us to new insights about the nature of the physical world. Most of us strongly expect that when we know enough, we shall be able to understand quasars and other peculiar galaxies in terms of known laws of physics. But consider the delight of the scientist when he finds something really new, and you will realize why many of us, deep inside, hope *not*.

SCENE 2

The Structure of the Universe

Here we sit on a rock revolving about one star—in fact, a very insignificant star among hundreds of thousands of millions of such stars in a rather ordinary galaxy. We estimate that there may be a thousand million galaxies that could be photographed with existing telescopes if we had the time and inclination to do so. Most of those galaxies, as we have seen, fit into a sort of pattern, but there are some aberrant ones among them—if aberrant is the proper word to describe such things as Seyfert galaxies and quasars. But these details seem quite trivial in the overall organization of the universe. It is the structure of the universe as a whole that we want to look at now.

Hubble, a man of no small perspective, was interested in the same question a few years after showing that galaxies even *existed*. He was interested, for example, in how galaxies are distributed in space. Are there as many in one direction of the sky as in any other? And if we count fainter and fainter galaxies, presumably farther and farther away, do we find that their numbers increase in the way they should if galaxies are distributed uniformly in depth?

Hubble's Faint Galaxy Survey

Hubble had at his disposal the world's largest telescopes—the 100-inch (2.5-m) and 60-inch (1.5-m) reflectors on Mount Wilson. But although those telescopes can probe to great depths, they can do so only in small fields of view. To photograph the entire sky with the 100-inch telescope would take not just a lifetime, but thousands of years. So instead, Hubble sampled the sky in many regions, much as Herschel did with his star gauging (see Act III Scene 2). In the 1930s Hubble photographed 1283 sample areas or fields with the telescopes, and on each photograph he carefully counted the numbers of galaxy images to various limits of brightness, the faintest corresponding to the greatest depth of space that could then be probed.

The results of Hubble's survey are shown in Figure VI.19, which is a map of the sky shown in what are called *galactic coordinates;* the Milky Way, across the middle of the plot, defines the galactic equator, and the top and bottom of

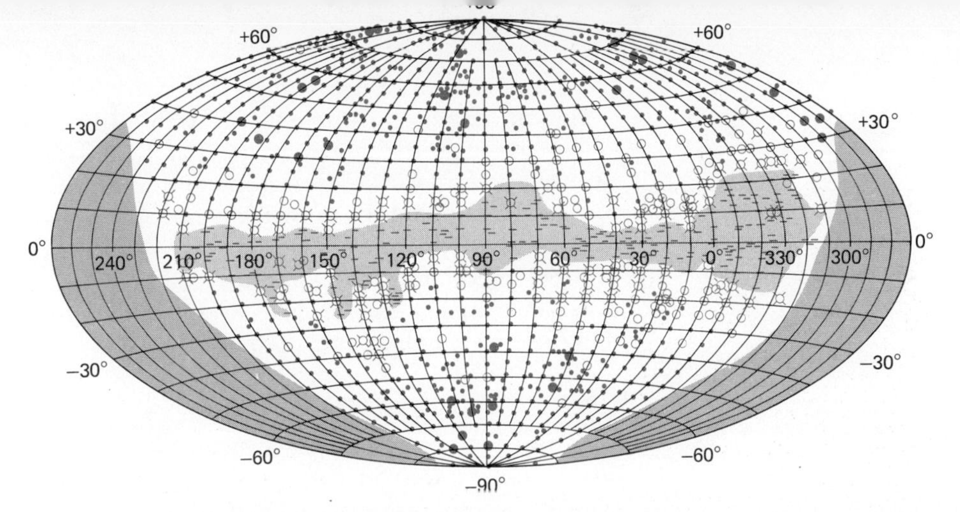

FIGURE VI.19
Numbers of galaxies in fields surveyed by Hubble.

the map—the galactic poles—are 90° away from the Milky Way. The empty sectors at the lower right and left are the parts of the sky too far south to observe from Mount Wilson. Each symbol represents one of the regions of the sky surveyed by Hubble, and the size of the symbol indicates the relative number of galaxies he could observe in that area.

The first obvious thing to notice is that we can't see galaxies in the direction of the Milky Way; the obscuring clouds of dust in our Galaxy hide what lies beyond in those directions. Hubble called this part of the sky the *zone of avoidance*. Near the Milky Way in direction, the counts of galaxies are below average and are denoted by open circles. The farther we look from the Milky Way, the less obscuring foreground dust lies in our line of sight and the more galaxies we see. From the counts of galaxies in different directions Hubble was actually able to figure out how much the dust obscured the light of distant galaxies, and he corrected for the effect.

After such correction, Hubble found that on the large scale the distribution of galaxies is *isotropic*, which means that if we look at a large enough area of the sky, we find as many galaxies in one direction as in any other. From the 44,000-odd galaxies Hubble counted in his selected regions, he calculated that about 100 million were potentially photographable with the 100-inch telescope. (This means that with the larger telescopes available today, we could, given enough time, be able to record about 1000 million.) Moreover, Hubble found that the numbers of galaxies increase with faintness about as we would expect if they were distributed uniformly in depth.

---

1. Assume for the sake of illustration that the dust in our Galaxy all lies in a relatively thin flat disk, with the sun in the central plane of that disk. Under these circumstances, show by a diagram that the obscuration of distant galaxies is less and less at greater and greater directions from the Milky Way (that is, at greater and greater galactic latitudes).

2. Suppose on one survey you count galaxies to a certain limiting faintness. On a second survey you count galaxies to a limit that is four times fainter.
   a) To how much greater distance does your second survey probe?
   b) How much greater is the volume of space you are reaching in your second survey?
   c) If galaxies are distributed homogeneously, how many times as many galaxies would you expect to count on your second survey?

These findings of Hubble were enormously important, for they indicated, at least to the precision of his data, that the universe is isotropic and homogeneous—the same in all directions and at all distances. In other words, his results indicate that the universe is not only about the same everywhere, but that the part we can see around us, aside from small-scale local differences, is representative of the whole. This idea of the uniformity of the universe is called the *cosmological principle,* and is the starting assumption for nearly all theories of cosmology (next scene).

Clustering of Galaxies

Now although the universe may be homogeneous on the *large scale,* it does not mean that it has to be exactly the same at every locality. I have never been in the Sahara Desert, but I am willing to suppose, for the sake of discussion, that large parts of it have a certain sameness. But that doesn't mean that every hill, every sand dune, and every clump of bushes is exactly the same as every other one. Your neighborhood may be more or less typical in the city in which you live, but that doesn't mean that every house is identical with your own. The cosmological principle allows local variations; it's on the large scale that things seem to be pretty much the same.

One kind of local variation, as we have already seen, is that galaxies tend to occur in clusters. In fact, when Hubble began his work, he was aware of several conspicuous clusters, and he made a point to avoid them in his photographic survey so that they would not bias his results. However, when he investigated the statistics of the numbers of galaxy images on the photographs, he realized that the galaxies are not distributed in space with perfect randomness, but show a clumpy distribution, as if they are clustered, even though in many cases sparse clusters at different distances, seen overlapping in projection, might pretty well wash themselves out from recognizability. The statistics, though, led Hubble to think that probably *all,* or at least *most,* galaxies are in groups or clusters, rather than being single in space. I think he was correct, but this hypothesis has not yet been absolutely proven.

---

3. If galaxies are distributed at random in space, statistical theory enables us to predict how many fields should contain a certain number of galaxies each. Hubble found that a far larger number of fields had too few galaxies, and also that a far larger number of fields had too many galaxies, than would be expected for a random distribution. Explain why this result suggests that galaxies tend to be clustered.

One such cluster is the one to which our own Galaxy belongs—the Local Group. It is spread over about a million parsecs, and contains at least 21 known members (some investigators count as members a somewhat larger number of galaxies at a somewhat greater distance). There are three large spiral galaxies (our own, the Andromeda galaxy, and M33), four irregulars, four intermediate ellipticals, and at least ten dwarf ellipticals, including a nearby dwarf discovered in 1977 at the Siding Spring Observatory in Australia. Dwarf ellipticals, you recall, are probably the most common kind of galaxy. Appendix 14 gives the properties of the galaxies that are generally ac-

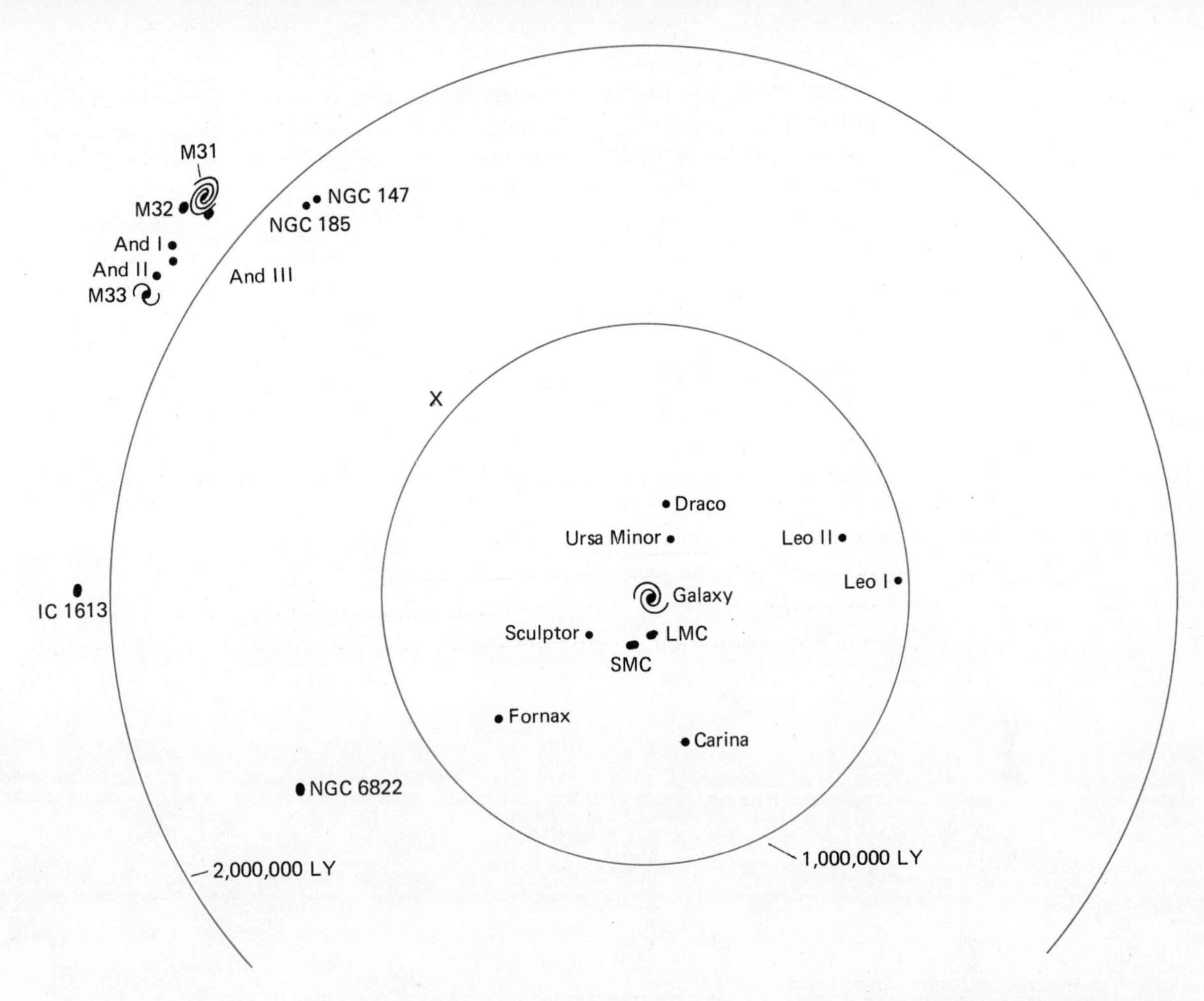

FIGURE VI.20
The Local Group.

cepted to be members of the Local Group. At distances of several million parsecs, we find several other groups of galaxies, more or less comparable to the Local Group. At a distance of several tens of millions of parsecs, however, something like the Local Group, seen in projection against the background of very many more distant galaxies, would never be noticed.

Shortly after World War II, two extensive photographic surveys were made of the sky. One was made with a 20-inch (50-cm) astrographic telescope at the Lick Observatory (in California) as part of a program to measure proper motions of stars. (A second series of photographs is now being taken with the same telescope; comparison of the first and second epoch photographs reveals the stars that have detectable proper motions.) It turns out that those photographs reveal an enormous number of images of galaxies, and astronomer C. D. Shane, at Lick, in collaboration with statisticians Jerzy Neyman and Elizabeth Scott, at the University of California, Berkeley, made a detailed analysis of their distribution.

Shane, Neyman, and Scott found, in agreement with the earlier hunch of Hubble, that the observed distribution of galaxies in the sky, at least as revealed on the Lick photographs, is compatible with the hypothesis that all galaxies are in clusters. The hypothesis is not *proven* by the analysis, but it seems likely to be correct. Of course most clusters are unrecognizable because of confusion with the background. On the other hand, a large number of very rich clusters (lots of members) are easy to spot—they stick out like sore thumbs.

The other photographic survey was the National Geographic Society-Palomar Observatory Sky Survey; it was made with the large 48-inch (1.2-m) Schmidt telescope on Palomar Mountain, and was financially supported by the National Geographic Society. The Palomar Survey reached a much greater depth in space than did the one at Lick.

As I mentioned earlier, while I was still completing my graduate studies, I was lucky enough to be hired as observer for that sky survey (which means I spent a lot of time at the observatory operating the telescope), and consequently had access to those wonderful photographs of the sky. They revealed tens of thousands of recognizable clusters of galaxies. I became terribly interested in them, and made it a project (in fact it became my doctoral dissertation) to survey the very richest of them. My thought was that the very rich clusters could be identified to very great depths in space, and serve as

FIGURE VI.21
The central part of the regular Coma cluster of galaxies. *(Kitt Peak National Observatory)*

markers of the large-scale distribution of matter in the universe. Moreover, with clusters, unlike with individual galaxies, we can estimate relative distances. I ended up cataloging 2712 clusters, but even more have been cataloged since.

Types of Clusters

Before telling you about the results of my study, let me tell you what clusters of galaxies are like. They fall into two main classes: regular and irregular. The regular clusters are very rich. On typical photographs hundreds of members can be identified, but if we include the numbers of dwarf ellipticals that also must be present, they could contain tens of thousands of galaxies. These clusters are roughly spherically symmetrical, and each has a high concentration of galaxies at its own center. Regular clusters often contain radio galaxies, and some of the richest are now known to emit X rays. We think the X rays are coming from hot gas in the clusters—possibly gas left over when the clusters were formed, or possibly matter ejected from stars, and which was subsequently swept out of the galaxies.

It is significant that the rich regular clusters contain few if any spiral galaxies—mostly only ellipticals and galaxies that look as though they *should* be spirals, but which have no spiral arms. It may be that spirals never formed in the rich clusters, but it is also possible that some simply have lost their interstellar matter by colliding with other galaxies. The stars in galaxies are so far apart that they have virtually no chance of colliding; galaxies just pass

FIGURE VI.22
The regular cluster of galaxies in Corona Borealis. *(Hale Observatories)*

through one another. But any interstellar matter in a galaxy *would* collide with that in another galaxy passing through it, and as the galaxies went on their respective ways, that gas and dust would be left high and dry in intergalactic space. Such may be the origin of the gas that is emitting X rays in some clusters. Anyway, that gas, whatever its origin, must be effective in sweeping gas out of other galaxies passing through it.

Irregular clusters are not as well organized as the regular ones. They are generally far less rich, and, in fact, range down to groups like the Local Group. The richest, though, may have as many galaxies as some regular clusters, and may be of comparable size—3 to 5 million parsecs across. They are not particularly symmetrical in appearance and do not have strong central concentration. They often have subclusters within them. And they contain galaxies of all types—spirals as well as ellipticals.

FIGURE VI.23
The irregular cluster of galaxies in Hercules. *(Hale Observatories)*

So up to now, we have this view of the universe: It is made up of galaxies, which range greatly in size and mass. The galaxies, in turn, are mostly (if not entirely) in groups and clusters. These systems of galaxies have diameters of from 1 to 5 million parsecs, or even a bit larger. The biggest clusters—the *great* clusters—are usually symmetrical and regular in structure, and have total masses comparable to a thousand million million (10^{15}) suns.

Large-scale Distribution of Matter in Space

As we look out into space (on our telescopic photographs) most groups and clusters of galaxies just seem to merge together in projection into a kind of irregular background of galaxies—in all directions in the sky. But here and there very rich clusters—the great clusters—stand out and can be recognized. Those are the fellows I studied. Figure VI.23 shows their distribution in the sky. The figure is a plot in galactic coordinates, as in Figure VI.19a, except that the part of the sky too far south to survey from Palomar happens to be shown in a different part of the map. Each symbol is a great cluster of galaxies, the larger symbols being the relatively nearer clusters and the small symbols the relatively more distant ones.

Well, my results more or less parallel Hubble's in two respects. First, just as he found galaxies, on the large scale, to be distributed more or less uniformly through space, so I found the large-scale distribution of clusters to be isotropic and homogeneous; there are as many in one direction as in any other (except, of course, for the obvious effects of absorption by dust in the Milky Way), and there is no evidence of clusters thinning out in distance. But on the *small* scale, just as Hubble found the galaxies to have a tendency to cluster, so I found that the clusters of galaxies tend to clump up in what many people call *superclusters*. Statistical tests show that the apparent clumping of clusters is not illusory, but that clusters of galaxies tend to be really associated with other clusters in space.

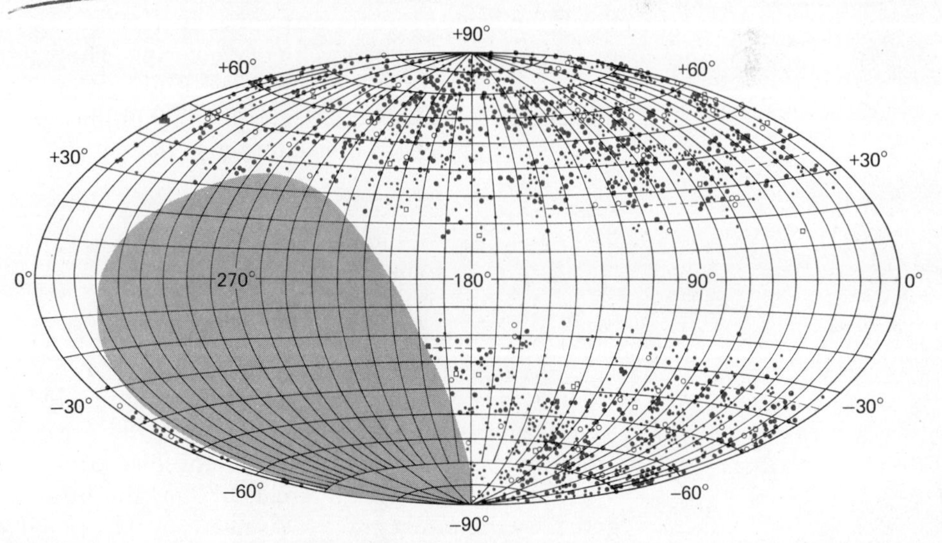

FIGURE VI.24
The distribution of rich clusters of galaxies. *(From a study by the author)*

Superclusters

The idea of superclusters was not new to me. In 1937 the Swedish astronomer Erik Holmberg, in investigating double and multiple galaxies, found that there seems to be a local concentration of them, and on a far larger scale than just the Local Group. Then, in the 1950s, Gerard de Vaucouleurs revived Holmberg's idea, and gave the matter an intensive study, which he is continuing to this day. De Vaucouleurs has amassed an immense amount of data to demonstrate that the Local Group, lots of other neighboring groups, and quite a few clusters of galaxies, including the nearest big (but *not great*) cluster—the Virgo cluster—are all part of a large system he calls the *Local Supercluster*. This local concentration of galaxies and groups and clusters has been confirmed by many independent observers, including two of my own graduate students. The diameter of the Local Supercluster is probably 50 to 75 million parsecs, and its mass probably lies in the range 10^{15} to 10^{16} times that of the sun.

Well, if our own Local Group is in a supercluster, the idea that there are other clusters of clusters of galaxies may not be so farfetched. It was controversial for some years, but now it is pretty well accepted by everybody. The general acceptance of the reality of superclusters had to await the availability of large computers. It is now possible to do very involved statistical studies of the distribution of galaxies and clusters of galaxies as they are revealed in photographic sky surveys. The present leader in this study is James Peebles, physicist and astronomer at Princeton. I enjoy mentioning Jim Peebles in this regard, because he started out as a skeptic, but now says he is a "true blue believer" in larger structures in the universe than clusters of galaxies.

Does this then mean that our universe may consist of a hierarchy of clustering—clusters of galaxies, clusters of clusters, clusters of clusters of clusters, ad infinitum? I do not believe so. We cannot be sure (at this writing), but I think the hierarchy stops at superclusters. And even superclusters are not really part of a true hierarchy of clustering. They, in my view, are simply the general regions of space where we find matter. They are the largest inhomogeneities we know of—typically 100 million parsecs across—but are not really *superclusters* in the same sense that clusters of galaxies are clusters.

---

4. Would an indefinite hierarchy of clustering (clusters upon clusters, without limit) be consistent or inconsistent with the cosmological principle? Why?

Clusters of galaxies are certainly gravitationally bound systems of galaxies. At least the rich, and almost certainly the great, clusters must have been around for most of the age of the universe, and are more or less permanent aggregates of matter. Superclusters, on the other hand, are not bound together by their own gravity. As nearly as we can tell, the Local Supercluster is expanding almost as fast as the universe as a whole. What flimsy evidence we have suggests that other superclusters are doing the same thing. They must be considered, I think, as regions where, in the early universe, matter collected and condensed into clusters and groups of galaxies, and possibly into individual intergroup galaxies as well.

It's sort of like a large metropolitan area. Of course I know Los Angeles better than any other megalopolis, but the idea must be the same elsewhere. Imagine looking at the city from space at night, and noticing the distribution of lights. There are several dense concentrations of lights — clusters — including the downtown city center, Pasadena, Hollywood, Long Beach, and some others. Then there are some smaller concentrtions — small clusters — including Sherman Oaks, Eagle Rock, and Bellflower. And then there are lots and lots of communities — groups. Whether there are individual houses, not parts of communities or towns or cities, is a matter of definition, I suppose. I suspect the same is true of individual galaxies in superclusters.

Just as the metropolitan area of Los Angeles (and Paris, too) is a real entity, so are superclusters. But they seem to be expanding with the universe, or at least almost as fast. Their existence must tell us something about the way the original universe formed into clusters and galaxies, but the details are not yet understood. In any case, embedded in the superclusters are occasional great clusters of galaxies, stable, bound systems that have probably been around almost from the beginning, and which serve as lampposts to tell us a great deal about how the universe is put together and how it may be tearing asunder — but that must wait for the next scene.

SCENE 3

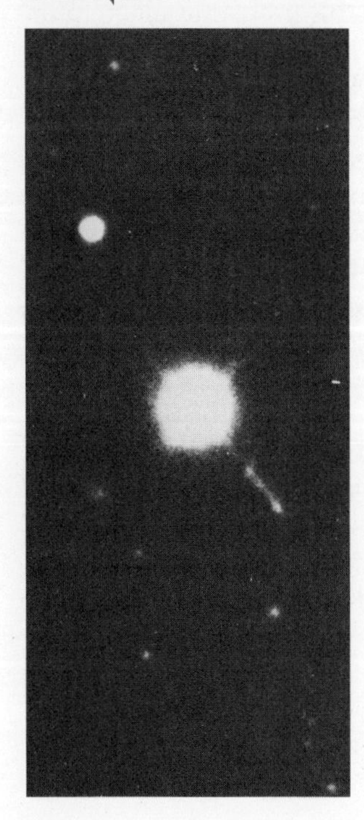

Beginnings and Endings

Two millenia ago our Greek forebears applied their best talents to understand the nature of the universe, and they achieved considerable advances, as we saw in Act I. To be sure, we have come a long way from them. Yet, in the field of modern *cosmology*—the study of the origin and evolution of the universe—we still bask in a lot of ignorance. We have discovered a great deal about the laws of physics that apply locally, and we freely apply them to the universe at large, sometimes with an almost religious conviction that they are the final and absolute truths of nature. Perhaps we are even right; but past experience suggests that new surprises may well await us. Future generations (should they exist) may well regard many of our ideas to be as naive as we regard the circles and epicycles of antiquity.

Still, as long as we keep things in humble perspective and don't take ourselves or our ideas too seriously, the study of cosmology can be a very fulfilling sport. We apply the best knowledge we have to try to answer the most profound questions man can ask. Next year, or in a decade, or in a hundred years, we will know more and may have to alter our thinking, but this need not deter us from searching for a new and better understanding of the universe.

The Cosmological Principle

In the laborabory we can do experiments many times and in many different ways to test the generality of a result. Even in astronomy there are usually very many objects available for observation to check out a hunch. But there is only *one* universe. What if the part we live in is somehow other than typical? In that case, unless someone were to tell us exactly *how* it is different from what is typical, there is no way we could learn the properties of the universe in general.

Of course you could sit down and *make up* all kinds of specifications about what the universe is like in the parts we cannot observe. You could say, for example, "Just beyond what we can observe, or ever hope to observe, the universe ends; but far away there are other similar universes, forever out of range of our telescopes." Or, you might say, "Far beyond the observable part of the universe, galaxies cease to exist; there are only isolated stars, all of them exactly like the sun, and each one with two planets that are inhabited." No one could prove your theories wrong, but neither could you ever prove them right. The same holds for a thousand more such ideas that I could advance in as many minutes (given time out for rest and refreshment). Some philosophers like to expound this way, but such speculation tells us nothing about nature, and it is not part of science.

Most cosmological theories, therefore, begin with the assumption (or hope, if you prefer) that the part of the universe we observe is representative of the whole thing. In the last scene we saw evidence that the observable region of space is homogeneous and isotropic. We simply assume that this homogeneity extends throughout the entire universe. This idea is the essence of the *cosmological principle.* Perhaps the principle is wrong, but if so, where do we begin? At least we can make some predictions on its basis, and see how well they match observations. Of course we'll never be able to know whether our theories are really *true,* and if so, whether they are true for the *entire* universe, but all science can ever do is represent the observable universe with models that describe its behavior. You'll have to ask your philosopher friends what the *real* truth is—or your guru, or an ancient Chinese oracle; those fellows may have private pipelines to absolute truth, but science does not.

Einstein's Static Universe

So to start with, we assume the cosmological principle. Next we need a physical theory. The most accurate applicable descritpion of nature yet found is Einstein's general theory of relativity. Shortly after Einstein introduced the theory in 1916, he attempted to apply it to a universe described by the cosmological principle. Now it turns out that to a considerable extent relativity predicts the same thing for the universe that Newtonian gravitational theory does: The universe is full of matter, and matter exerts gravitation; thus all matter in the universe should be falling together. But if that were the case, one would expect the universe to have collapsed long ago.

Well, Einstein found an out. You recall (possibly) in Act II Scene 5 how relativity describes gravitation in terms of a warping of spacetime by matter. The *field equations* describe that warping, and enable us to calculate how matter should move in spacetime. It turns out that in the derivation of the field equations that describe a universe obeying the cosmological principle, it is possible, without violating the premises of the theory, to introduce a term containing an arbitrary constant number. Often in the solution of physical problems, such constants can be introduced. (Those readers familiar with mathematics know, for example, how constants of integration crop up in the solution of differential equations.) Well, the constant that appears in the field equations of general relativity, which is called the *cosmological constant,* represents a force between all objects in the universe that is greater in proportion to their separation. If the cosmological constant is a number greater than zero, that force is one of repulsion. Now since no such force has ever been detected and since there is no evidence for it whatsoever, the natural inclination would be to assume that the value of the cosmological constant is zero. Then the field equations, in the regime where they can be tested, predict the right results.

Einstein, however, was presented with the problem of explaining why the universe does not fall together. So he assumed that the cosmological constant is *not* zero, and, in fact, he assigned it just the right value to counteract the effects of gravitation over large distances. In that case, the repulsion force would be too small to be noticed in the laboratory or even in the solar system. On the other hand there was no other evidence for such a force, and it seemed to many physicists that the need to introduce it just to make a new theory work was a serious weakness of that new theory.

The Friedmann Models

But then, in the early 1920s, other physicists, especially A. Friedmann, found that there exist solutions to the field equations that describe a homogeneous universe that does not collapse, and *without* the need of a cosmological constant. However, all such solutions require that the universe be *expanding*. It is as if some great explosion had started things off, and they are now coasting apart. If that explosion was strong enough to give all objects high enough initial speeds, they will separate forever; whereas if the explosion was not strong enough, the objects will someday fall together again. It is like a rocket shot away from the earth—if it has the escape speed of about 11 km/s, it will go on forever; if it has not, it will return to earth. In either case, we can understand the present universe, without involving any unknown forces of nature, if it is expanding. I emphasize that the expansion of the universe is *predicted theoretically* by the best description of gravitation we know—that provided by general relativity. It is not the expansion that has to be explained; on the contrary, if the universe were *not* expanding, *that* would require an explanation, say, by introducing an unknown force through the cosmological constant.

The Hubble Law

The story of the verification of that prediction begins at the Lowell Observatory in Arizona, where V. M. Slipher, from 1912 to 1925, observed spectra of more than 40 "nebulae." Slipher, of course, did not realize the true nature of those objects at the time, but of those later shown to be galaxies, all except a few nearby ones (in the Local Group) were found to have large positive radial velocities, ranging up to 1800 km/s.

After Hubble had demonstrated the extragalactic nature of the "nebulae," he began determining distances to those for which Slipher had found radial velocities. Meanwhile, Hubble's colleague, Milton Humason, began to observe spectra of additional galaxies for which Hubble had found distances. In 1929, Hubble made the first formal announcement of the *law of the redshifts*—now known as the *Hubble law*—that galaxies are receding from us at speeds that are proportional to their distances. (The constant of proportionality is now called the *Hubble constant*.) By 1931, Hubble and Humason had extended the law to greater distances with new observations, and they published full details of the newly established law in the *Astrophysical Journal* (Figure VI.25).

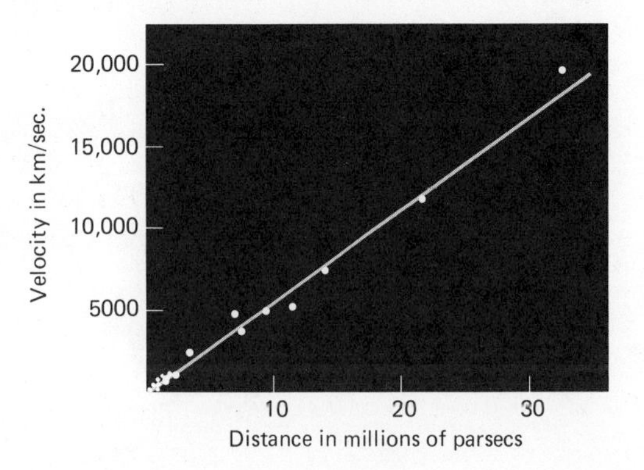

FIGURE VI.25
A copy of the diagram published in 1931 by Hubble and Humason, exhibiting a relation between the velocities and distances of galaxies. The scale of extragalactic distances has subsequently been modified.

1. Hubble originally assigned to the Hubble constant the value 550 km/s·10^6pc, which means that the speed of recession of a galaxy is 550 km/s for every million parsecs of its distance. Since Hubble's time, revisions in the extragalactic distance scale have reduced our estimate of the Hubble constant to about 50 km/s·10^6pc. How does such a revision affect our estimate of the age of the universe?

At this point, let's be sure that we all understand not only why the Hubble law shows the universe to be expanding, but *also* that it does *not* imply that we are at the center of the expansion. Imagine a ruler made of flexible rubber, with the usual lines marked off at each centimeter. Now suppose someone with strong arms grabs each end of the ruler and slowly stretches it, so that, say, it doubles in length in one minute (Figure VI.26). Consider an intelligent ant sitting on the mark at 2 cm—intentionally *not* at either end or in the middle. He measures how fast other ants, sitting at the 4-, 7-, and 12-cm marks move away from him as the ruler stretches. The one at 4 cm, originally 2 cm away, has doubled its distance; it has moved 2 cm/min. Similarly, the ones at 7 cm and 12 cm originally 5 and 10 cm distant, have had to move away at 5 and 10 cm/min, respectively. All ants move at speeds proportional to their distance. Now repeat the analysis, but put the intelligent ant on some other mark, say on 7 or 12, and you'll find that in all cases, as long as the ruler stretches uniformly, this ant finds that every other ant moves away at a speed proportional to its distance.

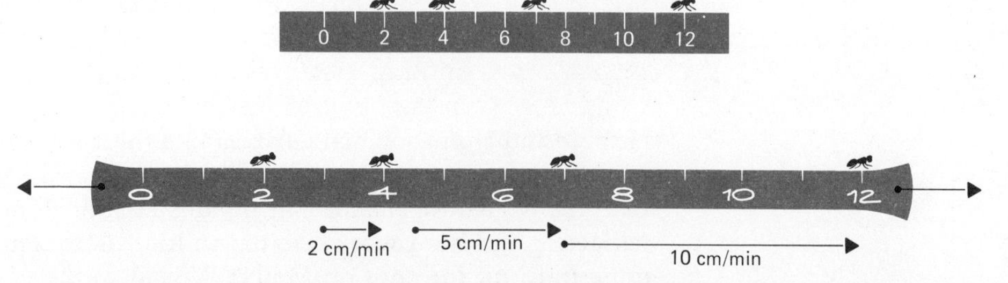

FIGURE VI.26
Stretching a ruler

For a more realistic, three-dimensional analogy, look at the raisin bread in Figure VI.27. The book has put too much yeast in the dough, and when he sets the bread out to rise, it doubles in size during the next hour and all the

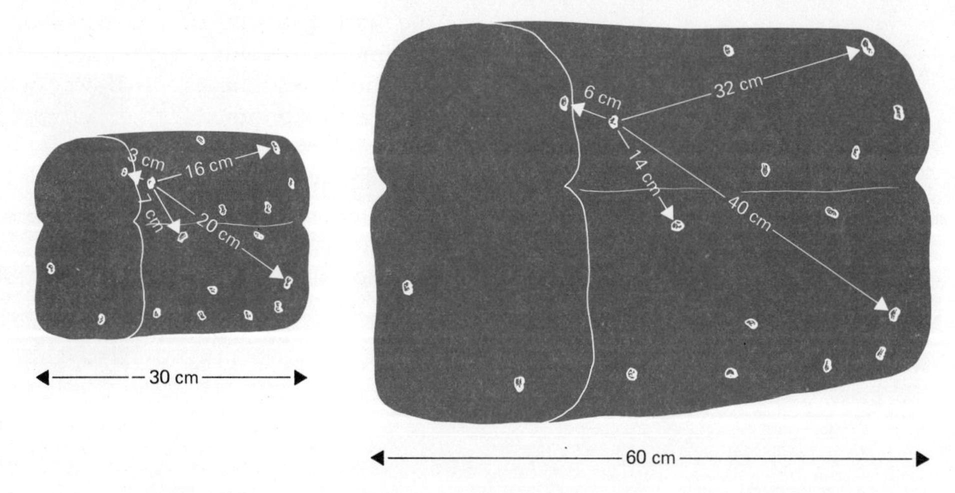

FIGURE VI.27
Expanding raisin bread.

raisins move farther apart. Some representative distances from one of the raisins (chosen arbitrarily, but not at the center) to several others are shown in the figure. Since each distance doubles during the hour, each raisin must move away from the one selected as origin at a speed proportional to its distance. The same is true, of course, no matter which raisin you start with. But don't carry the analogy too far; I use it here only to show that a uniform expansion of the universe must give the observed Hubble law. I do not, however, want to give the impression that some medium pervading space is carrying the galaxies apart from each other, as the bread dough does the raisins.

The expansion of the universe also does not mean that *everything* within it has to be expanding. Galaxies, and at least the richer clusters of galaxies, are bound together (we think) by their mutual gravitation, and merely separate from each other rather than themselves growing. Astronomer David Jenner likes to demonstrate this important point to his classes in elementary astronomy by gluing a number of ten-cent pieces to a large, tough balloon. He then blows up the balloon, so that the dimes all separate; but the coins themselves do not grow in size. As Dave points out, "The value of currency never increases during inflation."

The expansion of the universe is thus taken as a basic and essential postulate of nearly all theories of cosmology. It also saves the general theory of relativity without any need for the cosmological constant. When he learned of the Hubble law, Einstein is said to have remarked that his introduction of the cosmological constant was the biggest blunder of his life.

The Hubble Law Today

Hubble and Humason actively pursued their work on the redshift law for the rest of their careers. Humason, by the way, was an amazing man himself. Lacking formal education, he made a name for himself with native intelligence and hard work. He began his "astronomical" career by driving a mule train up the trail on Mount Wilson to the observatory; in those early days supplies had to be brought up that way. (Even astronomers hiked up to the mountaintop for their turns at the telescope; not too many of us are that dedicated anymore.) Anyway, Humason became interested in the astronomers and the work they were doing, so took on a job as janitor at the observatory. After a time, he became a night assistant, helping the astronomers run the telescope and take data. Eventually he made such a mark that he earned the rank of full astronomer at the observatory.

Among his many honors, Humason was awarded an honorary doctorate. It would be very hard today for a person to go so far in science without a university degree, because science has become increasingly specialized. Milton Humason may be one of the last of a rare breed. He was also a fine gentleman, whom we all counted ourselves privileged to know. I must add, too, that he was all but unbeatable at the observatory pool table, where astronomers would often spend cloudy nights studying ballistics.

Hubble and Humason are both dead now, but they were able to push their law out to include galaxies moving away from us at 60,000 km/s – 20 percent the speed of light. The work, of course, continues, and is now a major effort at several large observatories of the world. Radio astronomy has played an important role, too. We saw in Scene 1 how some galaxies emit radio energy with great strength. It is thought that many of the unidentified radio sources are probably giant elliptical galaxies in clusters, but so distant and faint that they haven't been recognized on photographic surveys. Thus one way to search for very remote clusters of galaxies of large redshift is to inspect fields containing unidentified radio sources.

FIGURE VI.28
Milton Humason. *(Courtesy of the Archives, California Institute of Technology)*

FIGURE VI.29
Rudolph Minkowski.
(Courtesy of the Archives, California Institute of Technology)

In this way Rudolph Minkowski made the first great stride in pushing out the redshift law since Humason's retirement. With the 200-inch telescope at Palomar, Minkowski searched the part of the sky corresponding to the position of a source of radiation recorded by the radio astronomers at Cambridge, England. There he found a very remote cluster of galaxies that appeared to be far more distant than any that had ever been observed before. But the galaxies in that cluster are so faint that it would take a Herculean effort to photograph their spectra.

Minkowski, however, had experience in photographing the spectra of faint objects. A former classmate of mine was once at a party at which Minkowski was also present. At one point, my friend was relating, to the amusement of his listeners, a story he had heard about an astronomer who had exposed a photograph with the telescope for all of one night to record the spectrum of a faint object, only to get so tired that when he went to the darkroom to develop his photographic plate he mistakenly put it into the hypo solution instead of the developer! (Photographers among you know that the hypo makes any development or further development impossible.) Well, Minkowski corrected my fellow student with the statement, "It vas *three* nights, und it vas me!"

But Minkowski did it right on Cambridge source 3C295! He got good spectra of two galaxies in the cluster, and found that it moves away from us at 36 percent the speed of light. Those were the last observations Minkowski made before his retirement from the Hale Observatories, and were also the most dramatic of his career. And I am very glad he succeeded; he was my boss on the Palomar Sky Survey, and was a truly great person. After retirement he remained active in research up to the end; he died early in 1976 at the age of 81. I shall always have a warm soft spot in my heart for Rudolph!

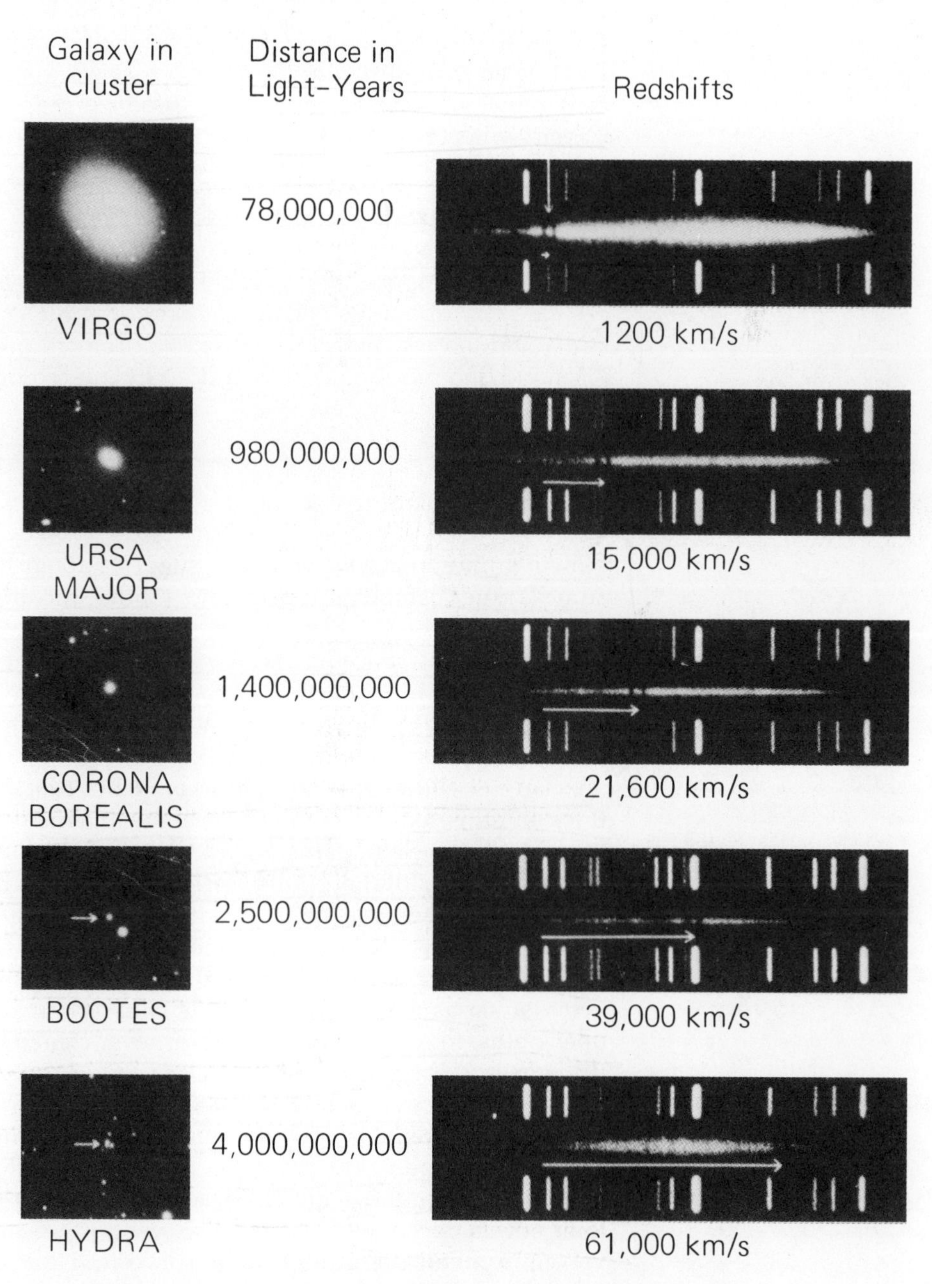

FIGURE VI.30 *(left)* Photographs of individual galaxies in successively more distant clusters; *(right)* the spectra of those galaxies, showing the Doppler shifts of the two strong absorption lines of ionized calcium. *(Hale Observatories)*

The invention of new electronic techniques in the 1960s has given our telescopes even more power. Today, other young astronomers, following in the footsteps of Minkowski, have discovered even more remote clusters of galaxies. The most distant whose radial velocity has been observed at this writing is moving away from us at more than 50 percent the speed of light. Quasars, as we have seen, have been observed with even far greater redshifts —indicating speeds up to 91 percent that of light—but as yet we have no independent way of determining the distances to quasars, so cannot use them to extend the Hubble law to greater depths.

I have already remarked (Scene 1) that some people worry about such high speeds for galaxies, and wonder if those large redshifts don't have some other explanation than that of a Doppler shift. But I hope you are convinced by now that the easiest way to interpret the redshifts is as Doppler shifts, indicating an expanding universe; otherwise we would have to come up with a new physical principle that could produce redshifts. Why make up new physics when existing theory *predicts* that the universe is expanding?

---

conserved

2. Some people have suggested that light "tires" in traversing intergalactic space, thereby increasing in wavelength. What does this idea say about the conservation of energy?

3. Could the redshift in the spectra of remote galaxies be due to dimming and reddening of their light by intergalactic dust? Why or why not?

Evolving Relativistic Models of the Universe

Now it's time to see what theory suggests about the universe. We will assume (1) the cosmological principle; (2) that the universe is expanding; and (3) the gravitational theory of general relativity, with zero cosmological constant. In general, spacetime is curved by the presence of matter, which means that the paths of light over long distances are not necessarily along the straight lines of Euclid, according to whose geometry the area of a circle is pi times the square of its radius (πR^2).

Relativity allows that the paths of light *could* follow Euclidean straight lines, but this requires a very critical mean density of matter in space. More likely, light travels in paths Euclid would have called curved. It might curve back on itself (although not necessarily in closed paths), in which case we could never observe, even in an infinite time, a light signal from a place infinitely far away. Then spacetime (or its geometry) is said to be *closed*—there is a *finite* region of it susceptible to even potential observation, and the universe, so far as we are concerned, is *finite*. Alternatively, light can curve on open paths, like hyperbolas, and can never come back; then spacetime (and its geometry) is *open*, and space can be *infinite*. In adopting the cosmological principle, we are assuming further that unless the geometry of spacetime is closed, the universe is, and always has been, infinite. Of course this is only an idealized model. The principle could also apply to a finite universe, but one whose boundaries are so far away that we can never observe them or their effects.

Well, with all these assumptions, we find that the universe must have started in an initial, very dense state (but one that even then could have been infinite). From that dense beginning everything has evolved according to the laws of general relativity; we call these models of the universe the *evolving relativistic models.* We presume that the expansion of the present universe started explosively with a "big bang." That's why these models of cosmology are also called, collectively, the *big-bang* theory.

The solution to the field equations of general relativity, with zero cosmological constant, allows three possibilities.

1. A *closed universe,* in which the expansion will eventually be overcome by the gravitation of matter, and the universe will then start to fall together again. As far as relativity theory is concerned, we could imagine another big bang occurring when all matter has fallen together again, resulting in a new expansion. Then, in principle, the universe could go on forever, alternating between expansions and contractions. Such a model has sometimes been called the *oscillating* theory of the universe. It is not a complete theory, though, because we do not know the appropriate physics to describe the details of what goes on at each contraction, new big bang, and expansion. In any case, in these models the geometry of spacetime is closed, circles have areas less than πR^2, and the observable universe is finite.

2. An *open universe,* in which, once started, the expansion goes on forever. In these models there is just one big bang, and it happened some thousands of millions of years in the past, but since then the universe has been, and always will be, expanding. As gravitation pulls between all the matter of the universe, the expansion is ever slowing, but will never quite stop. Here the geometry of spacetime is open, circles have areas greater than πR^2, and the universe is infinite.

3. The *Euclidean, or flat, universe,* in which the big bang was just right to make the universe expand forever, but with the matter in it only just barely escaping from itself. This model represents the boundary between the open and closed models. In it, space is Euclidean, or *flat.*

In all these models the cosmological principle is presumed to hold, so only the scale of the universe can change. Thus the whole problem of cosmology reduces to that of finding *how* the scale changes in time. Most cosmologists denote that scale of the universe by the capital letter R, which varies in time as the universe expands or contracts. We follow the history of the universe, then, by following how R changes with time.

Figure VI.31 shows a graph of the scale of the universe, R, as time goes on. The above three possibilities are all plotted, and labeled 1, 2, and 3, corre-

FIGURE VI.31
The change in scale with time for the evolutionary models discussed in the text.

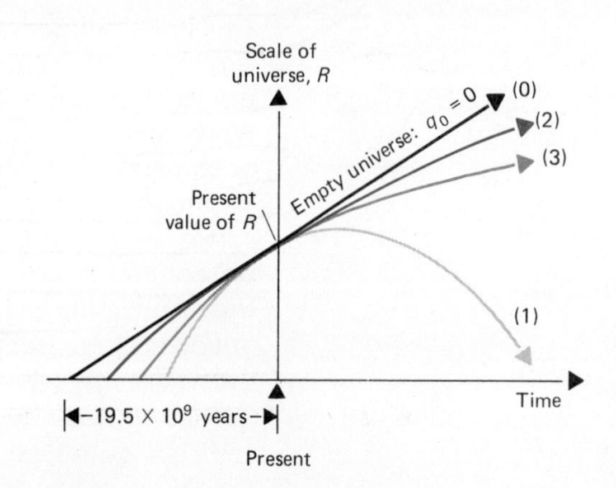

sponding to the closed, open, and flat models, respectively. All the models of cosmology shown in the figure are evolving relativistic universes, since they all assume general relativity, and they all change in time. And all started with some sort of singularity, or big bang. Now note that even that big bang did not have to occur in a small local place; there is no "site" for the start of the expansion. Because we are inside the universe *now*, we always were—it is, and was, always all around us—*everywhere*.

---

4. Some people have speculated that the universe began from a small ball—perhaps the size of a grapefruit. Is this idea consistent with the cosmological principle? Why or why not? In either case, have you grounds to absolutely rule out the idea?

Parameters of the Relativistic Models

Whereas there is no site for the "beginning," each model does predict a definite time in the past when the big bang occurred. In each model the expansion slows with time, or *decelerates*. The *rate* at which it is decelerating is all we need to know (if the whole idea is correct) to choose between the possible fates of the universe, for that rate of deceleration would tell us whether the universe will expand forever, in which case spacetime is open, or will eventually stop expanding and contract, in which case spacetime is closed.

The deceleration rate is so important that it is given a special symbol: q_0. All the relativistic evolving cosmologies with zero cosmological constant require that q_0 be greater than zero. If $q_0=1/2$, the universe will just barely expand forever. If q_0 is greater than $1/2$, the universe will one day stop expanding and start contracting, and possibly will even oscillate between alternate expansions and contractions. If, on the other hand, q_0 is less than $1/2$, the universe will expand forever.

Exactly when, in the past, the big bang occurred depends on the value of q_0. The smallest q_0 can be is zero, which would mean that there is no matter in the universe, and hence no gravitation at all, and no deceleration. This model of the universe is shown as curve *0* in Figure VI.31. Then the age of the universe (and the oldest age it *can* have according to our assumptions) is found by extrapolating the present rate of expansion backward, as though it had not slowed at all. Then, knowing how far away a galaxy is, we would need only to divide its present distance by its present speed to see how long it and we were close together in the beginning. According to the present best estimates of extragalactic distances, this greatest possible age for the universe is about 19.5 thousand million years.

If $q_0 = 1/2$, and the universe can just barely expand forever, the age of the universe is exactly two-thirds that maximum possible age—or about 13 thousand million years. Closed universes (q_0 greater than $1/2$) must have younger ages, and open universes (q_0 less than $1/2$) must have greater ages, but always less than 19.5 thousand million years.

These possibilities, all that exist with the models most frequently considered by cosmologists today, are summarized in the following table.

Kind of Universe	Deceleration Parameter, q_0	Age (millions of years)	Geometry of Spacetime	Future of the Universe
Closed	More than 1/2	Less than 13,000	Closed on itself; finite universe	Will stop expanding and contract someday
Open	0 to 1/2	13,000 to 19,500	Open and infinite	Will expand forever
Flat	Exactly 1/2	13,000	Flat (Euclidean and infinite)	Will barely expand forever

5. Are you perplexed with the task of imagining an infinite universe? What if it is finite; then what lies beyond? Do you suppose the experts can visualize infinities and eternities any better than you can? If not, can they simply accept these concepts, because they are used to talking and thinking about them? Collect opinions of such experts if you can find them. Who are the experts on infinities and eternities?
6. Suppose $q_0 > 1/2$, so that the universe is closed. Can there be other similar "universes" beyond the closed light paths in our own universe, with the same or even different properties? What might they be like, and how do they interrelate with our universe? Should cosmologists be concerned with studying such possibilities? Why or why not?
7. Resketch the part of Figure VI.31 that relates to the open universe. Show on the time axis of your sketch where we are, the approximate positions of the big bang, the most remote quasars, and the most remote clusters of galaxies for which we have measured redshifts.

Now, of course the universe does not really have to be like one of the models described in the table above. Those are the models which are possible with certain very restrictive assumptions, but no one can guarantee that the assumptions are correct. They are, in a sense, the simplest ones we can make, because they assume a large-scale uniformity of the universe, and that known laws of physics in their simplest form are the correct ones. Naturally, lots of people have considered alternatives, and lots of other models of cosmology exist. The trouble is, there are infinitely many possibilities. Most cosmologists want to try out the simplest of them first. I shall, however, mention briefly a few of the other suggestions.

Alternative Models

One possibility is that the cosmological constant is *not* zero, even though the universe *is* expanding. Models with a positive cosmological constant lead to universes that are older than those given in the table, and usually to universes that will expand forever. Figure VI.32 shows how R varies with time since the big bang in a typical model with a positive cosmological constant.

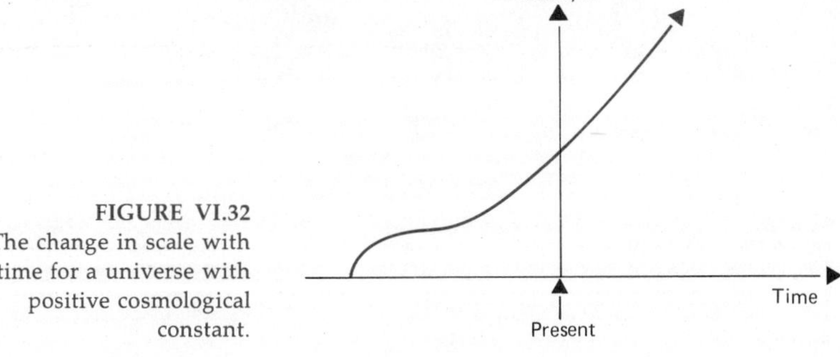

FIGURE VI.32
The change in scale with time for a universe with positive cosmological constant.

Another interesting theory of cosmology was introduced in the early 1950s by astrophysicists Hermann Bondi, Thomas Gold, and Fred Hoyle, the *steady-state theory*. The steady-state model can be derived from a single assumption: the *perfect cosmological principle,* which states that the universe is the same (on the large scale) not only at all places, but also *for all time*. This means if you could come back to life many thousands of millions of years in the future (or past) you would see the same general picture of everything. The universe is expanding, but still is infinite in age and infinite in extent. But if the universe expands, the matter in it must thin out in space, lowering the mean density of matter in the universe. This would violate the perfect cosmological principle (that nothing can change in time), unless new matter comes into being spontaneously to fill the void left by the separation of the existing galaxies. So the steady-state theory was also called the *continuous-creation theory*. The variation of R with time for the steady-state theory is shown in Figure VI.33.

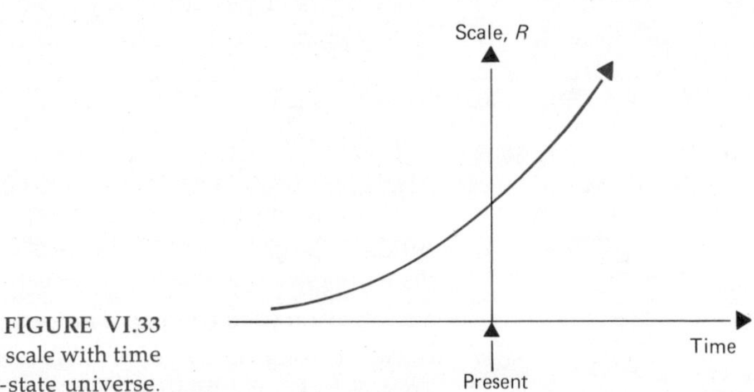

FIGURE VI.33
The change in scale with time for the steady-state universe.

Some people objected that matter cannot be created out of nothing—that such would be miraculous. But as Hoyle pointed out, it is also miraculous where the matter came from in the first place, and it is no more difficult to imagine a lot of little miracles occurring continuously all the time than one huge one all at once. There were at least two things going for the steady-state idea: it was philosophically nice in its symmetry and eternal nature, and it also stuck out its neck by making specific predictions. We shall see in a moment that those predictions have probably failed to be realized.

Another, newer theory proposed by Hoyle in collaboration with the Indian astronomer J. V. Narlikar is a new version of a static universe. In the Hoyle-Narlikar model, the universe does not expand and the redshift is not caused by a Doppler effect at all, but by all particles in the part of the universe in which we live gradually increasing in mass. This means if we look off into the distance and hence back into time, we are seeing galaxies at a time when their atoms were less massive. The model actually predicts all phenomena that we observe, but it supposes a new principle of physics—that masses of particles change with time.

The Hoyle-Narlikar theory says that if we go back in time far enough, to what corresponds to the big bang in the "conventional" models, we come to a time when particles all had zero mass. At an even earlier time, particles had nonzero mass again, but mass that interacted, gravitationally, in a negative way with mass in our region of the universe. In fact, the model supposes that all spacetime is divided into great regions of alternating mass like ours and mass that reacts negatively with ours. Within each region, the masses of particles increase for a while, and then decrease, eventually passing through a zero-mass boundary into a region where the mass increases again and interacts in the opposite direction.

Still another kind of unconventional cosmology is the *hierarchical* model, wherein the hierarchy of clustering goes on indefinitely. On the other hand, the evidence for large-scale homogeneity seems to be increasing, and this evidence is an argument against the hierarchical universe. The large-scale distribution of faint radio sources, for example, appears to rule out inhomogeneities larger than a few hundred million parsecs across. Also, counts of faint galaxies of various degrees of faintness made by George Rainey, a former student of mine, and also by Stanley Brown, a former student of de Vaucouleurs, are remarkably the same in different regions of the sky.

These are some of the cosmological ideas out of the mainstream. Space does not permit explaining them all thoroughly. Mainly, though, I wanted to be sure not to leave you with the idea that the universe necessarily *has* to be described by one of those models summarized in the foregoing table. A few years from now we may all be singing a very different tune indeed. But now, let's get back on the subject and see what evidence there is to support any of the notions we've described in this scene.

The Microwave Background

First, have we any evidence that there was even a big bang at all? The answer is yes. We described in the Prologue of this drama what we think the big bang might have been like. Calculations based on the physics we know today show that the primeval fireball should have been extremely hot and dense at one point, and should have consisted of only certain kinds of particles: protons, neutrons, electrons, neutrinos, and their antis. Then as the fireball expanded and cooled, some of these particles underwent fusion, and nucleogenesis should have built up many elements, but the only ones to survive should have been hydrogen, helium, a small trace of deuterium, and only the very tiniest traces of other atomic nuclei. The fireball continued to cool, and when it was roughly a million years old, the atoms in it essentially all hydrogen and helium) became neutral. At that point, it became transparent.

Now before that instant, the fireball was opaque, as it is inside a star. Thus there was a lot of radiation being radiated about from atom to atom. At the moment the fireball became transparent, though, when its temperature dropped to about 3000° and its density to about 1000 atoms per cubic centimeter, atoms in it no longer absorbed this radiation. What happened to all that light? Well, physicist George Gamow, back in 1948, predicted that the energy should still be floating about in the universe, but by now should be redshifted to radio wavelengths, for to see it we would have to look far back in time, and thus far off in space.

Gamow's idea was forgotten because there was no way, at that time, to observe such radiation in space. But in 1965 the idea was revived by a Princeton group of physicists, R. H. Dicke, Jim Peebles, P. G. Roll, and D. T. Wilkinson, who published a paper describing the nature of the microwave radiation that should be coming to us from all directions in space. Remember that we were inside the fireball when it happened (or at least the ancestral particles that eventually evolved to us were), so it was all around us then, and is now. Just before the fireball became transparent, we should have been bathed from all directions in the intense light of a hot gas.

Today, thousands of millions of years after the big bang, according to the Princeton physicists, now that the matter of the universe has cooled and condensed into superclusters, and clusters, and galaxies, and stars, and planets, and us, we should still be bathed in that radiation, and should see all around us that light of the fireball—the glowing embers, as it were, of the blast that started the expansion of the universe. But because of that expansion the radiation should have been increased in wavelength by a thousand times, and should now consist not of visible light, but of a background of microwave radio waves, like those emitted from a very cold body, just a few degrees above absolute zero.

The same year of that theoretical prediction, 1965, A. A. Penzias and R. W. Wilson, of the Bell Telephone Laboratories, were attempting to track down the sources of random noise in the laboratory's very sensitive radio antenna-receiver system. They found, though, that their equipment was picking up a faint signal from all directions, and they could not account for it in any other way than that it was coming from space. When the Princeton physicists heard of this, they were naturally very excited; they suspected that Penzias and Wilson had stumbled onto the microwave background radiation predicted from the big bang. Then followed a long series of observations that are still going on. What has been established by now is that the radiation is exceedingly like that expected from a black body just 2.7° above absolute zero. It is generally interpreted as direct evidence of the big bang.

All this has a good deal of additional significance. For one thing, if the interpretation of the microwave background radiation is correct, it rules out the steady-state theory, for in that theory the universe was always as it is now, and there could have been no big bang. Also, it turns out that the radiation is exceedingly uniform in all directions—that is, it is *very* isotropic. Its isotropy provides strong confirmation for the cosmological principle (again, if we interpret the radiation correctly). This probably pretty nearly rules out the hierarchical models. The Hoyle-Narlikar theory, on the other hand, also predicts the microwave background; in that theory it is radiation from a previous sector of spacetime, scattered about very efficiently by the particles passing through a boundary where their mass is temporarily zero. Well, as you can see, this discovery of the background radio waves has provided cosmologists with an absolute field day! Some are still arguing about what it means, but no matter what it means, it means *something,* and is a very exciting development in modern cosmology. And as I say, most cosmologists think that we are actually looking at the big bang itself—all around us!

FIGURE VI.34
Robert W. Wilson *(left)* and A. A. Penzias *(right)* with the horn-shaped antenna with which they discovered the microwave background radiation. *(Bell Laboratories)*

8. How would the microwave background radiation have been different if it were observed from 5 to 8 thousand million years ago?

The Velocity of the Galaxy in Space

The microwave background can tell us something else, too. Recall (Act II Scene 4) how Michelson and Morely tried to measure the absolute speed of the earth in space. They failed because absolute speed has no meaning; we can measure motion only relative to something else.

On the other hand, that hot gas from which the universe evolved would be about as absolute a standard as we could imagine. Of course it's not around anymore—most of it has long since condensed into stars and galaxies and other stuff—but we do still observe that gas by means of the microwave radiation all about us. Today, that radiation is like that emitted by a very cold black body. Now, if you approach a black body, its radiation is all Doppler shifted to shorter wavelengths and resembles that from a slightly hotter black body; and if you move away from it, the radiation is like that from a slightly cooler black body. Thus, in principle, we should be able to detect our speed with respect to the universe as a whole by observing a slight difference between the wavelength distributions of the microwave background radiation in opposite directions in the sky.

Just such a detection was announced late in 1977 by the University of California physicists Muller, Smoot, and Gorenstein, who observed the background radiation with instruments flown on a high-altitude U2 aircraft. The effect due to our motion is very hard to measure, and the results have a large uncertainty. However, they seem to indicate a speed for the sum of 390 km/s; after correcting for the rotation of the galaxy carrying the sun at a speed of 300 km/s, in a different direction, the Berkeley team find that our Galaxy as a whole is moving about 600 km/s in a direction toward the constellation Hydra—very roughly toward the center of the Local Supercluster. At this writing that result still awaits definitive confirmation, but it is nevertheless very exciting.

A Final Look Back

Now Figure VI.35 is the final spacetime diagram I'll show you in this drama. One of the three dimensions of space is, as before (Act II Scene 4), along the horizontal axis, and time increases vertically upward in the figure. The time at the horizontal axis is the time of the big bang, and the shaded region is that first million years (obviously not to scale) when the fireball (universe) was opaque. The times t_a, t_b, and so on, are various times in the past, and the time T is now. Our own path through spacetime—our world line—is along the time axis, moving upward into the future. The other vertical lines are the world lines of other galaxies. Actually, the galaxies are increasing in distance from one another (and us), but I have not tried to show the expansion of the universe because that would unduly complicate the picture.

As in previous spacetime diagrams, I have chosen units so that light travels along 45° lines. Thus we look off into space, and at the same time back into time, along those 45° lines. Notice how, as we look back along those light paths, we see other galaxies as they were in the past when light left them. The farther away and back we look, the greater the redshift of the light reaching us. We see relatively nearby galaxies by light not too different from that which left them, but remote ones that might look blue, if nearby, actually

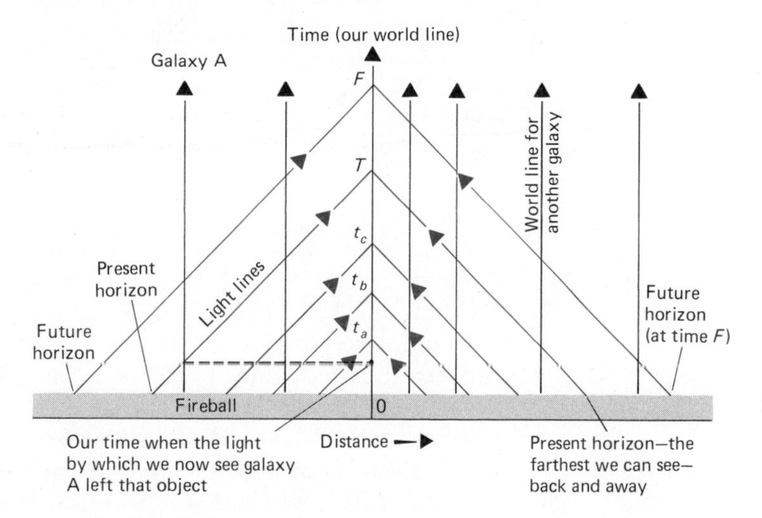

FIGURE VI.35 Spacetime diagram for the universe.

look red. We would expect to see the light of extremely remote galaxies largely in the infrared.

---

9. If the universe expands forever, what, eventually, will become of the 3°K microwave background radiation?

Finally, our line of sight intercepts the shaded zone in the diagram, and we can see no farther. That is the opaque fireball itself, and we cannot see into it any more than we can into a star, even though the bright light that left it reaches us as radio waves. Far in the future, when more time has gone by since the big bang and when we have reached point *F* on our world line, we will have to look farther before our line of sight stops at the fireball of radiation, and by than it will be more redshifted to even longer wavelengths. But there will always be a limit to how much of the universe we will be able to observe; to look off in space is to look back in time as well, and we can look back (and off) no farther than that fireball, from which we all came and which completely surrounds us in space.

And Then a Look Ahead

The observations, then, are at least compatible with a big-bang origin for the universe. But now, how might we determine whether the universe will expand forever, or eventually stop and collapse on itself? In the former case, recall, space is open and infinite; in the latter it is closed and finite—finite in the sense that light cannot escape to infinity. If light cannot escape, it is as if our entire universe is one vast black hole.

It turns out that there are, in principle, several ways in which we can choose between models. One of these involves the Hubble law itself. Obviously the proportionality between the speeds and distances of galaxies must break down eventually, because even if we could see the present speeds of very remote galaxies, we would find them getting closer and closer to the speed of light, but never quite reaching it, the farther away they are.

But we do not see the present speeds, or even the present distances, of the galaxies. We see other galaxies as they were in the past, when they were nearer to us than they are today. Moreover, the relativistic cosmologies (with zero cosmological constant) predict that the speeds of all galaxies must be decreasing, because of the gravitational deceleration of the universe. Therefore, for galaxies sufficiently far away that the look-back time to them is appreciable, both their observed distances and observed speeds must deviate from the simple proportionality of the Hubble law.

The astronomer who succeeded Hubble at the Hale Observatories is Allan Sandage. He has devoted much of his career to an enormously careful study of the Hubble law and all its ramifications, in the hope that we can determine its form for very remote clusters of galaxies accurately enough to distinguish between the predictions of various cosmological models. Figure VI.36 shows the Hubble law, from data gathered by Sandage up to 1976. The plotted points represent the redshifts of clusters of galaxies at various distances, and the magnitudes of their brightest members; the magnitude of the brightest cluster galaxy is, remember, an indicator of the cluster's distance.

Also shown on the figure are several solid lines and shaded zones that indicate the predictions of various cosmological models. The scatter of points is still too great to unequivocally select a model that best fits the data. When certain selection effects, as well as the probable effect of stellar evolution on the magnitudes of galaxies, are taken into account, the observations slightly favor the open universe, but no one would make any firm claims on the basis of the present data.

A second test of cosmology comes from the ages of the oldest systems of stars in our own Galaxy – the globular clusters. The best guess puts their ages at about 13 thousand million years; certainly they are at least 10 thousand million years old. On the other hand it must have taken the Galaxy at least 1000 million years – and probably longer – to form after the big bang. Now, according to present estimates of the scale of extragalactic distances, the oldest the universe can be is 19.5 thousand million years, which means if the universe is closed it cannot be more than 13 thousand million years old. The latter age is less than our best estimate for the age of our Galaxy since the big

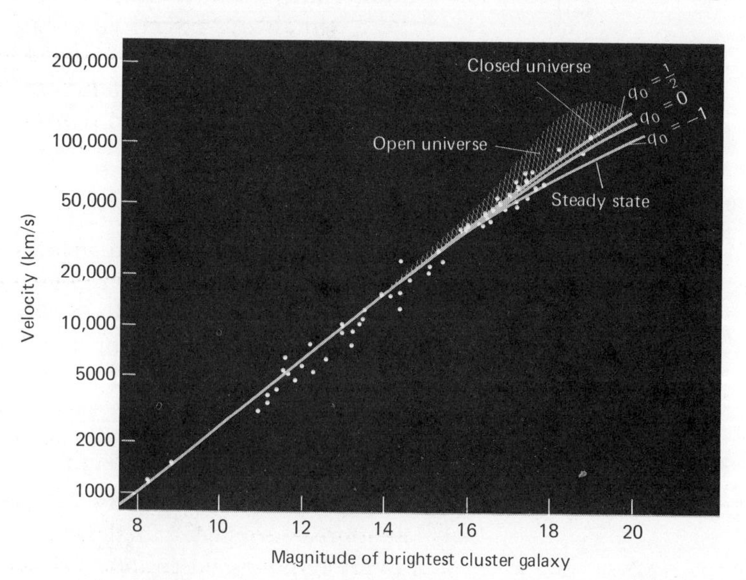

FIGURE VI.36
The Hubble diagram, showing the Hubble law, from data by A. Sandage.

bang. Therefore, the age of our Galaxy seems to require that the universe be old enough that it has to be open. This conclusion can be wrong, though, if our estimates either of extragalactic distances or of the age of our Galaxy are in error, and both estimates are, frankly, very shaky.

---

10. What would be the problem of postulating a closed universe if the Hubble constant were 100 km/s·10^6pc (twice the value assumed in this scene), and if the oldest star clusters in our Galaxy were known to be at least 10^{10} years old?
11. Suppose we were to count all galaxies to a certain distance in space. If the universe were not expanding, the total number counted should be proportional to the cube of the limiting distance of our counts. (Why?) Taking account of the finite time required for light to reach us, describe the relation that would be observed between the total count and the limiting distance for
    a) the steady-state theory
    b) the relativistic evolutionary theories of the universe.
    How might such observations help us choose between cosmological models?

Another hint about the past history of the universe is provided by that heavy isotope of hydrogen called deuterium—whose nucleus contains one proton and one neutron. Careful observations show that a tiny fraction—about one part in a hundred thousand—of the hydrogen in the interstellar space of our Galaxy is in the form of deuterium. Now we saw in Act IV Scene 3 how deuterium nuclei are formed from protons by nucleosynthesis in high-temperature stellar interiors. However, under the conditions that prevail in stars, *none of that deuterium* should survive, for its nuclei almost immediately absorb additional protons to make helium. The question is, then, where did the interstellar deuterium come from?

Well, one possibility is that it was formed in the big bang. On the other hand, calculations show that even in the primeval fireball any deuterium formed should fuse into helium, *unless* the density in the fireball was low enough for some of the deuterium to survive the high-temperature period of the fireball. But if the density were that low then, calculations show that it should be so low now that the universe cannot have sufficient mass to produce enough gravitation ever to stop the expansion, meaning that the universe would have to be open. Of course, we may still find some other origin for the deuterium.

I think the most definitive test to date (and it is not very definitive) is that of the mean density of matter today in the universe. What if we could take all the stars and galaxies apart, atom by atom, and spread those particles evenly throughout all of space, so that the density of matter in the universe were everywhere the same. That's what we mean by the *mean density* of matter in the universe. Since it is the matter that gives the universe gravitation, and gravitation that slows the expansion, if that mean density is too low, it means the expansion can never stop. We cannot, naturally, spread all the atoms about evenly, but we can calculate how much matter the stars and galaxies would make if we *could* do so, and we find that even allowing for lots of invisible gas within clusters of galaxies, the mean density of the universe is still too low by at least 10 or 20 times to stop the expansion and close the universe. Again, the evidence favors an open universe. But, also again, the conclusion can be wrong if, for example, there is a lot of uncondensed gas in space between the clusters of galaxies.

I've mentioned only a few observational tests that can help us choose between models of cosmology; there are quite a few others, and most of these give results that are compatible with a universe that will expand forever, or, as Sandage put it, "The universe happened only once." On the other hand, not a single test gives a very strong conclusion—the observations are still very difficult to make. Even a dozen marginal pieces of evidence do not prove a case. As Hubble put it, we are measuring shadows, and we "search among ghostly errors of measurement for landmarks that are scarcely more substantial." Even if the tests were definitive in themselves, they would still be based on the assumptions of our models. I hope you do not find it frustrating that I cannot tell you "how it is," but after all, the whole fun is in the quest.

There is just one other point I would like to raise. We have talked at some length about the exhaustion of energy—both for our use on earth and within stars. Yet, I have said that energy (including mass) is conserved. If something is conserved, how can it be used up? The point is that energy is of no use to us if we cannot transfer it from one place to another—to make it do something for us. As physicist Richard Feynman once put it, you can't dry yourself with a wet towel. If your body is wet and the towel is dry, only then can the water be transferred from your skin to the towel.

The universe has a lot of hot spots in it—stars, for example. And it has a lot of stored chemical energy—as in fossil fuels. And it has potential nuclear energy—in light elements that can be fused to heavier ones, and in radioactive elements. These energy sources are sort of ordered resources in the universe—they are like cosmic batteries, waiting to be drained. But in nature, things tend to go from an ordered state to a random state (where the towel and bather are equally wet). Hot spots tend to cool off and to warm the surroundings until everything is the same temperature. Batteries tend to discharge. A blob of cream in a cup of coffee disperses itself until it is evenly mixed in the brew. That cream will never reorganize itself into a blob spontaneously; the battery will not suck electricity out of the air to charge itself; stars will not drain heat out of space to warm up their insides again.

This transformation of the universe from an ordered to a random state is the essence of the second law of thermodynamics. It may provide us with the most fundamental distinction between past and future—the future has always evolved to a greater randomness. A measure of the state of this evolution is called *entropy;* we say that the entropy of the universe is always *increasing.*

If the universe were closed, and were one day to collapse on itself again, we could imagine that its great gravitational energy would once again reheat it to a colossal fireball, and force a reordering of matter and energy into stars and atoms. But what if, as observations suggest may be the case, the universe goes on expanding forever? Is there no way to ever reverse the upward march of entropy? Is the universe destined to end its life in a *thermodynamic götterdämmerung?* It's one thing to contemplate the end of mankind. Man may even deserve it. I can even accept (with sorrow) the end of all life on earth.

But to accept the end of the universe is to feel the deeper prick of a longer thorn. But perhaps there's something we don't know yet. After all, it *did* happen once. I wonder if the author of *Genesis* was fully aware of the profound significance of those four little words:

"Let there be light!"

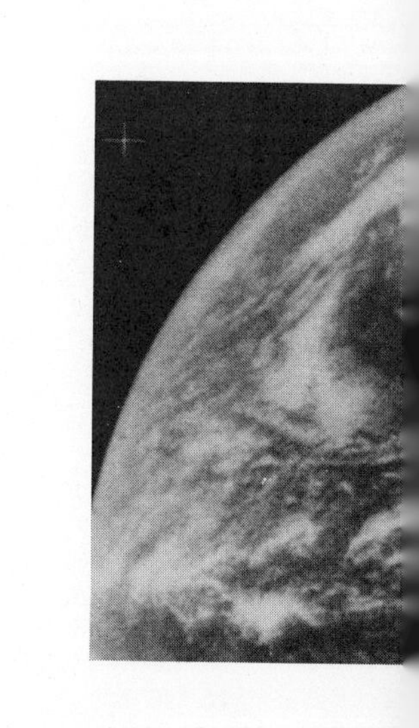

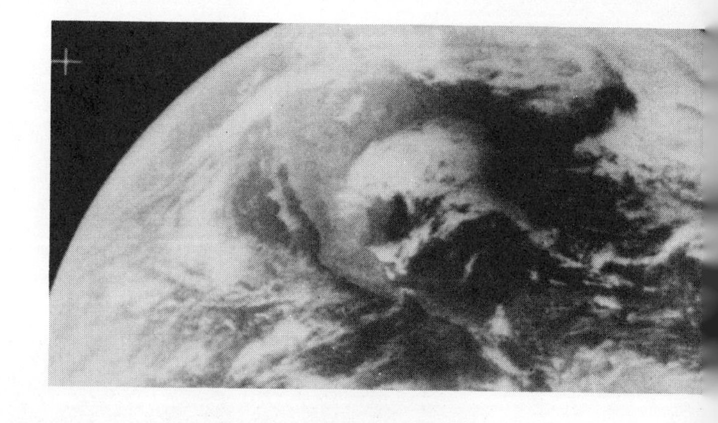

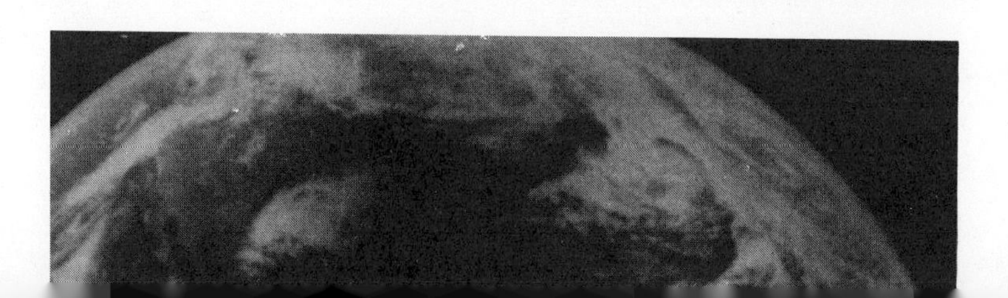

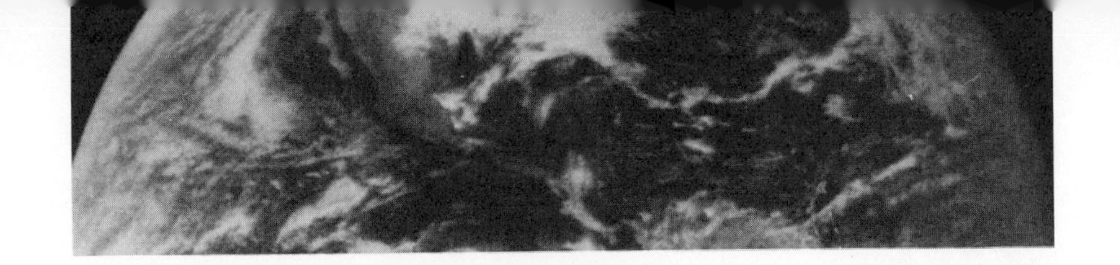

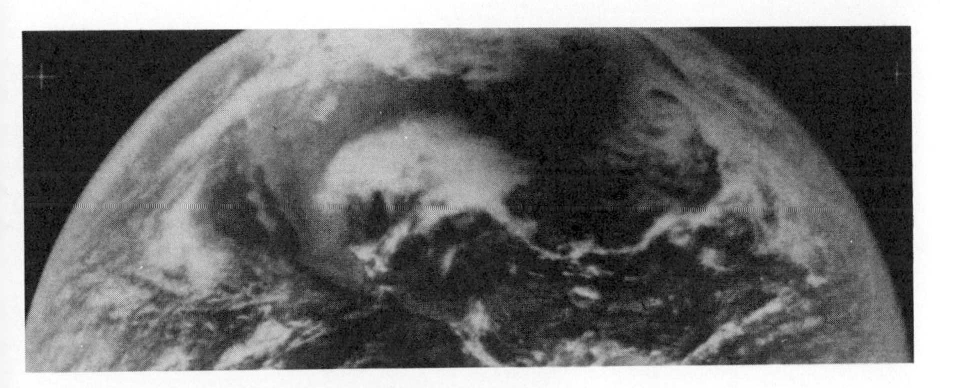

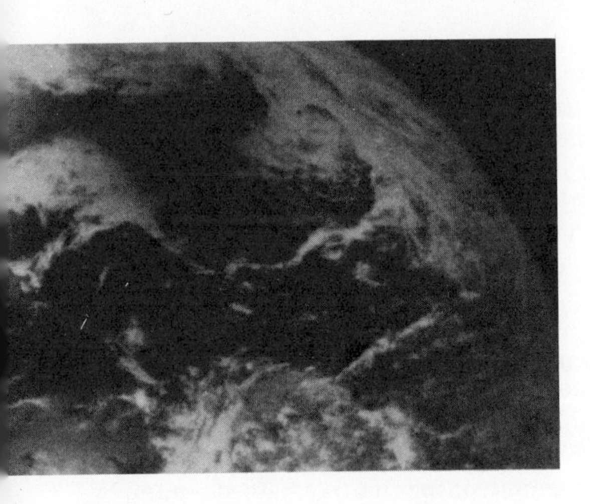

EPILOGUE

It was about noontime on 7 March 1970. We were on top of a small hill near the town of Miahuatlán in southern Mexico. A beautiful desert stretched out all around us, with hills and low mountains encircling the horizon. It was an absolutely cloudless day, with a clear blue sky. There were about 60 in our group from the United States, and many hundreds of Mexican families who had followed us into the shadow path to witness one of the most deeply moving experiences of their lives.

There was an attitude of friendship, anticipation, and excitement everywhere. Many of the Mexican children were helping us—a few (like me) professional, but most amateur, astronomers—set up our telescopes and cameras. Many had come from Mexico City and farther, but they knew we had come from farther yet, and that there must be an important reason for being there.

Soon the first bite appeared out of the limb of the sun. We passed around special filters—safe ones we had prepared ourselves so that no one would damage his eyes looking at the sun; through these, all could watch the partial phases. Fifty minutes after that first bite had appeared, the excitement had grown. The temperature had dropped noticeably, and a breeze had come up. The landscape had taken on eerie shades of red. Some of our people picked out the planet Venus in the still fairly bright sky. The effects heightened over the next few minutes. Soon Venus was shining brightly and Mercury could easily be seen. An awed quiet fell over the crowd.

Then we could see an indistinct shadow on the mountains across the valley, and a huge dark cone seemed to hang out of the sky. We could see that blurred shadow cone moving quickly nearer. On the ground around us appeared ghostly shimmering shadow bands, two or three meters apart, and moving about as fast as a man could walk. The landscape around us became very red, and the sky was deep purple. In the dry desert air the temperature had dropped about 10°C, and the breeze was by now quite strong.

Suddenly everything darkened, as if someone had quickly turned down a rheostat that controlled daylight. Although it was still as light as it is an hour into twilight, it seemed much darker, so suddenly had the illumination gone out. Everyone looked directly at the sun now—those special filters were no longer needed. As the approaching limb of the moon moved over to cover the sun's disk, the last rays of sunlight, shining through the lower lunar valleys, produced a string of red spots of light along the moon's edge—*Baily's beads*. Within seconds Baily's beads faded out, and as they went, there was a bright, starlike flash of the *diamond ring*—the final rays of the sun before it was completely hidden behind the moon. Then shining with the brilliance of a half-moon, the breathtaking, pearly-white solar corona appeared, stretching its streamers out a degree or more in some directions around the sun.

At first, you heard only the sounds of hundreds of intakes of breath; then expressions of wonder and amazement. After a few seconds, applause broke out from several groups of Mexicans, and shouts of glee and surprise. But mostly what you heard, even from the astronomers, was "It's incredible! It's fantastic! I just can't believe it!"

The three minutes of totality seemed to last both for an eternity and for only a fleeting instant—at the same time. One lost all perspective of time. The planets shone brightly in the sky, and the brightest stars could be seen. Bats, confused about the time of day, began to flutter around us. For those three minutes, we had transcended reality, or so it seemed, to witness something that should be reserved only for the gods.

Then it was over, and again we saw the diamond ring, Baily's beads, the shadow bands, and those eerie colors. Deeply moved, and elated, we returned to the city of Oaxaca, just outside the path of totality, where we spent the night. There was little sleep for us though, and no one cared, for the Mexicans in that beautiful city spent the whole night in celebration.

That was my first total eclipse of the sun. I have seen three more since, and if all goes well, it will be four before you read these words. And each in its way was different, but fully as exciting and moving as that first one. You don't seem to become jaded of total solar eclipses; on some of our expeditions people had seen as many as nine, and they treasured each one as they had their first.

Of course, long before ever having witnessed the event, I had told many thousands of people—students in classes, and members of the public at planetarium shows, in which we would try to recreate the spectacle—that it was the most moving of natural phenomena, and that, "If you ever have the opportunity to get into the path of a *total* eclipse (a partial or annular one is not the same—it must be *total*), do not fail to do so, or you will miss the experience of your life." I meant those words, because I believed them. I had talked to many people who *had* witnessed the event, and I had read about it in books. But after having *had* the experience, I now know that it cannot be described by words, verbally or in writing, nor reproduced in planetaria, on television, in motion pictures, or any other way. It is something one must experience for himself to believe.

Through the ages men must have been so awed. Surely many have considered it a religious experience. Surely total solar eclipses have inspired men to great accomplishments. What is it about the experience that is so absolutely incredible? I think it is that we, during those few brief moments, experience something in Nature that is obviously grander than man and his works, and even greater than all earthly affairs. It is, if you will, an introduction to the glory of Nature.

Of course there are grander phenomena: pulsating stars, supernovae, quasars, and, naturally, the big bang itself. We can *read* about those grander phenomena, and we can investigate them at first hand, but perhaps it takes the most profound *experience* Nature can offer *man* to help us really appreciate our universe. We cannot (today) *experience* the big bang, but we can, if we are very lucky, experience a total solar eclipse. I think it helps us understand better those basic forces of Nature (or at least to appreciate them better), including the big bang.

As I say, eclipses must have inspired man in his exploration of the universe and have played a positive role in his discovery of some of its secrets. But having learned those few secrets of Nature, what has man done with this God-given information? He has started in motion the machinery of his own destruction.

I believe we can *stop* that machinery—I believe we can reverse the trend to our own end. We know what must be done: We must marshal all our intellect; we must recognize our place on earth and our present plight; we must act. It will not do to leave it to our presidents, or to the United Nations, or to the "scientists" to solve our problems. It is *our* responsibility, and *now*. And we must, in some way, enlist the support of all four thousand million people on our very fragile planet if we want to survive. In another dozen years, it will be *five* thousand million people!

I hope we can do it; whether we do or not, the Drama of the Universe will go on. And, you know, the universe puts on a damned good show! I *so* hope someone is around to see how the next few acts come out.

APPENDIXES

APPENDIX 1

(Technical references are marked with an asterisk.)

General Texts

Abell, G. O., *Exploration of the Universe,* Third Edition. New York: Holt, Rinehart and Winston, 1975.

Hoyle, Fred, *Astronomy and Cosmology, A Modern Course,* San Francisco: W. H. Freeman and Company, 1975.

* Smith, Elske v.P., and Jacobs, K. C., *Introductory Astronomy and Astrophysics.* Philadelphia: Saunders, 1973.

Elementary Texts and Books on Astronomy

Abell, G. O., *Realm of the Universe,* New York: Holt, Rinehart and Winston, 1976.

Alter, D., Cleminshaw, C. H., and Phillips, J., *Pictorial Astronomy.* New York: Crowell, 1974.

Menzel, D. H., *Astronomy.* New York: Random House, 1970.

Verschuur, G. L., *The Invisible Universe,* New York: Springer-Verlag, 1974.

Histories of Astronomy and Astrology

Berendzen, R., Hart, R., and Seeley, D., *Man Discovers the Galaxies,* New York: Neale Watson Academic Publications, 1976.

Galileo, G., *Dialogue on the Great World Systems* (translated by T. Salisbury). Chicago: University of Chicago Press, 1953.

Gallant, Roy A., *Astrology, Sense or Nonsense?* Garden City, New York: Doubleday, 1974.

Gardner, *M., Fads and Fallacies in the Name of Science.* New York: Dover, 1957.

Hoyle, F., *Astronomy.* New York: Doubleday, 1962.

King, H. C., *Exploration of the Universe.* New York: New American Library, 1964.

Koestler, A., *The Sleepwalkers.* New York: Macmillan, 1959.

Newton, I., *Mathematical Principles of Natural Philosophy.* R. T. Crawford, ed. Berkeley, Calif.: University of California Press, 1966.

Pannekoek, A., *A History of Astronomy.* New York: Interscience, 1961.

Shapley, H., and Howarth, H. E., *A Source Book in Astronomy.* New York: McGraw-Hill, 1929.

Struve, O., and Zebergs, V., *Astronomy of the Twentieth Century.* New York: Macmillan, 1962.

Celestial Mechanics

Ahrendt, M. H., *The Mathematics of Space Exploration.* New York: Holt, Rinehart and Winston, 1965.

Ryabov, Y., *An Elementary Survey of Celestial Mechanics.* New York: Dover, 1961.

Van de Kamp, P., *Elements of Astromechanics.* San Francisco: Freeman, 1964.

Telescopes and Light

Christianson, W. N., and Hogborn, J. A., *Radio Telescopes.* London: Cambridge University Press, 1969.

Miczaika, G., and Sinton, W., *Tools of the Astronomer.* Cambridge, Mass.: Harvard University Press, 1961.

Minnaert, M., *The Nature of Light and Colour in the Open Air.* New York: Dover, 1954.

Page, T., and Page, L. W., *Telescopes; How to Make Them and Use Them.* New York: Macmillan, 1966.

Steinberg, J. L., and Lequeux, J., *Radio Astronomy.* New York: McGraw-Hill, 1963.

Earth and the Solar System

Hawkins, G. S., *Meteors, Comets, and Meteorites.* New York: McGraw-Hill, 1964.

Menzel, D., *Our Sun.* Cambridge, Mass.: Harvard University Press, 1959.

Watson, F., *Between the Planets.* Cambridge, Mass.: Harvard University Press, 1956.

Whipple, F., *Earth, Moon, and Planets.* Cambridge, Mass.: Harvard University Press, 1968.

Stellar Astronomy and Astrophysics

Aller, L. H., *Atoms, Stars and Nebulae* (rev. ed.). Cambridge, Mass.: Harvard University Press, 1971.

Bok, B. J., and Bok, P. E., *The Milky Way* (4th ed.). Cambridge, Mass.: Harvard University Press, 1973.

Brandt, J. C., *The Sun and Stars.* New York: McGraw-Hill, 1966.

Galaxies, Cosmology, and Relativity

* Couderc, P., *The Expansion of the Universe.* London: Faber and Faber, 1952.

Hodge, P. W., *Galaxies and Cosmology.* New York: McGraw-Hill, 1966.

Hubble, E., *The Realm of the Nebulae.* New Haven, Conn.: Yale University Press, 1936; also New York: Dover, 1958.

Kaufmann, W. J., *The Cosmic Frontiers of General Relativity,* Boston: Little Brown, 1977.

Sandage, A. R., *The Hubble Atlas of Galaxies.* Washington, D. C.: Carnegie Institution, 1961.

Sciama, D. W., *Modern Cosmology,* London: Cambridge University Press, 1971.

Sciama, D. W., *The Physical Foundations of General Relativity.* New York: Doubleday, 1969.

Sciama, D. W., *The Unity of the Universe.* New York: Doubleday, 1959.

Shapley, H., *Galaxies* (3rd ed.). Cambridge, Mass.: Harvard University Press, 1972.

Shipman, Harry L., *Black Holes, Quasars, and the Universe,* Boston: Houghton Mifflin, 1976.

Weinberg, Steven, *The First Three Minutes,* New York: Basic Books, 1977.

Life in the Universe

Berendzen, R. (ed.), *Life Beyond Earth and the Mind of Man.* Washington, D. C.: NASA, 1973.

Bracewell, R. N., *The Galactic Club.* Stratford, Calif.: Stanford Alumni Assn., 1974.

Drake, F., *Intelligent Life in Space.* New York: Macmillan, 1962.

Ponnamperuma, C., *The Origins of Life.* New York: Dutton, 1972.

Ponnamperuma, C., and Cameron, A. G. W., *Interstellar Communication: Scientific Perspectives.* Boston: Houghton Mifflin, 1974.

Sagan, C., *The Cosmic Connection.* Garden City, N.Y.: Anchor Press-Doubleday, 1973.

Shklovskii, I. S., and Sagan, C., *Intelligent Life in the Universe.* New York: Dell, 1966.

Star Atlases and Sky Guides

Allen, R. H., *Star Names.* New York: Dover, 1963.

Menzel, D. H., *A Field Guide to the Stars and Planets.* Boston: Houghton Mifflin, 1964.

Norton, W. W., *Sky Atlas.* Cambridge, Mass.: Sky Publishing Company, 1971.

Olcott, W. T., *Olcott's Field Book of the Skies.* New York: Putnam, 1954.

Journals and Periodicals

Astronomy, published monthly by AstroMedia Corp., Milwaukee, Wisconsin.

The Griffith Observer, published monthly by the Griffith Observatory, Los Angeles, California.

Mercury, published bimonthly by the Astronomical Society of the Pacific, San Francisco, California.

Scientific American, published monthly by Scientific American, New York.

Sky and Telescope, published monthly by the Sky Publishing Corporation, Harvard College Observatory, Cambridge, Massachusetts.

Career Information

Students interested in a career in astronomy can obtain an information leaflet from the Executive Officer, American Astronomical Society, 211 FitzRandolph Road, Princeton, New Jersey 08540.

APPENDIX 2

Glossary

aberration (of starlight). Apparent displacement in the direction of a star due to the earth's orbital motion.

absolute magnitude. Apparent magnitude a star would have at a distance of 10 pc.

absolute zero. A temperature of −273° C (or 0° K) where all molecular motion stops.

absorption spectrum. Dark lines superimposed on a continuous spectrum.

accelerate. To change velocity, either to speed up, slow down, or change direction.

acceleration of gravity. Numerical value of the acceleration produced by the gravitational attraction on an object at the surface of a planet or star.

active sun. The sun during times of unusual solar activity—spots, flares, and associated phenomena.

Age of Aquarius. Period (about 2000 years) during which the vernal equinox, moving because of precession, passes through the constellation of Aquarius.

albedo. The fraction of incident sunlight that a planet or minor planet reflects.

almanac. A book or table listing astronomical events.

alpha particle. The nucleus of a helium atom, consisting of two protons and two neutrons.

altitude. Angular distance above or below the horizon, measured along a vertical circle, to a celestial object.

amplitude. The range in variability, as in the light from a variable star.

angstrom (Å). A unit of length equal to 10^{-8} cm.

angular diameter. Angle subtended by the diameter of an object.

angular momentum. A measure of the momentum associated with motion about an axis or fixed point.

annular eclipse. An eclipse of the sun in which the moon is too distant to appear to cover the sun completely, so that a ring of sunlight shows around the moon.

anomalistic month. The period of revolution of the moon about the earth with respect to its line of apsides, or to the perigee point.

anomalistic year. The period of revolution of the earth about the sun with respect to its line of apsides, or to the perihelion point.

Antarctic Circle. Parallel of latitude 66½° S; at this latitude the noon altitude of the sun is 0° on the date of the summer solstice.

antimatter. Matter consisting of antiparticles: *antiprotons* (protons with negative rather than positive charge), *positrons* (positively charged electrons), and *antineutrons.*

apastron. The point of closest approach of two stars, as in a binary star orbit.

aperture. The diameter of an opening, or of the primary lens or mirror of a telescope.

aphelion. Point in its orbit where a planet is farthest from the sun.

apogee. Point in its orbit where an earth satellite is farthest from the earth.

apparent magnitude. A measure of the observed light flux received from a star or other object at the earth.

apparent solar day. The interval between two successive transits of the sun's center across the meridian.

apparent solar time. The hour angle of the sun's center *plus* 12 hours.

apse (or **apsis;** pl. **apsides).** The point in a body's orbit where it is nearest or farthest the object it revolves about.

Arctic Circle. Parallel of latitude 66½° N; at this latitude the noon altitude of the sun is 0° on the date of the winter solstice.

artificial satellite. A manmade object put into a closed orbit about the earth.

ascendant. (astrological term). The point on the zodiac that is on the eastern horizon, just rising at the moment of birth.

aspect. The situation of the sun, moon, or planets with respect to one another.

association. A loose cluster of stars whose spectral types, motions, or positions in the sky indicate that they have probably had a common origin.

asteroid. A synonym for "minor planet."

asthenosphere. The mantle beneath the lithosphere of the earth.

astigmatism. A defect in an optical system whereby pairs of light rays in different planes do not focus at the same place.

astrology. The pseudoscience that treats with supposed influences of the configurations and locations in the sky of the sun, moon, and planets on human destiny; a primitive religion having its origin in ancient Babylonia.

astrometric binary. A binary star in which one component is not observed, but its presence is deduced from the orbital motion of the visible component.

astrometry. That branch of astronomy that deals with the determination of precise positions and motions of celestial bodies.

astronautics. The science of the laws and methods of space flight.

astronomical unit (AU). Originally meant to be the semimajor axis of the orbit of the earth; now defined as the semimajor axis of the orbit of a hypothetical body with the mass and period that Gauss assumed for the earth. The semimajor axis of the orbit of the earth is 1.000 000 230 AU.

astronomy. The branch of science that treats of the physics and morphology of that part of the universe which lies beyond the earth's atmosphere.

astrophysics. That part of astronomy which deals principally with the physics of stars, stellar systems, and interstellar material. Astrophysics also deals, however, with the structures and atmospheres of the sun and planets.

atmospheric refraction. The bending, or refraction, of light rays from celestial objects by the earth's atmosphere.

atom. The smallest particle of an element that retains the properties which characterize that element.

atomic clock. A time-keeping device regulated by the natural frequency of the emission or absorption of radiation by a particular kind of atom.

atomic mass unit. *Chemical:* one-sixteenth of the mean mass of an oxygen atom. *Physical:* one-twelfth of the mass of an atom of the most common isotope of carbon. The atomic mass unit is approximately the mass of a hydrogen atom, 1.67×10^{-24}g.

atomic number. The number of protons in each atom of a particular element.

atomic time. The time kept by a cesium (atomic) clock, based on the **atomic second**—the time required for 9,192,631,770 cycles of the radiation emitted or absorbed in a particular transition of an atom of cesium-133.

atomic transition. A change in the state of energy of an atom; the atom may gain or lose energy by collision with another particle or by the emission or absorption of a photon.

atomic weight. The mean mass of an atom of a particular element in atomic mass units.

aurora. Light radiated by atoms and ions in the ionosphere, mostly in the polar regions.

autumnal equinox. The intersection of the ecliptic and celestial equator where the sun crosses the equator from north to south.

azimuth. The angle along the celestial horizon, measured eastward from the north point, to the intersection of the horizon with the vertical circle passing through an object.

Baily's beads. Small "beads" of sunlight seen passing through valleys along the limb of the moon in the instant preceding and following totality in a solar eclipse.

ballistic missile. A missile or rocket that is given its entire thrust during a brief period at the beginning of its flight, and that subsequently "coasts" to its target along an orbit.

Balmer lines. Emission or absorption lines in the spectrum of hydrogen that arise from transitions between the second (or first excited) and higher energy states of the hydrogen atoms.

barred spiral galaxy. Spiral galaxy in which the spiral arms begin from the ends of a "bar" running through the nucleus rather than from the nucleus itself.

barycenter. The center of mass of two mutually revolving bodies.

baryons (and antibaryons). The heavy atomic nuclear particles, such as protons and neutrons.

base line. That side of a triangle used in triangulation or surveying whose length is known (or can be measured), and which is included between two angles that are known (or can be measured).

"big bang" theory. A theory of cosmology in which the expansion of the universe is presumed to have begun with a primeval explosion.

billion. In the United States and France, one thousand million (10^9); in Great Britain and Germany, one million million (10^{12}).

binary star. A double star; two stars revolving about each other.

binding energy. The energy required to completely separate the constituent parts of an atomic nucleus.

biosphere. That part of the earth (its surface, atmosphere, and oceans) where life can exist.

black body. A hypothetical perfect radiator, which absorbs and reemits all radiation incident upon it.

black dwarf. A presumed final state of evolution for a star, in which all of its energy sources are exhausted and it no longer emits radiation.

black hole. A hypothetical body whose velocity of escape is equal to or greater than the speed of light; thus no radiation can escape from it.

blink microscope (or comparator). A microscope in which the user's view is shifted rapidly back and forth between the corresponding portions of two different photographs of the same region of the sky.

Bode's law. A scheme by which a sequence of numbers can be obtained that give the approximate distances of the planets from the sun in astronomical units.

Bohr atom. A particular model of an atom, invented by Niels Bohr, in which the electrons are described as revolving about the nucleus in circular orbits.

bolide. A very bright fireball or meteor; sometimes defined as a fireball accompanied by sound.

bolometric correction. The difference between the visual (or photovisual) and bolometric magnitudes of a star.

bolometric magnitude. A measure of the flux of radiation from a star or other object received just outside the earth's atmosphere, as it would be detected by a device sensitive to *all* forms of electromagnetic energy.

bremsstrahlung. Radiation from free-free transitions, in which electrons gain or lose energy while being accelerated in the field of an atomic nucleus or ion.

bubble chamber. A chamber in which bubbles form along the electrically charged path of a high-energy charged particle, rendering the track of that particle visible.

burnout. The instant when a rocket stops firing.

calculus. A branch of mathematics that permits computations involving rates of change (*differential* calculus) or of the contribution of an infinite number of infinitesimal quantities (*integral* calculus).

carbon cycle. A series of nuclear reactions involving carbon as a catalyst, by which hydrogen is transformed to helium.

cardinal points. The four principal points of the compass: North, East, South, and West.

Cassegrain focus. An optical arrangement

in a reflecting telescope in which light is reflected by a second mirror to a point behind the objective mirror.

cD galaxy. A supergiant elliptical galaxy frequently found in the centers of clusters of galaxies.

celestial equator. A great circle on the celestial sphere 90° from the celestial poles; the circle of intersection of the celestial sphere with the plane of the earth's equator.

celestial mechanics. That branch of astronomy which deals with the motions and gravitational influences of the members of the solar system.

celestial navigation. The art of navigation at sea or in the air from sightings of the sun, moon, planets, and stars.

celestial poles. Points about which the celestial sphere appears to rotate; intersections of the celestial sphere with the earth's polar axis.

celestial sphere. Apparent sphere of the sky; a sphere of large radius centered on the observer. Directions of objects in the sky can be denoted by the positions of those objects on the celestial sphere.

center of gravity. Center of mass.

center of mass. The mean position of the various mass elements of a body or system, weighted according to their distances from that center of mass; that point in an isolated system which moves with constant velocity, according to Newton's first law of motion.

centrifugal force (or acceleration). An imaginary force (or acceleration) that is often introduced to account for the illusion that a body moving on a curved path tends to accelerate radially from the center of curvature. The actual force present is the one that diverts the body's motion from a straight line and is directed *toward* the center of curvature. It is, however, legitimate to introduce a fictitious centrifugal force field in a rotating (and hence noninertial) coordinate system.

centripetal force (or acceleration). The force required to divert a body from a straight path into a curved path (or the acceleration experienced by the body); it is directed toward the center of curvature.

cepheid variable. A star that belongs to one of two classes (type I and type II) of yellow supergiant pulsating stars.

Ceres. Largest of the minor planets and first to be discovered.

cesium clock. An atomic clock that utilizes a transition in the atom cesium-133.

charm. The name given to a variety of quark; a quark is a hypothetical basic constituent of all nuclear particles.

chromatic aberration. A defect of optical systems whereby light of different colors is focused at different places.

chromosphere. That part of the solar atmosphere that lies immediately above the photospheric layers.

chronograph. A device for recording and measuring the times of events.

chronometer. An accurate clock.

circular velocity. The critical speed with which a revolving body can have a circular orbit.

circumpolar regions. Portions of the celestial sphere near the celestial poles that are either always above or always below the horizon.

cloud chamber. A chamber in which droplets of liquid condense along the electrically charged path of a high-energy charged particle, rendering the track of that particle visible.

Clouds of Magellan. Two neighboring galaxies visible to the naked eye from southern latitudes.

cluster of galaxies. A system of galaxies containing from several to thousands of member galaxies.

cluster variable (RR Lyrae variable). A member of a certain large class of pulsating variable stars, all with periods less than one day. These stars are often present in globular star clusters.

color index. Difference between the magnitudes of a star or other object measured in light of two different spectral regions, for example, photographic *minus* photovisual magnitudes.

color-magnitude diagram. Plot of the magnitudes (apparent or absolute) of the stars in a cluster against their color indices.

coma. A defect in an optical system in which off-axis rays of light striking different parts of the objective do not focus in the same place.

coma (of comet). The diffuse gaseous component of the head of a comet.

comet. A small body of icy and dusty matter, which revolves about the sun. When a comet comes near the sun, some of its material vaporizes, forming a large *coma* of tenuous gas, and often a *tail*.

comparison spectrum. The spectrum of a vaporized element (such as iron) photographed beside the image of a stellar spectrum, and with the same camera, for purposes of comparison of wavelengths.

compound. A substance composed of two or more chemical elements.

conduction. The transfer of energy by the direct passing of energy or electrons from atom to atom.

configuration. Any one of several particular orientations in the sky of the moon or a planet with respect to the sun.

conic section. The curve of intersection between a circular cone and a plane; these curves can be ellipses, circles, parabolas, or hyperbolas.

conjunction. The configuration of a planet when it has the same celestial longitude as the sun, or the configuration when any two celestial bodies have the same celestial longitude or right ascension.

conservation of angular momentum. The law that angular momentum is conserved in the absence of any force not directed toward or away from the point or axis about which the angular momentum is referred—that is, in the absence of a torque.

constellation. A configuration of stars named for a particular object, person, or animal; or the area of the sky assigned to a particular configuration.

contacts (of eclipses). The instants when certain stages of an eclipse begin.

continental drift. A gradual drift of the continents over the surface of the earth due to *plate tectonics*.

continuous spectrum. A spectrum of light comprised of radiation of a continuous range of wavelengths or colors rather than only certain discrete wavelengths.

convection. The transfer of energy by moving currents of a fluid containing that energy.

Coordinated Universal Time. Greenwich Mean Time standardized and regulated by an international agency, the *Bureau International de l'Heure,* on the basis of astronomical observations reported from around the world.

Copernicus satellite. An artificial satellite with scientific instrumentation especially designed for ultraviolet observations in space.

core (of earth). The central part of the earth, believed to be a liquid of high density.

corona. Outer atmosphere of the sun.

corona (or **halo) of Galaxy.** The outer portions of the Galaxy, especially on either side of the plane of the Milky Way.

coronagraph. An instrument of photographing the chromosphere and corona of the sun outside of eclipse.

corpuscular radiation. Charged particles, mostly atomic nuclei and electrons, emitted into space by the sun and possibly other objects.

cosmic rays. Atomic nuclei (mostly protons) that are observed to strike the earth's atmosphere with exceedingly high energies.

cosmogony. The study of the origin of the world or universe.

cosmological constant. A term that arises in the development of the field equations of general relativity, which represents a repulsive force in the universe. The cosmological constant is often assumed to be zero.

cosmological model. A specific model, or theory, of the organization and evolution of the universe.

cosmological principle. The assumption that, on the large scale, the universe at any given time is the same everywhere.

cosmology. The study of the organization and evolution of the universe.

coudé focus. An optical arrangement in a reflecting telescope whereby light is reflected by two or more secondary mirrors down the polar axis of the telescope to a focus at a place separate from the moving parts of the telescope.

crater (lunar). A more or less circular depression in the surface of the moon.

crater (meteoritic). A crater on the earth caused by the collision of a meteoroid with the earth, and a subsequent explosion.

crescent moon. One of the phases of the moon when its elongation is less than 90° from the sun and it appears less than half full.

crust (of earth). The outer layer of the earth.

Cyclops Project. A proposed system of radio antennae for detection of signals from extraterrestrial civilizations.

dark nebula. A cloud of interstellar dust that obscures the light of more distant stars, and appears as an opaque curtain.

daylight saving time. A time one hour more advanced than standard time, usually adopted in spring and summer to take advantage of long evening twilights.

deferent. A stationary circle in the Ptolemaic system along which moves the center of another circle (epicycle), along which moves an object or another epicycle.

degenerate gas. A gas in which the allowable states for the electrons have been filled; it behaves according to different laws from those that apply to "perfect" gases.

density. The ratio of the mass of an object to its volume.

deuterium. A "heavy" form of hydrogen, in which the nucleus of each atom consists of one proton and one neutron.

diamond ring. A flash of sunlight at the instants before and after totality in a solar eclipse while the corona is visible as a complete ring of light around the moon.

differential gravitational force. The difference between the respective gravitational forces exerted on two bodies near each other by a third, more distant body.

differentiation (geological). A separation or segregation of different kinds of material in different layers in the interior of a planet.

diffraction. The spreading out of light in passing the edge of an opaque body.

diffraction grating. A system of closely spaced equidistant slits or reflecting strips which, by diffraction and interference, produce a spectrum.

diffraction pattern. A pattern of bright and dark fringes produced by the interference of light rays, diffracted by different amounts, with each other.

diffuse nebula. A reflection or emission nebula produced by interstellar matter (not a planetary nebula).

disk (of planet or other object). The apparent circular shape that a planet (or the sun, or moon, or a star) displays when seen in the sky or viewed telescopically.

disk of Galaxy. The central disk or "wheel" of our Galaxy, superimposed on the spiral structure.

dispersion. Separation, from white light, of different wavelengths being refracted by different amounts.

diurnal. Daily.

diurnal circle. Apparent path of a star in the sky during a complete day due to the earth's rotation.

diurnal motion. Motion during one day.

diurnal parallax. Apparent change in direction of an object caused by a displacement of the observer due to the earth's rotation.

Doppler shift. Apparent change in wavelength of the radiation from a source due to its relative motion in the line of sight.

draconic month. The period of revolution of the moon about the earth with respect to the nodes of the moon's orbit.

dwarf (star). A main-sequence star (as opposed to a giant or supergiant).

dyne. The metric unit of force; the force required to accelerate a mass of 1 gram in the amount 1 centimeter per second per second.

east point. The point on the horizon 90° from the north point (measured clockwise as seen from the zenith).

eccentric. A point, about which an object revolves on a circular orbit, that is not at the center of the circle.

eccentricity (of ellipse). Ratio of the distance between the foci to the major axis.

eclipse. The cutting off of all or part of the light of one body by another passing in front of it.

eclipse path. The track along the earth's surface swept out by the tip of the shadow of the moon (or the extension of its shadow) during a total (or annular) solar eclipse.

eclipse season. A period during the year when an eclipse of the sun or moon is possible.

eclipsing binary star. A binary star in which the plane of revolution of the two stars is nearly edge-on to our line of sight, so that the light of one star is periodically diminished by the other passing in front of it.

ecliptic. The apparent annual path of the sun on the celestial sphere.

ecliptic limit. The maximum angular distance from a node where the moon can be for an eclipse to take place.

Einstein-Rosen bridge. A hypothetical connection in spacetime between two distinct regions of the universe, of which one may be a black hole and one a white hole; a worm hole.

electric charge. A quantity of electrons or of electrical charge of one sign.

electric current. The flow of electrons.

electric field. The region of space around an electric charge within which an electric force can act on another charged particle.

electromagnetic radiation. Radiation consisting of waves propagated through the building up and breaking down of electric and magnetic fields; these include radio, infrared, light, ultraviolet, X rays, and gamma rays.

electromagnetic spectrum. The whole array or family of electromagnetic waves.

electron. A negatively charged subatomic particle that normally moves about the nucleus of an atom.

electron volt. The kinetic energy acquired by an electron that is accelerated through an electric potential of 1 volt; 1 electron volt is 1.60207×10^{-12} erg.

electroscope. A device for measuring the amount of charge in the air.

element. A substance that cannot be decomposed, by chemical means, into simpler substances.

elements (of orbit). Any of several quantities that describe the size, shape, and orientation of the orbit of a body.

ellipse. A conic section: the curve of intersection of a circular cone and a plane cutting completely through the cone.

elliptical galaxy. A galaxy whose apparent photometric contours are ellipses and which contains no conspicuous interstellar material.

ellipticity. The ratio (in an ellipse) of the major axis *minus* the minor axis to the major axis.

emission line. A discrete bright spectral line.

emission nebula. A gaseous nebula that derives its visible light from the fluorescence of ultraviolet light from a star in or near the nebula.

emission spectrum. A spectrum consisting of emission lines.

energy. The ability to do work.

energy level (in an atom or ion). A particular level, or amount, of energy possessed by an atom or ion above the energy it possesses in its least energetic state.

energy spectrum. A table or plot showing the relative numbers of particles (in cosmic rays or corpuscular radiation) of various energies.

ephemeris. A table that gives the positions of a celestial body at various times, or other astronomical data.

ephemeris time. A kind of time that passes at a strictly uniform rate; used to compute the instants of various astronomical events.

epicycle. A circular orbit of a body in the Ptolemaic system, the center of which revolves about another circle (the deferent).

equant. A stationary point in the Ptolemaic system not at the center of a circular orbit about which a body (or the center of an epicycle) revolves with uniform angular velocity.

equation of state. An equation relating the pressure, temperature, and density of a substance (usually a gas).

equator. A great circle on the earth, 90° from its poles.

equatorial mount. A mounting for a telescope with one axis parallel to the earth's axis, so that a motion of the telescope about that axis compensates for the earth's rotation.

equinox. One of the intersections of the ecliptic and celestial equator.

erg. The metric unit of energy; the work done by a force of one dyne moving through a distance of one centimeter.

eruptive variable. A variable star whose changes in light are erratic or explosive.

Euclidean. Pertaining to Euclidean geometry, or *flat space.*

event. A point in four-dimensional space-time.

event horizon. The surface through which a collapsing star is hypothesized to pass when its velocity of escape is equal to the speed of light, that is, when the star becomes a black hole.

evolutionary cosmology. A theory of cosmology that assumes that all parts of the universe have a common age and evolve together.

eyepiece. A magnifying lens used to view the image produced by the objective of a telescope.

excitation. The process of imparting to an atom or an ion an amount of energy greater than that it has in its normal or least-energy state.

extinction. Attenuation of light from a celestial body produced by the earth's atmosphere, or by interstellar absorption.

extragalactic. Beyond the Galaxy.

faculus (pl. **faculae**). Bright region near the limb of the sun.

fermions. Subatomic particles, such as electrons, that obey certain laws formulated by Enrico Fermi.

field equations. A set of equations in general relativity that describe the curvature of spacetime in the presence of matter.

filtergram. A photograph of the sun (or part of it) taken through a special narrow-bandpass filter.

fireball. A spectacular meteor.

First Point of Aries. The vernal equinox.

fission. The breakup of a heavy atomic nucleus into two or more lighter ones.

flare. A sudden and temporary outburst of light from an extended region of the solar surface.

flare star. A member of a class of stars that show occasional, sudden, unpredicted increases in light.

flocculus (pl. **flocculi**). A bright region of the solar surface observed in the monochromatic light of some spectral line; flocculi are now usually called *plages*.

fluctions. Name given by Newton to the calculus.

fluorescence. The absorption of light of one wavelength and reemission of it at another wavelength; especially the conversion of ultraviolet into visible light.

focal length. The distance from a lens or mirror to the point where light converged by it comes to a focus.

focal ratio (speed). Ratio of the focal length of a lens or mirror to its aperture.

focus. Point where the rays of light converged by a mirror or lens meet.

focus of a conic section. Mathematical point associated with a conic section, whose distance to any point on the conic bears a constant ratio to the distance from that point to a straight line known as the *directrix*.

force. That which can change the momentum of a body; numerically, the rate at which the body's momentum changes.

Fraunhofer line. An absorption line in the spectrum of the sun or of a star.

Fraunhofer spectrum. The array of absorption lines in the spectrum of the sun or of a star.

free-free transition. An atomic transition in which the energy associated with an atom or ion and passing electron changes during the encounter, but without capture of the electron by the atom or ion.

frequency. Number of vibrations per unit time; number of waves that cross a given point per unit time (in radiation).

full moon. That phase of the moon when it is at opposition (180° from the sun) and its full daylight hemisphere is visible from the earth.

fusion. The building up of heavier atomic nuclei from lighter ones.

galactic cluster. An "open" cluster of stars located in the spiral arms or disk of the Galaxy.

galactic equator. Intersection of the principal plane of the Milky Way with the celestial sphere.

galactic latitude. Angular distance north or south of the galactic equator to an object, measured along a great circle passing through that object and the galactic poles.

galactic longitude. Angular distance, measured eastward along the galactic equator from the galactic center, to the intersection of the galactic equator with a great circle passing through the galactic poles and an object.

galactic poles. The poles of the galactic equator; the intersections with the celestial sphere of a line through the observer that is perpendicular to the plane of the galactic equator.

galactic rotation. Rotation of the Galaxy.

galaxy. A large assemblage of stars; a typical galaxy contains millions to hundreds of thousands of millions of stars.

Galaxy. The galaxy to which the sun and our neighboring stars belong; the Milky Way is light from remote stars in the Galaxy.

gamma rays. Photons (of electromagnetic radiation) of energy higher than those of X rays; the most energetic form of electromagnetic radiation.

gauss. A unit of magnetic flux density.

gegenschein (counterglow). A very faint, diffuse glow of light opposite the sun in the sky, believed to be caused by sunlight reflected from interplanetary particles.

Geiger counter. A device for counting high-energy charged particles and hence for measuring the intensity of corpuscular radiation.

geodesic. The path of a body in spacetime.

geodesic equations. A set of equations in general relativity by which the paths of objects in spacetime can be calculated.

geomagnetic. Referring to the geometrical center of the earth's magnetic field.

geomagnetic poles. The poles of a hypothetical bar magnet whose magnetic field most nearly matches that of the earth.

giant (star). A luminous star of large radius.

gibbous moon. One of the phases of the moon in which more than half, but not all, of the moon's daylight hemisphere is visible from the earth.

Gliese catalogue. A catalogue of nearby stars compiled by the astronomer W. Gliese.

globular cluster. One of about 120 larger star clusters that form a system of clusters centered on the center of the Galaxy.

globule. A small, dense, dark nebula; believed to be a possible protostar.

granulation. The "rice-grain"-like structure of the solar photosphere.

gravitation. The tendency of matter to attract itself.

gravitational constant, *G*. The constant of proportionality in Newton's law of gravitation; in metric units G has the value 6.668×10^{-8} dyne · cm^2/gm^2.

gravitational energy. Energy that can be released by the gravitational collapse, or partial collapse, of a system.

gravitational redshift. The redshift caused by a gravitational field. The slowing of clocks in a gravitational field.

gravitational waves. Oscillations in spacetime, propagated by changes in the distribution of matter.

great circle. Circle on the surface of a sphere that is the curve of intersection of the sphere with a plane passing through its center.

greatest elongation (east or west). The largest separation in celestial longitude (to the east or west) that an inferior planet can have from the sun.

greenhouse effect. The blanketing of infrared radiation near the surface of a planet by, for example, carbon dioxide in its atmosphere.

Greenwich meridian. The meridian of longitude passing through the site of the old Royal Greenwich Observatory, near London; origin of longitude on the earth.

Gregorian calendar. A calendar (now in common use) introduced by Pope Gregory XIII in 1582.

H I region. Region of neutral hydrogen in interstellar space.

H II region. Region of ionized hydrogen in interstellar space.

hadron. A subnuclear particle; one of hundreds now known to exist, of mass from somewhat less to somewhat more than that of the proton.

half-life. The time required for half of the radioactive atoms in a sample to disintegrate.

halo (around sun or moon). A ring of light around the sun or moon caused by refraction by the ice crystals of cirrus clouds.

halo (of galaxy). See corona.

harmonic law. Kepler's third law of planetary motion: the cubes of the semimajor axes of the planetary orbits are in proportion to the squares of the sidereal periods of the planets.

harvest moon. The full moon nearest the time of the autumnal equinox.

head (of comet). The main part of a comet, consisting of its nucleus and coma.

"heavy" elements. In astronomy, usually those elements of greater atomic number than helium.

Heisenberg uncertainty principle. A principle of quantum mechanics that places a limit on the precision with which the simultaneous position and momentum of a body or particle can be specified.

helio-. Prefix referring to the sun.

heliocentric. Centered on the sun.

helium flash. The nearly explosive ignition of helium in the triple-alpha process in the dense core of a red giant star.

Hertzsprung-Russell (H-R) diagram. A plot of absolute magnitude against temperature (or spectral class or color index) for a group of stars.

horary astrology. Astrology based on a horoscope drawn up for the place and instant at which a question or idea was first raised. Horary astrology purports to advise on the auspiciousness of an action based on the horoscope of the instant of its conception.

horizon (astronomical). A great circle on the celestial sphere 90° from the zenith.

horizon system. A system of celestial coordinates (altitude and azimuth) based on the astronomical horizon and the north point.

horoscope (astrological term). A chart showing the positions along the zodiac and in the sky of the sun, moon, and planets at some given instant and place on earth—generally corresponding to the time and place of a person's birth.

hour angle. The angle measured westward along the celestial equator from the local meridian to the hour circle passing through an object.

hour circle. A great circle on the celestial sphere passing through the celestial poles.

house. A division or segment of the sky numbered according to its position with respect to the horizon, and used by astrology in preparing a horoscope.

Hubble constant. Constant of proportionality in the relation between the velocities of remote galaxies and their distances. The Hubble constant is approximately 55 km/s · 10^6 pc.

Hubble law. The law of the redshifts.

hydrostatic equilibrium. A balance between the weights of various layers, as in a star or the earth's atmosphere, and *the pressures that support them.*

hyperbola. A conic section of eccentricity greater than 1.0; the curve of intersection between a circular cone and a plane that is at too small an angle with the axis of the cone to cut all of the way through it, and is not parallel to a line in the face of the cone.

hypothesis. A tentative theory or supposition, advanced to explain certain facts or phenomena, which is subject to further tests and verification.

image. The optical representation of an object produced by light rays from the object being refracted or reflected by a lens or mirror.

image tube. A device in which electrons, emitted from a photocathode surface exposed to light, are focused electronically.

inclination (of an orbit). The angle between the orbital plane of a revolving body and some fundamental plane—usually the plane of the celestial equator or of the ecliptic.

Index Catalogue, IC. The supplement to Dreyer's *New General Catalogue* of star clusters and nebulae.

index of refraction. A measure of the refracting power of a transparent substance; specifically, the ratio of the speed of light in a vacuum to its speed in the substance.

inertia. The property of matter that requires a force to act on it to change its state of motion; momentum is a measure of inertia.

inertial system. A system of coordinates that is not itself accelerated, but which is either stationary or is moving with constant velocity.

inferior conjunction. The configuration of an inferior planet when it has the same longitude as the sun, and is between the sun and earth.

inferior planet. A planet whose distance from the sun is less than the earth's.

infrared radiation. Electromagnetic radiation of wavelength longer than the longest (red) wavelengths that can be perceived by the eye, but shorter than radio wavelengths.

interferometer (stellar). An optical device, making use of the principle of interference of light waves, with which small angles can be measured.

International Date Line. An arbitrary line on the surface of the earth near longitude 180° across which the date changes by one day.

international magnitude system. The system of photographic and photovisual magnitudes, referring to the blue and yellow spectral regions, at one time adopted by international agreement, but now largely superseded by the *U, B, V* system.

interplanetary medium. The sparse distribution of gas and solid particles in the interplanetary space.

interstellar dust. Microscopic solid grains, believed to be mostly dielectric compounds of hydrogen and other common elements, in interstellar space.

interstellar gas. Sparse gas in interstellar space.

interstellar lines. Absorption lines superimposed on stellar spectra, produced by the interstellar gas.

interstellar matter. Interstellar gas and dust.

ion. An atom that has become electrically charged by the addition or loss of one or more electrons.

ionization. The process by which an atom gains or loses electrons.

ionization potential. The energy required to remove an electron from an atom.

ionosphere. The upper region of the earth's atmosphere in which many of the atoms are ionized.

ion tail (of comet). The relatively straight tail of a comet produced by the interaction of the solar wind with the ions in the comet.

irregular galaxy. The galaxy without rotational symmetry; neither a spiral nor elliptical galaxy.

irregular variable. A variable star whose light variations do not repeat with a regular period.

island universe. Historical synonym for galaxy.

isotope. Any of two or more forms of the same element, whose atoms all have the same number but different masses.

isotropic. The same in all directions.

Jovian planet. Any of the planets Jupiter, Saturn, Uranus, and Neptune.

Julian Calendar. A calendar introduced by Julius Caesar in 45 B.C.

Julian day. The number of the day in a running sequence beginning January 1, 4713 B.C.

Jupiter. The fifth planet from the sun in the solar system.

Kepler's laws. The three laws, discovered by Kepler, that describe the motions of the planets.

kiloparsec (kpc). 1000 parsecs, or about 3260 LY.

kinetic energy. Energy associated with motion; the kinetic energy of a body is one-half the product of its mass and the square of its velocity.

kinetic theory (of gases). The science that treats the motions of the molecules that compose gases.

laser. An acronym for *light amplification by stimulated emission of radiation;* a device for amplifying a light signal at a particular wavelength into a coherent beam.

latitude. A north-south coordinate on the surface of the earth; the angular distance north or south of the equator measured along a meridian passing through a place.

launch window. A range of dates during which a space vehicle can be launched for a specific mission without exceeding the fuel capabilities of that system.

law. A statement of order or relation between phenomena that, under given conditions, is presumed to be invariable.

law of areas. Kepler's second law: the radius vector from the sun to any planet sweeps out equal areas in the planet's orbital plane in equal intervals of time.

law of the redshifts. The relation between the radial velocity and distance of a remote galaxy: the radial velocity is proportional to the distance of the galaxy.

lead sulfide cell. A device used to measure infrared radiation.

leap year. A calendar year with 366 days, intercalated approximately every four years to make the average length of the calendar year as nearly equal as possible to the tropical year.

lepton. A light subatomic particle, such as an electron, neutrino, or muon.

light. Electromagnetic radiation that is visible to the eye.

light curve. A graph that displays the variation in light or magnitude of a variable or eclipsing binary star.

light-year. The distance light travels in a vacuum in one year; 1 LY = 9.46×10^{17} cm, or about 6×10^{12} mi.

limb (of sun or moon). Apparent eage of the sun or moon as seen in the sky.

limb darkening. The phenomenon whereby the sun is less bright near its limb than near the center of its disk.

limiting magnitude. The faintest magnitude that can be observed with a given instrument or under given conditions.

line of apsides. The line connecting the apsides of an orbit (the perifocus and farthest-from-focus points); or the line along the major axis of the orbit.

line of nodes. The line connecting the nodes of an orbit.

line profile. A plot of the intensity of light versus wavelength across a spectral line.

linear diameter. Actual diameter in units of length.

lithosphere. The upper layer of the earth, to a depth of 50 to 100 km, involved in plate tectonics.

Local Group. The cluster of galaxies to which our Galaxy belongs.

local oscillator. The old (classical) idea of an atom absorbing or emitting radiation by setting itself in oscillation or by reducing that oscillation. The local oscillator has been replaced by a different model in the modern quantum theory.

local standard of rest A coordinate system that shares the average motion of the sun and its neighboring stars about the galactic center.

local supercluster (or **supergalaxy**). A proposed cluster of clusters of galaxies, to which the Local Group belongs.

longitude. An east-west coordinate on the earth's surface; the angular distance, measured east or west along the equator from the Greenwich meridian, to the meridian passing through a place.

Lorentz transformation. A mathematical way of transforming lengths, masses, speeds, and so on, from one system to another in uniform relative motion, while preserving known physical laws.

low-velocity star (or **object**). A star (or object) that has low space velocity; generally an object that shares the sun's high orbital speed about the galactic center.

luminosity. The rate of radiation of electromagnetic energy into space by a star or other object.

luminosity class. A classification of a star according to its luminosity for a given spectral class.

luminous energy. Light.

lunar. Referring to the moon.

lunar eclipse. An eclipse of the moon.

Lyman lines. A series of absorption or emission lines in the spectrum of hydrogen that arise from transitions to and from the lowest energy states of the hydrogen atoms.

Magellanic Clouds. See Clouds of Magellan.

magnetic field. The region of space near a magnetized body within which magnetic forces can be detected.

magnetic pole. One of two points on a magnet (or the earth) at which the greatest density of lines of force emerge. A compass needle aligns itself along the local lines of force on the earth and points more or less toward the magnetic poles of the earth.

magnetometer. A device for measuring magnetic fields.

magnetosphere. The region around the earth occupied by its magnetic field.

magnifying power. The number of times larger (in angular diameter) an object appears through a telescope than with the naked eye.

magnitude. A measure of the amount of light flux received from a star or other luminous object.

main sequence. A sequence of stars on the Hertzsprung-Russell diagram, containing the majority of stars, that runs diagonally from the upper left to the lower right.

major axis (of ellipse). The maximum diameter of an ellipse.

major planet. A Jovian planet.

mantle (of earth). The greatest part of the earth's interior, lying between the crust and the core.

mare. Latin for "sea"; name applied to many of the "sealike" features on the moon or Mars.

Mariner space probes. A series of space probes launched in the 1960s and early 1970s to explore the planets Mercury, Venus, and Mars.

Mars. Fourth planet from the sun in the solar system.

maser. An acronym for *microwave amplification of stimulated emission radiation;* a device for amplifying a microwave (radio) signal at a particular wavelength into a coherent beam.

mass. A measure of the total amount of material in a body; defined either by the inertial properties of the body or by its gravitational influence on other bodies.

mass defect. The amount by which the mass of an atomic nucleus is less than the sum of the masses of the individual nucleons that compose it.

mass-luminosity relation. An empirical

relation between the masses and luminosities of many (principally main-sequence) stars.

Maxwell's equations. A set of four equations that describe the fields around magnetic and electric charges, and how changes in those fields produce forces and electromagnetic radiation.

mean solar day. Interval between successive meridian passages of the mean sun; average length of the apparent solar day.

mean solar time. Local hour angle of the mean sun *plus* 12 hours.

mean sun. A fictitious body that moves eastward with uniform angular velocity along the celestial equator, completing one circuit of the sky with respect to the vernal equinox in a tropical year.

mechanics. That branch of physics which deals with the behavior of material bodies under the influence of, or in the absence of, forces.

medical astrology. That branch of astrology that deals with supposed connections between planets and zodiacal signs and bodily organs and their diseases.

megaparsec (Mpc). One million (10^6) pc.

Mercury. Nearest planet to the sun in the solar system.

meridian (celestial). The great circle on the celestial sphere that passes through an observer's zenith and the north (or south) celestial pole.

meridian (terrestrial). The great circle on the surface of the earth that passes through a particular place and the north and south poles of the earth.

mesosphere. The layer of the ionosphere immediately above the stratosphere.

Messier catalogue. A catalogue of nonstellar objects compiled by Charles Messier in 1787.

meteor. The luminous phenomenon observed when a meteoroid enters the earth's atmosphere and burns up; popularly called a "shooting star."

meteor shower. Many meteors appearing to radiate from a common point in the sky caused by the collision of the earth with a swarm of meteoritic particles.

meteorite. A portion of a meteoroid that survives passage through the atmosphere and strikes the ground.

meteorite fall. The occurrence of a meteorite striking the ground.

meteoroid. A meteoritic particle in space before any encounter with the earth.

micrometeorite. A meteoroid so small that, on entering the atmosphere of the earth, it is slowed quickly enough that it does not burn up or ablate but filters through the air to the ground.

microwave. Short-wave radio wavelengths.

Milky Way. The band of light encircling the sky, which is due to the many stars and diffuse nebulae lying near the plane of the Galaxy.

minor axis (of ellipse). The smallest or least diameter of an ellipse.

minor planet. One of several tens of thousands of small planets, ranging in size from a few hundred kilometers to less than one kilometer in diameter.

Mira Ceti-type variable star. Any of a large class of red-giant long-period or irregular pulsating variable stars, of which the star Mira is a prototype.

missile. A projectile, especially a rocket.

model atmosphere (or **photosphere**). The result of a theoretical calculation of the run of temperature, pressure, density, and so on, through the outer layers of the sun or a star.

molecule. A combination of two or more atoms bound together; the smallest particle of a chemical compound or substance that exhibits the chemical properties of that substance.

momentum. A measure of the inertia or state of motion of a body; the momentum of a body is the product of its mass and velocity. In the absence of a force, momentum is conserved.

monochromatic. Of one wavelength or color.

mundane astrology. Astrology applied to nations and kings, rather than to individuals.

muon. A particle that behaves like an electron but is about 200 times as massive.

nadir. The point on the celestial sphere 180° from the zenith.

nanosecond. One thousand-millionth (10^{-9}) second.

natal astrology. Astrology based on the horoscope drawn up for the place and moment of one's birth.

nautical mile. The mean length of one minute of arc on the earth's surface along a meridian.

navigation. The art of finding one's position and course at sea or in the air.

neap tide. The lowest tides in the month which occur when the moon is near first or third quarter.

nebula. Cloud of interstellar gas or dust.

nebular hypothesis. The basic idea that the sun and planets formed from the same cloud of gas and dust in interstellar space.

Neptune. Eighth planet from the sun in the solar system.

neutrino. A particle that has no mass or charge but that carries energy away in the course of certain nuclear transformations.

neutron. A subatomic particle with no charge and with mass approximately equal to that of the proton.

neutron star. A star of extremely high density composed entirely of neutrons.

New General Catalogue (NGC). A catalogue of star clusters, nebulae, and galaxies compiled by J. L. E. Dreyer in 1888.

new moon. Phase of the moon when its longitude is the same as that of the sun.

Newtonian focus. An optical arrangement in a reflecting telescope where the light is reflected by a flat mirror to a focus at the side of the telescope tube just before it reaches the focus of the objective.

Newton's laws. The laws of mechanics and gravitation formulated by Isaac Newton.

night sky light. The faint illumination of the night sky; the main source is usually fluorescence by atoms high in the atmosphere.

node. The intersection of the orbit of a body with a fundamental plane—usually the plane of the celestial equator or of the ecliptic.

nodical month. The period of revolution of the moon about the earth with respect to the line of nodes of the moon's orbit.

nodical (eclipse) year. Period of revolution of the earth about the sun with respect to the line of nodes of the moon's orbit.

nonthermal radiation. See synchrotron radiation.

north point. That intersection of the celestial meridian and astronomical horizon lying nearest the north celestial sphere.

north polar sequence. A group of stars in the vicinity of the north celestial pole whose magnitudes serve as standards for the international magnitude system.

nova. A star that experiences a sudden outburst of radiant energy, temporarily increasing its luminosity by hundreds to thousands of times.

nuclear. Referring to the nucleus of the atom.

nuclear bulge. Central part of our Galaxy.

nuclear transformation. Transformation of one atomic nucleus into another.

nucleogenesis. The formation and the evolution of the chemical elements by nuclear reactions.

nucleon. Any one of the subatomic particles that compose a nucleus.

nucleus (of atom). The heavy part of an atom, composed mostly of protons and neutrons, and about which the electrons revolve.

nucleus (of comet). A swarm of solid particles in the head of a comet.

nucleus (of galaxy). Central concentration of stars and gas at the center of a galaxy.

null geodesic. The path of a light ray in four-dimensional spacetime.

objective. The principal image-forming component of a telescope or other optical instrument.

objective prism. A prismatic lens that can be placed in front of a telescope objective to transform each star image into an image of its spectrum.

oblate spheroid. A solid formed by rotating an ellipse about its minor axis.

oblateness. A measure of the "flattening" of an oblate spheroid; numerically, the ratio of the difference between the major and minor diameters (or axes) to the major diameter (or axis).

obliquity of the ecliptic. Angle between the planes of the celestial equator and the ecliptic; about $23\frac{1}{2}°$.

obscuration (interstellar). Absorption of starlight by interstellar dust.

occultation. An eclipse of a star or planet by the moon or a planet.

opacity. Absorbing power; capacity to impede the passage of light.

open cluster. A comparatively loose or "open" cluster of stars, containing from a few dozen to a few thousand members, located in the spiral arms or disk of the Galaxy; galactic cluster.

opposition. Configuration of a planet when its elongation is 180°.

optical binary. Two stars at different distances nearly lined up in projection so that they appear close together, but which are not really dynamically associated.

optics. The branch of physics that deals with light and its properties.

orbit. The path of a body that is in revolution about another body or point.

outgassing. The process by which the gasses of a planetary atmosphere work their way out from the crust of the planet.

Pallas. Second minor planet to be discovered.

Pangaea. Name given to the hypothetical continent from which the present continents of the earth separated.

parabola. A conic section of eccentricity 1.0; the curve of intersection between a circular cone and a plane parallel to a straight line in the surface of the cone.

parabolic speed. See velocity of escape.

paraboloid. A parabola of revolution; a curved surface of parabolic cross section. Especially applied to the surface of the primary mirror in a standard reflecting telescope.

parallactic ellipse. A small ellipse that a comparatively nearby star appears to trace out in the sky, which results from the orbital motion of the earth about the sun.

parallax. An apparent displacement of an object due to a motion of the observer.

parallax (stellar). An apparent displacement of a nearby star that results from the motion of the earth around the sun; numerically, the angle subtended by 1 AU at the distance of a particular star.

parsec. The distance of an object that would have a stellar parallax of one second of arc; 1 parsec = 3.26 light-years.

partial eclipse. An eclipse of the sun or moon in which the eclipsed body does not appear completely obscured.

Pauili exclusion principle. The principle that states that no two subatomic particles of certain types can exist in the same place and time with the same state (or condition).

penumbra. The portion of a shadow from which only part of the light source is occulted by an opaque body.

penumbral eclipse. A lunar eclipse in which the moon passes through the penumbra, but not the umbra, of the earth's shadow.

perfect cosmological principle. The assumption that, on the large scale, the universe appears the same from every place and at all times.

perfect gas. An "ideal" gas that obeys the perfect gas laws.

perfect gas laws. Certain laws that describe the behavior of an ideal gas; Charles' law, Boyle's law, and the equation of state for a perfect gas.

perfect radiator. Black body; a body that absorbs and subsequently reemits all radiation incident upon it.

periastron. The place in the orbit of a star in a binary star system where it is closest to its companion star.

perifocus. The place on an elliptical orbit that is closest to the focus occupied by the central force.

perigee. The place in the orbit of an earth satellite where it is closest to the center of the earth.

perihelion. The place in the orbit of an object revolving about the sun where it is closest to the center of the sun.

period. A time interval; for example, the time required for one complete revolution.

period-density relation. Proportionality between the period and the inverse square root of the mean density for a pulsating star.

period-luminosity relation. An empirical relation between the periods and luminosities of cepheid-variable stars.

periodic comet. A comet whose orbit has been determined to have an eccentricity of less than 1.0.

perturbation. The disturbing effect, when small, on the motion of a body as predicted by a simple theory, produced by a third body or other external agent.

phases of the moon. The progression of changes in the moon's appearance during the month that results from the moon's turning different portions of its illuminated hemisphere to our view.

photocell (photoelectric cell). An electron tube in which electrons are dislodged from the cathode when it is exposed to light and are accelerated to the anode, thus producing a current in the tube, whose strength serves as a measure of the light striking the cathode.

photoelectric effect. The emission of an electron by the absorption of a photon by a substance.

photographic magnitude. The magnitude of an object, as measured on the traditional, blue- and violet-sensitive photographic emulsions.

photometry. The measurement of light intensities.

photomultiplier. A photoelectric cell in which the electric current generated is amplified at several stages within the tube.

photon. A discrete unit of electromagnetic energy.

photosphere. The region of the solar (or a stellar) atmosphere from which radiation escapes into space.

photosynthesis. The formation of carbohydrates in the chlorophyll-containing tissues of plants exposed to sunlight. In the process, oxygen is released to the atmosphere.

photovisual magnitude. A magnitude corresponding to the spectral region to which the human eye is most sensitive, but measured by photographic methods with suitable green- and yellow-sensitive emulsions and filters.

pion. A particular kind of *meson*, or subatomic particle of mass intermediate between that of the proton and electron.

Pioneer spacecraft. A series of spacecraft launched to Jupiter and more distant planets in the early 1970s.

Planck's constant. The constant of proportionality relating the energy of a photon to its frequency.

Planck's radiation law. A formula from which can be calculated the intensity of radiation at various wavelengths emitted by a black body.

planet. Any of nine solid bodies revolving about the sun.

planetarium. An optical device for projecting on a screen or domed ceiling the stars and planets and their apparent motions in the sky.

planetary nebula. A shell of gas ejected from, and enlarging about, a certain kind of extremely hot star.

planetoid. Synonym for minor planet.

plasma. A hot ionized gas.

plate tectonics. The motion of segments or plates of the outer layer of the earth over the underlying mantle.

Pluto. Ninth planet from the sun in the solar system.

polar axis. The axis of rotation of the earth; also, an axis in the mounting of a telescope that is parallel to the earth's axis.

polarization. A condition in which the planes of vibration (or the *E* vectors) of the various rays in a light beam are at least partially aligned.

polarized light. Light in which polarization is present.

Polaroid. Trade name for a transparent substance that produces polarization in light.

Population I and II. Two classes of stars (and systems of stars), classified according to their spectra, chemical compositions, radial velocities, ages, and locations in the Galaxy.

positron. An electron with a positive rather than a negative charge.

postulate. An essential prerequisite to a hypothesis or theory.

potential energy. Stored energy that can be converted into other forms; especially gravitational energy.

precession (of earth). A slow, conical motion of the earth's axis of rotation, caused principally by the gravitational torque of the moon and sun on the earth's equatorial bulge. *Lunisolar precession,* precession caused by the moon and sun only; *planetary precession,* a slow change in the orientation of the plane of the earth's orbit caused by planetary perturbations; *general precession,* the combination of these two effects on the motion of the earth's axis with respect to the stars.

precession of the equinoxes. Slow westward motion of the equinoxes along the ecliptic that results from precession.

primary cosmic rays. The cosmic-ray particles that arrive at the earth from beyond its atmosphere, as opposed to the secondary particles that are produced by collisions between primary cosmic rays and air molecules.

prime focus. The point in a telescope where the objective focuses the light.

prime meridian. The terrestrial meridian passing through the site of the old Royal Greenwich Observatory; longitude 0°.

primeval atom. A single mass whose explosion (in some cosmological theories) has been postulated to have resulted in all the matter now present in the universe.

primeval fireball. The extremely hot opaque gas that is presumed to have comprised the entire mass of the universe at the time of or immediately following the "big bang"; the exploding primeval atom.

Principia. Contraction of *Philosophiae Naturalis Principia Mathematica,* the great book by Newton in which he set forth his laws of motion and gravitation in 1687.

principle of equivalence. A principle of general relativity that states that forces and accelerations are equivalent, and that, in particular, the force of gravitation can be replaced by a suitable acceleration in the coordinate system.

principle of relativity. The assumption basic to special relativity that the laws of physics are the same in all systems in uniform motion with respect to each other.

prism. A wedge-shaped piece of glass that is used to disperse white light into a spectrum.

prolate spheroid. The solid produced by the rotation of an ellipse about its major axis.

prominence. A phenomenon in the solar corona that commonly appears like a flame above the limb of the sun.

proper motion. The angular change in direction of a star per year.

proton. A heavy subatomic particle that carries a positive charge, and one of the two principal constituents of the atomic nucleus.

proton-proton chain. A chain of thermonuclear reactions by which nuclei of hydrogen are built up into nuclei of helium.

protoplanet (or **-star** or **-galaxy**). The original material from which a planet (or a star or galaxy) condensed.

pulsar. A variable radio source of small angular size that emits radio pulses in very regular periods that range from 0.03 to 5 seconds.

pulsating variable. A variable star that pulsates in size and luminosity.

quadrature. A configuration of a planet in which its elongation is 90°.

quantum mechanics. The branch of physics that deals with the structure of atoms and their interactions with each other and with radiation.

quark. A hypothetical fundamental subatomic particle. Quarks of from 1 to 6 different kinds, in various combinations, are presumed to make up all other particles.

quarter moon. Either of the two phases of the moon when its longitude differs by 90° from that of the sun; the moon appears half full at these phases.

quasar. A quasi-stellar source.

quasi-stellar galaxy (QSG). A stellar-appearing object of very large redshift presumed to be extragalactic and highly luminous.

quasi-stellar source (QSS). A stellar-appearing object of very large redshift that is a strong source of radio waves; presumed to be extragalactic and highly luminous.

RR Lyrae variable. One of a class of giant pulsating stars with periods less than one day; a cluster variable.

radar. A technique for observing the reflection of radio waves from a distant object.

radial velocity. The component of relative velocity that lies in the line of sight.

radial velocity curve. A plot of the variation of radial velocity with time for a binary or variable star.

radiant (of meteor shower). The point in the sky from which the meteors belonging to a shower seem to radiate.

radiation. A mode of energy transport whereby energy is transmitted through a vacuum; also the transmitted energy itself, either electromagnetic or corpuscular.

radiation pressure. The transfer of momentum carried by electromagnetic radiation to a body that the radiation impinges upon.

radioactive dating. The science of determining the ages of rocks or other specimens by the amount of radioactive decay of certain radioactive elements contained therein.

radioactivity (radioactive decay). The process by which certain kinds of atomic nuclei naturally decompose with the spontaneous emission of subatomic particles and gamma rays.

radio astronomy. The technique of making astronomical observations in radio wavelengths.

radio galaxy. A galaxy that emits greater amounts of radio radiation than average.

radio telescope. A telescope designed to make observations in radio wavelengths.

ray (lunar). Any of a system of bright elongated streaks, sometimes associated with a crater on the moon.

recurrent nova. A nova that has been known to erupt more than once.

reddening (interstellar). The reddening of starlight passing through interstellar dust, caused by the dust scattering blue light more effectively than red.

red giant. A large, cool star of high luminosity; a star occupying the upper right portion of the Hertzsprung-Russell diagram.

redshift. A shift to longer wavelengths of the light from remote galaxies; presumed to be produced by a Doppler shift.

reflecting telescope. A telescope in which the principal optical component (objective) is a concave mirror.

reflection. The return of light rays by an optical surface.

reflection nebula. A relatively dense dust cloud in interstellar space that is illuminated by starlight.

refracting telescope. A telescope in which the principal optical component (objective) is a lens or system of lenses.

refraction. The bending of light rays passing from one transparent medium (or a vacuum) to another.

regression of nodes. A consequence of certain perturbations on the orbit of a revolving body whereby the nodes of the orbit slide westward in the fundamental plane (usually the plane of the ecliptic or of the celestial equator).

relative orbit. The orbit of one of two mutually revolving bodies referred to the other body as origin.

relativistic particle (or electron). A particle (or electron) moving at nearly the speed of light.

relativity. A theory formulated by Einstein that describes the relations between measurements of physical phenomena by two different observers who are in relative motion at constant velocity (the *special theory of relativity*), or that describes how a gravitational field can be replaced by a curvature of spacetime (the *general theory of relativity*).

resolution. The degree to which fine details in an image are separated or resolved.

resolving power. A measure of the ability of an optical system to resolve or separate fine details in the image it produces; in astronomy, the angle in the sky that can be resolved by a telescope.

rest mass. The mass of an object or particle as measured when it is at rest in the laboratory.

retrograde motion. An apparent westward motion of a planet on the celestial sphere or with respect to the stars.

retrorockets. Rockets fired from a spacecraft in the direction of its motion to slow it down.

revolution. The motion of one body around another.

right ascension. A coordinate for measuring the east-west positions of celestial bodies; the angle measured eastward along the celestial equator from the vernal equinox to the hour circle passing through a body.

rille (or rill). A crevasse or trenchlike depression in the moon's surface.

Roche's limit. The smallest distance from a planet or other body at which purely gravitational forces can hold together a satellite or secondary body of the same mean density as the primary; within this distance the tidal forces of the primary would break up the secondary.

rotation. Turning of a body about an axis running through it.

satellite. A body that revolves about a larger one; for example, a moon of a planet.

Saturn. The sixth planet from the sun in the solar system.

scale (of telescope). The linear distance in the image corresponding to a particular angular distance in the sky; say, so many centimeters per degree.

Schmidt telescope. A type of reflecting telescope invented by B. Schmidt, in which certain aberrations produced by a spherical concave mirror are compensated for by a thin objective correcting lens.

Schwarzschild radius. See event horizon.

science. The attempt to find order in nature or to find laws that describe natural phenomena.

scientific method. A specific procedure in science: (1) the observation of phenomena or the results of experiments; (2) the formulation of hypotheses that describe these phenomena, and that are consistent with the body of knowledge available; (3) the testing of these hypotheses by noting whether or not they adequately predict and describe new phenomena or the results of new experiments.

Sculptor-type system. A dwarf elliptical galaxy, of which the system in Sculptor is a typical example.

secondary cosmic rays. Secondary particles produced by interactions between primary cosmic rays from space and the atomic nuclei in molecules of the earth's atmosphere.

second-order cluster of galaxies. A cluster of clusters of galaxies.

secular. Not periodic.

seeing. The unsteadiness of the earth's atmosphere, which blurs telescopic images.

seismic waves. Vibrations traveling through the earth's interior that result from earthquakes.

seismograph. An instrument used to record and measure seismic waves.

seismology. The study of earthquakes and the conditions that produce them, and of the internal structure of the earth as deduced from analyses of seismic waves.

seleno-. Prefix referring to the moon.

semimajor axis. Half the major axis of a conic section.

semiregular variable. A variable star, usually a red giant or supergiant, whose period of pulsation is far from constant.

separation (in a visual binary). The angular separation of the two components of a visual binary star.

Seyfert galaxy. A spiral galaxy whose nucleus shows bright emission lines; one of a class of galaxies first described by C. Seyfert. (They are sometimes radio sources.)

shadow cone. The umbra of the shadow of a spherical body (such as the earth) in sunlight.

shell star. A type of star, usually of spectral-type B to F, surrounded by a gaseous ring or shell.

shower (of cosmic rays). A large "rain" of secondary cosmic-ray particles produced by a very energetic primary particle impinging on the earth's atmosphere.

shower (meteor). Many meteors, all seeming to radiate from a common point in the sky, caused by the encounter by the earth of a swarm of meteoroids moving together through space.

sidereal astrology. Astrology in which the horoscope is based on the positions of the planets with respect to the fixed stars rather than with respect to signs that, as a consequence of precession, slide through the zodiac.

sidereal day. The interval between two successive meridian passages of the vernal equinox.

sidereal month. The period of the moon's revolution about the earth with respect to the stars.

sidereal period. The period of revolution of one body about another with respect to the stars.

sidereal time. The local hour angle of the vernal equinox.

sidereal year. Period of the earth's revolution about the sun with respect to the stars.

sign (of zodiac). Astrological term for any of twelve equal sections along the ecliptic, each of length 30°. Starting at the vernal equinox, and moving eastward, the signs are Aries, Taurus, Gemini, Cancer, Leo, Virgo, Libra, Scorpio, Sagittarius, Capricorn, Aquarius, and Pisces.

simultaneity. The occurrence of two events at the same time. In relativity, absolute simultaneity is seen not to have meaning.

Skylab. An orbiting scientific laboratory occupied by several successive teams of astronauts in the late 1960s and early 1970s.

solar activity. Phenomena of the solar atmosphere: sunspots, plages, and related phenomena.

solar antapex. Direction away from which the sun is moving with respect to the local standard of rest.

solar apex. The direction toward which the sun is moving with respect to the local standard of rest.

solar constant. Mean amount of solar radiation received per unit time, by a unit area, just outside the earth's atmosphere, and perpendicular to the direction of the sun; the numerical value is 1.37×10^6 ergs/cm$^2 \cdot$ s.

solar motion. Motion of the sun, or the velocity of the sun, with respect to the local standard of rest.

solar nebula. The cloud of gas and dust from which the solar system is presumed to have formed.

solar parallax. Angle subtended by the equatorial radius of the earth at a distance of 1 AU.

solar system. The system of the sun and the planets, their satellites, the minor planets, comets, meteoroids, and other objects revolving around the sun.

solar time. A time based on the sun; usually the hour angle of the sun *plus* 12 hours.

solar wind. A radial flow of corpuscular radiation leaving the sun.

solstice. Either of two points on the celestial sphere where the sun reaches its maximum distances north and south of the celestial equator.

south point. Intersection of the celestial meridian and astronomical horizon 180° from the north point.

space motion. The velocity of a star with respect to the sun.

space probe. An unmanned interplanetary rocket carrying scientific instruments to obtain data on other planets or on the interplanetary environment.

space technology. The applied science of the immediate space environment of the earth.

spacetime. A system of one time and three spatial coordinates, with respect to which the time and place of an *event* can be specified; also called *spacetime continuum.*

specific gravity. The ratio of the density of a body or substance to that of water.

spectral class (or type). A classification of a star according to the characteristics of its spectrum.

spectral sequence. The sequence of spectral classes of stars arranged in order of decreasing temperatures of stars of those classes.

spectrogram. A photograph of a spectrum.

spectrograph. An instrument for photographing a spectrum; usually attached to a telescope to photograph the spectrum of a star.

spectroheliogram. A photograph of the sun obtained with a spectroheliograph.

spectroheliograph. An instrument for photographing the sun, or part of the sun, in the monochromatic light of a particular spectral line.

spectrophotometry. The measurement of the intensity of light from a star or other source at different wavelengths.

spectroscope. An instrument for directly viewing the spectrum of a light source.

spectroscopic binary star. A binary star in which the components are not resolved optically, but whose binary nature is indicated by periodic variations in radial velocity, indicating orbital motion.

spectroscopic parallax. A parallax (or distance) of a star that is derived by comparing the apparent magnitude of the star with its absolute magnitude as deduced from its spectral characteristics.

spectroscopy. The study of spectra.

spectrum. The array of colors or wavelengths obtained when light from a source is dispersed, as in passing it through a prism or grating.

spectrum analysis. The study and analysis of spectra, especially stellar spectra.

spectrum binary. A binary star whose binary nature is revealed by spectral characteristics that can only result from the composite of the spectra of two different stars.

speed. The rate at which an object moves without regard to its direction of motion; the numerical or absolute value of velocity.

spherical aberration. A defect of optical systems whereby on-axis rays of light striking different parts of the objective do not focus at the same place.

spicule. A narrow jet of rising material in the solar chromosphere.

spiral arms. Arms of interstellar material and young stars that wind out in a plane from the central nucleus of a spiral galaxy.

spiral galaxy. A flattened, rotating galaxy with pinwheel-like arms of interstellar material and young stars winding out from its nucleus.

sporadic meteor. A meteor that does not belong to a shower.

spring tide. The highest tide of the month, produced when the sun and moon have longitudes that differ from each other by nearly 0° or 180°.

Sputnik. Russian for "satellite," or "fellow traveler"; the name given to the first Soviet artificial satellite.

stadium. A Greek unit of length, based on the Olympic Stadium; roughly $^1/_6$ km.

standard time. The local mean solar time of a standard meridian, adopted over a large region to avoid the inconvenience of continuous time changes around the earth.

star. A self-luminous sphere of gas.

star cluster. An assemblage of stars held together by their mutual gravitation.

steady state (theory of cosmology). A theory of cosmology embracing the perfect cosmological principle, and involving the continuous creation of matter.

Stefan's law or **Stefan-Boltzmann Law.** A formula from which the rate at which a black body radiates energy can be computed; the total rate of energy emission from a unit area of a black body is proportional to the fourth power of its absolute temperature.

stellar evolution. The changes that take place in the sizes, luminosities, structures, and so on, of stars as they age.

stellar model. The result of a theoretical calculation of the run of physical conditions in a stellar interior.

stellar parallax. The angle subtended by 1 AU at the distance of a star; usually measured in seconds of arc.

Stonehenge. An assemblage of upright stones in Salisbury Plain, England, believed to have been constructed by early people for astronomical observations connected with timekeeping and the calendar.

stratosphere. The layer of the earth's atmosphere above the troposphere, where most weather takes place, and below the ionosphere.

strong nuclear force. The force that binds together the parts of the atomic nucleus.

subdwarf. A star of luminosity lower than that of main-sequence stars of the same spectral type.

subgiant. A star of luminosity intermediate between those of main-sequence stars and normal giants of the same spectral type.

summer solstice. The point on the celestial sphere where the sun reaches its greatest distance north of the celestial equator.

sun. The star about which the earth and other planets revolve.

sundial. A device for keeping time by the shadow a marker (gnomon) casts in sunlight.

sunspot. A temporary cool region in the solar photosphere that appears dark by contrast against the surrounding hotter photosphere.

sunspot cycle. The semiregular 11-year period with which the frequency of sunspots varies.

supergiant. A star of very high luminosity.

superior conjunction. The configuration of a planet in which it and the sun have the same longitude, with the planet being more distant than the sun.

superior planet. A planet more distant from the sun than the earth.

supernova. A stellar outburst or explosion in which a star suddenly increases its luminosity by from hundreds of thousands to hundreds of millions of times.

surface gravity. The weight of a unit mass at the surface of a body.

surveying. The technique of measuring distances and relative positions of places over the surface of the earth (or elsewhere); generally accomplished by triangulation.

synchrotron radiation. The radiation emitted by charged particles being accelerated in magnetic fields and moving at speeds near that of light.

synodic month. The period of revolution of the moon with respect to the sun, or its cycle of phases.

synodic period. The interval between successive occurrences of the same configuration of a planet; for example, between successive oppositions or successive superior conjunctions.

tachyon. A hypothetical particle that always moves with a speed greater than that of light. (There is no evidence that tachyons exist.)

tail (of comet). Gases and solid particles ejected from the head of a comet and forced away from the sun by radiation pressure or corpuscular radiation.

tangential (transverse) velocity. The component of a star's space velocity that lies in the plane of the sky.

tectonics. See plate tectonics.

tektites. Rounded glassy bodies that are suspected to be of meteoritic origin.

telescope. An optical instrument used to aid the eye in viewing or measuring, or to photograph distant objects.

telluric. Of terrestrial origin.

temperature (absolute). Temperature measured in centigrade (Celsius) degrees from absolute zero.

temperature (Celsius; formerly centigrade). Temperature measured on a scale where water freezes at 0° and boils at 100°.

temperature (color). The temperature of a star as estimated from the intensity of the stellar radiation at two or more colors or wavelengths.

temperature (effective). The temperature of a black body that would radiate the same total amount of energy that a particular body does.

temperature (excitation). The temperature of a star as estimated from the relative strengths of lines in its spectrum that originate from atoms in different stages of excitation.

temperature (Fahrenheit). Temperature measured on a scale where water freezes at 32° and boils at 212°.

temperature (ionization). The temperature of a star as estimated from the relative strengths of lines in its spectrum that originate from atoms in different stages of ionization.

temperature (Kelvin). Absolute temperature measured in centigrade degrees.

temperature (kinetic). A measure of the speeds or mean energy of the molecules in a substance.

temperature (radiation). The temperature of a black body that radiates the same amount of energy in a given spectral region as does a particular body.

tensor. A generalization of the concept of the vector, consisting of an array of numbers or quantities that transform according to specific rules.

terminator. The line of sunrise or sunset on the moon or a planet.

terrestrial planet. Any of the planets Mercury, Venus, Earth, Mars, and sometimes Pluto.

Tetrabiblos. A standard and widely used treatise on astrology by Ptolemy.

theory. A set of hypotheses and laws that have been well demonstrated as applying to a wide range of phenomena associated with a particular subject.

thermal energy. Energy associated with the motions of the molecules in a substance.

thermal equilibrium. A balance between the input and outflow of heat in a system.

thermal radiation. The radiation emitted by any body or gas that is not at absolute zero.

thermodynamics. The branch of physics that deals with heat and heat transfer among bodies.

thermonuclear energy. Energy associated with thermonuclear reactions or that can be released through thermonuclear reactions.

thermonuclear reaction. A nuclear reaction or transformation that results from encounters between nuclear particles that are given high velocities (by heating them).

thermosphere. The region of the earth's atmosphere lying between the mesosphere and the exosphere.

tidal force. A differential gravitational force that tends to deform a body.

tide. Deformation of a body by the differential gravitational force exerted on it by another body; in the earth, the deformation of the ocean surface by the differential gravitational forces exerted by the moon and sun.

ton (American short). 2000 lb.

ton (English long). 2240 lb.

ton (metric). One million grams.

topography. The configuration or relief of the surface of the earth, moon or a planet.

total eclipse. An eclipse of the sun in which the sun's photosphere is entirely hidden by the moon, or an eclipse of the moon in which it passes completely into the umbra of the earth's shadow.

train (of meteor). A temporarily luminous trail left in the wake of a meteor.

transit. An instrument for timing the exact instant a star or other object crosses the local meridian. Also, the passage of a celestial body across the meridian; or the passage of a small body (say, a planet) across the disk of a larger one (say, the sun).

triangulation. The operation of measuring some of the elements of a triangle so that other ones can be calculated by the methods of trigonometry, thus determining distances to remote places without having to span them directly.

trigonometry. The branch of mathematics that deals with the analytical solutions of triangles.

trillion. In the United States, 1 thousand billions or million millions (10^{12}); in Great Britain, 1 million billions or million-million millions (10^{18}).

triple-alpha process. A series of two nuclear reactions by which three helium nuclei are built up into one carbon nucleus.

Trojan minor planet. One of several minor planets that share Jupiter's orbit around the sun, but located approximately 60° around the orbit from Jupiter.

Tropic of Cancer. Parallel of latitude 23½° N.

Tropic of Capricorn. Parallel of latitude 23½° S.

tropical astrology. The conventional practice of astrology, in which the horoscope is based on the signs that move through the zodiac with precession.

tropical year. Period of revolution of the earth about the sun with respect to the vernal equinox.

troposphere. Lowest level of the earth's atmosphere, where most weather takes place.

tsunami. A series of very fast waves of seismic origin traveling through the ocean; popularly called "tidal waves."

21-cm line. A spectral line of neutral hydrogen at the radio wavelength of 21 centimeters.

Uhuru. An earth satellite equipped for observations at X-ray wavelengths.

ultraviolet radiation. Electromagnetic radiation of wavelengths shorter than the shortest (violet) wavelengths to which the eye is sensitive; radiation of wavelengths in the approximate range 100 to 4000 angstroms.

umbra. The central, completely dark part of a shadow.

uncertainty principle. See Heisenberg uncertainty principle.

universal time. The local mean time of the prime meridian.

universe. The totality of all matter and radiation and the space occupied by same.

Uranus. Seventh planet from the sun in the solar system.

Van Allen layer. Doughnut-shaped region surrounding the earth where many rapidly moving charged particles are trapped in its magnetic field.

variable star. A star that varies in luminosity.

variation of latitude. A slight semiperiodic change in the latitudes of places on the earth that results from a slight shifting of the body of the earth with respect to its axis of rotation.

vector. A quantity that has both magnitude and direction.

velocity. A vector that denotes both the speed and direction a body is moving.

velocity of escape. The speed with which an object must move in order to enter a parabolic orbit about another body (such as the earth), and hence move permanently away from the vicinity of that body.

Venus. The second planet from the sun in the solar system.

vernal equinox. The point on the celestial sphere where the sun crosses the celestial equator passing from south to north.

Vikings. A series of spacecrafts that landed laboratories on Mars in 1976. The Viking landers were particularly designed to search for life on Mars.

visual binary star. A binary star in which the two components are telescopically resolved.

volume. A measure of the total space occupied by a body.

Voyagers. A series of spacecrafts that were launched by the United States toward Jupiter and more distant planets in 1977.

Vulcan. A hypothetical planet once believed to exist and have an orbit between that of Mercury and the sun; the existence of Vulcan is now generally discredited.

walled plain. A large lunar crater.

wandering of the poles. A semiperiodic shift of the body of the earth relative to its axis of rotation; responsible for variation of latitude.

watt. A unit of power; 10 million ergs expended per second.

wavelength. The spacing of the crests or troughs in a wave train.

weak nuclear force. The nuclear force involved in radioactive decay. The weak force is characterized by the slow rate of certain nuclear reactions—such as the decay of the neutron, which occurs at the average rate of 17 min.

weight. A measure of the force due to gravitational attraction.

west point. The point on the horizon 270° around the horizon from the north point, measured in a clockwise direction as seen from the zenith.

white dwarf. A star that has exhausted most or all of its nuclear fuel and has collapsed to a very small size; believed to be near its final stage of evolution.

white hole. The hypothetical time reversal of a black hole, in which matter and radiation gush up.

Widmanstätten figures. Crystalline structure that can be observed in cut and polished meteorites.

Wien's law. Formula that relates the temperature of a black body to the wavelength at which it emits the greatest intensity of radiation.

winter solstice. Point on the celestial sphere where the sun reaches its greatest distance south of the celestial equator.

worm hole. See **Einstein-Rosen Bridge.**

X rays. Photons of wavelengths intermediate between those of ultraviolet radiation and gamma rays.

X-ray stars. Stars (other than the sun) that emit observable amounts of radiation at X-ray frequencies.

year. The period of revolution of the earth around the sun.

Zeeman effect. A splitting or broadening of spectral lines due to magnetic fields.

zenith. The point on the celestial sphere opposite to the direction of gravity; or the direction opposite to that indicated by a plumb bob.

zenith distance. Arc distance of a point on the celestial sphere from the zenith; 90° minus the altitude of the object.

zero-age main sequence. Main sequence for a system of stars that have completed their contraction from interstellar matter, are now deriving all their energy from nuclear reactions, but whose chemical composition has not yet been altered by nuclear reactions.

zodiac. A belt around the sky centered on the ecliptic.

zodiacal light. A faint illumination along the zodiac, believed to be sunlight reflected and scattered by interplanetary dust.

zone of avoidance. A region near the Milky Way where obscuration by interstellar dust is so heavy that few or no exterior galaxies can be seen.

zone time. The time, kept in a zone 15° wide in longitude, that is the local mean time of the central meridian of that zone. Zone time is used at sea, but over land the boundaries are irregular to conform to political boundaries, and it is called *standard time*.

APPENDIX 3

Metric and English Units

In the English system of measure the fundamental units of length, mass, and time are the yard, pound, and second, respectively. There are also, of course, larger and smaller units, which include the ton (2000 lb), the mile (1760 yd), the rod ($16\frac{1}{2}$ ft), the inch ($\frac{1}{36}$ yd), the ounce ($\frac{1}{16}$ lb), and so on. Such units are inconvenient for conversion and arithmetic computation.

In science, therefore, it is more usual to use the metric system, which has been adopted universally in nearly all countries. The fundamental units of the metric system are:

length: 1 meter (m)
mass: 1 kilogram (kg)
time: 1 second (sec)

A meter was originally intended to be 1 ten-millionth of the distance from the equator to the North Pole along the surface of the earth. It is about 1.1 yd. A kilogram is about 2.2 lb. The second is the same in metric and English units. The most commonly used quantities of length and mass of the metric system are the following:

length

1 km	= 1 kilometer	= 1000 meters	= 0.6214 mile	
1 m	= 1 meter	= 1.094 yards	= 39.37 inches	
1 cm	= 1 centimeter	= 0.01 meter	= 0.3937 inch	
1 mm	= 1 millimeter	= 0.001 meter	= 0.1 cm	= 0.03937 inch
1μ	= 1 micron	= 0.000 001 meter	= 0.0001 cm	= 3.937×10^{-5} inch

also: 1 mile = 1.6093 km
1 inch = 2.5400 cm

mass

1 metric ton	= 10^6 grams	= 1000 kg	= 2.2046×10^3 lb
1 kg	= 1000 grams	= 2.2046 lb	
1 g	= 1 gram	= 0.0022046 lb	= 0.0353 oz
1 mg	= 1 milligram	= 0.001 g	= 2.2046×10^{-6} lb

also: 1 lb = 453.6 g
1 oz = 28.3495 g

APPENDIX 4

Temperature Scales

Three temperature scales are in general use:

1 Fahrenheit (F); water freezes at 32° F and boils at 212° F.

2 Celsius or centigrade* (C); water freezes at 0° C and boils at 100° C.

3 Kelvin or absolute (K); water freezes at 273° K and boils at 373° K.

All molecular motion ceases at $-459°F = -273°C = 0°K$. Thus Kelvin temperature is measured from this lowest possible temperature, called *absolute zero*. It is the temperature scale most often used in astronomy. Kelvin degrees have the same value as centigrade or Celsius degrees, since the difference between the freezing and boiling points of water is 100 degrees in each.

On the Fahrenheit scale, water boils at 212 degrees and freezes at 32 degrees; the difference is 180 degrees. Thus to convert Celsius or Kelvin degrees to Fahrenheit it is necessary to multiply by $180/100 = 9/5$. To convert from Fahrenheit to Celsius or Kelvin degrees, it is necessary to multiply by $100/180 = 5/9$.

Example 1: What is 68° F in Celsius and in Kelvin?

$$68° \text{ F} - 32° \text{ F} = 36° \text{ F above freezing.}$$

$$\frac{5}{9} \times 36° = 20°;$$

thus,

$$68° \text{ F} = 20° \text{ C} = 293° \text{ K}.$$

Example 2: What is 37° C in Fahrenheit and in Kelvin?

$$37° \text{ C} = 273° + 37° = 310° \text{ K};$$

$$\frac{9}{5} \times 37° = 66.6°;$$

thus,

$$37° \text{ C is } 66.6° \text{ F above freezing}$$

or

$$37° \text{ C} = 32° + 66.6° = 98.6° \text{ F}.$$

*Celsius is now the name used for centigrade temperature; it has a more modern standardization, but differs from the old centigrade scale by less than 0.1°.

APPENDIX 5

Mathematical, Physical, and Astronomical Constants

MATHEMATICAL CONSTANTS:

$$\pi = 3.1415926536$$

$$1 \text{ radian} = 57^{\circ}.2957795 = 3437'.74677 = 206264''.806$$

Number of square degrees on a sphere = 41 252.96124

PHYSICAL CONSTANTS:

velocity of light	$c = 2.997925 \times 10^{10}$ cm/s
constant of gravitation	$G = 6.668 \times 10^{-8}$ dyne $\cdot$ cm^2/g^2
Planck's constant	$h = 6.626 \times 10^{-27}$ erg $\cdot$ s
Boltzmann's constant	$k = 1.380 \times 10^{-16}$ erg/deg
mass of hydrogen atom	$m_H = 1.673 \times 10^{-24}$ g
mass of electron	$m_e = 9.1096 \times 10^{-28}$ g
charge on electron	$\epsilon = 4.803 \times 10^{-10}$ electrostatic units
Stefan-Boltzmann constant	$\sigma = 5.669 \times 10^{-5}$ erg/cm$^2 \cdot$ deg$^4 \cdot$ s
constant in Wien's law	$\lambda_{max}T = 0.28979$ cm/ deg
Rydberg's constant	$R = 1.09737 \times 10^5$ per cm
1 electron volt	eV $= 1.60207 \times 10^{-12}$ erg
1 angstrom	Å $= 10^{-8}$ cm

ASTRONOMICAL CONSTANTS:

astronomical unit	AU $= 1.49597892 \times 10^{13}$ cm
parsec	pc $= 206265$ AU
	$= 3.262$ LY
	$= 3.086 \times 10^{18}$ cm
light-year	LY $= 9.4605 \times 10^{17}$ cm
	$= 6.324 \times 10^4$ AU
tropical year	$= 365.242199$ ephemeris days
sidereal year	$= 365.256366$ ephemeris days
	$= 3.155815 \times 10^7$ s
mass of earth	$M_{\oplus} = 5.977 \times 10^{27}$ g
mass of sun	$M_{\odot} = 1.989 \times 10^{33}$ g
equatorial radius of earth	$R_{\oplus} = 6378$ km
radius of sun	$R_{\odot} = 6.960 \times 10^{10}$ cm
luminosity of sun	$L_{\odot} = 3.90 \times 10^{33}$ erg/s
solar constant	$S = 1.37 \times 10^6$ erg/cm$^2 \cdot$ s
obliquity of ecliptic (1900)	$\epsilon = 23^{\circ}27'8''.26$
direction of galactic center (1950)	$\alpha = 17^{h}42^{m}.4$
	$\delta = -28^{\circ}55'$
direction of north galactic pole (1950)	$\alpha = 12^{h}49^{m}$
	$\delta = +27^{\circ}.4$

APPENDIX 6

Orbital Data For the Planets

PLANET	SYMBOL	SEMIMAJOR AXIS AU	SEMIMAJOR AXIS 10^6 KM	SIDEREAL PERIOD TROPICAL YEARS	SIDEREAL PERIOD DAYS	SYNODIC PERIOD (DAYS)	MEAN ORBITAL SPEED (KM/S)	ORBITAL ECCENTRICITY	INCLINATION OF ORBIT TO ECLIPTIC
Mercury	☿	0.3871	57.9	0.24085	87.97	115.88	47.8	0.206	7.°004
Venus	♀	0.7233	108.2	0.61521	224.70	583.92	35.0	0.007	3.394
Earth	⊕	1.0000	149.6	1.000039	365.26	—	29.8	0.017	0.0
Mars	♂	1.5237	227.9	1.88089	686.98	779.94	24.2	0.093	1.850
(Ceres)	①	2.7673	414	4.604		466.6	17.9	0.077	10.615
Jupiter	♃	5.2028	778	11.86223		398.88	13.1	0.048	1.305
Saturn	♄	9.5388	1427	29.4577		378.09	9.7	0.056	2.490
Uranus	⛢ or ♅	19.182	2870	84.013		369.66	6.8	0.047	0.773
Neptune	♆	30.058	4497	164.793		367.49	5.4	0.009	1.774
Pluto	♇	39.5177	5912	248.43		366.74	4.7	0.249	17.170

For the mean equator and equinox of 1960.

APPENDIX 7

Physical Data For the Planets

PLANET	DIAMETER KM	DIAMETER EARTH = 1	MASS (EARTH = 1)	MEAN DENSITY (G/CM³)	PERIOD OF ROTATION	INCLINATION OF EQUATOR TO ORBIT	OBLATENESS	SURFACE GRAVITY (EARTH = 1)	ALBEDO	VISUAL MAGNITUDE AT MAXIMUM LIGHT	VELOCITY OF ESCAPE (KM/S)
Mercury	4,878	0.38	0.055	5.44	58.6 days	~0°	0	0.38	0.06	−1.9	4.3
Venus	12,112	0.95	0.82	5.3	242.9 days	~180°	0	0.91	0.76	−4.4	10.3
Earth	12,756	1.00	1.00	5.52	$23^h56^m04^s$	23°27′	1/298.2	1.00	0.39	—	11.2
Mars	6,800	0.53	0.107	3.9	$24^h37^m23^s$	24°	1/192	0.38	0.15	−2.8	5.1
Jupiter	143,000	11.2	317.9	1.3	9^h50^m to 9^h55^m	3°	1/15	2.64	0.51	−2.5	60
Saturn	121,000	9.49	95.2	0.7	10^h14^m to 10^h38^m	27°	1/9.5	1.13	0.50	−0.4	35
Uranus	47,000	3.69	14.6	1.6	$20^h - 25^h$	98°	1/14	1.07	0.66	+5.6	22
Neptune	45,000	3.50	17.2	2.3	$15^h - 20^h$	29°	1/40	1.41	0.62	+7.9	25
Pluto	<6,000	<0.47	0.11(?)	?	6.387 days		0	?	?	+14.9	?

APPENDIX 8

Satellites of Planets

PLANET	SATELLITE	DISCOVERED BY	MEAN DISTANCE FROM PLANET (KM)	SIDEREAL PERIOD (DAYS)	ORBITAL ECCEN-TRICITY	DIAMETER OF SATELLITE (KM)*	MASS (PLANET = 1)†	APPROXIMATE MAGNITUDE AT OPPOSITION
Earth	Moon	—	384,404	27.322	0.055	3476	0.0123	−12.5
Mars	Phobos	A. Hall (1877)	9,380	0.319	0.021	25	(2.7×10^{-8})	+12
	Deimos	A. Hall (1877)	23,500	1.262	0.003	13	4.8×10^{-9}	13
Jupiter	V	Barnard (1892)	180,500	0.498	0.003	(150)	(2×10^{-9})	13
	I Io	Galileo (1610)	421,600	1.769	0.000	3640	4×10^{-5}	5
	II Europa	Galileo (1610)	670,800	3.551	0.000	3050	2.5×10^{-5}	6
	III Ganymede	Galileo (1610)	1,070,000	7.155	0.002	5270	8×10^{-5}	5
	IV Callisto	Galileo (1610)	1,882,000	16.689	0.008	5000	5×10^{-5}	6
	VI	Perrine (1904)	11,470,000	250.57	0.158	(120)	(8×10^{-10})	14
	VII	Perrine (1905)	11,800,000	259.65	0.207	(40)	(4×10^{-11})	18
	X	Nicholson (1938)	11,850,000	263.55	0.130	(10)	(1×10^{-12})	19
	XIII	Kowal (1974)	11,110,000	239.2	0.147	(8)	(5×10^{-13})	20
	XII	Nicholson (1951)	21,200,000	631.1	0.169	(10)	(7×10^{-13})	18
	XI	Nicholson (1938)	22,600,000	692.5	0.207	(15)	(2×10^{-12})	19
	VIII	Melotte (1908)	23,500,000	738.9	0.378	(25)	(8×10^{-12})	17
	IX	Nicholson (1914)	23,700,000	758	0.275	(15)	(2×10^{-12})	19
Saturn	Janus	A. Dollfus (1966)	157,500	0.749		(350)	(3×10^{-8})	14
	Mimas	W. Herschel (1789)	185,400	0.942	0.020	(500)	6.6×10^{-8}	12
	Enceladus	W. Herschel (1789)	237,900	1.370	0.004	(500)	1.5×10^{-7}	12
	Tethys	Cassini (1684)	294,500	1.888	0.000	(1000)	1.1×10^{-6}	11
	Dione	Cassini (1684)	377,200	2.737	0.002	(1000)	2×10^{-6}	11
	Rhea	Cassini (1672)	526,700	4.518	0.001	1600	3×10^{-6}	10
	Titan	Huygens (1655)	1,221,000	15.945	0.029	5800	2.5×10^{-4}	8
	Hyperion	Bond (1848)	1,479,300	21.277	0.104	(400)	2×10^{-7}	14
	Iapetus	Cassini (1671)	3,558,400	79.331	0.028	(1200)	4×10^{-6}	11
	Phoebe	W. Pickering (1898)	12,945,500	550.45	0.163	(300)	5×10^{-8}	14
Uranus	Miranda	Kuiper (1948)	123,000	1.414	0	(200)	1×10^{-6}	17
	Ariel	Lassell (1851)	191,700	2.520	0.003	(600)	1.5×10^{-5}	15
	Umbriel	Lassell (1851)	267,000	4.144	0.004	(400)	6×10^{-6}	15
	Titania	W. Herschel (1787)	438,000	8.706	0.002	(1000)	5×10^{-5}	14
	Oberon	W. Herschel (1787)	585,960	13.463	0.001	(900)	3×10^{-5}	14
Neptune	Triton	Lassell (1846)	353,400	5.877	0.000	6000	3×10^{-3}	14
	Nereid	Kuiper (1949)	5,560,000	359.881	0.749	(500)	(10^{-6})	19

*A diameter of a satellite given in parentheses is estimated from the amount of sunlight it reflects.

†A mass of a satellite given in parentheses is estimated from its size and an assumed density.

APPENDIX 9

Important Total Solar Eclipses in the Second Half of the Twentieth Century

DATE	*DURATION OF TOTALITY (MIN)*	*WHERE VISIBLE*
1952 Feb. 25	3.0	Africa, Asia
1954 June 30	2.5	North-Central U.S. (Great Lakes), Canada, Scandinavia, U.S.S.R., Central Asia
1955 June 20	7.2	Southeast Asia
1958 Oct. 12	5.2	Pacific, Chile, Argentina
1959 Oct. 2	3.0	Northern and Central Africa
1961 Feb. 15	2.6	Southern Europe
1962 Feb. 5	4.1	Indonesia
1963 July 20	1.7	Japan, Alaska, Canada, Maine
1965 May 30	5.3	Pacific Ocean, Peru
1966 Nov. 12	1.9	South America
1970 March 7	3.3	Mexico, Florida, parts of U.S. Atlantic coastline
1972 July 10	2.7	Alaska, Northern Canada
1973 June 30	7.2	Atlantic Ocean, Africa
1974 June 20	5.3	Indian Ocean, Australia
1976 Oct. 23	4.9	Africa, Indian Ocean, Australia
1977 Oct. 12	2.8	Northern South America
1979 Feb. 26	2.7	Northwest U.S., Canada
1980 Feb. 16	4.3	Central Africa, India
1981 July 31	2.2	Siberia
1983 June 11	5.4	Indonesia
1984 Nov. 22	2.1	Indonesia, South America
1987 March 29	0.3	Central Africa
1988 March 18	4.0	Philippines, Indonesia
1990 July 22	2.6	Finland, Arctic Regions
1991 July 11	7.1	Hawaii, Central America, Brazil
1992 June 30	5.4	South Atlantic
1994 Nov. 3	4.6	South America
1995 Oct. 24	2.4	South Asia
1997 March 9	2.8	Siberia, Arctic
1998 Feb. 26	4.4	Central America
1999 Aug. 11	2.6	Central Europe, Central Asia

APPENDIX 10 *The Nearest Stars*

STAR	RIGHT ASCENSION (1980)		DECLINATION (1980)		DISTANCE (PC)	PROPER MOTION	RADIAL VELOCITY (KM/S)	SPECTRA OF COMPONENTS			VISUAL MAGNITUDES OF COMPONENTS			ABSOLUTE VISUAL MAGNITUDES OF COMPONENTS		
								A	B	C	A	B	C	A	B	C
	h	m	°	′		″										
Proxima Centauri	14	28.6	−62	36	1.31	3.86	−16	M5V			+11.05			+15.4		
α Centauri	14	38.5	−60	46	1.35	3.68	−22	G2V	K0V		−0.01	+1.33		+4.4	+5.7	
Barnard's Star	17	56.9	+4	33	1.81	10.34	−108	M5V			+9.54			+13.2		
Wolf 359	10	55.7	+7	09	2.35	4.70	+13	M8V			+13.53			+16.7		
Lalande 21185	10	02.4	+36	09	2.52	4.78	−84	M2V			+7.50			+10.5		
Sirius	6	44.2	−16	41	2.65	1.33	−8	A1V	wd		−1.46	+8.68		+1.4	+11.6	
Luyten 726-8	1	37.8	−18	04	2.72	3.36	+29	M5.5V	M5.5V		+12.45	+12.95		+15.3	+15.8	
Ross 154	18	48.6	−23	51	2.90	0.72	−4	M4.5V			+10.6			+13.3		
Ross 248	23	40.9	+44	05	3.14	1.60	−81	M6V			+12.29			+14.8		
Luyten 789-6	22	37.4	−15	27	3.28	3.26	−60	M7V			+12.18			+14.6		
ε Eridani	3	32.0	−9	32	3.31	0.98	+16	K2V			+3.73			+6.1		
Ross 128	11	46.7	+0	56	3.32	1.38	−13	M5V			+11.10			+13.5		
61 Cygni	21	05.9	+38	37	3.38	5.22	−64	K5V	K7V		+5.22	+6.03		+7.6	+8.4	
ε Indi	22	01.6	−56	51	3.44	4.33	−40	K5V			+4.68			+7.0		
Procyon	7	38.3	+5	17	3.51	1.25	−3	F5IV-V	wd		+0.37	+10.7		+2.6	+13.0	
BD+43°44	0	17.1	+43	54	3.55	2.90	+13	M1V	M6V		+8.07	+11.04		+10.3	+13.3	
BD+59°1915	18	42.6	+59	35	3.55	2.29	+10	M4V	M5V		+8.90	+9.69		+11.2	+11.9	
CD−36°15693	23	04.3	−35	58	3.58	6.90	+10	M2V			+7.36			+9.6		
τ Ceti	1	43.3	−16	03	3.61	1.92	−16	G8V			+3.50			+5.7		
BD+5°1668	7	26.3	+5	19	3.70	3.77	+26	M5V			+9.82			+12.0		
CD−39°14192	21	16.2	−38	56	3.85	3.46	+21	M0V			+6.67			+8.8		
Kapteyn's Star	5	10.6	−44	58	3.91	8.72	+245	M0V			+8.81			+10.8		
Kruger 60	22	27.3	+57	36	3.95	0.86	−26	M3V	M4.5V		+9.85	+11.3		+11.9	+13.3	
Ross 614	6	28.3	−2	47	3.97	1.00	+24	M7V	?		+11.07	+14.8		+13.1	+16.8	
BD−12°4523	16	29.2	−12	36	4.02	1.18	−13	M5V			+10.12			+12.1		
van Maanen's Star	0	48.0	+5	19	4.18	2.99	+54	wd			+12.37			+14.3		
Wolf 424	12	32.3	+9	08	4.33	1.76	−5	M5.5V	M6V		+13.16	+13.4		+15.0	+15.2	
CD−37°15492	0	04.0	−37	26	4.44	6.11	+23	M4V			+8.63			+10.4		
BD+50°1725	10	10.2	+49	33	4.50	1.45	−26	K7V			+6.59			+8.3		
CD−46°11540	17	27.1	−46	52	4.63	1.06	–	M4V			+9.36			+11.0		
CD−49°13515	21	32.2	−49	05	4.67	0.81	+8	M1V			+8.67			+10.3		
Luyten 1159-16	1	59.1	+12	59	4.69	2.09	–	M8V			+12.27			+13.9		
CD−44°11909	17	35.7	−44	18	4.69	1.16	–	M5V			+11.2			+12.8		
BD+68°946	17	36.6	+68	22	4.69	1.31	−22	M3.5V			+9.15			+10.8		
Ross 780	22	52.2	−14	21	4.78	1.14	+9	M5V			+10.17			+11.8		
Luyten 145-141	11	44.4	−64	44	4.85	2.68	–	wd			+11.44			+13.0		
40 Eridani	4	14.6	−7	40	4.88	4.08	−42	K1V	wd	M4.5V	+4.43	+9.53	+11.17	+6.0	+11.1	+12.7
BD+20°2465	10	18.5	+19	58	4.90	0.49	+11	M4.5V			+9.43			+11.0		
Lalande 25372	13	44.7	+15	01	4.98	2.30	+15	M4V			+8.50			+10.0		

APPENDIX 11

The Twenty Brightest Stars

STAR	RIGHT ASCENSION (1980)		DECLI-NATION (1980)		DIS-TANCE (PC)†	PROPER MOTION	SPECTRA OF COMPONENTS			VISUAL MAGNITUDES OF COMPONENTS			ABSOLUTE VISUAL MAGNITUDES OF COMPONENTS		
							A	B	C	A	B	C	A	B	C
	h	m	°	′		″									
Sirius	6	44.2	−16	41	2.7	1.33	A1V	wd		−1.46	+8.7		+1.4	+11.6	
Canopus	6	23.5	−52	41	30	0.02	F01b-II			−0.72			−3.1		
α Centauri	14	38.5	−60	46	1.3	3.68	G2V	K0V		−0.01	+1.3		+4.4	+5.7	
Arcturus	14	14.8	+19	19	11	2.28	K2IIIp			−0.06			−0.3		
Vega	18	36.2	+38	46	8.0	0.34	A0V			+0.04			+0.5		
Capella	5	15.2	+45	59	14	0.44	GIII	M1V	M5V	+0.05	+10.2	+13.7	−0.7	+9.5	+13
Rigel	5	13.5	−8	13	250	0.00	B8 Ia	B9		+0.14	+6.6		−6.8	−0.4	
Procyon	7	38.3	+5	17	3.5	1.25	F5IV-V	wd		+0.37	+10.7		+2.6	+13.0	
Betelgeuse	5	54.1	+7	24	150	0.03	M2Iab			+0.41v			−5.5		
Achernar	1	37.0	−57	20	20	0.10	B5V			+0.51			−1.0		
β Centauri	14	02.4	−60	17	90	0.04	B1III	?		+0.63	+4		−4.1	−0.8	
Altair	19	49.7	+8	49	5.1	0.66	A7IV-V			+0.77			+2.2		
α Crucis	12	25.5	−62	59	120	0.04	B1IV	B3		+1.39	+1.9		−4.0	−3.5	
Aldebaran	4	34.7	+16	29	16	0.20	K5III	M2V		+0.86	+13		−0.2	+12	
Spica	13	24.2	−11	03	80	0.05	B1V			+0.91v			−3.6		
Antares	16	28.1	−26	23	120	0.03	M1Ib	B4eV		+0.92v	+5.1		−4.5	−0.3	
Pollux	7	44.2	+28	05	12	0.62	K0III			+1.16			+0.8		
Fomalhaut	22	56.5	−29	43	7.0	0.37	A3V	K4V		+1.19	+6.5		+2.0	+7.3	
Deneb	20	40.7	+45	12	430	0.00	A2Ia			+1.26			−6.9		
β Crucis	12	46.6	−59	34	150	0.05	B0.5IV			+1.28v			−4.6		

† Distances of the more remote stars have been estimated from their spectral types and apparent magnitudes, and are only approximate.

Note: Several of the components listed are themselves spectroscopic binaries. A "v" after a magnitude denotes that the star is variable, in which case the magnitude at median light is given. A "p" after a spectral type indicates that the spectrum is peculiar. An "e" after a spectral type indicates that emission lines are present. When the luminosity classification is rather uncertain a range is given.

APPENDIX 12

Pulsating Variable Stars

TYPE OF VARIABLE	SPECTRA	PERIOD (DAYS)	MEDIAN MAGNITUDE (ABSOLUTE)	AMPLITUDE (MAGNITUDES)	DESCRIPTION	EXAMPLE	NUMBER KNOWN IN GALAXY†
Cepheids (type I)	F to G supergiants	3 to 50	−1.5 to −5	0.1 to 2	Regular pulsation; period-luminosity relation exists	δ Cep	706
Cepheids (type II)	F to G supergiants	5 to 30	0 to −3.5	0.1 to 2	Regular pulsation; period-luminosity relation exists	W Vir	(About 50; included with type I)
RV Tauri	G to K yellow and red bright giants	30 to 150	−2 to −3	Up to 3	Alternate large and small maxima	RV Tau	104
Long-period (Mira-type)	M red giants	80 to 600	+2 to −2	>2.5	Brighten more or less periodically	o Cet (Mira)	4566
Semiregular	M giants and supergiants	30 to 2000	0 to −3	1 to 2	Periodicity not dependable; often interrupted	α Ori	2221
Irregular	All types	Irregular	<0	Up to several magnitudes	No known periodicity; many may be semiregular, but too little data exist to classify them as such	π Gru	1687
RR Lyrae or cluster-type	A to F blue giants	<1	0 to +1	<1 to 2	Very regular pulsations	RR Lyr	4433
β Cephei or β Canis Majoris	B blue giants	0.1 to 0.3	−2 to −4	0.1	Maximum light occurs at time of highest compression	β Cep	23
δ Scuti	F subgiants	<1	0 to +2	<0.25	Similar to, and possibly related to, RR Lyrae variables	δ Sct	17
Spectrum variables	A main sequence	1 to 25	0 to +1	0.1	Anomalously intense lines of Si, Sr, and Cr vary in intensity with same period as light; most have strong variable magnetic fields	α^2C Vn	28

†According to the 1968 edition of the Soviet *General Catalogue of Variable Stars.*

APPENDIX 13

Eruptive Variable Stars

TYPE OF VARIABLE	SPECTRA	DURATION OF INCREASED BRIGHTNESS	NORMAL ABSOLUTE MAGNITUDE	AMPLITUDE (MAGNITUDES)	DESCRIPTION	EXAMPLE	NUMBER KNOWN IN GALAXY†
Novae	O to A hot subdwarfs	Months to years	>0	7 to 16	Rapid rise to maximum; slow decline; ejection of gas shell	GK Per	166
Novalike variables or P Cygni stars	Hot B stars	Erratic	−3 to −6	Several magnitudes	Slow, erratic, and novalike variations in light; may be unrelated to novae. Gas shell ejected	P Cyg	39
Supernovae	?	Months to years	?	15 or more	Sudden, violent flareup, followed by decline and ejection of gas shell	CM Tau (Crab nebula)	7
R Coronae Borealis	F to K supergiant	10 to several hundred days	−5	1 to 9	Sudden and irregular drops in brightness. Low in hydrogen abundance, but high in carbon abundance	R CrB	32
T Tauri or RW Aurigae	B to M main sequence and subgiants	Rapid and erratic	0 to +8	Up to a few magnitudes	Rapid and irregular light variations. Generally associated with interstellar material. Subtypes from G to M are called T Tauri variables	RW Aur, T Tau	1109
U Geminorum or SS Cygni or "dwarf novae"	A to F hot subdwarfs	Few days to few weeks	>0	2 to 6	Novalike outbursts at mean intervals which range from 20 to 600 days. Those with longer intervals between outbursts tend to have greater amplitudes. many, if not all, are members of binary-star systems	SS Cyg, U Gem	215
Flare stars	M main sequence	Few minutes	>8	Up to 6	Sudden flareups in light; probably localized flares on surface of star	UV Cet	28
Z Camelopardalis variables	A to F hot subdwarfs	Few days	>0	2 to 5	Similar to U Geminorium, except that variations are sometimes interrupted by constant light for several cycles. Intervals between outbursts normally range from 10 to 40 days.	Z Cam	20

†According to the 1968 edition of the Soviet *General Catalogue of Variable Stars.*

APPENDIX 14

The Local Group

The table lists the galaxies that are generally agreed to be members of the Local Group. Uncertain data are in parentheses.

GALAXY	*TYPE*	*RIGHT ASCENSION (1980)*		*DECLINATION (1980)*		*VISUAL MAGNITUDE* (m_v)	*DISTANCE*		*DIAMETER*		*ABSOLUTE MAGNITUDE* (M_v)	*RADIAL VELOCITY (KM/S)*	*MASS (SOLAR MASSES)*
							(KPC)	*(1000 LY)*	*(KPC)*	*(1000 LY)*			
Our Galaxy	Sb	—	—	—		——	—	——	30	100	(−21)	—	2×10^{11}
Large Magellanic Cloud	Irr I	5^h	26^m	−69°		0.9	48	160	10	30	−17.7	+276	2.5×10^{10}
Small Magellanic Cloud	Irr I	0	51	−73		2.5	56	180	8	25	−16.5	+168	
Ursa Minor system	E4 (dwarf)	15	8.6	+67	11′		70	220	1	3	(−9)		
Sculptor system	E3 (dwarf)	0	58.9	−33	48	8.0	83	270	2.2	7	−11.8		(2 to 4 × 10^6)
Draco system	E2 (dwarf)	17	19.9	+57	56		100	330	1.4	4.5	(−10)		
Carina system	E3 (dwarf)	6	41.1	−50	57		(170)	(550)	1.5	4.8	(−10)		
Fornax system	E3 (dwarf)	2	38.9	−34	36	8.4	250	800	4.5	15	−13.6	+39	(1.2 to 2 × 10^7)
Leo II system	E0 (dwarf)	11	12.4	+22	16		230	750	1.6	5.2	−10.0		(1.1 × 10^6)
Leo I system	E4 (dwarf)	10	7.4	+12	24	12.0	280	900	1.5	5	−10.4		
NGC 6822	Irr I	19	43.8	−14	50	8.9	460	1500	2.7	9	−14.8	−32	
NGC 147	E6	0	32.0	+48	23	9.73	570	1900	3	10	−14.5		
NGC 185	E2	0	37.8	+48	14	9.43	570	1900	2.3	8	−14.8	−305	
NGC 205	E5	0	39.2	+41	35	8.17	680	2200	5	16	−16.5	−239	
NGC 221 (M32)	E3	0	41.6	+40	46	8.16	680	2200	2.4	8	−16.5	−214	
IC 1613	Irr I	1	2.1	+1	51	9.61	680	2200	5	16	−14.7	−238	
Andromeda galaxy (NGC 224; M31)	Sb	0	41.6	+41	10	3.47	680	2200	40	130	−21.2	−266	3×10^{11}
And I	E0 (dwarf)	0	44.6	+37	54	(14)	(680)	(2200)	0.5	1.6	(−11)		
And II	E0 (dwarf)	1	15.2	+33	18	(14)	(680)	(2200)	0.7	2.3	(−11)		
And III	E3 (dwarf)	0	34.2	+36	24	(14)	(680)	(2200)	0.9	0.9	(−11)		
NGC 598 (M33)	Sc	1	32.7	+30	33	5.79	720	2300	17	60	−18.9	−189	8×10^9

APPENDIX 15 The Messier Catalogue of Nebulae and Star Clusters

M	NGC OR (IC)	RIGHT ASCENSION (1980)		DECLI-NATION (1980)		APPARENT VISUAL MAGNITUDE	DESCRIPTION
		h	m	°	′		
1	1952	5	33.3	+22	01	8.4	"Crab" nebula in Taurus; remains of SN 1054
2	7089	21	32.4	−0	54	6.4	Globular cluster in Aquarius
3	5272	13	41.2	+28	29	6.3	Globular cluster in Canes Venatici
4	6121	16	22.4	−26	28	6.5	Globular cluster in Scorpio
5	5904	15	17.5	+2	10	6.1	Globular cluster in Serpens
6	6405	17	38.8	−32	11	5.5	Open cluster in Scorpio
7	6475	17	52.7	−34	48	3.3	Open cluster in Scorpio
8	6523	18	02.4	−24	23	5.1	"Lagoon" nebula in Sagittarius
9	6333	17	18.1	−18	30	8.0	Globular cluster in Ophiuchus
10	6254	16	56.1	−4	05	6.7	Globular cluster in Ophiuchus
11	6705	18	50.0	−6	18	6.8	Open cluster in Scutum Sobieskii
12	6218	16	46.3	−1	55	6.6	Globular cluster in Ophiuchus
13	6205	16	41.0	+36	30	5.9	Globular cluster in Hercules
14	6402	17	36.6	−3	14	8.0	Globular cluster in Ophiuchus
15	7078	21	28.9	+12	05	6.4	Globular cluster in Pegasus
16	6611	18	17.8	−13	47	6.6	Open cluster with nebulosity in Serpens
17	6618	18	19.6	−16	11	7.5	"Swan" or "Omega" nebula in Sagittarius
18	6613	18	18.7	−17	08	7.2	Open cluster in Sagittarius
19	6273	17	01.4	−26	14	6.9	Globular cluster in Ophiuchus
20	6514	18	01.2	−23	02	8.5	"Trifid" nebula in Sagittarius
21	6531	18	03.4	−22	30	6.5	Open cluster in Sagittarius
22	6656	18	35.2	−23	56	5.6	Globular cluster in Sagittarius
23	6494	17	55.8	−19	00	5.9	Open cluster in Sagittarius
24	6603	18	17.3	−18	26	4.6	Open cluster in Sagittarius
25	(4725)	18	30.5	−19	16	6.2	Open cluster in Sagittarius
26	6694	18	44.1	−9	25	9.3	Open cluster in Scutum Sobieskii
27	6853	19	58.8	+22	40	8.2	"Dumbbell" planetary nebula in Vulpecula
28	6626	18	23.2	−24	52	7.6	Globular cluster in Sagittarius
29	6913	20	23.3	+38	27	8.0	Open cluster in Cygnus
30	7099	21	39.2	−23	16	7.7	Globular cluster in Capricornus

M	*NGC OR (IC)*	*RIGHT ASCENSION (1980)*		*DECLI-NATION (1980)*		*APPARENT VISUAL MAGNITUDE*	*DESCRIPTION*
		h	m	°	′		
31	224	0	41.6	+41	10	3.5	Andromeda galaxy
32	221	0	41.6	+40	46	8.2	Elliptical galaxy; companion to M31
33	598	1	32.7	+30	33	5.8	Spiral galaxy in Triangulum
34	1039	2	40.7	+42	43	5.8	Open cluster in Perseus
35	2168	6	07.5	+24	21	5.6	Open cluster in Gemini
36	1960	5	35.0	+34	05	6.5	Open cluster in Auriga
37	2099	5	51.1	+32	33	6.2	Open cluster in Auriga
38	1912	5	27.3	+35	48	7.0	Open cluster in Auriga
39	7092	21	31.5	+48	21	5.3	Open cluster in Cygnus
40		12	21	+59			Close double star in Ursa Major
41	2287	6	46.2	−20	43	5.0	Loose open cluster in Canis Major
42	1976	5	34.4	−5	24	4	Orion nebula
43	1982	5	34.6	−5	18	9	Northeast portion of Orion nebula
44	2632	8	39	+20	04	3.9	Praesepe; open cluster in Cancer
45		3	46.3	+24	03	1.6	The Pleiades; open cluster in Taurus
46	2437	7	40.9	−14	46	6.6	Open cluster in Puppis
47	2422	7	35.7	−14	26	5	Loose group of stars in Puppis
48	2548	8	12.8	−5	44	6	"Cluster of very small stars"; identifiable
49	4472	12	28.8	+8	06	8.5	Elliptical galaxy in Virgo
50	2323	7	02.0	−8	19	6.3	Loose open cluster in Monoceros
51	5194	13	29.1	+47	18	8.4	"Whirlpool" spiral galaxy in Canes Venatici
52	7654	23	23.3	+61	30	8.2	Loose open cluster in Cassiopeia
53	5024	13	12.0	+18	16	7.8	Globular cluster in Coma Berenices
54	6715	18	53.8	−30	30	7.8	Globular cluster in Sagittarius
55	6809	19	38.7	−30	59	6.2	Globular cluster in Sagittarius
56	6779	19	15.8	+30	08	8.7	Globular cluster in Lyra
57	6720	18	52.8	+33	00	9.0	"Ring" nebula; planetary nebula in Lyra
58	4579	12	36.7	+11	55	9.9	Spiral galaxy in Virgo
59	4621	12	41.0	+11	46	10.0	Spiral galaxy in Virgo
60	4649	12	42.6	+11	40	9.0	Elliptical galaxy in Virgo
61	4303	12	20.8	+4	35	9.6	Spiral galaxy in Virgo
62	6266	16	59.9	−30	05	6.6	Globular cluster in Scorpio
63	5055	13	14.8	+42	07	8.9	Spiral galaxy in Canes Venatici
64	4826	12	55.7	+21	39	8.5	Spiral galaxy in Coma Berenices
65	3623	11	17.9	+13	12	9.4	Spiral galaxy in Leo
66	3627	11	19.2	+13	06	9.0	Spiral galaxy in Leo; companion to M65
67	2682	8	50.0	+11	53	6.1	Open cluster in Cancer
68	4590	12	38.4	−26	39	8.2	Globular cluster in Hydra
69	6637	18	30.1	−32	23	8.0	Globular cluster in Sagittarius
70	6681	18	42.0	−32	18	8.1	Globular cluster in Sagittarius

M	NGC OR (IC)	RIGHT ASCENSION (1980)		DECLI-NATION (1980)		APPARENT VISUAL MAGNITUDE	DESCRIPTION
		h	m	°	′		
71	6838	19	52.8	+18	44	7.6	Globular cluster in Sagitta
72	6981	20	52.3	−12	38	9.3	Globular cluster in Aquarius
73	6994	20	57.8	−12	43	9.1	Open cluster in Aquarius
74	628	1	35.6	+15	41	9.3	Spiral galaxy in Pisces
75	6864	20	04.9	−21	59	8.6	Globular cluster in Sagittarius
76	650	1	41.0	+51	28	11.4	Planetary nebula in Perseus
77	1068	2	41.6	−0	04	8.9	Spiral galaxy in Cetus
78	2068	5	45.7	0	03	8.3	Small emission nebula in Orion
79	1904	5	23.3	−24	32	7.5	Globular cluster in Lepus
80	6093	16	15.8	−22	56	7.5	Globular cluster in Scorpio
81	3031	9	54.2	+69	09	7.0	Spiral galaxy in Ursa Major
82	3034	9	54.4	+69	47	8.4	Irregular galaxy in Ursa Major
83	5236	13	35.4	−29	31	7.6	Spiral galaxy in Hydra
84	4374	12	24.1	+13	00	9.4	Elliptical galaxy in Virgo
85	4382	12	24.3	+18	18	9.3	Elliptical galaxy in Coma Berenices
86	4406	12	25.1	+13	03	9.2	Elliptical galaxy in Virgo
87	4486	12	29.7	+12	30	8.7	Elliptical galaxy in Virgo
88	4501	12	30.9	+14	32	9.5	Spiral galaxy in Coma Berenices
89	4552	12	34.6	+12	40	10.3	Elliptical galaxy in Virgo
90	4569	12	35.8	+13	16	9.6	Spiral galaxy in Virgo
91	omitted						
92	6341	17	16.5	+43	10	6.4	Globular cluster in Hercules
93	2447	7	43.7	−23	49	6.5	Open cluster in Puppis
94	4736	12	50.0	+41	14	8.3	Spiral galaxy in Canes Venatici
95	3351	10	42.9	+11	49	9.8	Barred spiral galaxy in Leo
96	3368	10	45.7	+11	56	9.3	Spiral galaxy in Leo
97	3587	11	13.7	+55	07	11.1	"Owl" nebula; planetary nebula in Ursa Major
98	4192	12	12.7	+15	01	10.2	Spiral galaxy in Coma Berenices
99	4254	12	17.8	+14	32	9.9	Spiral galaxy in Coma Berenices
100	4321	12	21.9	+15	56	9.4	Spiral galaxy in Coma Berenices
101	5457	14	02.5	+54	27	7.9	Spiral galaxy in Ursa Major
102	5866(?)	15	05.9	+55	50	10.5	Spiral galaxy (identification as M102 in doubt)
103	581	1	31.9	+60	35	6.9	Open cluster in Cassiopeia
104*	4594	12	39.0	−11	31	8.3	Spiral galaxy in Virgo
105*	3379	10	46.8	+12	51	9.7	Elliptical galaxy in Leo
106*	4258	12	18.0	+47	25	8.4	Spiral galaxy in Canes Venatici
107*	6171	16	31.4	−13	01	9.2	Globular cluster in Ophiuchus
108*	3556	11	10.5	+55	47	10.5	Spiral galaxy in Ursa Major
109*	3992	11	56.6	+53	29	10.0	Spiral galaxy in Ursa Major
110*	205	0	39.2	+41	35	9.4	Elliptical galaxy (companion to M31)

* Not in Messier's original (1781) list; added later by others.

APPENDIX 16

The Chemical Elements

ELEMENT	SYMBOL	ATOMIC NUMBER	ATOMIC WEIGHT† (CHEMICAL SCALE)	NUMBER OF ATOMS PER 10^{12} HYDROGEN ATOMS‡
Hydrogen	H	1	1.0080	1×10^{12}
Helium	He	2	4.003	8×10^{10}
Lithium	Li	3	6.940	<3
Beryllium	Be	4	9.013	1.2×10^{1}
Boron	B	5	10.82	$<1.6 \times 10^{2}$
Carbon	C	6	12.011	4.2×10^{8}
Nitrogen	N	7	14.008	8.7×10^{7}
Oxygen	O	8	16.0000	6.9×10^{8}
Fluorine	F	9	19.00	3.6×10^{4}
Neon	Ne	10	20.183	3.7×10^{7}
Sodium	Na	11	22.991	1.9×10^{6}
Magnesium	Mg	12	24.32	3.2×10^{7}
Aluminum	Al	13	26.98	3.3×10^{6}
Silicon	Si	14	28.09	4.0×10^{7}
Phosphorus	P	15	30.975	3.9×10^{5}
Sulfur	S	16	32.066	1.6×10^{7}
Chlorine	Cl	17	35.457	2.2×10^{5}
Argon	Ar(A)	18	39.944	1.0×10^{6}
Potassium	K	19	39.100	1.2×10^{5}
Calcium	Ca	20	40.08	2.5×10^{6}
Scandium	Sc	21	44.96	1.1×10^{3}
Titanium	Ti	22	47.90	5.6×10^{4}
Vanadium	V	23	50.95	1.0×10^{4}
Chromium	Cr	24	52.01	6.9×10^{5}
Manganese	Mn	25	54.94	2.6×10^{5}
Iron	Fe	26	55.85	2.5×10^{7}
Cobalt	Co	27	58.94	3.2×10^{4}
Nickel	Ni	28	58.71	2.1×10^{6}
Copper	Cu	29	63.54	1.1×10^{4}
Zinc	Zn	30	65.38	2.8×10^{4}
Gallium	Ga	31	69.72	6.3×10^{2}
Germanium	Ge	32	72.60	3.2×10^{3}
Arsenic	As	33	74.91	2.6×10^{2}
Selenium	Se	34	78.96	2.7×10^{3}
Bromine	Br	35	79.916	5.4×10^{2}
Krypton	Kr	36	83.80	1.9×10^{3}
Rubidium	Rb	37	85.48	4.1×10^{2}
Strontium	Sr	38	87.63	7.6×10^{2}
Yttrium	Y	39	88.92	2.1×10^{2}
Zirconium	Zr	40	91.22	6.0×10^{2}
Niobium (Columbium)	Nb(Cb)	41	92.91	2.0×10^{2}
Molybdenum	Mo	42	95.95	1.5×10^{2}
Technetium	Tc(Ma)	43	(99)	—
Ruthenium	Ru	44	101.1	66
Rhodium	Rh	45	102.91	26
Palladium	Pd	46	106.4	20
Silver	Ag	47	107.880	7

ELEMENT	SYMBOL	ATOMIC NUMBER	ATOMIC WEIGHT† (CHEMICAL SCALE)	NUMBER OF ATOMS PER 10^{12} HYDROGEN ATOMS‡
Cadmium	Cd	48	112.41	72
Indium	In	49	114.82	40
Tin	Sn	50	118.70	25
Antimony	Sb	51	121.76	8
Tellurium	Te	52	127.61	2.6×10^2
Iodine	I(J)	53	126.91	44
Xenon	Xe(X)	54	131.30	2.14×10^2
Cesium	Cs	55	132.91	16
Barium	Ba	56	137.36	1.3×10^2
Lanthanum	La	57	138.92	66
Cerium	Ce	58	140.13	76
Praseodymium	Pr	59	140.92	35
Neodymium	Nd	60	144.27	71
Promethium	Pm	61	(147)	—
Samarium	Sm(Sa)	62	150.35	63
Europium	Eu	63	152.0	5
Gadolinium	Gd	64	157.26	13
Terbium	Tb	65	158.93	2
Dysprosium	Dy(Ds)	66	162.51	13
Holmium	Ho	67	164.94	3
Erbium	Er	68	167.27	7
Thulium	Tm(Tu)	69	168.94	4
Ytterbium	Yb	70	173.04	6
Lutecium	Lu(Cp)	71	174.99	7
Hafnium	Hf	72	178.50	8
Tantalum	Ta	73	180.95	1
Tungsten	W	74	183.86	3×10^2
Rhenium	Re	75	186.22	2
Osmium	Os	76	190.2	6
Iridium	Ir	77	192.2	1.6×10^2
Platinum	Pt	78	195.09	1×10^2
Gold	Au	79	197.0	5
Mercury	Hg	80	200.61	$<10^2$
Thallium	Tl	81	204.39	8
Lead	Pb	82	207.21	71
Bismuth	Bi	83	209.00	<80
Polonium	Po	84	(209)	—
Astatine	At	85	(210)	—
Radon	Rn	86	(222)	—
Francium	Fr(Fa)	87	(223)	—
Radium	Ra	88	226.05	—
Actinium	Ac	89	(227)	—
Thorium	Th	90	232.12	7
Protoactinium	Pa	91	(231)	—
Uranium	U(Ur)	92	238.07	<4
Neptunium	Np	93	(237)	—
Plutonium	Pu	94	(244)	—
Americium	Am	95	(243)	—
Curium	Cm	96	(248)	—
Berkelium	Bk	97	(247)	—
Californium	Cf	98	(251)	—
Einsteinium	E	99	(254)	—
Fermium	Fm	100	(253)	—
Mendeleevium	Mv	101	(256)	—
Nobelium	No	102	(253)	—

†Where mean atomic weights have not been well determined, the atomic mass numbers of the most stable isotopes are given in parentheses.

‡Provided by L. H. Aller.

APPENDIX 17 *The Constellations*

CONSTELLATION (LATIN NAME)	GENITIVE CASE ENDING	ENGLISH NAME OR DESCRIPTION	ABBRE-VIA-TION	APPROXIMATE POSITION α	δ
				h	°
Andromeda	Andromedae	Princess of Ethiopia	And	1	+40
Antlia	Antliae	Air pump	Ant	10	−35
Apus	Apodis	Bird of Paradise	Aps	16	−75
Aquarius	Aquarii	Water bearer	Aqr	23	−15
Aquila	Aquilae	Eagle	Aql	20	+5
Ara	Arae	Altar	Ara	17	−55
Aries	Arietis	Ram	Ari	3	+20
Auriga	Aurigae	Charioteer	Aur	6	+40
Boötes	Boötis	Herdsman	Boo	15	+30
Caelum	Caeli	Graving tool	Cae	5	−40
Camelopardus	Camelopardis	Giraffe	Cam	6	+70
Cancer	Cancri	Crab	Cnc	9	+20
Canes Venatici	Canum Venaticorum	Hunting dogs	CVn	13	+40
Canis Major	Canis Majoris	Big dog	CMa	7	−20
Canis Minor	Canis Minoris	Little dog	CMi	8	+5
Capricornus	Capricorni	Sea goat	Cap	21	−20
†Carina	Carinae	Keel of Argonauts' ship	Car	9	−60
Cassiopeia	Cassiopeiae	Queen of Ethiopia	Cas	1	+60
Centaurus	Centauri	Centaur	Cen	13	−50
Cephus	Cephei	King of Ethiopia	Cep	22	+70
Cetus	Ceti	Sea monster (whale)	Cet	2	−10
Chamaeleon	Chamaeleontis	Chameleon	Cha	11	−80
Circinus	Circini	Compasses	Cir	15	−60
Columba	Columbae	Dove	Col	6	−35
Coma Berenices	Comae Berenices	Berenice's hair	Com	13	+20
Corona Australis	Coronae Australis	Southern crown	CrA	19	−40
Corona Borealis	Coronae Borealis	Northern crown	CrB	16	+30
Corvus	Corvi	Crow	Crv	12	−20
Crater	Crateris	Cup	Crt	11	−15
Crux	Crucis	Cross (southern)	Cru	12	−60
Cygnus	Cygni	Swan	Cyg	21	+40
Delphinus	Delphini	Porpoise	Del	21	+10
Dorado	Doradus	Swordfish	Dor	5	−65
Draco	Draconis	Dragon	Dra	17	+65
Equuleus	Equulei	Little horse	Equ	21	+10
Eridanus	Eridani	River	Eri	3	−20
Fornax	Fornacis	Furnace	For	3	−30
Gemini	Geminorum	Twins	Gem	7	+20
Grus	Gruis	Crane	Gru	22	−45
Hercules	Herculis	Hercules, son of Zeus	Her	17	+30
Horologium	Horologii	Clock	Hor	3	−60
Hydra	Hydrae	Sea serpent	Hya	10	−20
Hydrus	Hydri	Water snake	Hyi	2	−75
Indus	Indi	Indian	Ind	21	−55
Lacerta	Lacertae	Lizard	Lac	22	+45
Leo	Leonis	Lion	Leo	11	+15
Leo Minor	Leonis Minoris	Little lion	LMi	10	+35
Lepus	Leporis	Hare	Lep	6	−20
Libra	Librae	Balance	Lib	15	−15

CONSTELLATION (LATIN NAME)	GENITIVE CASE ENDING	ENGLISH NAME OR DESCRIPTION	ABBRE-VIA-TION	APPROXIMATE POSITION	
				α	δ
				h	°
Lupus	Lupi	Wolf	Lup	15	−45
Lynx	Lyncis	Lynx	Lyn	8	+45
Lyra	Lyrae	Lyre or harp	Lyr	19	+40
Mensa	Mensae	Table Mountain	Men	5	−80
Microscopium	Microscopii	Microscope	Mic	21	−35
Monoceros	Monocerotis	Unicorn	Mon	7	−5
Musca	Muscae	Fly	Mus	12	−70
Norma	Normae	Carpenter's level	Nor	16	−50
Octans	Octantis	Octant	Oct	22	−85
Ophiuchus	Ophiuchi	Holder of serpent	Oph	17	0
Orion	Orionis	Orion, the hunter	Ori	5	+5
Pavo	Pavonis	Peacock	Pav	20	−65
Pegasus	Pegasi	Pegasus, the winged horse	Peg	22	+20
Perseus	Persei	Perseus, hero who saved Andromeda	Per	3	+45
Phoenix	Phoenicis	Phoenix	Phe	1	−50
Pictor	Pictoris	Easel	Pic	6	−55
Pisces	Piscium	Fishes	Psc	1	+15
Piscis Austrinus	Piscis Austrini	Southern fish	PsA	22	−30
†Puppis	Puppis	Stern of the Arognauts' ship	Pup	8	−40
†Pyxis (= Malus)	Pyxidis	Compass on the Argonauts' ship	Pyx	9	−30
Reticulum	Reticuli	Net	Ret	4	−60
Sagitta	Sagittae	Arrow	Sge	20	+10
Sagittarius	Sagittarii	Archer	Sgr	19	−25
Scorpius	Scorpii	Scorpion	Sco	17	−40
Sculptor	Sculptoris	Sculptor's tools	Scl	0	−30
Scutum	Scuti	Shield	Sct	19	−10
Serpens	Serpentis	Serpent	Ser	17	0
Sextans	Sextantis	Sextant	Sex	10	0
Taurus	Tauri	Bull	Tau	4	+15
Telescopium	Telescopii	Telescope	Tel	19	−50
Triangulum	Trianguli	Triangle	Tri	2	+30
Triangulum Australe	Trianguli Australis	Southern triangle	TrA	16	−65
Tucana	Tucanae	Toucan	Tuc	0	−65
Ursa Major	Ursae Majoris	Big bear	UMa	11	+50
Ursa Minor	Ursae Minoris	Little bear	VMi	15	+70
†Vela	Velorum	Sail of the Argonauts' ship	Vel	9	−50
Virgo	Virginis	Virgin	Vir	13	0
Volans	Volantis	Flying fish	Vol	8	−70
Vulpecula	Vulpeculae	Fox	Vul	20	+25

†The four constellations Carina, Puppis, Pyxis, and Vela originally formed the single constellation, Argo Navis.

APPENDIX 18

Star Maps

The star maps, one for each month, are printed on six removable sheets at the back of this book. To learn the stars and constellations, the sheet containing the map for the current month should be taken outdoors and compared directly with the sky. The maps were designed for a latitude of about 35° N but are useful anywhere in the continental United States. Each map shows the appearance of the sky at about 9:00 P.M. (Standard Time) near the middle of the month for which it is intended; near the beginning and end of the month, it shows the sky as it appears about 10:00 P.M. and 8:00 P.M., respectively. To use a map, hold the sheet vertically, and turn it so that the direction you are facing is shown at the bottom.

These star maps were originally prepared by C. H. Cleminshaw for the *Griffith Observer* (published by the Griffith Observatory, P.O. Box 27787, Los Angeles, California, 90027), and are reproduced here by the very kind permission of the Griffith Observatory, Edwin C. Krupp, Director.

INDEX

Index

Page numbers in boldface refer to principal discussions.
Page numbers in parentheses refer to figures.

F

H

K

L

N

P

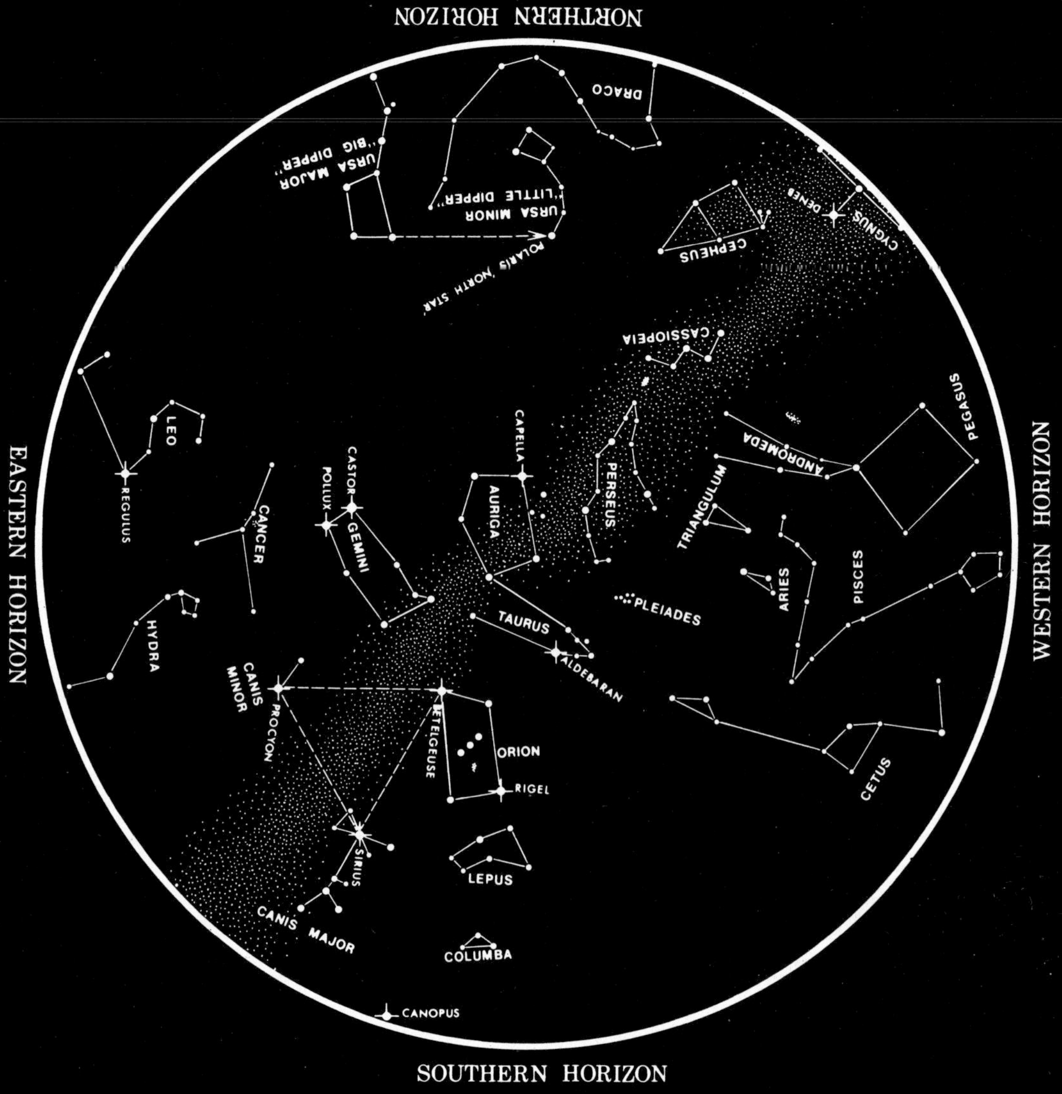

NORTHERN HORIZON

EASTERN HORIZON

WESTERN HORIZON

SOUTHERN HORIZON

THE NIGHT SKY IN JANUARY

Latitude of chart is 34°N, but it is practical throughout the continental United States.

To use: Hold chart vertically and turn it so the direction you are facing shows at the bottom.

Chart time (Local Standard):
10 p.m. First of month
9 p.m. Middle of month
8 p.m. Last of month

Star Chart from *GRIFFITH OBSERVER*, Griffith Observatory, Los Angeles

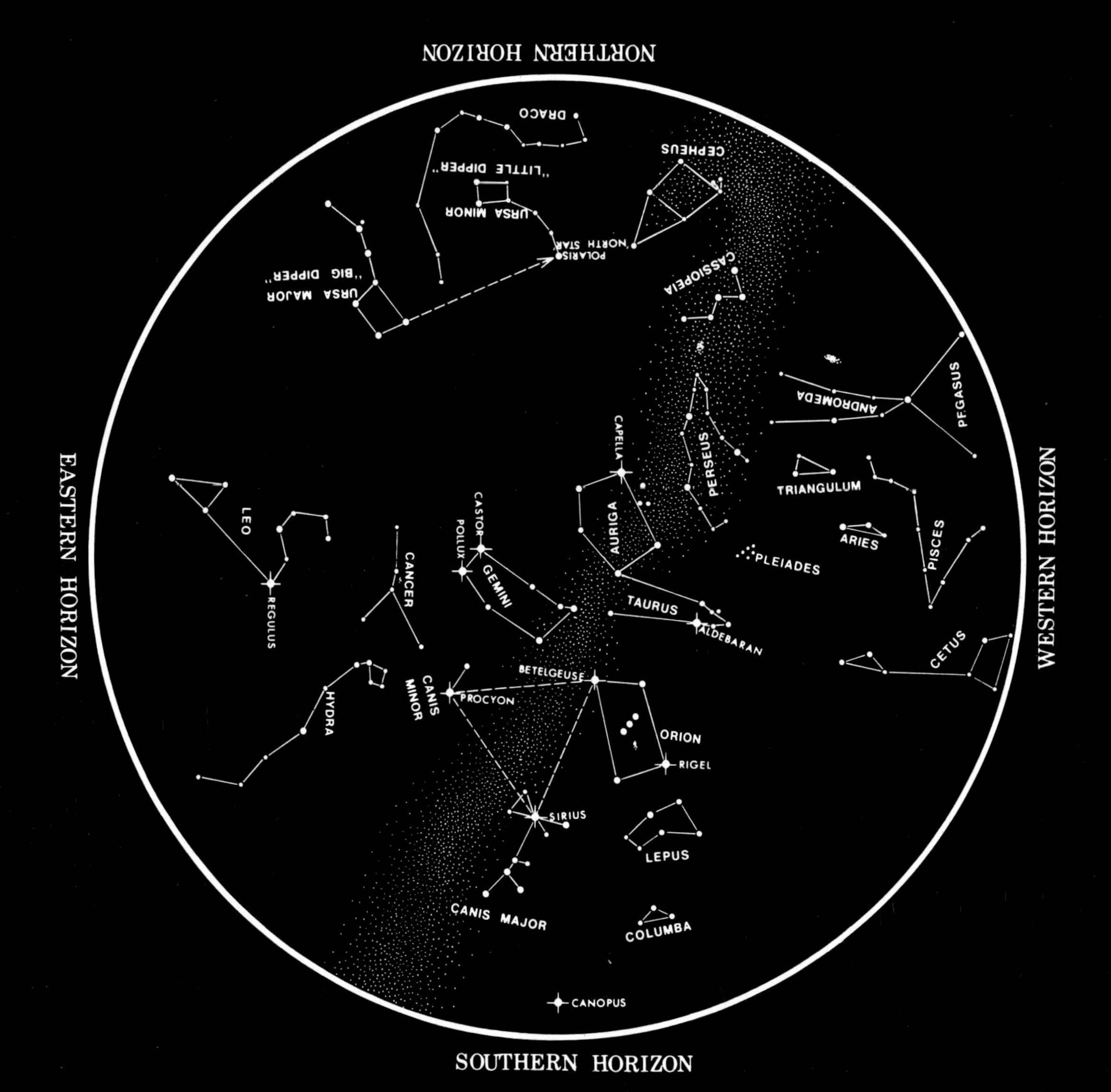

THE NIGHT SKY IN FEBRUARY

Latitude of chart is 34°N, but it is practical throughout the continental United States.

To use: Hold chart vertically and turn it so the direction you are facing shows at the bottom.

Chart time (Local Standard):
10 p.m. First of month
9 p.m. Middle of month
8 p.m. Last of month

Star Chart from *GRIFFITH OBSERVER*, Griffith Observatory, Los Angeles

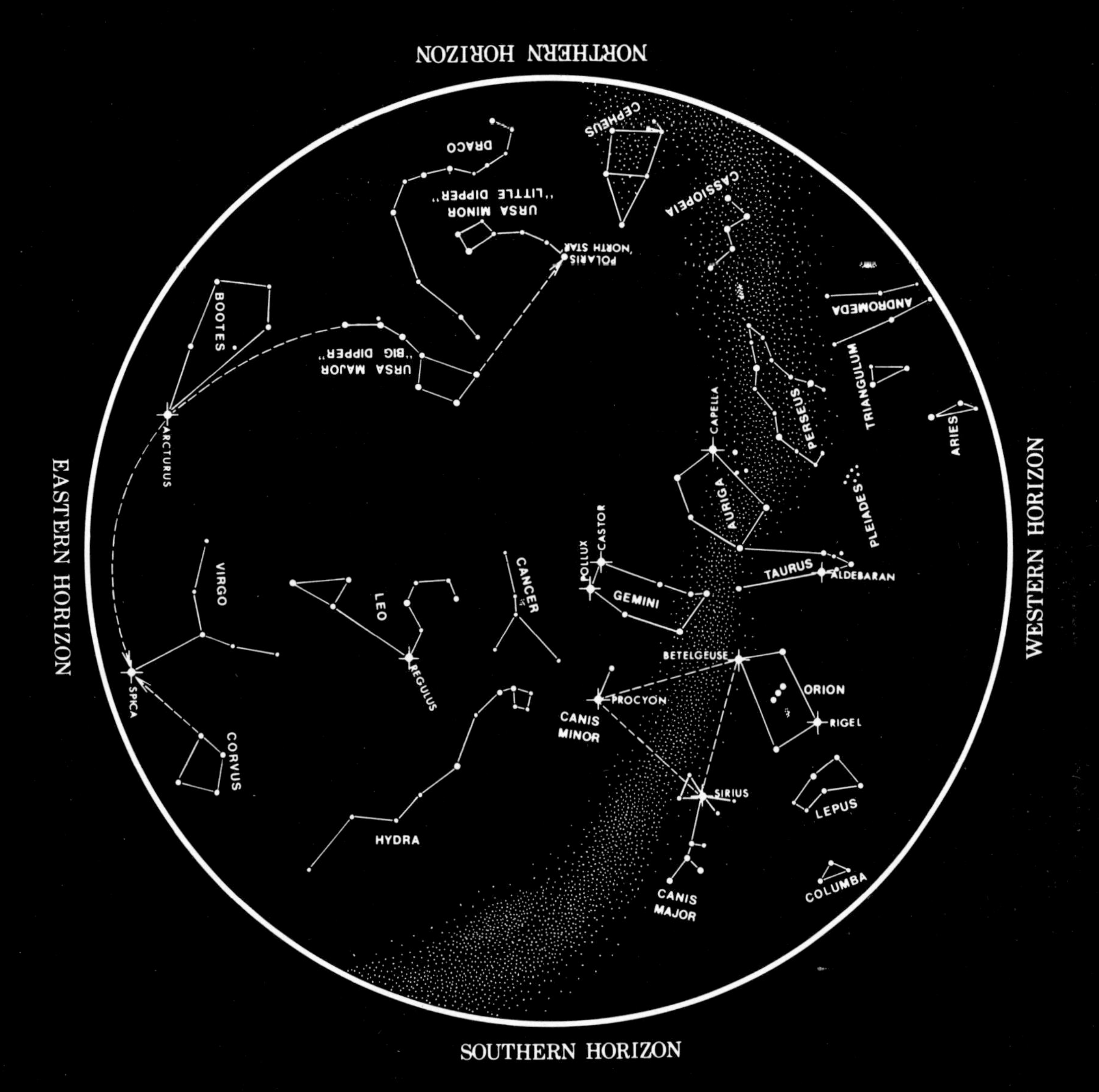

THE NIGHT SKY IN MARCH

Latitude of chart is 34°N, but it is practical throughout the continental United States.

To use: Hold chart vertically and turn it so the direction you are facing shows at the bottom.

Chart time (Local Standard):
10 p.m. First of month
9 p.m. Middle of month
8 p.m. Last of month

Star Chart from *GRIFFITH OBSERVER*, Griffith Observatory, Los Angeles

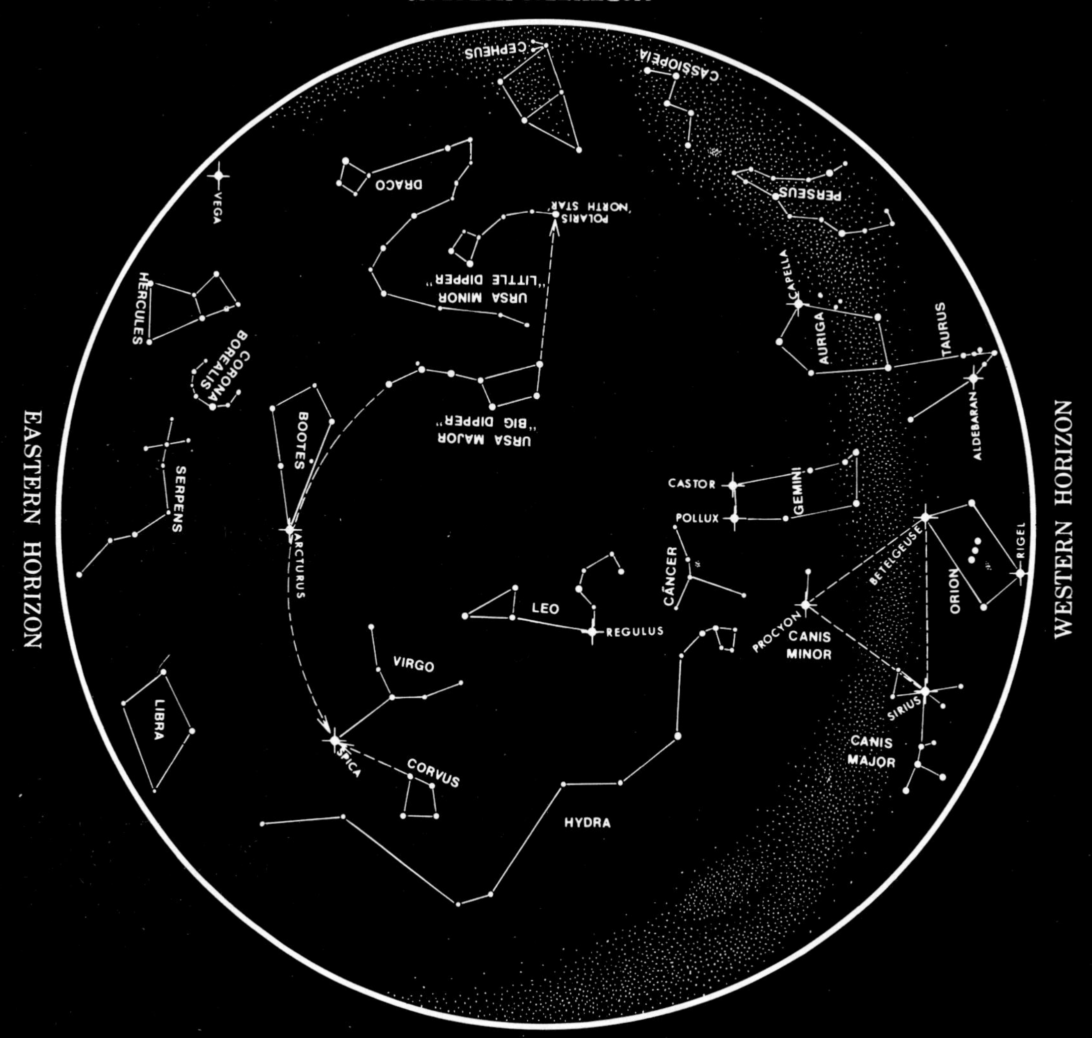

THE NIGHT SKY IN APRIL

Latitude of chart is 34°N, but it is practical throughout the continental United States.

To use: Hold chart vertically and turn it so the direction you are facing shows at the bottom.

Chart time (Local Standard):
10 p.m. First of month
9 p.m. Middle of month
8 p.m. Last of month

Star Chart from *GRIFFITH OBSERVER*, Griffith Observatory, Los Angeles

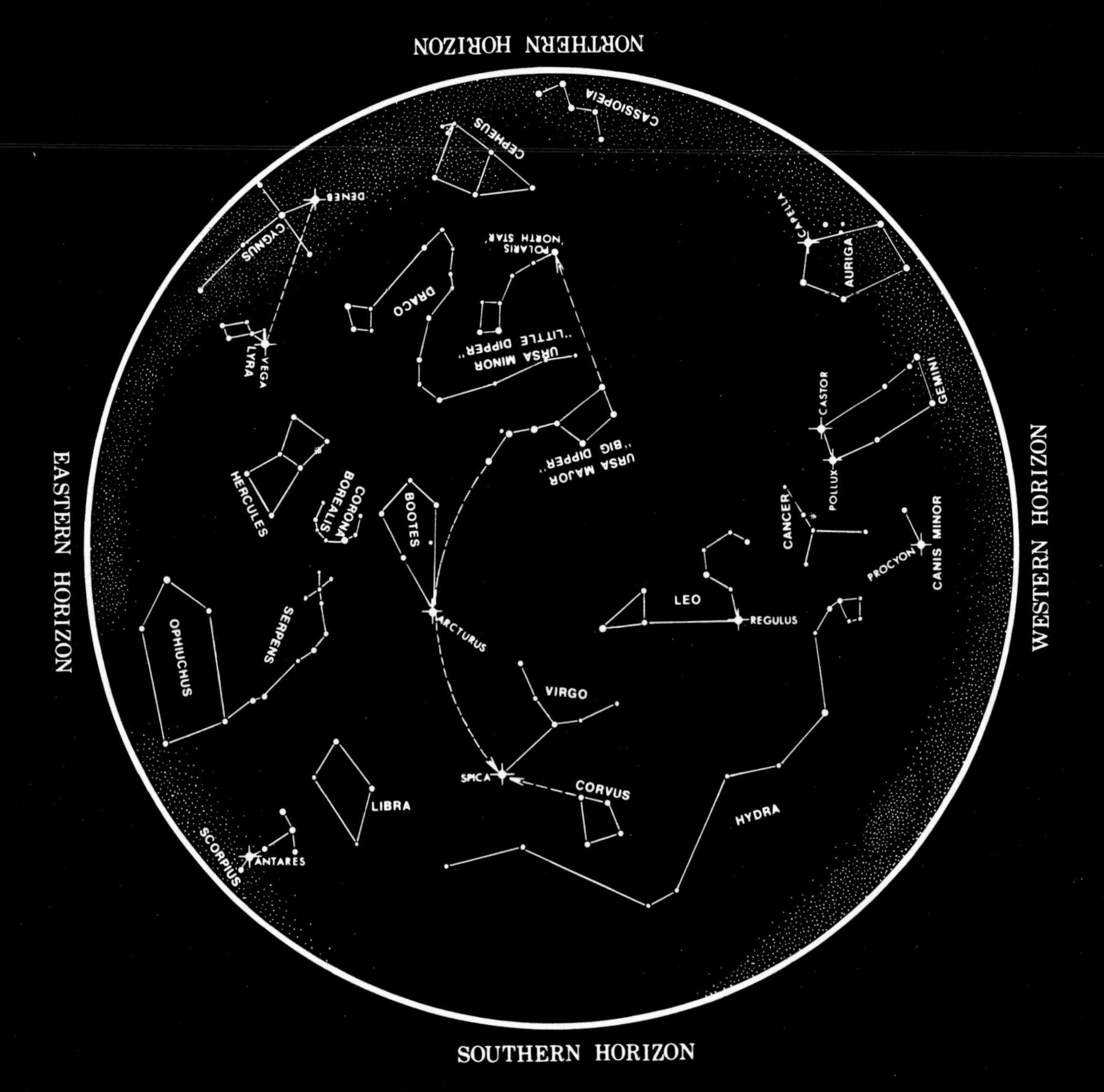

THE NIGHT SKY IN MAY

Latitude of chart is 34°N, but it is practical throughout the continental United States.

To use: Hold chart vertically and turn it so the direction you are facing shows at the bottom.

Chart time (Local Standard):
10 p.m. First of month
9 p.m. Middle of month
8 p.m. Last of month

Star Chart from *GRIFFITH OBSERVER*, Griffith Observatory, Los Angeles

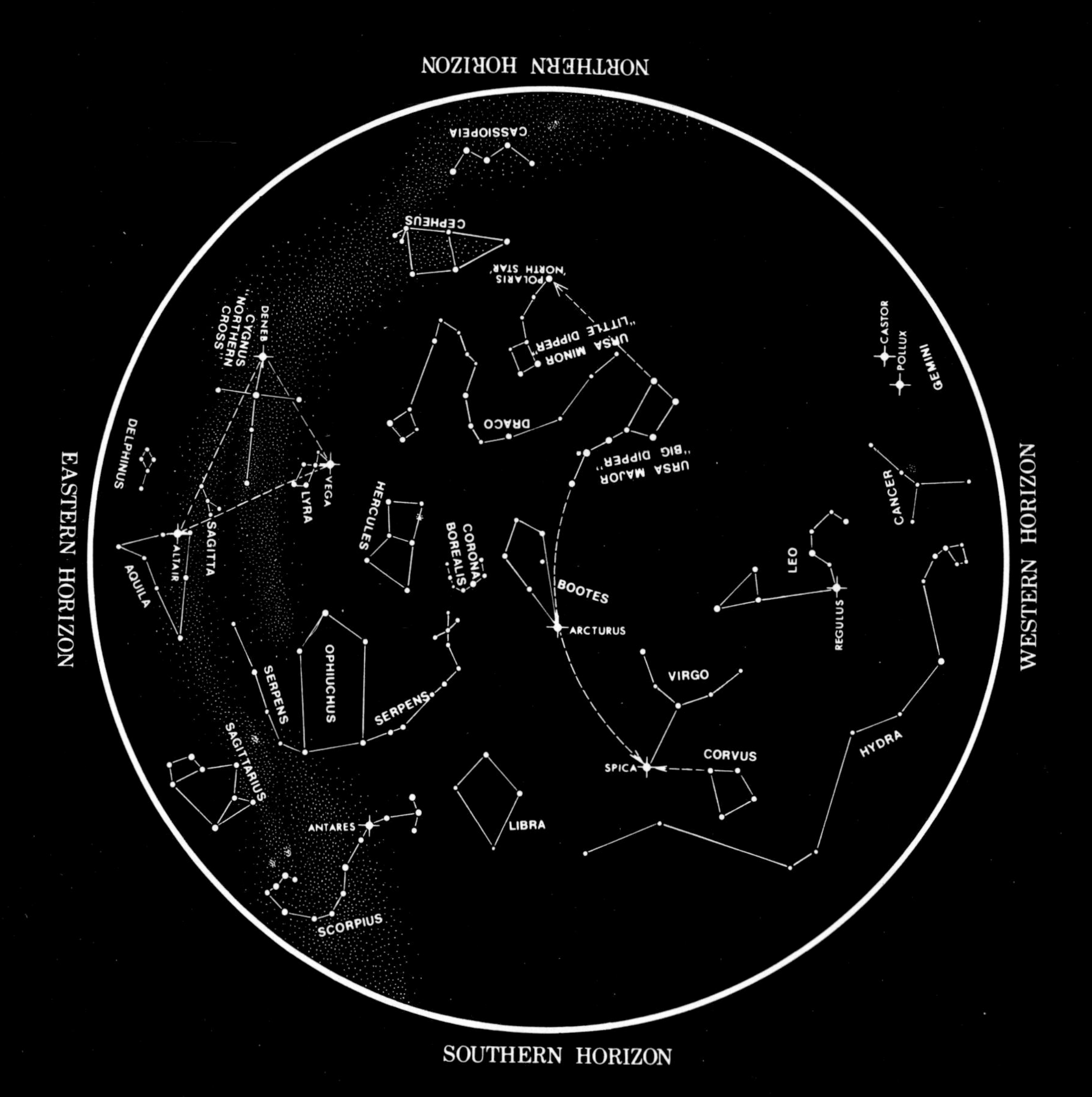

THE NIGHT SKY IN JUNE

Latitude of chart is 34°N, but it is practical throughout the continental United States.

To use: Hold chart vertically and turn it so the direction you are facing shows at the bottom.

Chart time (Local Standard):
10 p.m. First of month
9 p.m. Middle of month
8 p.m. Last of month

Star Chart from *GRIFFITH OBSERVER*, Griffith Observatory, Los Angeles

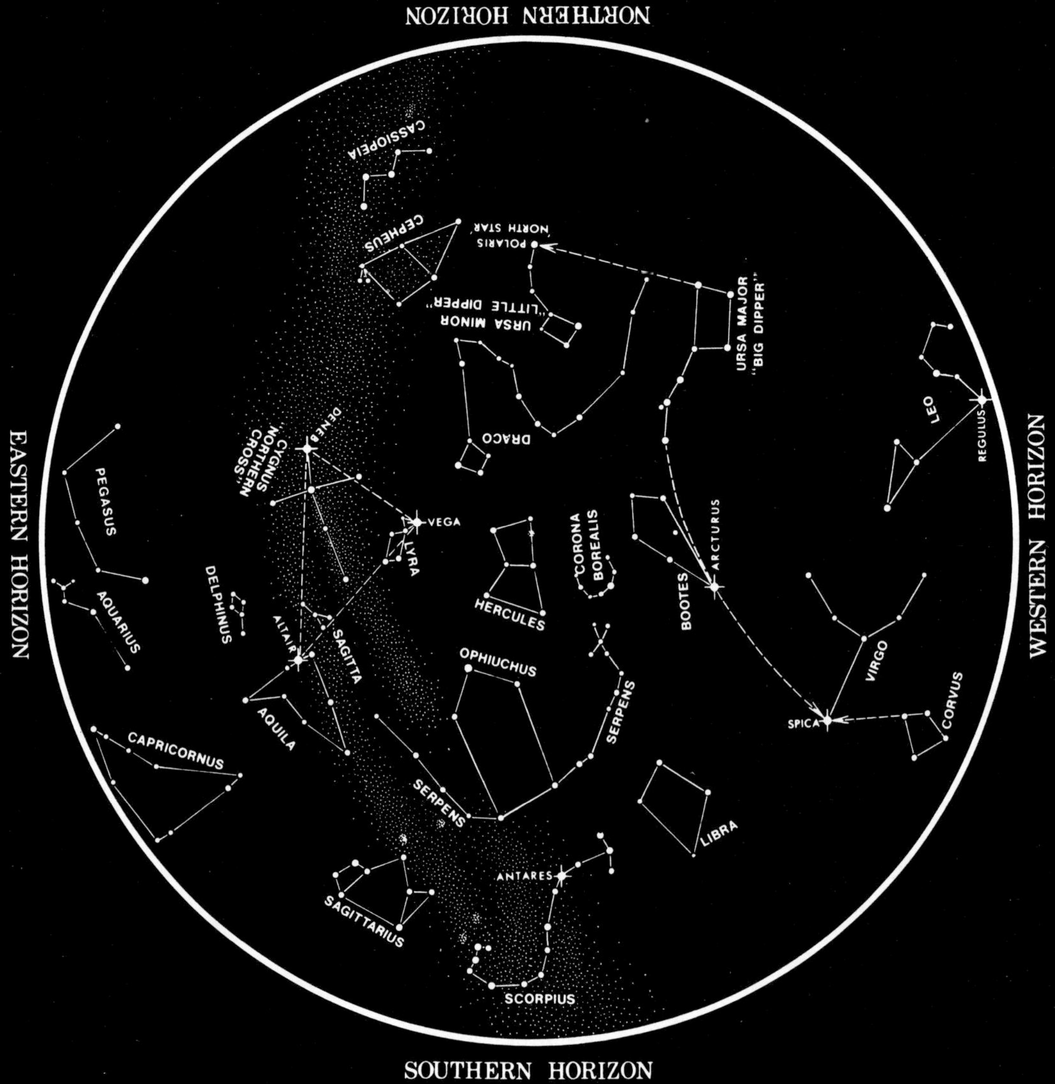

THE NIGHT SKY IN JULY

Latitude of chart is 34°N, but it is practical throughout the continental United States.

To use: Hold chart vertically and turn it so the direction you are facing shows at the bottom.

Chart time (Local Standard):
10 p.m. First of month
9 p.m. Middle of month
8 p.m. Last of month

Star Chart from *GRIFFITH OBSERVER*, Griffith Observatory, Los Angeles

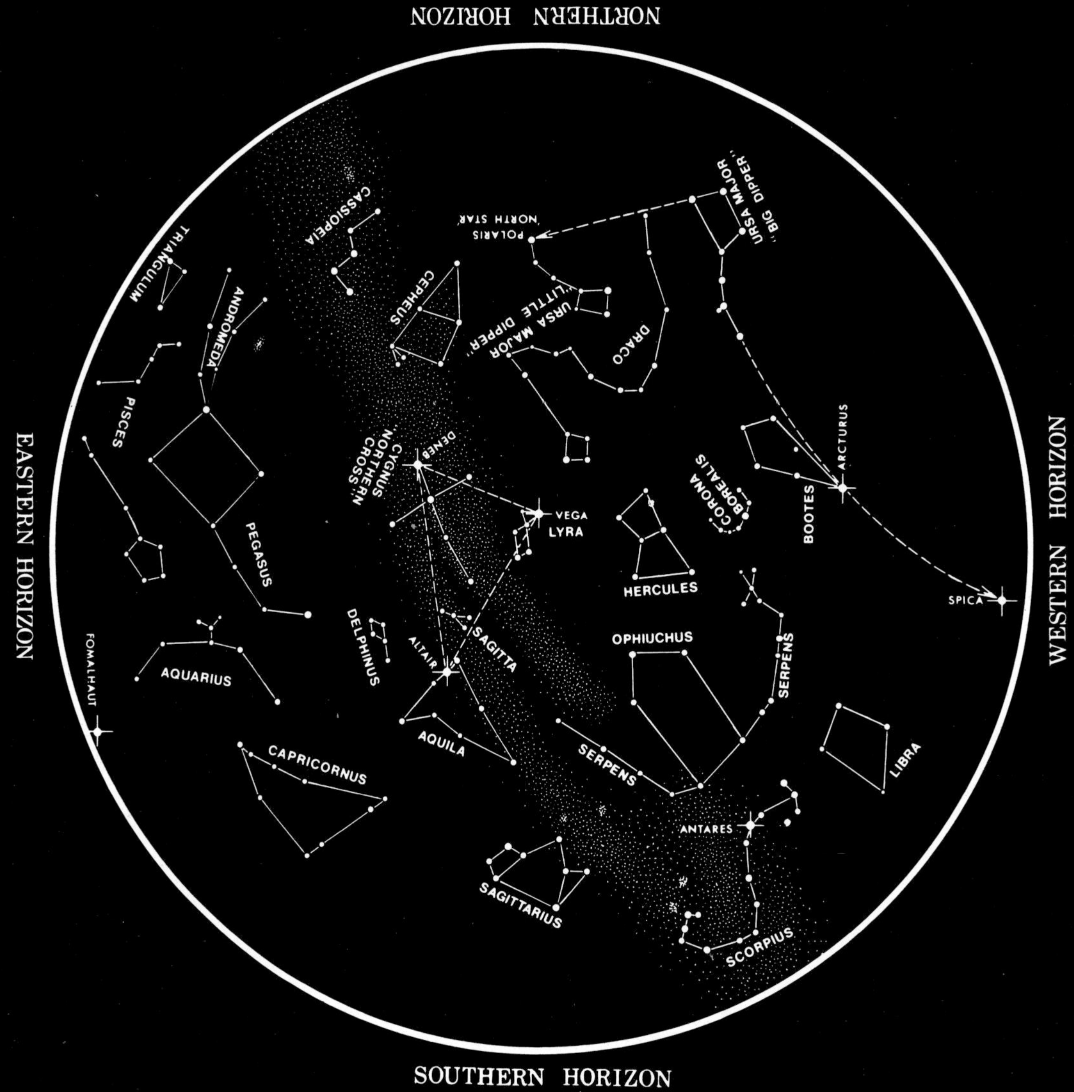

THE NIGHT SKY IN AUGUST

Latitude of chart is 34°N, but it is practical throughout the continental United States.

To use: Hold chart vertically and turn it so the direction you are facing shows at the bottom.

Chart time (Local Standard):
10 p.m. First of month
9 p.m. Middle of month
8 p.m. Last of month

Star Chart from *GRIFFITH OBSERVER*, Griffith Observatory, Los Angeles

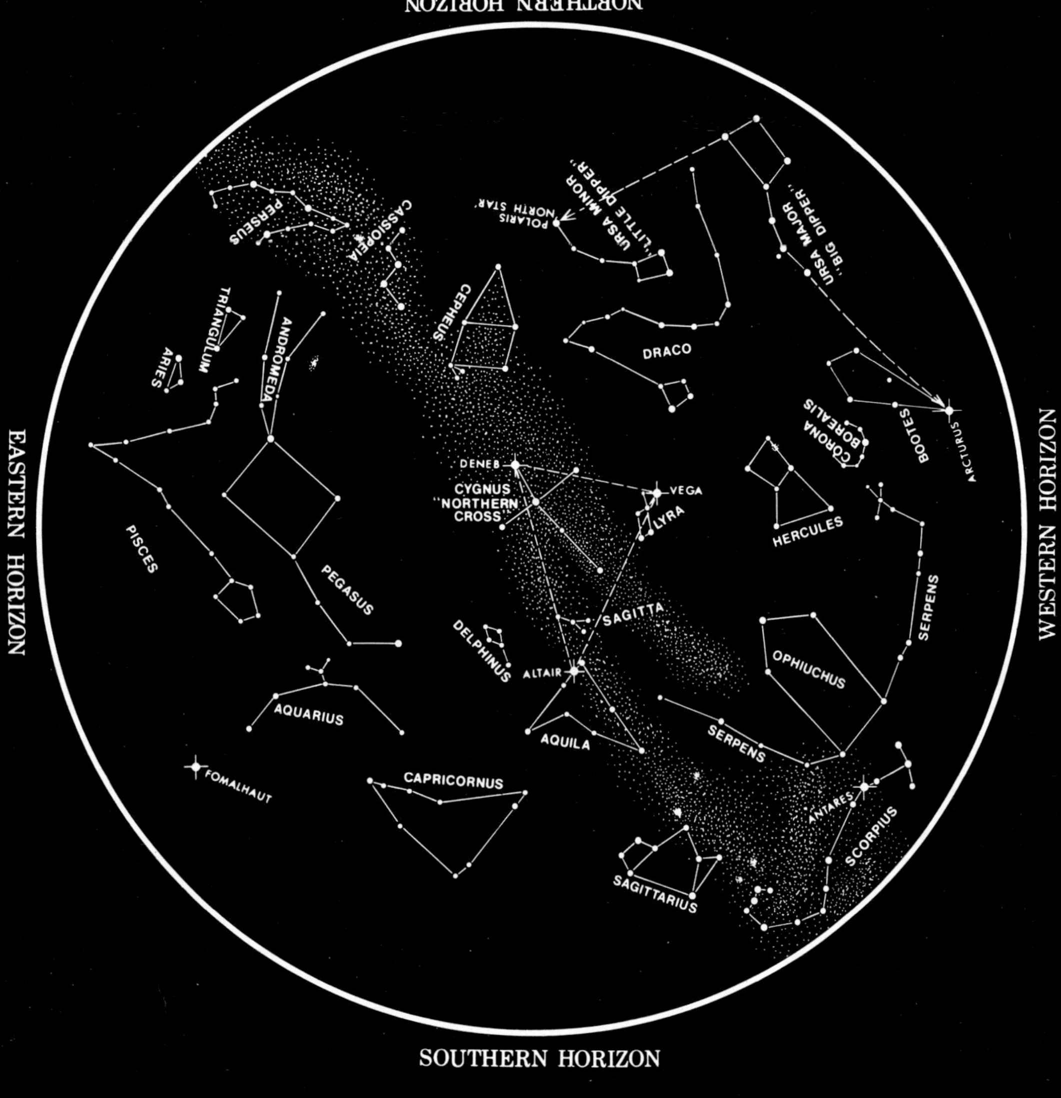

THE NIGHT SKY IN SEPTEMBER

Latitude of chart is 34°N, but it is practical throughout the continental United States.

To use: Hold chart vertically and turn it so the direction you are facing shows at the bottom.

Chart time (Local Standard):

10 p.m. First of month

9 p.m. Middle of month

8 p.m. Last of month

Star Chart from *GRIFFITH OBSERVER*, Griffith Observatory, Los Angeles

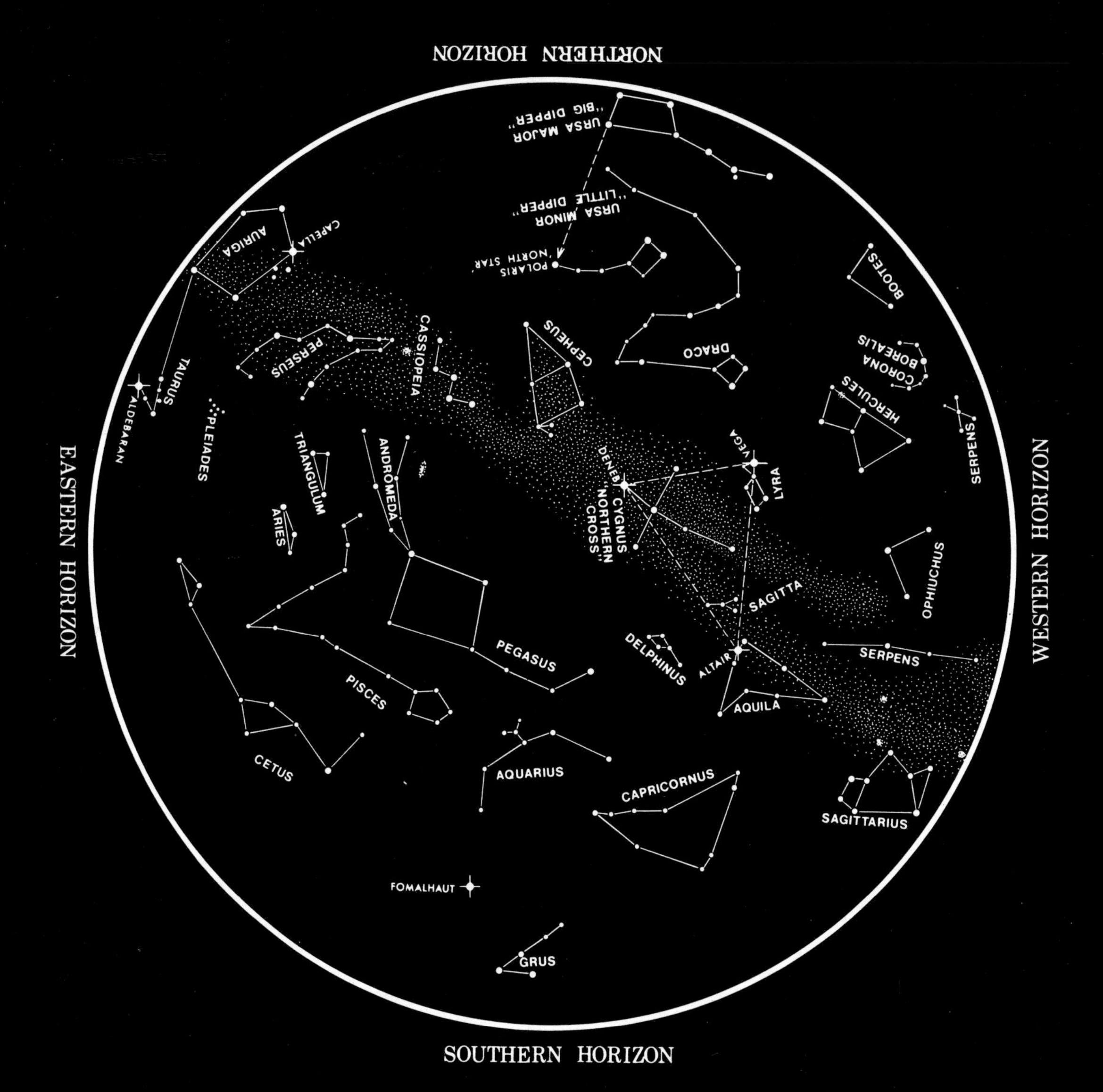

THE NIGHT SKY IN OCTOBER

Latitude of chart is 34°N, but it is practical throughout the continental United States.

To use: Hold chart vertically and turn it so the direction you are facing shows at the bottom.

Chart time (Local Standard):
10 p.m. First of month
9 p.m. Middle of month
8 p.m. Last of month

Star Chart from *GRIFFITH OBSERVER*, Griffith Observatory, Los Angeles

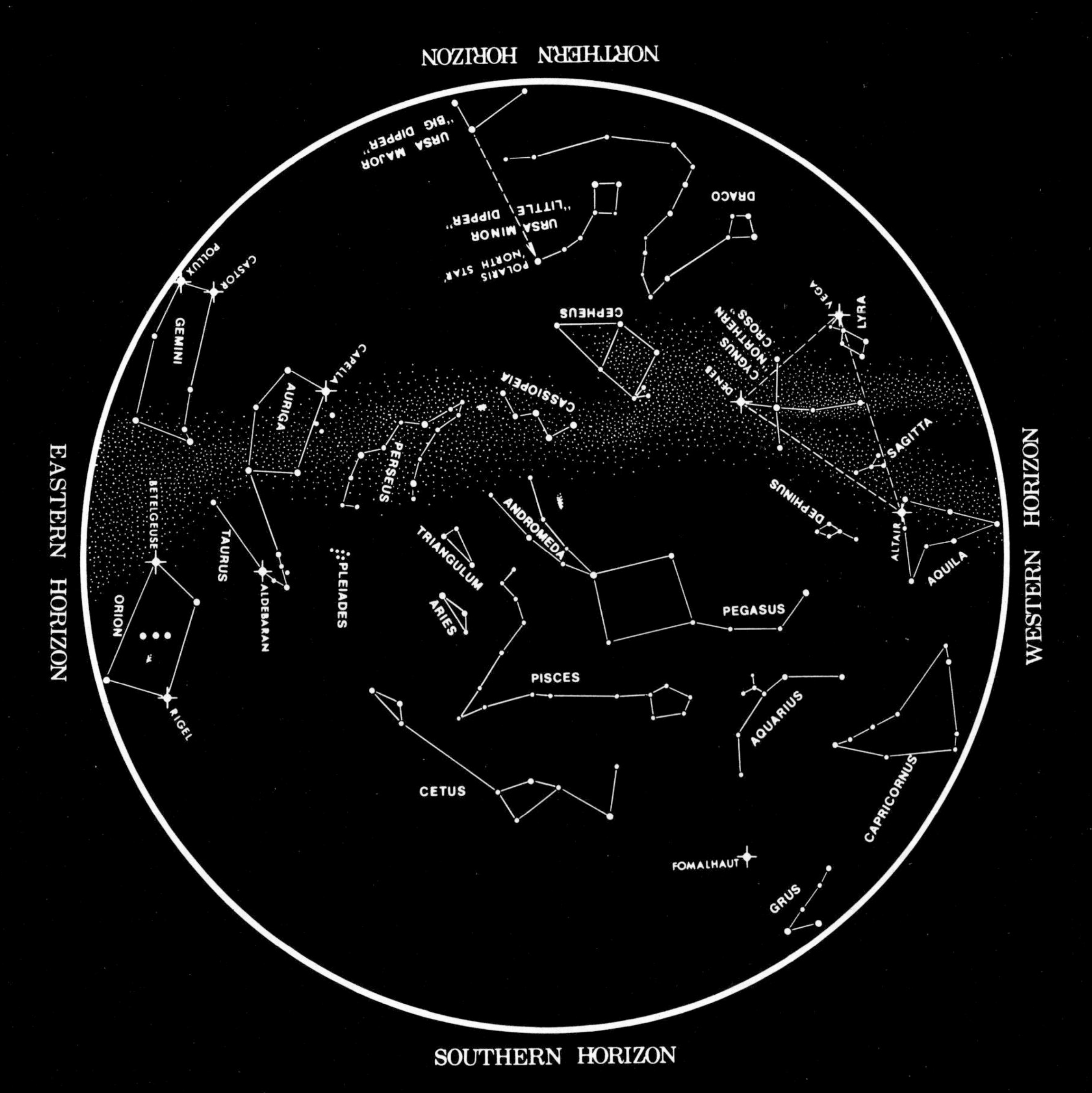

THE NIGHT SKY IN NOVEMBER

Latitude of chart is 34°N, but it is practical throughout the continental United States.

To use: Hold chart vertically and turn it so the direction you are facing shows at the bottom.

Chart time (Local Standard):
10 p.m. First of month
9 p.m. Middle of month
8 p.m. Last of month

Star Chart from *GRIFFITH OBSERVER,* Griffith Observatory, Los Angeles

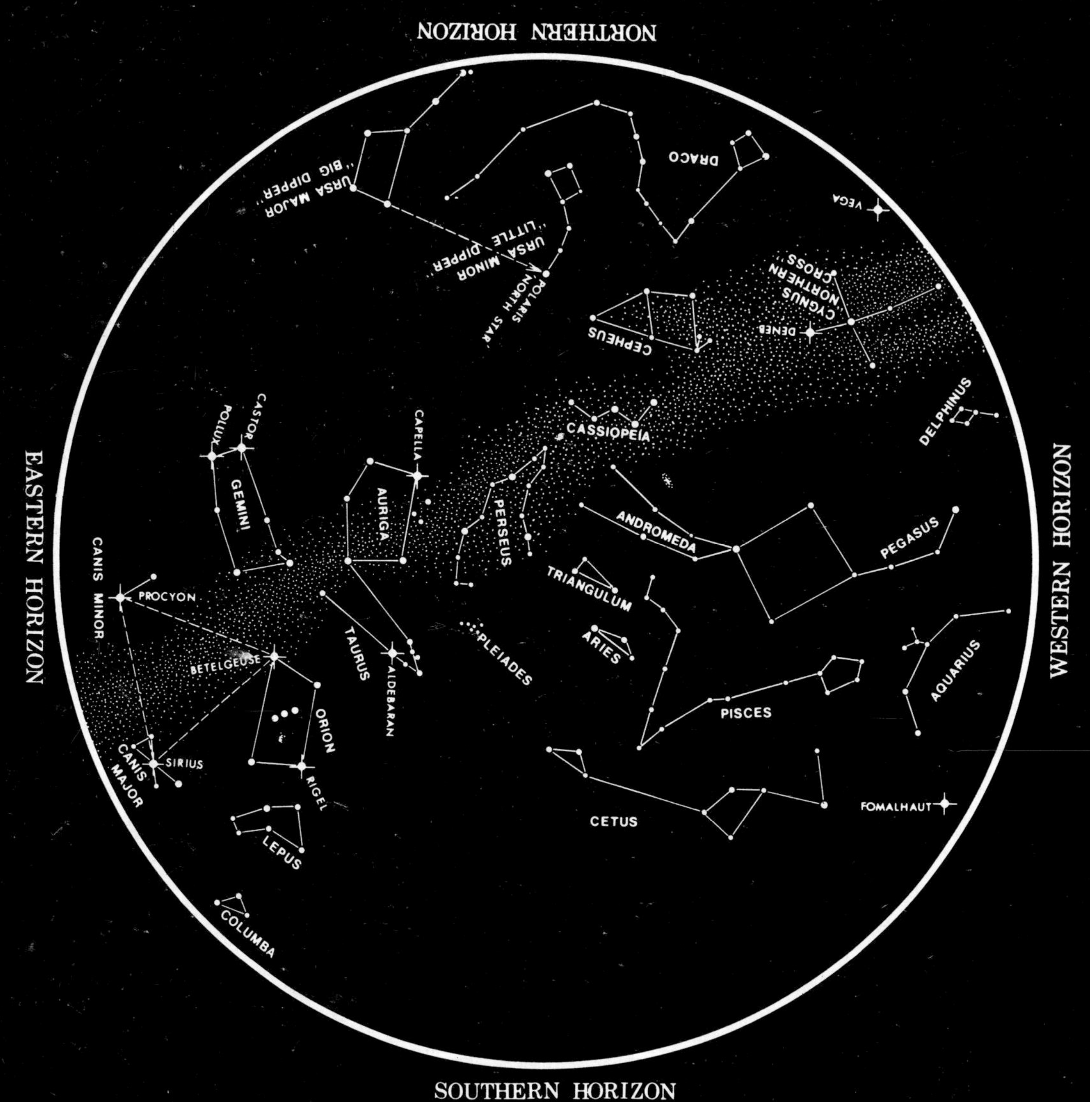

THE NIGHT SKY IN DECEMBER

Latitude of chart is 34°N, but it is practical throughout the continental United States.

To use: Hold chart vertically and turn it so the direction you are facing shows at the bottom.

Chart time (Local Standard):
10 p.m. First of month
9 p.m. Middle of month
8 p.m. Last of month

Star Chart from *GRIFFITH OBSERVER*, Griffith Observatory, Los Angeles